About the book . . .

Molecular Vib-Rotors presents the theory of the vibrating-rotating mole-
-le and describes in detail ba
Although basic the

MOLECULAR VIB-ROTORS

Molecular Vib-Rotors

THE THEORY AND INTERPRETATION OF HIGH RESOLUTION INFRARED SPECTRA

Harry C. Allen, Jr. National Bureau of Standards, Washington, D.C.

Paul C. Cross Mellon Institute, Pittsburgh, Pennsylvania

JOHN WILEY AND SONS, INC., NEW YORK AND LONDON

Library of Congress Catalog Card Number: 63-11426
Printed in the United States of America

PREFACE

Since 1945 there has been a vigorous development in infrared instrumentation. Commercial spectrometers are now available that can resolve the rotational structure of vibrational bands of lighter molecules, and many research instruments are available that can resolve the rotational structure in the spectra of moderately heavy molecules. Much of the basic theory for the interpretation of these spectra was developed during the early years of quantum mechanics. The available knowledge up to 1945 is well covered in the excellent book by Herzberg. In the past fifteen years much of this theory has been correlated and extended. Unfortunately, most of this work was scattered throughout the literature.

It is our purpose to present this material in an orderly manner, to relate the diverse notations, and to summarize some of the accumulated experience in applying the theory to the analysis of observed spectra.

The treatment in the first five chapters is intended to give a consistent development of the basic theory. The aim has been to present the theory with sufficient completeness to enable the reader to understand its derivation and to be prepared to extend and adapt the methods involved to new problems. No attempt has been made to include all the esoteric extensions of the theory that have been made in the past few years, since these treatments are so detailed that the notation becomes ponderous and confusing to the beginner.

The last three chapters deal with illustrative applications of the theory to linear, symmetric, and asymmetric rotors. Individual molecules and their spectra have been introduced only as examples, and no exhaustive survey of the literature is included.

The examples have been chosen not only to demonstrate the straightforward application of the theory but also to illustrate some of the complications which can arise in the complete analysis of observed spectra.

The mathematical demands on the reader are moderate. He is expected to have an acquaintance with the methods of quantum mechanics and of the principles of group theory. The necessary background is no greater than that acquired by a study of the vibrational problem. The demands are less in the early chapters where the derivations are presented in considerable detail. In later chapters, where the derivations require only repeated applications of previously presented techniques, only the key steps are given or the results merely quoted.

In preparing the manuscript we have drawn heavily on many sources. It is not feasible to give specific credit in every instance, since a given source may have been used unwittingly. However, sources to which we are especially indebted are *Atomic Spectra*, by E. U. Condon and G. H. Shortley; *Infrared and Raman Spectra*, by G. Herzberg; *Molecular Vibrations*, by E. B. Wilson, Jr., J. C. Decius, and P. C. Cross; and *Microwave Spectroscopy*, by C. H. Townes and A. H. Schawlow.

Of the recent contributions to the scientific literature, we have relied chiefly upon those of Professor E. Bright Wilson, Jr., and of his colleagues. Others upon whose work we have drawn extensively include Professors D. M. Dennison, H. C. Longuet-Higgins, H. H. Nielsen, and E. K. Plyler.

We have had much valuable assistance in the final preparation of the manuscript, especially from Dr. Bernard Kirtman, whose thorough review and criticism we greatly appreciate. Dr. Bryce L. Crawford, Jr., also made many useful suggestions. Others whose assistance with parts of the task we wish to acknowledge include W. G. Fately, R. N. Kortzeborn, R. L. Redington, W. B. Olson, and R. E. Wilde.

HARRY C. ALLEN, JR.
PAUL C. CROSS

January 1963

CONTENTS

1

HAMILTONIANS FOR THE MOLECULAR MODELS

1a INTRODUCTION

The observable vibrational spectra of most molecules occur in the spectral region between 200 and 10,000 cm^{-1}. The availability of vastly improved detectors for this region in the last few years has made possible the observation of vibrational bands under very high resolution. With the improved resolution, we can now observe the rotational fine structure of these vibrational bands for a large number of molecules. The analysis of this fine structure reveals a considerable amount of information about the structure of the molecules giving rise to the spectrum. The effective moments of inertia in the two vibrational states and the frequency of the vibrational transition are readily determined with a precision which was unattainable a few years ago. From the moments of inertia of the simple molecules it is possible to deduce structural parameters such as bond lengths and bond angles. If a sufficient number of vibrational frequencies are determined precisely, the anharmonicity and interaction terms in the vibrational energy expression may be deduced. These terms can give, in many instances, information about the shape of the potential energy function. The force constants of the various bonds can be calculated giving quantitative data on bond properties.

The rapid development of the study of pure rotational spectra in the microwave region has also contributed vast numbers of structural parameters for simple molecules. This does not mean that a study of the infrared vibrational-rotational spectrum is precluded; on the contrary, the two methods should be thought of as complementary.

The microwave study usually gives effective inertial constants in the ground vibrational state. An infrared analysis makes possible the determination of the variation of the inertial constants with vibrational state from which equilibrium inertial constants can be determined. Molecules which do not have a permanent dipole moment cannot be studied in the microwave region, and molecules with small moments of inertia are very difficult to study because of the limited frequency range that is presently accessible.

The theory of the purely vibrational aspects of infrared spectra has been covered in the recent book of Wilson, Decius, and Cross [9]. The purpose of this work is to develop only those aspects of the theory which arise in the high resolution work. The excellent treatise of Herzberg [4] contains examples illustrating the application of most of these results.

A fairly large number of perturbations of the vibrational-rotational levels can be observed in the microwave region that cannot be observed with present techniques in the infrared region. As a result only those effects which can reasonably be expected to be observed in infrared spectra are treated. For details especially applicable in microwave studies the excellent books by Gordy, Smith, and Trambarulo [3] and by Townes and Schawlow [8] should be consulted.

Ib MOLECULAR MODEL AND CLASSICAL KINETIC ENERGY

In order to develop a mathematical formalism with which to treat vibrational-rotational spectra a molecular model must be developed. The commonly used model is that of point masses moving in a potential field provided by the averaged motion of the electrons. The atoms can be thought of as being held together by rather rigid springs giving rise to a molecule that can be described mathematically as a semirigid body.

The energy levels of such a model are the eigenvalues of the Schrödinger equation

$$(\mathbf{H} - E)\psi = 0, \tag{1}$$

where $\mathbf{H}$ is the quantum-mechanical operator for the rotating-vibrating model. The quantum-mechanical operator will be derived by the method first used by Wilson and Howard [125]. This derivation is carried out in great detail by Wilson, Decius, and Cross [9] and is partially repeated here only to give the reader easy access to definitions.

Ic CLASSICAL KINETIC ENERGY

A rigorous treatment of the problem starts with a determination of the kinetic energy in a suitable set of coordinates. Such a set of coordinates consists of the Cartesian coordinates X, Y, Z of the center of mass of body, the three Eulerian angles θ, ϕ, χ (shown in Fig. 4k1; see also [9] Appendix I) which describe the orientation in space of a set of rotating coordinate axes x, y, z whose origin is at the center of mass of the molecule, and the 3N-6 normal coordinates which define the relative positions of the atoms in the rotating axis system (for a nonlinear molecule).

Let the position of the αth particle with respect to the center of mass be given by the vector $\mathbf{r}_\alpha$. In the moving coordinate system, which is fully defined later, this vector has the components x_α, y_α, z_α. The center of mass is located by the vector $\mathbf{R}$ from the origin of the space-fixed axis system. The equilibrium position of the αth particle is given by the vector $\mathbf{a}_\alpha$ which is fixed in the moving axis system. The displacement vector $\boldsymbol{\rho}_\alpha$ can now be defined by the relation

$$\boldsymbol{\rho}_\alpha = \mathbf{r}_\alpha - \mathbf{a}_\alpha. \tag{1}$$

If at any instant the rotating axes have the angular velocity $\boldsymbol{\omega}$ and if the vector $\mathbf{v}_\alpha$ has the components $\dot{x}_\alpha$, $\dot{y}_\alpha$, $\dot{z}_\alpha$, the velocity of the αth particle in space is

$$\mathbf{v}_{t\alpha} = \dot{\mathbf{R}} + (\boldsymbol{\omega} \times \mathbf{r}_\alpha) + \mathbf{v}_\alpha. \tag{2}$$

The kinetic energy of the whole molecule is then

$$\begin{aligned} 2T = \sum_\alpha m\mathbf{v}_{t\alpha}{}^2 &= \dot{\mathbf{R}}^2 \sum_\alpha m_\alpha + \sum_\alpha m_\alpha(\boldsymbol{\omega} \times \mathbf{r}_\alpha)\cdot(\boldsymbol{\omega} \times \mathbf{r}_\alpha) + \sum_\alpha m_\alpha \mathbf{v}_\alpha^2 \\ &+ 2\dot{\mathbf{R}}\cdot\boldsymbol{\omega} \times \sum_\alpha m_\alpha \mathbf{r}_\alpha + 2\dot{\mathbf{R}}\cdot\sum_\alpha m_\alpha \mathbf{v}_\alpha + 2\boldsymbol{\omega}\cdot\sum_\alpha (m_\alpha \mathbf{r}_\alpha \times \mathbf{v}_\alpha). \end{aligned} \tag{3}$$

The meaning of each of the terms is readily ascertained. The first term is the translational kinetic energy, whereas the second and third terms are, respectively, the purely rotational and purely vibrational kinetic energies. The fourth and fifth terms are the interaction of translation with rotation and with vibration, and the last term is the kinetic energy of interaction between rotation and vibration.

Since by definition, the origin O of the rotating coordinate system is at the center of mass of the molecule, it must always be true that

$$\sum_\alpha m_\alpha \mathbf{r}_\alpha = 0. \tag{4}$$

By differentiating (4), we obtain

$$\begin{aligned} \sum_\alpha m_\alpha \dot{\mathbf{r}}_\alpha &= \sum_\alpha m_\alpha[(\boldsymbol{\omega} \times \mathbf{r}_\alpha) + \mathbf{v}_\alpha] \\ &= \boldsymbol{\omega} \times \sum_\alpha m_\alpha \mathbf{r}_\alpha + \sum_\alpha m_\alpha \mathbf{v}_\alpha = \sum_\alpha m_\alpha \mathbf{v}_\alpha = 0. \end{aligned} \tag{5}$$

Eckart [41, 101] shows that three more conditions are necessary to define completely the rotating coordinate system, which should be chosen to insure that the axes rotate with the molecule. The additional conditions are necessary because the vibrating atoms are in motion relative to each other. These conditions derive from the relation

$$\sum_\alpha m_\alpha \mathbf{a}_\alpha \times \mathbf{r}_\alpha = 0. \tag{6}$$

Upon differentiating (6), we find

$$\begin{aligned} 0 &= \sum m_\alpha \dot{\mathbf{a}}_\alpha \times \mathbf{r}_\alpha + \sum m_\alpha \mathbf{a}_\alpha \times \dot{\mathbf{r}}_\alpha \\ &= \sum m_\alpha(\boldsymbol{\omega} \times \mathbf{a}_\alpha) \times \mathbf{r}_\alpha + \sum m_\alpha \mathbf{a}_\alpha \times (\boldsymbol{\omega} \times \mathbf{r}_\alpha) + \sum m_\alpha \mathbf{a}_\alpha \times \mathbf{v}_\alpha \\ &= \sum m_\alpha \mathbf{a}_\alpha \times \mathbf{v} \end{aligned} \tag{7}$$

The final equation is the one to be used in defining the rotating coordinate system. It states that, to the extent that $\mathbf{r}_\alpha$ may be replaced by $\mathbf{a}_\alpha$, there will be no resultant angular momentum with respect to the rotating coordinate system.

Returning to (3) and substituting $\mathbf{r}_\alpha = \boldsymbol{\rho}_\alpha + \mathbf{a}_\alpha$ in the last term as well as the conditions (4), (5), and (7), we obtain

$$2T = \dot{\mathbf{R}}^2 M + \sum m_\alpha(\boldsymbol{\omega} \times \mathbf{r}_\alpha) \cdot (\boldsymbol{\omega} \times \mathbf{r}_\alpha) + \sum m_\alpha \mathbf{v}_\alpha^2 + 2\boldsymbol{\omega} \cdot \sum m_\alpha(\boldsymbol{\rho}_\alpha \times \mathbf{v}_\alpha), \quad (8)$$

where $M = \sum_\alpha m_\alpha$.

The translational energy is completely separable as is shown by (8); henceforth it will not be considered since it is not important in field-free problems. Thus we are left with terms arising from kinetic energy of rotation, vibration, and rotational-vibrational interaction.

By standard methods of classical mechanics [6] (8) can be expanded to give

$$\begin{aligned} 2T = {} & I_{xx}\omega_x^2 + I_{yy}\omega_y^2 + I_{zz}\omega_z^2 - 2I_{xy}\omega_x\omega_y - 2I_{yz}\omega_y\omega_z - 2I_{xz}\omega_x\omega_z \\ & + \sum m_\alpha \mathbf{v}_\alpha^2 + 2\omega_x \sum m_\alpha(\boldsymbol{\rho}_\alpha \times \mathbf{v}_\alpha)_x + 2\omega_y \sum m_\alpha(\boldsymbol{\rho}_\alpha \times \mathbf{v}_\alpha)_y \\ & + 2\omega_z \sum m_\alpha(\boldsymbol{\rho}_\alpha \times \mathbf{v}_\alpha)_z. \end{aligned} \quad (9)$$

In (9) I_{xx}, I_{yy}, and I_{zz} are the instantaneous moments of inertia with respect to the moving axes x, y, z, and I_{xy}, I_{yz}, and I_{xz} are the instantaneous products of inertia. It should be emphasized that these quantities are not constants but functions of the positions of the particles. The ω_x, ω_y, and ω_z are the components of the angular velocity $\boldsymbol{\omega}$ of the rotating coordinate system.

It is now convenient to introduce a set of $3N - 6$ internal coordinates, which are best chosen to be the normal coordinates Q_k of the vibrational motion. These are defined in terms of the mass-weighted components $\xi_\alpha = \sqrt{m_\alpha}\, \Delta x_\alpha$, $\eta_\alpha = \sqrt{m_\alpha}\, \Delta y_\alpha$, $\zeta_\alpha = \sqrt{m_\alpha}\, \Delta z_\alpha$ of the displacement vector $\boldsymbol{\rho}_\alpha$ by the equations

$$\begin{aligned} \xi_\alpha &= \sum_k \frac{\partial \xi_\alpha}{\partial Q_k} Q_k = \sum_k l_{\alpha k} Q_k, \\ \eta_\alpha &= \sum_k \frac{\partial \eta_\alpha}{\partial Q_k} Q_k = \sum_k m_{\alpha k} Q_k, \\ \zeta_\alpha &= \sum_k \frac{\partial \zeta_\alpha}{\partial Q_k} Q_k = \sum_k n_{\alpha k} Q_k. \end{aligned} \quad (10)$$

in which ξ_α, η_α, ζ_α measure the displacement from the equilibrium position in terms of a mass-adjusted scale and $l_{\alpha k}$, $m_{\alpha k}$, $n_{\alpha k}$ are constant coefficients. Since the normal coordinates are normalized and orthogonal,

$$\sum m_\alpha v_\alpha^2 = \sum_\alpha (\dot{\xi}_\alpha^2 + \dot{\eta}_\alpha^2 + \dot{\zeta}_\alpha^2) = \sum_k \dot{Q}_k^2. \quad (11)$$

In terms of the normal coordinates the parts of the coupling terms become

$$\begin{aligned} \sum_\alpha m_\alpha(\boldsymbol{\rho}_\alpha \times \mathbf{v}_\alpha)_x &= \sum (\eta_\alpha \dot{\zeta}_\alpha - \zeta_\alpha \dot{\eta}_\alpha) = \sum_{k=1}^{3N-6} \mathfrak{X}_k \dot{Q}_k \\ \sum_\alpha m_\alpha(\boldsymbol{\rho}_\alpha \times \mathbf{v}_\alpha)_y &= \sum \mathfrak{Y}_k \dot{Q}_k \\ \sum m_\alpha(\boldsymbol{\rho}_\alpha \times \mathbf{v}_\alpha)_z &= \sum \mathfrak{Z}_k \dot{Q}_k, \end{aligned} \quad (12)$$

where

$$-\mathfrak{X}_k = \sum_{\alpha l} \frac{\partial(\zeta_\alpha, \eta_\alpha)}{\partial(Q_k, Q_l)} Q_l = \sum_{\alpha l} (n_{\alpha k} m_{\alpha l} - m_{\alpha k} n_{\alpha l}) Q_l$$

$$-\mathfrak{Y}_k = \sum_{\alpha l} \frac{\partial(\xi_\alpha, \zeta_\alpha)}{\partial(Q_k, Q_l)} Q_l = \sum_{\alpha l} (l_{\alpha k} n_{\alpha l} - n_{\alpha k} l_{\alpha l}) Q_l \tag{13}$$

$$-\mathfrak{Z}_k = \sum_{\alpha l} \frac{\partial(\eta_\alpha, \xi_\alpha)}{\partial(Q_k, Q_l)} Q_l = \sum_{\alpha l} (m_{\alpha k} l_{\alpha l} - l_{\alpha k} m_{\alpha l}) Q_l,$$

where $[\partial(\zeta_\alpha, \eta_\alpha)]/[\partial(Q_k, Q_l)]$ is the usual abbreviation for

$$\begin{vmatrix} \dfrac{\partial \zeta_\alpha}{\partial Q_k} & \dfrac{\partial \eta_\alpha}{\partial Q_k} \\ \dfrac{\partial \zeta_\alpha}{\partial Q_l} & \dfrac{\partial \eta_\alpha}{\partial Q_l} \end{vmatrix}. \tag{14}$$

The kinetic energy is thus given by

$$\begin{aligned} 2T = {} & I_{xx}\omega_x^2 + I_{yy}\omega_y^2 + I_{zz}\omega_z^2 - 2I_{xy}\omega_x\omega_y - 2I_{yz}\omega_y\omega_z \\ & - 2I_{xz}\omega_x\omega_z + 2\omega_x \sum_k \mathfrak{X}_k \dot{Q}_k + 2\omega_y \sum_k \mathfrak{Y}_k \dot{Q}_k + 2\omega_z \sum_k \mathfrak{Z}_k \dot{Q}_k + \sum_k \dot{Q}_k^2. \end{aligned} \tag{15}$$

Id THE CLASSICAL HAMILTONIAN

To obtain the quantum-mechanical operator **H**, it is necessary to have the kinetic energy expressed in terms of the angular momenta instead of the angular velocities. The angular momentum is a vector which may be defined as

$$\mathbf{P} = \sum_\alpha m_\alpha \mathbf{r}_\alpha \times \dot{\mathbf{r}}_\alpha = \sum_\alpha m_\alpha [\mathbf{r}_\alpha \times (\boldsymbol{\omega} \times \mathbf{r}_\alpha)] + \sum_\alpha m_\alpha (\mathbf{r}_\alpha \times \mathbf{v}_\alpha) \tag{1}$$

Equation (1) can be expanded in an analogous manner to 1c9 to give

$$P_x = I_{xx}\omega_x - I_{xy}\omega_y - I_{xz}\omega_z + \sum \mathfrak{X}_k \dot{Q}_k = \frac{\partial T}{\partial \omega_x}$$

$$P_y = -I_{xy}\omega_x + I_{yy}\omega_y - I_{yz}\omega_z + \sum \mathfrak{Y}_k \dot{Q}_k = \frac{\partial T}{\partial \omega_y} \tag{2}$$

$$P_z = -I_{xz}\omega_x - I_{yz}\omega_y + I_{zz}\omega_z + \sum \mathfrak{Z}_k \dot{Q}_k = \frac{\partial T}{\partial \omega_z}$$

The momentum p_k conjugate to Q_k is

$$p_k = \frac{\partial T}{\partial \dot{Q}_k} = \dot{Q}_k + \mathfrak{X}_k\omega_x + \mathfrak{Y}_k\omega_y + \mathfrak{Z}_k\omega_z. \tag{3}$$

Equation (3) is solved for $\dot{Q}_k$ and substituted into (2) to give

$$P_x = I_{xx}\omega_x - I_{xy}\omega_y - I_{xz}\omega_z + \sum \mathfrak{X}_k(p_k - \mathfrak{X}_k\omega_x - \mathfrak{Y}_k\omega_y - \mathfrak{Z}_k\omega_z), \text{ etc.} \tag{4}$$

It is now convenient to define the following quantities:

$$
\begin{aligned}
I_{xx}' &= I_{xx} - \sum \mathfrak{X}_k^2 & I_{xy}' &= I_{xy} + \sum \mathfrak{X}_k \mathfrak{Y}_k & p_x &= \sum \mathfrak{X}_k p_k \\
I_{yy}' &= I_{yy} - \sum \mathfrak{Y}_k^2 & I_{yz}' &= I_{yz} + \sum \mathfrak{Y}_k \mathfrak{Z}_k & p_y &= \sum \mathfrak{Y}_k p_k \\
I_{zz}' &= I_{zz} - \sum \mathfrak{Z}_k^2 & I_{xz}' &= I_{xz} + \sum \mathfrak{Z}_k \mathfrak{X}_k & p_z &= \sum \mathfrak{Z}_k p_k,
\end{aligned} \tag{5}
$$

where p_x, p_y, and p_z arise from vibration alone as may be seen from their definition; they are called components of internal angular momentum.

On substituting these definitions into (4), we find

$$
\begin{aligned}
(P_x - p_x) &= I_{xx}'\omega_x - I_{xy}'\omega_y - I_{xz}'\omega_z \\
(P_y - p_y) &= -I_{xy}'\omega_x + I_{yy}'\omega_y - I_{yz}'\omega_z \\
(P_z - p_z) &= -I_{xz}'\omega_x - I_{yz}'\omega_y + I_{zz}'\omega_z.
\end{aligned} \tag{6}
$$

The inverse transformation of (6) may be written as

$$
\begin{aligned}
\omega_x &= \mu_{xx}(P_x - p_x) + \mu_{xy}(P_y - p_y) + \mu_{xz}(P_z - p_z) \\
\omega_y &= \mu_{yx}(P_x - p_x) + \mu_{yy}(P_y - p_y) + \mu_{yz}(P_z - p_z) \\
\omega_z &= \mu_{zx}(P_x - p_x) + \mu_{zy}(P_y - p_y) + \mu_{zz}(P_z - p_z),
\end{aligned} \tag{7}
$$

in which

$$
\begin{aligned}
\mu_{xx} &= \frac{I_{yy}'I_{zz}' - I_{yz}'^2}{\Delta} & \mu_{xy} &= \frac{I_{zz}'I_{xy}' + I_{yz}'I_{xz}'}{\Delta} \\
\mu_{yy} &= \frac{I_{xx}'I_{zz}' - I_{xz}'^2}{\Delta} & \mu_{xz} &= \frac{I_{xy}'I_{yz}' + I_{yy}'I_{xz}'}{\Delta} \\
\mu_{zz} &= \frac{I_{xx}'I_{yy}' - I_{xy}'^2}{\Delta} & \mu_{yz} &= \frac{I_{xx}'I_{yz}' + I_{xy}'I_{xz}'}{\Delta}
\end{aligned} \tag{8}
$$

$$
\Delta = \begin{vmatrix} I_{xx}' & -I_{xy}' & -I_{xz}' \\ -I_{xy}' & I_{yy}' & -I_{yz}' \\ -I_{xz}' & -I_{yz}' & I_{zz}' \end{vmatrix}.
$$

Now substituting (5), (6), and (7) into 1c15 the kinetic energy becomes

$$
\begin{aligned}
2T = {} & \mu_{xx}(P_x - p_x)^2 + \mu_{yy}(P_y - p_y)^2 + \mu_{zz}(P_z - p_z)^2 \\
& + 2\mu_{xy}(P_x - p_x)(P_y - p_y) + 2\mu_{yz}(P_y - p_y)(P_z - p_z) \\
& + 2\mu_{zx}(P_z - p_z)(P_x - p_x) + \sum_k p_k^2.
\end{aligned} \tag{9}
$$

The coefficients $\mu_{gg'}$ are functions only of the normal coordinates. The Hamiltonian is obtained by adding the potential energy V to (9). The form of V is discussed more fully in WDC [9].

1e THE QUANTUM-MECHANICAL HAMILTONIAN

The transformation from the classical Hamiltonian to the quantum-mechanical Hamiltonian is not a straightforward process. It is described in detail by Wilson,

Decius, and Cross [9]. The derivation is not important for our work since no new quantities arise; only the results are quoted here, and the interested reader is referred to the mentioned source for the complete details. Thus the complete quantum-mechanical Hamiltonian is

$$\mathbf{H} = \tfrac{1}{2}\mu^{1/4}\sum_{gg'} (\mathbf{P}_g - \mathbf{p}_g)\mu_{gg'}\mu^{-1/2}(\mathbf{P}_{g'} - \mathbf{p}_{g'})\mu^{1/4} + \tfrac{1}{2}\mu^{1/4}\sum_k \mathbf{p}_k\mu^{-1/2}\mathbf{p}_k\mu^{1/4} + \mathbf{V}, \tag{1}$$

in which gg' denote x, y, or z and μ is the determinant of the coefficients $\mu_{gg'}$ which are defined in 1d8. The $\mu_{gg'}$ are functions of the normal coordinates, so at this point no limitations have been placed on the amplitude of vibrations. However, the general solution of (1) is not possible; hence for practical applications it is necessary to introduce approximations.

If THE RIGID ROTOR-HARMONIC-OSCILLATOR APPROXIMATION

For some molecules the vibrational potential energy has a deep minimum at the equilibrium position. Classically, this would mean that the atoms vibrate with small amplitudes about their equilibrium positions. If this is so then it should be a good approximation to neglect the dependence of μ and $\mu_{gg'}$ on the normal coordinates. If μ and $\mu_{gg'}$ are constants, they are not affected by the operators, and the Hamiltonian reduces to

$$\mathbf{H} = \tfrac{1}{2}\sum_{gg'} \mu_{gg'}(\mathbf{P}_g - \mathbf{p}_g)(\mathbf{P}_{g'} - \mathbf{p}_{g'}) + \tfrac{1}{2}\sum_k \mathbf{p}_k^2 + \mathbf{V}. \tag{1}$$

If the x, y, z axes moving with the molecule are chosen to coincide with the principal inertial axes of the molecule in the equilibrium configuration, then the terms depending on products of inertia (I_{xy}, I_{yz}, I_{xz}) will vanish. As a further approximation the $\sum_k \mathfrak{X}_k\mathfrak{Y}_k$ are neglected because they depend on the squares of the displacements. Then, the coefficients I_{xy}' I_{yz}' I_{xz}' in 1d5 vanish and I_{xx}' becomes I_x, I_{yy}' becomes I_y, and I_{zz}' becomes I_z. The single subscripts x, y, z designate the principal inertial axes. The coefficients μ_{gg} become the reciprocals of the rigid moments of inertia and the Hamiltonian operator becomes, in this approximation,

$$\mathbf{H} = \left[\frac{(\mathbf{P}_x - \mathbf{p}_x)^2}{2I_x} + \frac{(\mathbf{P}_y - \mathbf{p}_y)^2}{2I_y} + \frac{(\mathbf{P}_z - \mathbf{p}_z)^2}{2I_z}\right] + \tfrac{1}{2}\sum_k \mathbf{p}_k^2 + \mathbf{V}. \tag{2}$$

The I_g's are the equilibrium values of the principal moments of inertia. The term in the brackets represents the rotational energy together with the interaction of the rotational and internal angular momentum. The last two terms represent purely vibrational motion.

In the simplest approximation the internal angular momenta are neglected, leading to

$$\mathbf{H} = \frac{1}{2}\left[\frac{\mathbf{P}_x^2}{I_x} + \frac{\mathbf{P}_y^2}{I_y} + \frac{\mathbf{P}_z^2}{I_z}\right] + \frac{1}{2}\sum_k \mathbf{p}_k^2 + \mathbf{V}. \tag{3}$$

In this form the Hamiltonian consists of the energy of a rigid, rotating body plus the vibrational energy of a nonrotating molecule. The rotational part of (3) can be solved precisely. If V is a Hooke's Law potential, the vibrational problem is that of an harmonic oscillator which can also be solved completely. The solution for this problem serves as a starting point for almost all problems of molecular spectroscopy.

1g LINEAR POLYATOMIC MOLECULES

An examination of the Hamiltonian given in 1e1 reveals that it is not appropriate for a linear molecule on the basis of the molecular model used here. Our model requires that $I_z \to 0$ $I_y \to I_x$ for a linear molecule of point masses. This, of course, gives an infinite term in the Hamiltonian. This problem has been discussed at length by Nielsen [89].

We restrict ourselves to pointing out the method proposed by Nielsen. He has shown that the linear molecule may be regarded as a limiting case of the axially symmetric molecule. Hence in the classical Hamiltonian the quantity $P_z - p_z = 0$. This also carries over into quantum mechanics. There are now only five coordinates necessary to specify the rigid motion, x, y, z, θ, and ϕ, and there must be $3N - 5$ normal coordinates. This results in only five equations of the type 1c5 and 1c7, which are not sufficient to define the moving axes uniquely. One proceeds just as in the more complicated case, where χ enters as an arbitrary parameter that may be set equal to zero, thus completing the definition of the axes. The problem can then be solved by techniques to be discussed later.

2

THE RIGID ROTOR

2a THE SIMPLEST MODEL

The rotational energy levels of a vibrating-rotating molecule in the simplest approximation are given by the solution of the wave equation resulting from the operator given in 1f3. In this approximation the rotational and vibrational energies are completely separable and the energy levels are those of the rigid rotor and the harmonic-oscillator problems. The harmonic-oscillator problem is solved and fully discussed in most elementary quantum mechanics books and is not discussed here [2, 7].

The problem of more immediate interest here is the solution for the rigid rotor. The Hamiltonian for this problem is given by the first term of 1f3, that is,

$$\mathbf{H}_R = \frac{1}{2}\left(\frac{\mathbf{P}_x^{\ 2}}{I_x} + \frac{\mathbf{P}_y^{\ 2}}{I_y} + \frac{\mathbf{P}_z^{\ 2}}{I_z}\right). \tag{1}$$

2b THE ANGULAR MOMENTUM OPERATORS

Among the angular momentum operators there exists a set of rules of combination called commutation rules. If the operators $\mathbf{P}_g$ are written out explicitly in wave mechanical form, these commutation rules may be deduced. On the other hand, in matrix mechanics these rules are basic postulates [2]. The latter viewpoint is accepted here, and the derivation from the wave mechanical operators is left as an exercise for the reader. For instance, see [2], p. 39].

Since much of the asymmetric rotor knowledge has been systemized and extended by King, Hainer, and Cross, [36, 67], the conventions which they have set up are used here. In their notation the commutation rules are*

$$\begin{aligned} \mathbf{P}_x\mathbf{P}_y - \mathbf{P}_y\mathbf{P}_x &= [\mathbf{P}_x, \mathbf{P}_y] = -i\hbar\mathbf{P}_z \\ \mathbf{P}_y\mathbf{P}_z - \mathbf{P}_z\mathbf{P}_y &= [\mathbf{P}_y, \mathbf{P}_z] = -i\hbar\mathbf{P}_x \\ \mathbf{P}_z\mathbf{P}_x - \mathbf{P}_x\mathbf{P}_z &= [\mathbf{P}_z, \mathbf{P}_x] = -i\hbar\mathbf{P}_y\,. \end{aligned} \tag{1}$$

2c THE MATRIX ELEMENTS† OF P^2 AND P_z

The square of the angular momentum is

$$\mathbf{P}^2 = \mathbf{P}_x^{\,2} + \mathbf{P}_y^{\,2} + \mathbf{P}_z^{\,2}. \tag{1}$$

From this definition and a little operator manipulation it can be seen that

$$[\mathbf{P}^2, \mathbf{P}] = [\mathbf{P}\cdot\mathbf{P}, \mathbf{P}] = \mathbf{P}\cdot[\mathbf{P}, \mathbf{P}] - [\mathbf{P}, \mathbf{P}]\cdot\mathbf{P}. \tag{2}$$

An operator always commutes with itself; that is, $[\mathbf{P}, \mathbf{P}] = \mathbf{PP} - \mathbf{PP} = 0$, and it follows that $[\mathbf{P}^2, \mathbf{P}] = 0$. Since $\mathbf{P}^2$ commutes with $\mathbf{P}$, it commutes with all its components and any polynomial of $\mathbf{P}$ such as $\mathbf{H}$. Hence, there exist simultaneous eigenstates of $\mathbf{P}^2$ and one of the components, say $\mathbf{P}_z$. The operator $\mathbf{P}^2$ must be Hermitian, which requires that its eigenvalues be real. (For properties of Hermitian operators see [2, p. 37]. Thus the eigenvalues of $\mathbf{P}^2 \geqslant 0$; let us call them P^2. It is now possible to find the allowed values of $\mathbf{P}^2$ and $\mathbf{P}_z$.

Let $\psi(P^2, P_z, \mathscr{P}_Z)$ be an eigenfunction of the operators $\mathbf{P}^2$ and $\mathbf{P}_z$ and a set of observables which we shall call, for convenience, $\mathscr{P}_Z$, which make up with $\mathbf{P}^2$ and $\mathbf{P}_z$ a complete set of independent commuting observables for this system. We represent the eigenvalues of these observables by P^2, P_z, and $\mathscr{P}_Z$. It is required that $\mathscr{P}_Z$ commute with $\mathbf{P}_x$ and $\mathbf{P}_y$ as well as $\mathbf{P}^2$ and $\mathbf{P}_z$.

Now

$$\begin{aligned} \mathbf{P}^2\psi(P^2, P_z, \mathscr{P}_Z) = (\mathbf{P}_x^{\,2} + \mathbf{P}_y^{\,2} + \mathbf{P}_z^{\,2})\psi(P^2, P_z, \mathscr{P}_Z) &= P^2\psi(P^2, P_z, \mathscr{P}_Z) \\ \mathbf{P}_z\psi(P^2, P_z\mathscr{P}_Z) &= P_z\psi(P^2, P_z, \mathscr{P}_Z), \end{aligned} \tag{3}$$

from which it is apparent that

$$(\mathbf{P}_x^{\,2} + \mathbf{P}_y^{\,2})\psi(P^2, P_z, \mathscr{P}_Z) = (P^2 - P_z^{\,2})\psi(P^2, P_z, \mathscr{P}_Z). \tag{4}$$

* The commutation relations for the components of the angular momentum relative to the space-fixed axis system X, Y, Z are normally written with a positive sign; that is, $\mathscr{P}_X\mathscr{P}_Y - \mathscr{P}_Y\mathscr{P}_X = i\hbar\mathscr{P}_Z$, etc. The total angular momentum is a constant of the motion and hence has the same value regardless of the system of axes to which it is referred. The components of the angular momentum relative to the molecular fixed axes are related to the components relative to the space-fixed system through the relations $\mathbf{P}_g = \sum_F \phi_{Fg}\mathscr{P}_F$. A simple transformation on the commutator $\mathscr{P}_X\mathscr{P}_Y - \mathscr{P}_Y\mathscr{P}_X$ using the relating equations and the fact that the direction cosines transform like vectors, hence satisfying the commutators $\phi_{Xx}\mathbf{P}_y - \mathbf{P}_y\phi_{Xx} = -i\hbar\phi_{Xz}$, shows that the corresponding commutators in the molecule-fixed system must have a negative sign. For instance, see J. H. Van Vleck [117].

† This development follows that given by Condon and Shortley [1].

Since $\mathbf{P}_x$ and $\mathbf{P}_y$ correspond to real observables and since the allowed values of $(\mathbf{P}_x^2 + \mathbf{P}_y^2)$ and of $\mathbf{P}^2$ are real and positive, we can say that

$$P_z \leqslant (P^2)^{1/2}, \tag{5}$$

which shows that P_z is bounded and lies between the values $+(P^2)^{1/2}$ and $-(P^2)^{1/2}$.

From the commutation relations 2b1, it is found that

$$\begin{aligned}\mathbf{P}_z(\mathbf{P}_y \pm i\mathbf{P}_x) &= \mathbf{P}_z\mathbf{P}_y \pm i\mathbf{P}_z\mathbf{P}_x \\ &= \mathbf{P}_y\mathbf{P}_z + [\mathbf{P}_z, \mathbf{P}_y] \pm i\mathbf{P}_x\mathbf{P}_z \pm i[\mathbf{P}_z, \mathbf{P}_x] \\ &= (\mathbf{P}_y \pm i\mathbf{P}_x)(\mathbf{P}_z \pm \hbar).\end{aligned} \tag{6}$$

Operating on $\psi(P^2, P_z, \mathscr{P}_Z)$ by both sides of (6), we obtain

$$\mathbf{P}_z(\mathbf{P}_y \pm i\mathbf{P}_x)\psi(P^2, P_z, \mathscr{P}_Z) = (P_z \pm \hbar)(\mathbf{P}_y \pm i\mathbf{P}_x)\psi(P^2, P_z \mathscr{P}_Z). \tag{7}$$

This is the equation of an eigenvalue problem which states that unless

$$(\mathbf{P}_y \pm i\mathbf{P}_x)\psi(P^2, P_z, \mathscr{P}_Z) = 0, \tag{8}$$

it is an eigenfunction of $\mathbf{P}_z$ belonging to the eigenvalue $P_z \pm \hbar$. It is also an eigenfunction of $\mathbf{P}^2$ belonging to the eigenvalue P^2 and of $\mathscr{P}_Z$ belonging to the eigenvalue $\mathscr{P}_Z$ since both $\mathbf{P}^2$ and $\mathscr{P}_Z$ commute with $\mathbf{P}_x$ and $\mathbf{P}_y$. Therefore, starting with a given pair of allowed simultaneous eigenvalues P^2 and P_z, in general, a whole series of allowed simultaneous eigenvalues are found.

$$\ldots P^2, P_z - \hbar; \qquad P^2, P_z; \qquad P^2, P_z + \hbar \ldots. \tag{9}$$

It has already been shown that the values of P_z which may occur in simultaneous eigenstates with P^2 are bounded. Therefore the series must have a lowest member, P^2, P_z^0 and a highest member P^2, P_z^1.

From (8) it is seen that these states must satisfy the relations

$$\begin{aligned}&(a) \qquad (\mathbf{P}_y - i\mathbf{P}_x)\psi(P^2, P_z^0, \mathscr{P}_Z) = 0 \\ &(b) \qquad (\mathbf{P}_y + i\mathbf{P}_x)\psi(P^2, P_z^1, \mathscr{P}_Z) = 0;\end{aligned} \tag{10}$$

otherwise the left side of these relations would be eigenfunctions belonging to $P_z^0 - \hbar$ and $P_z^1 + \hbar$, which is contrary to the hypothesis.

Operation on (10*a*) by $(\mathbf{P}_y + i\mathbf{P}_x)$ gives

$$(\mathbf{P}_x^2 + \mathbf{P}_y^2 + \hbar\mathbf{P}_z)\psi(P^2, P_z^0, \mathscr{P}_Z) = (P^2 - (P_z^0)^2 + \hbar P_z^0)\psi(P^2, P_z^0, \mathscr{P}_Z) = 0. \tag{11}$$

Since $\psi(P^2, P_z^0, \mathscr{P}_Z) \neq 0$ by hypothesis,

$$P^2 - (P_z^0)^2 + \hbar P_z^0 = 0. \tag{12}$$

Operation on (10*b*) by $(\mathbf{P}_y - i\mathbf{P}_x)$ leads in a similar way to the relation

$$P^2 - (P_z^1)^2 - \hbar P_z^1 = 0. \tag{13}$$

Solving (12) and (13) for P^2 and setting the results equal gives

$$(P_z^1 + P_z^0)(P_z^0 - P_z^1 - \hbar) = 0. \tag{14}$$

By hypothesis $P_z^1 \geqslant P_z^0$; hence the first factor shows that $P_z^0 = -P_z^1$. From (14) and (9) it is seen that the difference $P_z^1 - P_z^0$ must be zero or a positive integer times $\hbar$. Call this integer $2J$, where J has the values 0, $\frac{1}{2}$, 1, $\frac{3}{2}$, 2, etc. Thus we have

$$P_z^0 = -J\hbar, \qquad P_z^1 = J\hbar. \tag{15}$$

Substitution of these values into (12) and (13) gives the result that

$$P^2 = J(J+1)\hbar^2 \tag{16}$$

For any one value of P^2 the values of P_z are

$$J\hbar,\ (J-1)\hbar, \qquad (J-2)\hbar \ldots -(J-1)\hbar, \qquad -J\hbar. \tag{17}$$

The different constants in (17) are denoted by K, and M is used for the present to denote the quantum number associated with $\mathscr{P}_Z$. Hence the matrix elements of $\mathbf{P}_z$ are

$$(P^2, P_z, \mathscr{P}_Z\,|\mathbf{P}_z|\,P^2, P_z, \mathscr{P}_Z) = K\hbar, = (J, K, M\,|\mathbf{P}_z|\,J, K, M), \tag{18}$$

whereas for P^2 one finds

$$(P^2, P_z, \mathscr{P}_Z\,|\mathbf{P}^2|\,P^2, P_z, \mathscr{P}_Z) = J(J+1)\hbar^2. \tag{19}$$

A brief discussion about the allowed values of K is in order at this time. In future discussions it is assumed that the integral values of K are those appropriate to our problem. This selection, which is necessary so that the wave function may be single valued, is discussed in Appendix I where it is proved that the half integral quantum numbers are unsuitable for the model, the rigid rotor, being discussed. The half-integral values of K are important in problems of spin-angular momentum.

2d THE MATRIX ELEMENTS OF $\mathbf{P}_x$ AND $\mathbf{P}_y$

We now want to find the matrix elements of $\mathbf{P}_x$ and $\mathbf{P}_y$ in the representation in which $\mathbf{P}^2$ and $\mathbf{P}_z$ are diagonal. To simplify the notation, the eigenvalues P^2 and P_z are denoted by their quantum numbers J and K, respectively. The quantum number M for $\mathscr{P}_Z$ is chosen in anticipation of the later result that $\mathscr{P}_Z$ is the Z component of the angular momentum in the space-fixed coordinate system.

The first step in this process is to normalize the wave functions found in 2c7. This equation tells us that

$$\psi(J, K \pm 1, M) = N_{\pm}(\mathbf{P}_y \pm i\mathbf{P}_x)\psi(J, K, M), \tag{1}$$

where $N_{\pm}$ is the factor necessary to normalize $\psi(J, K \pm 1, M)$ when

$$\int \bar{\psi}(J, K, M)\psi(J, K, M)\, d\tau = 1, \tag{2}$$

where $\bar{\psi}(J, K, M)$ is the complex conjugate of $\psi(J, K, M)$.

Since $\overline{(\mathbf{P}_y \pm i\mathbf{P}_x)} = (\mathbf{P}_y \mp i\mathbf{P}_x)$, we find that

$$\begin{aligned}
&\int \bar{\psi}(J, K \pm 1, M)\psi(J, K \pm 1, M)\, d\tau \\
&= \int \overline{N_\pm(\mathbf{P}_y \pm i\mathbf{P}_x)\psi(J, K, M)} N_\pm(\mathbf{P}_y \pm i\mathbf{P}_x)\psi(J, K, M)\, d\tau \\
&= \int \bar{N}_\pm N_\pm \bar{\psi}(J, K, M)(\mathbf{P}_y \mp i\mathbf{P}_x)(\mathbf{P}_y \pm i\mathbf{P}_x)\psi(J, K, M)\, d\tau \\
&= \int \bar{N}_\pm N_\pm \bar{\psi}(J, K, M)(\mathbf{P}_x^2 + \mathbf{P}_y^2 \mp \hbar\mathbf{P}_z)\psi(J, K, M)\, d\tau \qquad (3) \\
&= \int \bar{N}_\pm N_\pm \bar{\psi}(J, K, M)[P^2 - P_z(P_z \pm \hbar)]\psi(J, K, M)\, d\tau \\
&= \bar{N}_\pm N_\pm \hbar^2[J(J+1) - K(K \pm 1)] \\
&= \bar{N}_\pm N_\pm[(J \mp K)(J \pm K + 1)]\hbar^2,
\end{aligned}$$

where use has been made of the Hermitian character of the operators. For this to equal unity.

$$N_\pm = \frac{e^{i\delta}}{\hbar[(J \mp K)(J \pm K + 1)]^{1/2}}, \qquad (4)$$

where δ is an arbitrary real number.

This is the arbitrary phase which occurs in any ψ because two states are not distinct unless they are linearly independent. Such indeterminacy of phase can have no effect on any results of physical significance. For this reason we shall therefore set $\delta = 0$, and thus from (1) and (4) we obtain

$$(\mathbf{P}_y \pm i\mathbf{P}_x)\psi(J, K, M) = \hbar[(J \mp K)(J \pm K + 1)]^{1/2}\psi(J, K \pm 1, M). \qquad (5)$$

This corresponds to the phase chosen by King, Hainer, and Cross [67].

The matrix elements of $\mathbf{P}_x$ and $\mathbf{P}_y$ can now be written down immediately.

$$\begin{aligned}
(J', K', M' |\mathbf{P}_y| J, K, M) = \tfrac{1}{2}\int \bar{\psi}(J', K', M')(\mathbf{P}_y + i\mathbf{P}_x)\psi(J, K, M)\, d\tau \\
+ \tfrac{1}{2}\int \bar{\psi}(J', K', M')(\mathbf{P}_y - i\mathbf{P}_x)\psi(J, K, M)\, d\tau.
\end{aligned} \qquad (6)$$

It is seen from (5) that this element vanishes unless $J' = J$, $M' = M$, $K' = K \pm 1$. Consequently, the nonvanishing elements are

$$(J, K \pm 1, M |\mathbf{P}_y| J, K, M) = \frac{\hbar}{2}[(J \mp K)(J \pm K + 1)]^{1/2}. \qquad (7)$$

By replacing the positive sign between the integrals on the right of (6) with a negative sign, we obtain the elements of $\mathbf{P}_x$ immediately.

$$(J, K \pm 1, M |\mathbf{P}_x| J, K, M) = \mp\frac{i\hbar}{2}[(J \mp K)(J \pm K + 1)]^{1/2}. \qquad (8)$$

A more general way of connecting these results is

$$(J, K', M |\mathbf{P}_y| J, K, M) = -e^{i(\pi/2)(K'-K)}(J, K', M |\mathbf{P}_x| J, K, M). \qquad (9)$$

In these expressions M represents the quantum numbers of a set of observables which commute with $\mathbf{P}$. Consider in the J, K, M representation an observable $\mathscr{P}_Z$ which commutes with $\mathbf{P}$. The matrix is diagonal with respect to J and K and can be shown to be entirely independent of K by the following method:

$$\mathscr{P}_Z\psi(J, K, M') = \sum_M \psi(J, K, M)(J, K, M\,|\mathscr{P}_Z|\,J, K, M'). \tag{10}$$

As

$$(\mathbf{P}_y \pm i\mathbf{P}_x)\mathscr{P}_Z\psi(J, K, M') = \sum_M \psi(J, K \pm 1, M)\hbar[(J \mp K)(J \pm K + 1)]^{1/2} \times (J, K, M\,|\mathscr{P}_Z|\,J, K, M') \tag{11}$$

$$\mathscr{P}_Z(\mathbf{P}_y \pm i\mathbf{P}_x)\psi(J, K, M') = \sum_M \psi(J, K \pm 1, M)\hbar[(J \mp K)(J \pm K + 1)]^{1/2} \times (J, K \pm 1, M\,|\mathscr{P}_Z|\,J, K \pm 1, M'). \tag{12}$$

These two expressions are equal since $\mathscr{P}_Z$ and $\mathbf{P}$ commute by hypothesis. Comparing coefficients of $\psi(J, K \pm 1, M)$ it is seen that

$$(J, K, M\,|\mathscr{P}_Z|\,J, K, M') = (J, K \pm 1, M\,|\mathscr{P}_Z|\,J, K \pm 1, M'),$$

if

$$[\mathscr{P}_Z, \mathbf{P}] = 0, \tag{13}$$

for all values of K such that the coefficient $[(J \mp K)(J \pm K + 1)]^{1/2} \neq 0$. This theorem is important when the Hamiltonian commutes with various angular momenta.

2e THE MATRIX ELEMENTS OF $\mathbf{P}_x^2$ AND $\mathbf{P}_y^2$

To obtain the matrix elements of the Hamiltonian it is necessary to have the matrix elements of $\mathbf{P}_x^2$ and $\mathbf{P}_y^2$. These elements are readily obtained from the matrix elements of $\mathbf{P}_x$ and $\mathbf{P}_y$, 2d7 and 2d8, by matrix multiplication. It is found that $\mathbf{P}_x^2$ and $\mathbf{P}_y^2$ have elements on the main diagonal and two off the main diagonal; thus

$$(J, K, M\,|\mathbf{P}_y^2|\,J, K, M) = (J, K, M\,|\mathbf{P}_x^2|\,J, K, M) = \frac{\hbar^2}{2}[J(J + 1) - K^2], \tag{1}$$

$$(J, K \pm 2, M\,|\mathbf{P}_y^2|\,J, K, M) = -(J, K \pm 2, M\,|\mathbf{P}_x^2|\,J, K, M) = \frac{\hbar^2}{4}[(J \mp K)(J \pm K + 1)(J \mp K - 1)(J \pm K + 2)]^{1/2}. \tag{2}$$

2f THE MATRIX ELEMENTS OF THE ANGULAR MOMENTUM IN THE NONROTATING COORDINATE SYSTEM: SPACE QUANTIZATION

Although the energy levels, in the absence of external fields, are independent of space quantization, the elements of the angular momentum in the space-fixed coordinate system will be needed later for the solution of the intensity problem. For convenience they are deduced now.

The angular momentum operators in the space-fixed coordinate system, $\mathscr{P}_F$ ($F = X, Y, Z$), have the following commutation rules:

$$(\mathscr{P}_X\mathscr{P}_Y) = i\hbar\mathscr{P}_Z; \qquad (\mathscr{P}_Y, \mathscr{P}_Z) = i\hbar\mathscr{P}_X; \qquad [\mathscr{P}_Z\mathscr{P}_X] = i\hbar\mathscr{P}_Y. \tag{1}$$

Furthermore, since the square of the angular momentum must be independent of the reference frame

$$P^2 = (P_x^{\,2} + P_y^{\,2} + P_z^{\,2}) = (\mathscr{P}_X^{\,2} + \mathscr{P}_Y^{\,2} + \mathscr{P}_Z^{\,2}) = \mathscr{P}^2, \tag{2}$$

which leads to the commutation rules

$$[\mathbf{P}^2, \mathscr{P}_F] = 0 \qquad [\mathbf{P}_g, \mathscr{P}_F] = 0 \tag{3}$$

for all F and g.

We can readily see now that the operator $\mathscr{P}_Z$ carried along through the previous derivations is the operator corresponding to the Z-component of the angular momentum in the space-fixed coordinate system.

Using methods entirely analogous to those used in the preceding sections, the nonvanishing elements are found to be

$$(J, K, M\,|\mathscr{P}_Z|\,J, K, M) = M\hbar$$

$$(J, K, M\,|\mathscr{P}_Y|\,J, K, M+1) = -i(J, K, M\,|\mathscr{P}_X|\,J, K, M+1) \tag{4}$$

$$= \frac{\hbar}{2}\,[J(J+1) - M(M+1)]^{1/2}.$$

2g THE ENERGY MATRIX H

The energy matrix $\mathbf{H}$ is readily determined from the form of the Hamiltonian operator 2a1 and the matrix elements already determined. This is not, in general, a diagonal matrix but has the following nonvanishing elements:

$$(J, K, M\,|\mathbf{H}|\,J, K, M) = \frac{\hbar^2}{4}\left\{\left(\frac{1}{I_x} + \frac{1}{I_y}\right)[J(J+1) - K^2] + \frac{2K^2}{I_z}\right\}$$

$$(J, K+2, M\,|\mathbf{H}|\,J, K, M) = (J, K, M\,|\mathbf{H}|\,J, K+2, M) \tag{1}$$

$$= \frac{\hbar^2}{8}\left\{\left(\frac{1}{I_y} - \frac{1}{I_x}\right)[(J-K)(J-K-1)(J+K+1)(J+K+2)]^{1/2}\right\}.$$

In (1) x, y, and z refer to the principal axes of the inertial dyadic. It is customary in molecular spectroscopy to designate these axes as a, b, and c with the convention that $I_a < I_b < I_c$. There are $n!$ or six ways in which the a, b, c axes can be identified with the x, y, z axes. The six possibilities and their identities in the King, Hainer, and Cross [67] notation are given in Table 2g1. Depending on the values of the moments of inertia, certain identifications are more convenient in special cases.

It is also convenient to incorporate the constants and the moments of inertia into new constants. These new constants are the quantities determined from experiment and are characteristic of the molecule under study; thus

$$A = \frac{\hbar^2}{2I_a}; \qquad B = \frac{\hbar^2}{2I_b}; \qquad C = \frac{\hbar^2}{2I_c}; \tag{2}$$

with $A > B > C$.

The quantized energy levels are the characteristic roots of the matrix **H**. The "probability amplitudes" corresponding to each rotational level define the transformation which diagonalizes **H**; that is, transforms **H** from a J, K-representation to a J, τ-representation. The number τ is not a true quantum number but serves to index

Table 2g I—The Six Ways of Identifying a, b, c with x, y, z

	I^r	II^r	III^r	I^l	II^l	III^l
x	B	C	A	C	A	B
y	C	A	B	B	C	A
z	A	B	C	A	B	C

the energy levels of the most general case, the asymmetric rotor. This index runs from $-J$ to $+J$, giving $2J + 1$ levels.

2h THE SYMMETRIC ROTOR

If two moments of inertia for a given molecule are equal, it is called a symmetric rotor. Obviously there are two possibilities for such a situation. In Case 1 the two

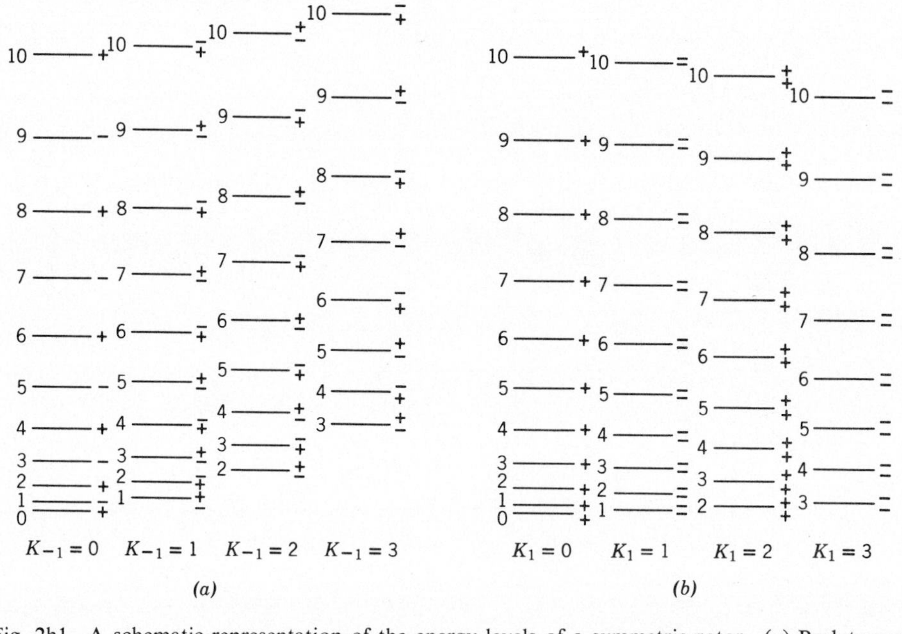

Fig. 2h1 A schematic representation of the energy levels of a symmetric rotor. (*a*) Prolate symmetric rotor, $\kappa = -1$. (*b*) Oblate symmetric rotor, $\kappa = +1$. For nonplanar molecules the plus and minus signs represent the behavior with respect to inversion of only the upper level of each pair of nearly degenerate levels. (See [4], pp. 25–27.) (From Herzberg, *Infrared and Raman Spectra*, copyright 1945, D. van Nostrand Company, Princeton, New Jersey.)

equal moments of inertia are smaller than the unique moment of inertia, and in Case 2 the two equal moments of inertia are larger than the unique moment of inertia. In either case the off-diagonal terms in the Hamiltonian matrix become zero since $I_x = I_y$. The energy is given by

$$F(J, K) = \frac{\hbar^2}{2}\left[\left(\frac{1}{I_x}\right)J(J+1) + \left(\frac{1}{I_z} - \frac{1}{I_x}\right)K^2\right]. \qquad (1)$$

Referring to Table 2g1, if a III (unique initial axis is *c*-axis) representation is chosen, the energy equation for Case 1 is obtained; that is,*

$$F(J, K) = AJ(J+1) + (C - A)K_1^2. \qquad (2)$$

This case is called the oblate symmetric limit. Since $C - A < 0$ the lowest energy level in a given J is that for $K_1 = J$. The energies of the various K_1 levels decrease as K_1 increases. Since K_1 enters into the energy expression as a square, each level, except for $K_1 = 0$, is doubly degenerate.

If one chooses a Type I (unique inertial axis is *a*-axis) representation, the energy equation takes the form appropriate to Case 2.

$$F(J, K) = CJ(J+1) + (A - C)K_{-1}^2. \qquad (3)$$

This case is called the prolate symmetric limit. Here $A - C > 0$ and the level with $K_{-1} = 0$ is the level of lowest energy in a given J. As K_{-1} increases within a given J the energy increases. Again each level with $K_{-1} \neq 0$ is doubly degenerate. The levels appropriate to these two limits are shown in Fig. 2h1. Beside the K_{-1} degeneracy each level has a $2J + 1$ degeneracy because of the space quantization.

2i THE LINEAR AND SPHERICAL ROTOR

The similarity of the energy expression for the spherical rotor and the linear molecule makes it convenient to consider them together since a good plausibility argument can be made which indicates the difference in the two cases. For a linear molecule $I_x = I_y$, $I_z = 0$. The off-diagonal terms are zero, whereas the coefficient of K^2 becomes infinite, which leads to infinite energies for the levels of a linear molecule. The only time a finite energy is possible is for $K = 0$. Since the linear molecule does have finite levels, they must occur for $K = 0$; hence for a linear molecule the vibrational-rotational levels are given by

$$F(J) = BJ(J+1). \qquad (1)$$

The symbol B is used to denote the inertial constant in the energy expressions for linear and spherical rotors to agree with established convention since no confusion can arise from its use. The only degeneracy of these states is $2J + 1$ arising from the space-quantization.

The spherical rotor has $I_x = I_y = I_z$. Again the off-diagonal elements of **H** become zero. The coefficient of K^2 in this instance becomes zero, leading to

$$F(J) = BJ(J+1); \qquad (2)$$

hence all K values lead to the same energies. There is also an additional degeneracy of $2J + 1$, which together with the space degeneracy gives a total degeneracy of

* The significance of the K subscripts will become apparent when the asymmetric case is discussed. For the present they may be regarded as differentiating the two symmetric limits.

$(2J + 1)^2$. This K degeneracy indicates that there is no preferred direction in the molecule-fixed axis system, along which one component of the angular momentum is quantized.

2j THE ASYMMETRIC ROTOR

The asymmetric rotor, $I_x \neq I_y \neq I_z$, is the most general case. As might be expected the energy level pattern is more complex since the K degeneracy is removed. The only meaningful quantum number affecting the energy is J, the quantum number for the total angular momentum. It was common practice to index the different levels of a given J with a pseudo-quantum number τ. This index took on integral values between $-J$ and $+J$, the value $\tau = -J$ being assigned to the lowest energy state in a given J. This notation has been largely superseded by a more meaningful double index notation which is described in the following paragraph.

Remembering that there are six possible ways in which the a, b, c axes can be identified with the x, y, z axes, the rigid rotor energy is now written in the form

$$E(A, B, C) = \frac{\mathbf{P}_a^{\,2}}{2I_a} + \frac{\mathbf{P}_b^{\,2}}{2I_b} + \frac{\mathbf{P}_c^{\,2}}{2I_c} = \frac{A\mathbf{P}_a^{\,2} + B\mathbf{P}_b^{\,2} + C\mathbf{P}_c^{\,2}}{\hbar^2}. \tag{1}$$

The actual identification of axes can be made from Table 2g1 for the specific problem at hand. The calculation of the energy levels is greatly facilitated by the introduction of a change of variables. Through the years several such parameters have been introduced. One of the most useful and most widely used is that due to Ray [99].

Let σ and ρ be scalar factors and let us introduce the following change of variables:

$$\begin{aligned} E(\sigma A + \rho, \sigma B + \rho, \sigma C + \rho) &= [(\sigma A + \rho)\mathbf{P}_a^{\,2} + (\sigma B + \rho)\mathbf{P}_b^{\,2} + (\sigma C + \rho)\mathbf{P}^2]/\hbar_c^{\,2} \\ &= [\sigma(A\mathbf{P}_a^{\,2} + B\mathbf{P}_b^{\,2} + C\mathbf{P}_c^{\,2}) + \rho(\mathbf{P}_a^{\,2} + \mathbf{P}_b^{\,2} + \mathbf{P}_c^{\,2})]/\hbar^2. \end{aligned} \tag{2}$$

With the help of (1) and 2c19 this can be reduced to

$$E(\sigma A + \rho, \sigma B + \rho, \sigma C + \rho) = \sigma E(A, B, C) + \rho J(J + 1). \tag{3}$$

Now choose

$$\sigma = \frac{2}{A - C}, \qquad \rho = -\frac{A + C}{A - C}, \tag{4}$$

so that

$$\begin{aligned} \sigma A + \rho &= 1 \\ \sigma B + \rho &= \frac{2B - A - C}{A - C} \\ \sigma C + \rho &= -1. \end{aligned} \tag{5}$$

An asymmetry parameter κ is now defined as

$$\kappa = \frac{2B - A - C}{A - C}. \tag{6}$$

The parameter κ results naturally from Ray's choice of diagonalizing $\mathbf{P}_b$, the angular momentum about the intermediate inertial axis. The chief value of using κ can be seen from the relation proved by Ray

$$E_\tau{}^J(\kappa) = -E_{-\tau}{}^J(-\kappa), \tag{7}$$

which gives the energies of positive κ from the energies of negative κ. The limit of $\kappa = -1$, or $B = C$ is the prolate symmetric rotor 2h3, whereas $\kappa = +1$, or $B = A$ is the oblate symmetric rotor 2h2.

Substituting (4), (5), and (6) into (2), we obtain

$$E(1, \kappa, -1) = E(\kappa) = \frac{2}{A - C} E(A, B, C) - \left(\frac{A + C}{A - C}\right) J(J + 1), \tag{8}$$

which leads to the energy of any asymmetric rotor on rearrangement

$$E(A, B, C) = F(J_\tau) = \frac{A + C}{2} J(J + 1) + \frac{A - C}{2} E_\tau(\kappa). \tag{9}$$

Equations (9) and (7) show that if the energy levels $E(\kappa)$ for $0 < \kappa > -1$ are determined, the energy levels for any rigid asymmetric rotor can be calculated by a simple arithmetic procedure.

The energy $E(\kappa)$ is essentially the energy of a molecule with inertial constants 1, κ, and -1. As such, $E(\kappa)$ can be calculated from the matrix elements of the Hamiltonian 2g1 by making the appropriate substitutions. Depending on the representation chosen from Table 2g1, the coefficients will have different values. The matrix elements of $E(\kappa)$ may be written as

$$\begin{aligned}(J, K, M\,|E(\kappa)|\,J, K, M) &= F[J(J + 1) - K^2] + GK^2 \\ &= FJ(J + 1) + (G - F)K^2\end{aligned} \tag{10}$$

$$\begin{aligned}(J, K, M\,|E(\kappa)|\,J, K + 2, M) &= (J, K + 2, M\,|H|\,J, K, M) \\ &= H[f(J, K + 1)]^{1/2},\end{aligned} \tag{11}$$

where the $f(J, K + 1)$ are given by

$$f(J, n) = f(J, -n) = \tfrac{1}{4}[J(J + 1) - n(n + 1)][J(J + 1) - n(n - 1)]. \tag{12}$$

Values of $f(J, n)$ may be found for J up to 30 in Appendix II.

The six forms resulting from the particular identifications of $\mathbf{P}_a$, $\mathbf{P}_b$, $\mathbf{P}_c$ with $\mathbf{P}_x$, $\mathbf{P}_y$, $\mathbf{P}_z$ may be arranged in two sets of three, each differing only in the sign of H; that is, the off-diagonal elements. The difference in sign of nondiagonal elements does not affect the roots of Hermitian matrices, but the sign of H must be retained to assign symmetry properly.

The expressions for F, G, and H in terms of κ for the three right-handed representations are given in Table 2j1.

For an oblate spheroid, $I_b = I_a$, $B = A$, $\kappa = 1$ and $z = c$ (Type III), the energy matrix becomes diagonal; that is, $H = 0$. The matrix elements of $E(\kappa) = E(1)$ become

$$(J, K, M\,|E(\kappa)|\,J, K, M) = J(J + 1) - 2K_1{}^2. \tag{13}$$

Substitution of (13) into (9) yields 2h2.

For κ in the neighborhood of $+1$, Type III representations are nearly diagonal, and hence the most convenient to use in determining the energy levels.

For the prolate limit $I_b = I_c$, $B = C$, $\kappa = -1$, and $a = z$ (Type I), the energy matrix is again diagonal, the elements being

$$(J, K, M\,|E(\kappa)|\,J, K, M) = -J(J + 1) + 2K_{-1}^2, \tag{14}$$

giving 2h3 when (14) is substituted into (9). Type I, being nearly diagonal for κ nearly -1, is the most convenient representation to use in this region of κ.

Table 2j1—The Coefficients for the Matrix Elements of E(κ)

REPRESENTATION	I^r	II^r	III^r
F	$\frac{1}{2}(\kappa - 1)$	0	$\frac{1}{2}(\kappa + 1)$
G	1	κ	-1
H	$-\frac{1}{2}(\kappa + 1)$	1	$\frac{1}{2}(\kappa - 1)$
$G - F$	$-\frac{1}{2}(\kappa - 3)$	κ	$-\frac{1}{2}(\kappa + 3)$
$F + G - H$	$\kappa + 1$	$\kappa - 1$	0
$F + G + H$	0	$\kappa + 1$	$\kappa - 1$

Type II representations never become diagonal. Their main use is for maximum asymmetry $\kappa = 0$. Here its main diagonal elements are all zero.

2k THE WANG ASYMMETRY PARAMETER

In early spectroscopic work another parameter of asymmetry was commonly used, which is due to Wang [118]. Although the usefulness of this parameter has been greatly reduced since extensive tables of asymmetric rotor energies have become available in terms of κ, no treatment of the problem would be complete without a consideration of this parameter. It might be said that this parameter is still useful in certain perturbation calculations and in some ways is a more natural parameter than κ.

This parameter is introduced most easily if we return to 2j4 and redefine σ and ρ as

$$\begin{aligned} \sigma &= \frac{1}{A - \frac{1}{2}(B + C)} \\ \rho &= -\frac{B + C}{2[A - \frac{1}{2}(B + C)]}, \end{aligned} \tag{1}$$

from which it is found that

$$\begin{aligned} \sigma A + \rho &= 1 \\ \sigma B + \rho &= -\frac{C - B}{2[A - \frac{1}{2}(B + C)]} \\ \sigma C + \rho &= \frac{C - B}{2[A - \frac{1}{2}(B + C)]}. \end{aligned} \tag{2}$$

The parameter of asymmetry b is now defined as

$$b = \frac{C - B}{2[A - \frac{1}{2}(B + C)]} = \frac{\kappa + 1}{\kappa - 3}, \tag{3}$$

which has the property $-1 < b < 0$, the limit $b = 0$, $B = C$ being the prolate symmetric limit, whereas $b = -1$, $A = B$, is the oblate symmetric limit. With these definitions the energy equation for any asymmetric rotor becomes

$$E(A, B, C) = \frac{B + C}{2} J(J + 1) + [A - \tfrac{1}{2}(B + C)]E(b). \tag{4}$$

The matrix elements of $E(b)$ may also be written as shown in 2j10 and 2j11 if the proper definitions of F, G, and H are used. These definitions are given in Table 2k1. $E(b)$ corresponds to the energy of a rotor having the reduced parameters $(1, -b, b)$.

Table 2kl—The Coefficients of the Matrix Elements of E(b)

b^*	III^l	II^l	I^l
b	I^r	II^r	III^r
F	0	$\frac{1}{2}(1 + b)$	$\frac{1}{2}(1 - b)$
G	1	$-b$	b
H	b	$\frac{1}{2}(1 - b)$	$-\frac{1}{2}(1 + b)$

The most asymmetric case corresponds to $b = -\frac{1}{3}$. Thus it can be seen that b does not have the same symmetric character as κ. However, if b is redefined as

$$b^* = \frac{A - B}{2[C - \frac{1}{2}(A + B)]} = \frac{\kappa - 1}{\kappa + 3}, \tag{5}$$

the energy equation becomes

$$E(A, B, C) = \frac{A + B}{2} J(J + 1) + [C - \tfrac{1}{2}(A + B)]E(b^*). \tag{6}$$

This redefinition shows the symmetry of b and b^*. They both have the same range; hence if the values of $E(b)$ are calculated for one-half the asymmetry range, say from -1 to $-\frac{1}{3}$, these values may be used for $E(b^*)$ if b^* is defined as in (5) and the energy is calculated from (6). We choose the range -1 to $-\frac{1}{3}$ since b changes less slowly with asymmetry, making interpolation between known energies more reliable.

2l SYMMETRY PROPERTIES OF THE ASYMMETRIC ROTOR WAVE FUNCTIONS

Before continuing with the calculation of asymmetric rotor energy levels, it will be helpful to investigate the symmetry properties of the wave functions. These symmetry properties greatly reduce the tediousness of evaluating the roots $E(\kappa)$ or $E(b)$. For introduction to group theory see WDC [9].

In group theoretical nomenclature, all rigid rotor wave functions belong to representations of the continuous three-dimensional rotation group. This group is called the external rotation group. These representations, characterized by the quantum numbers J and M, correspond to the infinity of possible rotations of the space-fixed axis system that leave the energy of the rotor invariant. The external rotation group

Table 2/1—Internal Species Classification for Symmetric Rotor Function $D_\infty(z)$

D_∞	E	C_∞^z	∞C_2
Σ_1 (even J)	1	1	1
Σ_2 (odd J)	1	1	−1
Π	2	$2\cos(2\pi/n)$	0
Δ	2	$2\cos(4\pi/n)$	0
Φ	2	$2\cos(6\pi/n)$	0

has infinitely many irreducible representations, one for each J value, each with a $(2J + 1)$-fold degeneracy. In the absence of external fields we do not need to investigate this point any further other than to note the $(2J + 1)$ degeneracy of the energy levels.

The symmetric rotor wave functions can be further classified under the continuous two-dimensional rotation group $\mathbf{D}_\infty$, the internal rotation group of the symmetric rotor. This group also has infinitely many representations characterized by the quantum numbers J and K.

There is one representation for each K value, except for $K = 0$ for which there are two. These levels are nondegenerate, but all other levels, $K > 0$, have a twofold degeneracy corresponding to $\pm K$. Table 2/1 gives the details of the classification. The species symbols $\Sigma, \Pi, \Delta, \Phi, \ldots$ for $K = 0, 1, 2, 3, \ldots$ are used in analogy to the nomenclature for the electronic states of diatomic molecules.

For the asymmetric rotor, the internal rotation group is the finite group V composed of four operations whose four, nondegenerate representations define four species of wave functions that may conveniently be labeled as in Table 2/2. The notation for the

Table 2/2—Character Table of V Group

	E	C_2^c	C_2^b	C_2^a
A	1	1	1	1
B_c	1	1	−1	−1
B_b	1	−1	1	−1
B_a	1	−1	−1	1

representations shows directly the axis of rotation for which the character is +1.

It is now necessary to turn to the detailed form of the wave functions. The symmetric top wave functions may be written as (1). See, for example, Pauling and Wilson, [7].

$$\psi_{JKM}(\theta, \phi, \chi) = \Theta_{JKM}(\theta) e^{iK\chi} e^{iM\phi}, \tag{1}$$

where

$$\Theta_{JKM}(\theta) = N_{JKM}x^{\frac{1}{2}|K-M|}(1-x)^{\frac{1}{2}|K+M|}\sum_{0}^{n}\alpha_\nu x^\nu$$

with $n = J - \frac{1}{2}|K + M| - \frac{1}{2}|K - M|$ and $x = \frac{1}{2}(1 - \cos\theta)$.

The ψ's of (1) exist only for $J \geqslant K$, $J \geqslant M$. In the following it is to be understood that the positive, numerical factor $N_{J,K,M}$ is such that $\psi_{J,K,M}$ is normalized, and the leading term of the Jacobi polynomial $\theta_{JKM}(\theta)$ has been taken as $+1$.

In order always to have the correct phase, a suggestion originally made by Van Vleck [116, 118] is followed by defining for $K \neq 0$

$$\overset{\times}{\psi}_{J,K,M} = (-1)^\beta \psi_{J,K,M} \tag{2}$$

where β is the larger of the two quantities K or M, or another way of saying the same thing $\beta = \frac{1}{2}|K + M| + \frac{1}{2}|K - M|$.

Furthermore we define

$$\begin{aligned} S(J, K, M, \gamma) &= 2^{-\frac{1}{2}}[\overset{\times}{\psi}_{J,K,M} + (-1)^\gamma \overset{\times}{\psi}_{J,-K,M}] \\ S(J, 0, M, 0) &= \psi(J, 0, M, 0), \end{aligned} \tag{3}$$

where γ may be even or odd, say 0 or 1, and for $K = 0$ only γ even exists.

If $K = 0$, there is one symmetric top wave function $\psi_{J,O,M}$ for each value of J, M. This belongs to the symmetric rotor species Σ_1, or Σ_2, depending on whether J is even or odd. If $K > 0$, species Π, Δ, etc., there are two wave functions of equal energy for each J, M. These may most conveniently be taken either as $\psi_{J,K,M}$ and $\psi_{J,-K,M}$ of (1) or as the two functions $S(J, K, M, \gamma)$ of (3).

Each asymmetric rotor wave function $A_{J,\tau,M}$ may be expressed as a linear combination of the symmetric rotor functions. These linear combinations assume the simplest form if we build them from the $S(J, K, M, \gamma)$'s. This is true because each $S(J, K, M, \gamma)$ satisfies the requirements for classification under a definite asymmetric rotor species which is not true for the $\psi(J, K, M)$'s. Under these conditions we may write

$$A(J, \tau, M) = \sum_{K\gamma} a_{K\gamma}^{JM\tau} S(J, K, M, \gamma). \tag{4}$$

Since any $A(J, \tau, M)$ belongs to one of the four species of the **V** group, only $S(J, K, M, \gamma)$'s of that particular species have nonvanishing a's in (4). In general, there will be several different $A(J, \tau, M)$'s of the same species characterized by different sets of a's, hence the additional index τ. If two moments of inertia of an asymmetric rotor approach equality, and if the $S(J, K, M, \gamma)$'s of the nearest symmetric rotor case are used in (4), then one $a_{K\gamma}^{J,M,\tau}$ approaches unity and the others zero. Thus for every $A(J, \tau, M)$ of a slightly asymmetric top, there is one $S(J, K, M, \gamma)$ which it closely approximates.

These new basis functions $S(J, K, M, \gamma)$ have been constructed relative to arbitrary axes x, y, z and not relative to the axes a, b, c, of the molecule as defined previously. They are therefore characterized by the representations A, B_z, B_y, B_x of the Four-group $\mathbf{V}(x, y, z)$. Here also the representations have been labeled to show directly the axis of rotation for which the character is $+1$. This makes easy the correlation of the

Table 2l3—Correlation of the Species Classification of Symmetric Rotor and Asymmetric Rotor Wave Functions

		PARITY OF		REPRESENTATION OF	REPRESENTATIONS OF $\mathbf{V}(a, b, c)$					
K	$\mathbf{D}_\infty(z)$	K	$J + \gamma$	$\mathbf{V}(x, y, z)$	I^r	I^l	II^r	II^l	III^r	III^l
0	Σ_1	e	e	A	A	A	A	A	A	A
0	Σ_2	e	o	B_z	B_a	B_a	B_b	B_b	B_c	B_c
1	Π	o	e	B_y	B_c	B_b	B_a	B_c	B_b	B_a
1	Π	o	o	B_x	B_b	B_c	B_c	B_a	B_a	B_b
2	Δ	e	e	A	A	A	A	A	A	A
2	Δ	e	o	B_z	B_a	B_a	B_b	B_b	B_c	B_c

representations of $\mathbf{V}(a, b, c)$ with those of $\mathbf{V}(x, y, z)$. A always corresponds to A, and B_a, B_b, B_c correspond to B_x, B_y, B_z according to the same permutation as identifies a, b, c, with x, y, z. These correlations are given in Table 2/3, which also includes the identification of the representations of the group $\mathbf{D}_\infty(z)$, to which the $S(J, K, M, \gamma)$ also belong, as well as the representations of the group $\mathbf{V}(x, y, z)$.

2m FACTORS OF THE REDUCED ENERGY MATRIX IN THE $S(J,K,M,\gamma)$ REPRESENTATION

For a given J the energy $E(\kappa)$ or $E(b)$ in any representation based on the $\psi^\times(J, K, M)$'s is of the order $2J + 1$ since $-J \leqslant K \leqslant J$. Since the matrix has elements only of the form K, K or $K \pm 2$, K it may be displayed as two submatrices whose indices involve, respectively, only even and only odd K's. These matrices may be set up by inspection from 2j10, 2j11, 2j12, Table 2j1, and Table 2k1. The transformation to a representation based on the $S(J, K, M, \gamma)$'s enables further factoring of the $E(\kappa)$ or $E(b)$ matrix into four submatrices, that is, a submatrix for each of the four symmetry species in the $\mathbf{V}$ group. This transformation may be represented symbolically as

$$X'E(\kappa)X = E^+ + E^- + 0^+ + 0^-, \tag{1}$$

where

$$X = X' = 2^{-\frac{1}{2}} \begin{vmatrix} \cdots & & & & \cdots \\ -1 & & & & 1 \\ & -1 & & 1 & \\ & & 2^{\frac{1}{2}} & & \\ & 1 & & 1 & \\ 1 & & & & 1 \\ \cdots & & & & \cdots \end{vmatrix} \tag{2}$$

is the Wang transformation, $\psi^\times = XS$. In order to use X in the form shown in (2) the order of K in setting up the $E(\kappa)$ matrix is extremely important. The orders from the top left corner are $\psi^\times(J, -J, M)$, $\psi^\times(J, -J + 1, M)$, $\ldots$, $\psi^\times(J, J, M)$ and $S(J, J, M, 1)$, $S(J, J - 1, M, 1)$, $S(J, 0, M, 0)$, $\ldots$, $S(J, J, M, 0)$. The submatrices, in terms of the original elements of $E(\kappa)$, may be displayed in the form

$$E^+ = \left|\begin{array}{c:cccc} E_{00} & 2^{1/2}E_{02} & 0 & \cdots \\ \hdashline 2^{1/2}E_{02} & E_{22} & E_{24} & \cdots \\ 0 & E_{24} & E_{44} & E_{46} \\ \cdots & \cdots & \cdots & \cdots \end{array}\right|. \qquad (3)$$

E^- has the same elements as E^+ after removal of the first row and first column as indicated by the dotted lines.

$$0^{\pm} = \left|\begin{array}{cccc} E_{11} \pm E_{-11} & E_{13} & 0 & \cdots \\ E_{13} & E_{33} & E_{35} & \cdots \\ 0 & E_{35} & E_{55} & \cdots \\ \cdots & \cdots & \cdots & \cdots \end{array}\right|. \qquad (4)$$

In (3) and (4) use has been made of the fact that

$$E_{K,K} = E_{-K,-K}, \qquad E_{K,K+2} = E_{K+2,K} = E_{-K,-K-2} = E_{-K-2,-K}.$$

These submatrices are identified by the symbols E^+, E^-, 0^+, 0^-, in which E and 0 refer to the even or oddness of the K values in the matrix elements, and plus and minus to the even or oddness of γ. These four species together with the six original representations of $E(\kappa)$ give twenty-four different submatrices,

$$\begin{array}{cccc} \mathrm{I}^rE^+ & \mathrm{I}^rE^- & \mathrm{I}^r0^+ & \mathrm{I}^r0^- \\ \mathrm{I}^lE^+ & \mathrm{I}^lE^- & \mathrm{I}^l0^+ & \mathrm{I}^l0^- \\ \cdots & \cdots & \cdots & \cdots \\ \mathrm{III}^lE^+ & \mathrm{III}^lE^- & \mathrm{III}^l0^+ & \mathrm{III}^l0^- \end{array}$$

which belong to the four symmetry species. Each submatrix of a given type, that is, I^rE^+ corresponds to the same set of energy levels as some submatrix of each of the other types. This correlation is accomplished by determining the symmetry classification of each submatrix.

2n SYMMETRY OF THE SUBMATRICES

The functions $S(J, K, M, \gamma)$ that occur in any submatrix must all belong to the same irreducible representation of $V(x, y, z)$. These representations may therefore be used to classify the submatrices.

To carry out this classification we must determine the effect of each of the operations E, C_2^x, C_2^y, C_2^z on the functions $S(J, K, M, \gamma)$. On detailed investigation we find [84]*

$$\begin{aligned} E \quad S(J, K, M, \gamma) &= S(J, K, M, \gamma) \\ C_2^z \quad S(J, K, M, \gamma) &= (-1)^K S(J, K, M, \gamma) \\ C_2^y \quad S(J, K, M, \gamma) &= (-1)^{J+\gamma} S(J, K, M, \gamma) \\ C_2^x \quad S(J, K, M, \gamma) &= (-1)^{J+K+\gamma} S(J, K, M, \gamma). \end{aligned} \qquad (1)$$

* The details of this calculation are carried out in Appendix III.

Hence by knowing the parity of K and $J + \gamma$, one knows the parity of any level. The parities of $J + \gamma$ and K are included in Table 2/3.

Now from the submatrix designation one knows the parity of K and $J + \gamma$ for a given J. Thus from this information and Table 213 one can construct Table 2n1. The corresponding classification of E^+, E^-, 0^+; 0^- for the Types I^r to III^l under the group $V(a, b, c)$ follows from Table 2/3.

The symmetry species of any given energy level can be found from the classification of the submatrix from which the level is obtained. To distinguish one level from the other levels of a given J it has been found very convenient to label it by the values of K

Table 2n1—Symmetry Classification of the Submatrices in V(x, y, z) (Note that as $J \rightarrow J \pm 1$ the $\pm$ matrices interchange symmetry)

			$J + \gamma$		SPECIES REPRESENTATION	
SUBMATRIX	K	γ	J EVEN	J ODD	J EVEN	J ODD
E^+	e	e	e	o	A	B_z
E^-	e	o	o	e	B_z	A
0^+	o	e	e	o	B_y	B_x
0^-	o	o	o	e	B_x	B_y

to which it corresponds in the limiting cases $\kappa = -1$, prolate symmetric rotor, and $\kappa = +1$, oblate symmetric rotor. The notation has the advantage that the symmetry classification under $V(a, b, c)$ may be found directly from the even or oddness of the two K's of a given level. That is, the four species A, B_c, B_b, B_a are indicated, respectively by the bipartite indices ee, oe, eo, oo in which the first symbol gives the parity of K for $\kappa = -1$ and the second for $\kappa = +1$.

That the symmetry classification in $V(a, b, c)$ is given uniquely in terms of the parity of K_{-1}, K_1 follows from the assignments of a, b, c to x, y, z for the cases $z = a$ and $z = c$. From Tables 2/3 and 2n1 we find for the $\kappa = -1$ limit, Type I, that for K_{-1} even requires the symmetry A or $B_z \rightarrow A$ or B_a, for K_{-1} odd the symmetry must be B_y or $B_x \rightarrow B_b$ or B_c. On the other hand, for K_1 even ($\kappa = 1$, $z = c$, Type III) requires A or $B_z \rightarrow A$ or B_c, K_1 odd requires B_x or $B_y \rightarrow B_a$ or B_b. Hence it must follow that $A = ee$, $B_c = oe$, $B_b = oo$ and $B_a = eo$. It is immaterial whether right- or left-hand Types I and III are chosen for the prolate and oblate representations. The classification of the submatrices to the symmetry species A, B_c, B_b, B_a in terms of the parities of the limiting K_{-1}, K_1 symmetric rotor cases is given in Table 2n2.

An early method of labeling energy levels is by the notation J_τ, where J specifies the J-set and τ takes on the $2J + 1$ values $-J \leqslant \tau \geqslant J$, the lowest energy level being J_{-J}, and τ increases so that the highest energy level is J_J. Figure 2n1 is a plot of the values of $E(\kappa)$ for low J's starting on the left for the case $\kappa = -1$ with κ increasing to the right until the limit $\kappa = +1$ is reached. Such a plot shows that there is no crossing of levels between the two symmetric limits within a given J; hence the τ labeling is unique. It, however, gives no indication of the symmetry. The bipartite index, the first K_{-1}, being 0, 1, 1, 2, 2, . . . from the lowest to the highest energy levels

Table 2n2—Symmetry Classification of the Submatrices in V(a, b, c) by the Parity of K_{-1}, K. ee = A, oe = B_c, oo = B_b, eo = B_a

	I^r		I^l		II^r		II^l		III^r		III^l	
SUB-MATRIX	J EVEN	J ODD	J EVEN	J ODD	J EVEN	J ODD	J EVEN	J ODD	J EVEN	J ODD	J EVEN	J ODD
E^+	ee	eo	ee	eo	ee	oo	ee	oo	ee	oe	ee	oe
E^-	eo	ee	eo	ee	oo	ee	oo	ee	oe	ee	oe	ee
0^+	oe	oo	oo	oe	eo	oe	oe	eo	oo	eo	eo	oo
0^-	oo	oe	oe	oo	oe	eo	eo	oe	eo	oo	oo	eo

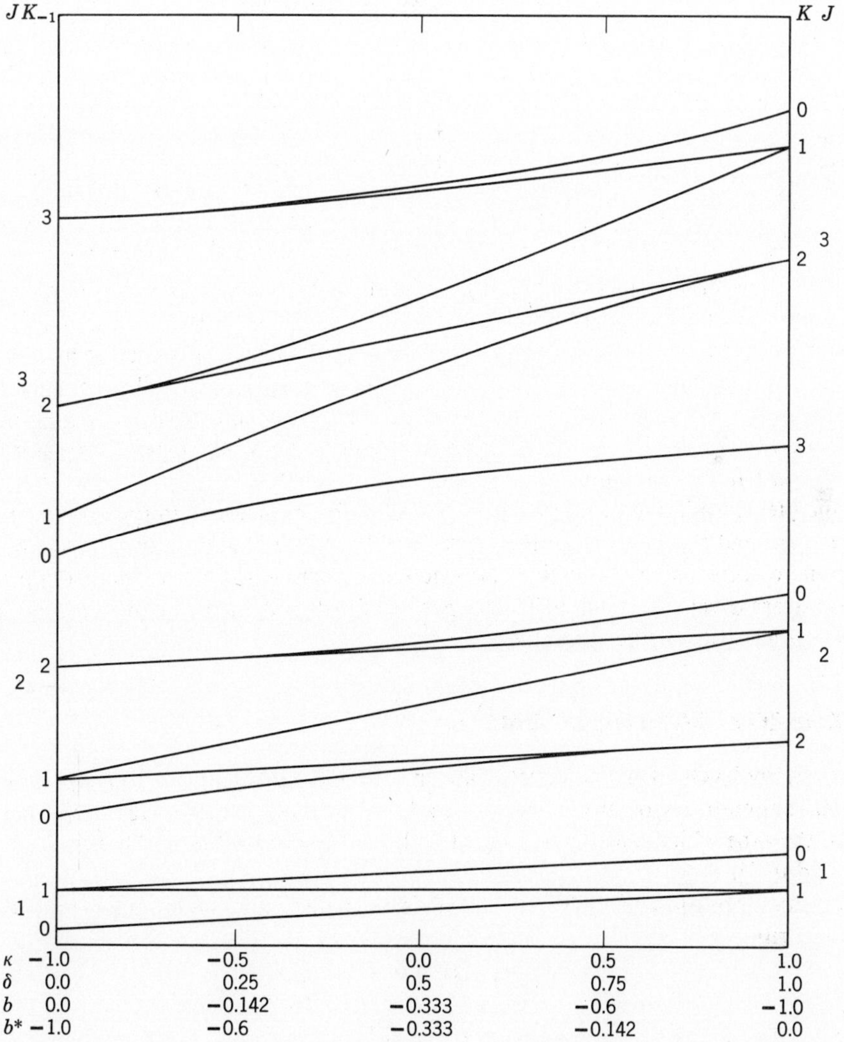

Fig. 2n1 A plot of $E(\kappa)$ versus κ and other asymmetry parameters for low J's. $\delta = (\kappa + 1)/2$. $b = (\kappa + 1)/(\kappa - 3)$. $b^* = (\kappa - 1)/(\kappa + 3)$.

and the second K_1, being 0, 1, 1, 2, 2 from the highest to the lowest energy levels, gives not only the symmetry through the parity of the indices but also τ, or the rank by means of the relation

$$\tau = K_{-1} - K_1 \tag{2}$$

The submatrix which contains any designated level in any type representation $I^r, \ldots, III^l$ is readily determined from the parity of the subscripts and Table 2n2. All the levels from a given submatrix of a given type of representation may be listed by writing the first subscripts as a descending series of numbers having the parity of the first index, $J \geqslant K \geqslant 0$, then writing the second subscripts as an ascending series having the parity of the second index $0 \leqslant K \leqslant J$, with the exception that the zero must be omitted from an even series unless it is paired with K_{-1} or $K_1 = J$.

Table 2n3—Species Classification

GROUP THEORY MULLIKEN [84]	KK (K_{-1}, K_1) [67]	DENNISON [38]	MECKE [80]	RAY [99]	$J + \gamma$
A	ee	$++$	(ABC)	abc	e
B_c	oe	$+-$	(AB)	c	o
B_b	oo	$--$	(AC)	b	e
B_a	eo	$-+$	(BC)	a	o

To illustrate this consider the level 7_{43}. It has the symmetry eo or B_a; $\tau = 4 - 3 = 1$; thus this level is the seventh-high for $J = 7$. If a III^r representation is used from Table 2n2, we see that the level is in the 0^+ submatrix. The levels occurring in 0^+ from the rules outlined above are 7_{61}, 7_{43}, 7_{25}, 7_{07}. The zero subscript is allowed since it is paired with $K_1 = J = 7$. The levels for E^-, which has the symmetry ee in this representation for $J = 7$, are 7_{62}, 7_{44}, 6_{26}. The zero subscript does not appear here since ee cannot contain the maximum K for J odd.

In the literature numerous authors have used different methods to specify symmetry species. In reading the original literature one should be able to relate the symmetry notations of other authors to that used here. A correlation of the more important notations are given in Table 2n3. The parity of $J + \gamma$ is included so it can be seen that it is the parity of the sum of $K_{-1} + K_1$.

2o ENERGY CALCULATIONS

The energy levels of the rigid rotor are the roots of the characteristic equations of the four submatrices of any Type $I^r \ldots III^l$. The type is chosen for convenience, that is, the type which converges fastest for the particular problem. These cases were discussed in 2j.

From an inspection of 2m3 and 2m4 it is seen that these submatrices are of the general form

$$\mathbf{E} = \begin{vmatrix} k_0 & b_1^{1/2} & 0 & \cdots \\ b_1^{1/2} & k_1 & b_2^{1/2} & \cdots \\ 0 & b_2^{1/2} & k_2 & \cdots \\ \cdots & \cdots & \cdots & \cdots \end{vmatrix}. \tag{1}$$

These can be put in a more convenient form by a transformation

$$\mathbf{D} = \mathbf{L}^{-1}\mathbf{E}\mathbf{L} = \begin{vmatrix} k_0 & 1 & 0 & \cdots \\ b_1 & k_1 & 1 & \cdots \\ 0 & b_2 & k_2 & \cdots \\ \cdots & \cdots & \cdots & \cdots \end{vmatrix} \tag{2}$$

where the matrix $\mathbf{L}$ has the form

$$\mathbf{L} = |l_{ii}| = \begin{vmatrix} 1 & 0 & 0 & \cdots \\ 0 & b_1^{-1/2} & 0 & \cdots \\ 0 & 0 & b_1^{-1/2}\, b_2^{-1/2} & \cdots \\ \cdots & \cdots & \cdots & \cdots \end{vmatrix}. \tag{3}$$

The values of k_i's and b_i's for each submatrix may be obtained from the proper forms of 2m3 and 2m4.

$$\mathbf{E}^{\pm};\ \mathbf{D} = \begin{vmatrix} F\mathcal{J} & 1 & 0 & \cdots \\ 2H^2 f(J,1) & 4(G-F)+F\mathcal{J} & 1 & \cdots \\ 0 & H^2 f(J,3) & 16(G-F)+F\mathcal{J} & \cdots \\ \cdots & \cdots & \cdots & \cdots \end{vmatrix} \tag{4}$$

$$0^{\pm};\ \mathbf{D} = \begin{vmatrix} (G-F)+F\mathcal{J}+Hf^{1/2}(J,0) & 1 & 0 & \cdots \\ H^2 f(J,2) & 9(G-F)+F\mathcal{J} & 1 & \cdots \\ 0 & H^2 f(J,4) & 25(G-F)+F\mathcal{J} & \cdots \\ \cdots & \cdots & \cdots & \cdots \end{vmatrix} \tag{5}$$

where $\mathcal{J} = J(J+1)$. The values of F, G, and H are found from Table 2j1 and the values $f(J, n)$ are given in Appendix II.

The usual polynomial form of a characteristic equation is most easily obtained by expanding the determinant $|D - \lambda I| = 0$ in terms of the first principal minors p_i, of order $i + 1$ by means of the recursion relations

$$\begin{aligned} p_0 &= k_0 - \lambda \\ p_1 &= (k_1 - \lambda)p_0 - b_1 \\ p_i &= (k_i - \lambda)p_{i-1} - b_i p_{i-2}. \end{aligned} \tag{6}$$

If the order of D is $n + 1$,

$$(-1)^{n+1} p_n = 0 \tag{7}$$

is the usual form of the characteristic equation expressed as a polynomial in λ, with λ^{n+1} as the leading term.

$$\begin{gathered} \lambda - k_0 = 0 \\ \lambda^2 - \lambda(k_1 + k_0) + k_1 k_0 - b_1 = 0 \\ \lambda^3 - \lambda^2(k_2 + k_1 + k_0) + \lambda(k_2 k_1 + k_2 k_0 + k_1 k_0 - b_2 - b_1) \\ - k_2 k_1 k_0 + k_2 b_1 + k_0 b_2 = 0, \quad \text{etc.} \end{gathered} \tag{8}$$

With the aid of (2), (4), and (5) the characteristic equations of any submatrix may be systematically developed. Type II representations give the equations most readily and in their simplest form since the k_i and b_i have the simplest forms in these representations. The true energies of a given rotor are then determined by the substitution of the roots $\lambda[E(\kappa)]$ of these equations into 2j9. Although, in general, $E(\kappa)$ cannot be given in explicit form, roots that can be derived from linear or quadratic factors may be expressed explicitly and are given in Table 2o1.

Table 2o1—Explicit Solutions of E(κ) and E(b)

$J_{K_{-1}K_1}$	$E(\kappa)$	$E(b)$
0_{00}	0	0
1_{10}	$\kappa + 1$	$1 - b$
1_{11}	0	$1 + b$
1_{01}	$\kappa - 1$	0
2_{20}	$2[\kappa + (\kappa^2 + 3)^{1/2}]$	$2[1 + (1 + 3b^2)^{1/2}]$
2_{21}	$\kappa + 3$	4
2_{11}	4κ	$1 - 3b$
2_{12}	$\kappa - 3$	$1 + 3b$
2_{02}	$2[\kappa - (\kappa^2 + 3)^{1/2}]$	$2[1 - (1 + 3b^2)^{1/2}]$
3_{30}	$5\kappa + 3 + 2(4\kappa^2 - 6\kappa + 6)^{1/2}$	$5 - 3b + 2(4 + 6b + 6b^2)^{1/2}$
3_{31}	$2[\kappa + (\kappa^2 + 15)^{1/2}]$	$5 + 3b + 2(4 - 6b + 6b^2)^{1/2}$
3_{21}	$5\kappa - 3 + 2(4\kappa^2 + 6\kappa + 6)^{1/2}$	$2[1 + (1 + 15b^2)^{1/2}]$
3_{22}	4κ	4
3_{12}	$5\kappa + 3 - 2(4\kappa^2 - 6\kappa + 6)^{1/2}$	$5 - 3b - 2(4 + 6b + 6b^2)^{1/2}$
3_{13}	$2[\kappa - (\kappa^2 + 15)^{1/2}]$	$5 + 3b - 2(4 - 6b + 6b^2)^{1/2}$
3_{03}	$5\kappa - 3 - 2(4\kappa^2 + 6\kappa + 6)^{1/2}$	$2[1 - (1 + 15b^2)^{1/2}]$
4_{41}	$5\kappa + 5 + 2(4\kappa^2 - 10\kappa + 22)^{1/2}$	$10 + 2(9 + 7b^2)^{1/2}$
4_{31}	$10\kappa + 2(9\kappa^2 + 7)^{1/2}$	$5(1 - b) + 2(4 + 10b + 22b^2)^{1/2}$
4_{32}	$5\kappa - 5 + 2(4\kappa^2 + 10\kappa + 22)^{1/2}$	$5(1 + b) + 2(4 - 10b + 22b^2)^{1/2}$
4_{23}	$5\kappa + 5 - 2(4\kappa^2 - 10\kappa + 22)^{1/2}$	$10 - 2(9 + 7b^2)^{1/2}$
4_{13}	$10\kappa - 2(9\kappa^2 + 7)^{1/2}$	$5(1 - b) - 2(4 + 10b + 22b^2)^{1/2}$
4_{14}	$5\kappa - 5 - 2(4\kappa^2 + 10\kappa + 22)^{1/2}$	$5(1 + b) - 2(4 - 10b + 22b^2)^{1/2}$
5_{42}	$10\kappa + 6(\kappa^2 + 3)^{1/2}$	$10 + 6(1 + 3b^2)^{1/2}$
5_{24}	$10\kappa - 6(\kappa^2 + 3)^{1/2}$	$10 - 6(1 + 3b^2)^{1/2}$

The solution of the secular determinant for cases that cannot be given explicitly must be solved by some iterative process by which the root can be approximated as precisely as one pleases. The solution is greatly facilitated by the fact that the secular determinant is a continuant equivalent to the continued fraction

$$(k_0 - \lambda) - \frac{b_1}{(k_1 - \lambda) - [b_2/(k_2 - \lambda) - \ldots]} = 0. \qquad (9)$$

By the use of (4) and (5), Table 2j1, and Appendix II the continued fraction form of any submatrix in terms of J and κ may be written down.

Since the secular equations must be solved by approximation methods, it is important to choose the Type I, II, or III representation which gives the most rapid convergence of the successive approximations. Furthermore, it is advantageous to approximate the mth root of a given equation by a form of the continued fraction which has k_m as the leading term [34].

$$\lambda_m = k_m - \frac{b_m}{(k_{m-1} - \lambda_m) - [b_{m-1}/(k_{m-2} - \lambda_m) - \ldots]} - \frac{b_{m+1}}{(k_{m+1} - \lambda_m) - [b_{m+2}/(k_{m+2} - \lambda_m) - \ldots]}. \quad (10)$$

Since the trace of a matrix remains unchanged under a similarity transformation, the relation $\Sigma\lambda_m = \Sigma k_m$ may be used to obtain the root for which covergence is the poorest or to check the numerical accuracy of the solutions.

Table 2o2—Values * of K_1 and τ for $K_{-1} = n$

SUBMATRIX	REPRESENTATION J EVEN	J ODD	K_1	τ
$\mathrm{I}^r\ E^+$	$A\ ee$	$B_a\ eo$	$J - n$	$2n - J$
$\mathrm{I}^r\ E^-$	$B_a\ eo$	$A\ ee$	$J - n + 1$	$2n - J - 1$
$\mathrm{I}^r\ 0^+$	$B_c\ oe$	$B_b\ oo$	$J - n + 1$	$2n - J - 1$
$\mathrm{I}^r\ 0^-$	$B_b\ oo$	$B_c\ oe$	$J - n$	$2n - J$

* In first row, $n = 0, 2, \ldots, J$ (or $J - 1$); in second row, $n = 2, 4, \ldots, J$ (or $J - 1$); in last two rows, $n = 1, 3, \ldots, J$ (or $J - 1$).

The τ value of the level λ_m can be determined from the K_{-1} and K_1 values. In Type I^r, the K_{-1} value is that of the K which enters as K^2 in the main diagonal element. The values of K_1 and τ for $K_{-1} = n$ are given in Table 2o2.

2p TRANSFORMATIONS WHICH DIAGONALIZE THE ENERGY MATRICES

To determine transition intensities, or to apply perturbation theory to energy calculations over small ranges of the energy parameter κ, it is necessary to find the matrix T of the similarity transformation which diagonalizes each energy submatrix, that is,

$$T'ET = |\lambda_\tau| = \Lambda, \quad (1)$$

where Λ is diagonal. Here τ is not a running index but refers to the actual τ values occurring in the submatrix under consideration.

The asymmetric rotor wave functions $A(J, \tau, M, \gamma)$ are expressed as linear combinations of the $S(J, K, M, \gamma)$ as in 2*l*4, that is,

$$A(J\tau M\gamma) = \sum_K a_K S(J, K, M, \gamma). \quad (2l4)$$

The summation over K is over only those K's which occur as squares in the main diagonals of the submatrix. The value of γ is fixed for each submatrix.

The matrix **T** can be found if we examine the diagonalization of D. There is a matrix **V** diagonalizing D, that is,

$$V^{-1}DV = \Lambda, \quad (2)$$

which can be computed from the last columns of adj $|D - I\lambda|$. Let such a column, which is the column of V corresponding to λ_τ, be denoted by v_τ. The components of v_τ can be evaluated by substituting λ_τ into the recursion formula 2o6. Denoting $p_i(\lambda_\tau)$ as p_{i_τ} we obtain

$$v_\tau = (1, -p_{0\tau}, p_{1\tau}, -p_{2\tau}, \ldots). \tag{3}$$

Thus, if v_{i_τ} is the $(i + 1)$st element of this column,

$$V = |v_{i\tau}|, \qquad i = 0 \text{ to } n$$

with

$$v_{0\tau} = 1, \qquad v_{i\tau} = (-1)^i p_{i-1,\tau}. \tag{4}$$

The relation between V and T can be found from the relation between the Hamiltonian E and the continuant matrix D, that is, from 2o2

$$D = L^{-1}EL. \tag{2o2}$$

Substituting 2o2 into (2) we find

$$V^{-1}L^{-1}ELV = \Lambda. \tag{5}$$

The transformation LV can be normalized by post-multiplication by the diagonal matrix

$$N = n_{\tau\tau} = [\Sigma_i (l_{ii}v_{i\tau})^2]^{-1/2}. \tag{6}$$

The transformation required in (1) is then

$$T = LVN = \begin{vmatrix} 1 & 0 & 0 & \cdots \\ 0 & b_1^{-1/2} & 0 & \cdots \\ 0 & 0 & b_1^{-1/2}b_2^{-1/2} & \cdots \\ \cdots & \cdots & \cdots & \cdots \end{vmatrix} \begin{vmatrix} 1 & 1 & 1 & \cdots \\ -p_{0\tau} & -p_{0\tau'} & -p_{0\tau''} & \cdots \\ p_{1\tau} & p_{1\tau'} & p_{1\tau''} & \cdots \\ \cdots & \cdots & \cdots & \cdots \end{vmatrix}$$

$$\times \begin{vmatrix} n_{\tau\tau} & 0 & 0 & \cdots \\ 0 & n_{\tau'\tau'} & 0 & \cdots \\ 0 & 0 & n_{\tau''\tau''} & \cdots \\ \cdots & \cdots & \cdots & \cdots \end{vmatrix}. \tag{7}$$

Mecke [80] also derived certain sum rule relationships between the energy levels and the inertial constants which hold regardless of the value of κ. These sum rules are also given in Appendix V.

Certain approximate methods for obtaining rigid rotor energies have been developed. These methods are not of great importance for infrared work since the actual values of $E(\kappa)$ are known at intervals of 0.1 in κ up through $J = 40$. For completeness these methods are discussed in Appendix IV.

3

HIGHER APPROXIMATIONS TO THE ENERGIES OF AN ASYMMETRIC ROTOR

3a PERTURBATION TREATMENT

The rigid rotor approximation discussed in Chapter 2 is not sufficient by itself to describe the energy levels that are observed in real molecules. It is now necessary to return to the complete Hamiltonian 1e1, in order to find a closer approximation.

In the rigid rotor approximation the dependence of the $\mu_{gg'}$ on the vibrational coordinates was completely neglected. The exact solution of the wave equation resulting from the operator in 1e1 is not possible. However, a better approximation to the actual energy levels and a better understanding of molecular dynamics can be obtained with the help of perturbation theory [125].

A more convenient form of the Hamiltonian can be obtained by expanding 1e1, noting that $\mathbf{P}_g$ commutes with $\mathbf{p}_g$ and that the $\mu_{gg'}$ are functions of only the vibrational coordinates. The resulting Hamiltonian is

$$\mathbf{H} = \tfrac{1}{2}\sum_{gg'} \mu_{gg'}\mathbf{P}_g\mathbf{P}_{g'} - \sum_{g} \mathbf{h}_g\mathbf{P}_g + \tfrac{1}{2}\sum_{gg'} \mu^{1/4}\mathbf{p}_g\mu_{gg'}\mu^{-1/2}\mathbf{p}_{g'}\mu^{1/4} + \tfrac{1}{2}\sum_{k} \mu^{1/4}\mathbf{p}_k\mu^{-1/2}\mathbf{p}_k\mu^{1/4} + \mathbf{V}, \tag{1}$$

in which

$$\mathbf{h}_g = \tfrac{1}{2}\sum_{g'} (\mu^{-1/4}\mu_{gg'}\mathbf{p}_{g'}\mu^{1/4} + \mu^{1/4}\mathbf{p}_{g'}\mu_{g'g}\mu^{-1/4}). \tag{1a}$$

This form of the Hamiltonian may now be written as the sum of three terms: a purely vibrational part that includes the last three terms of (1), $\mathbf{H}_R{}^0$ that consists of those parts of the first term that are diagonal in the vibrational quantum numbers V, and the perturbing term $\lambda\mathbf{H}'$ that encompasses the second term and those parts of the first term not included in $\mathbf{H}_R{}^0$. λ is a small positive parameter which is much less than one. $\mathbf{H}^0$ includes the vibrational part plus $\mathbf{H}_R{}^0$.

If we assume that the wave functions ψ_{RV} may be written as a product function $\psi_R\psi_V$ in which ψ_R depends only on the rotational coordinates and ψ_V depends only on the vibrational coordinates, we can then set up the matrices of $\mathbf{H}_R{}^0$ and $\lambda\mathbf{H}'$ in terms of these basis functions. We further assume that the ψ_V are the normalized and

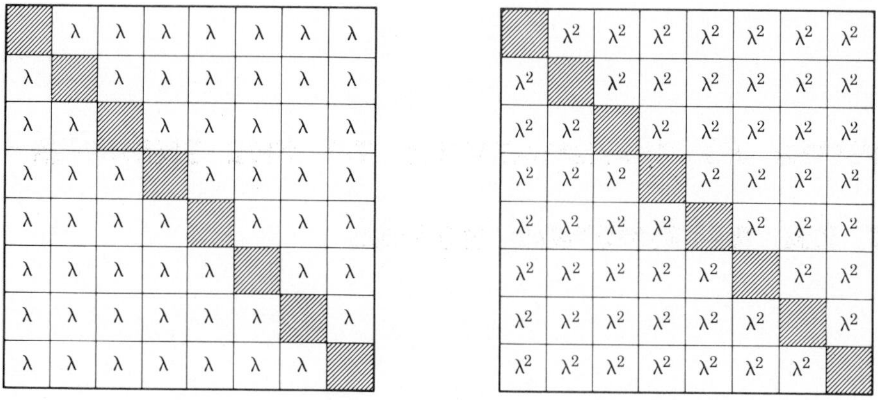

(*a*) (*b*)

Fig. 3a1 A section of the matrix **H** before (*a*) and after (*b*) transformation. Each block is labeled by two vibrational quantum numbers V and V′. The blocks are further subdivided and labeled by the rotational quantum numbers R, R′. The unshaded blocks in both (*a*) and (*b*) show the order of the perturbation being neglected.

orthogonal solutions of the vibrational wave equation. In this event $\mathbf{H}^0$ is diagonal in V, but, as has been seen, is not necessarily diagonal in R. The ψ_R for an asymmetric rotor is the $S(J, K, M, \gamma)$ defined in 2*l*3. Such a matrix is obviously not necessarily diagonal in K. The perturbation $\lambda\mathbf{H}'$ will by hypothesis have small elements.

For a problem such as this, it has been shown [5, 63] that it is possible to find a transformation which will transform the matrix $\mathbf{H} = \mathbf{H}^0 + \lambda\mathbf{H}'$ into a matrix where the only terms nondiagonal in V are of the order λ^2 or higher. These nondiagonal terms may be neglected in obtaining the energy correctly to the second order. By neglecting these higher order, off-diagonal terms, the transformed matrix factors into smaller matrices $\mathbf{E}$, one for each vibrational state, whose elements, $(R, V\,|\mathbf{H}|\,R'V)$, are labeled by the rotational quantum numbers only, it being understood that these elements are diagonal in V. $\mathbf{H}$ is given by

$$\mathbf{H} = \mathbf{H}_0 + \lambda\mathbf{H}_1 + \lambda^2\mathbf{H}_2, \tag{2}$$

in which the elements $\mathbf{H}_0 + \lambda\mathbf{H}_1$ are given by

$$(VR\,|\mathbf{H}^0|\,VR') + \lambda(VR\,|\mathbf{H}'|\,VR') = (VR\,|\mathbf{H}|\,VR'), \tag{3}$$

whereas the elements of $\lambda^2(\mathbf{H}_2)$ are

$$\lambda^2(\mathbf{H}_2)_{R,R'} = \lambda^2 \sum_{R''V''}' \frac{(VR\,|\mathbf{H}'|\,V''R'')(V''R''\,|\mathbf{H}'|\,VR')}{E_V - E_{V''}},$$

$$(V, R\,|\mathbf{H}'|\,V''R'') = \int \psi_V{}^*\psi_R{}^*\mathbf{H}'\psi_{R''}\psi_{V''}\,d\tau. \tag{4}$$

These are merely the results of a second-order perturbation calculation.

In (4) it has been assumed that the rotational spacing is small compared with that of the vibrational levels so that $E_{RV}{}^0 - E^0_{R''V''}$ can be replaced by $E_V{}^0 - E_{V''}{}^0$, the difference of the vibrational energies. The prime on the summation indicates that terms in which $V'' = V$ are omitted in the summation. $\mathbf{H}$ is the complete Hamiltonian $\mathbf{H}^0 + \lambda\mathbf{H}'$.

Figure 3a1 shows schematically the nature of a section of the matrix $\mathbf{H}$ before and after the transformation. $\mathbf{H}$ is really an infinite matrix. Before the transformation $\mathbf{H}$ can be factored into a problem of rotation and vibration only by neglecting terms of the order λ as was done in Chapter 2. After the transformation the factoring is possible when terms in V of the order λ^2 are neglected. This approximation giving the energy levels correct to terms of the order of λ^2 also differs from the rigid approximation in that the small rotational matrices $\mathbf{H}$ are not the same for each vibrational state since the elements of $\mathbf{H}_0 + \lambda\mathbf{H}_1$ and $\lambda^2\mathbf{H}_2$ are functions of V.

3b NATURE OF THE ROTATIONAL MATRIX

Through the use of the transformation discussed in 3a, the infinite energy matrix has been factored into rotational matrices when elements of the order of λ^2 are neglected. The eigenvalues of these matrices are the rotational energy levels for a particular vibrational state. The general nature of these rotational matrices is now considered.

The elements of zeroth and first order in λ are given by

$$(RV\,|\mathbf{H}^0 + \lambda\mathbf{H}'|\,R'V) = \int \psi_V{}^*\psi_R{}^*\mathbf{H}\psi_{R'}\psi_V\,d\tau, \tag{1}$$

where $\mathbf{H}$ is the complete Hamiltonian and the $\psi_R\psi_V = \psi_{RV}$ are the product type wave functions assumed in 3a. Each of the terms of $\mathbf{H}$ except the purely vibrational terms (the last three terms in 3a1) is the product of a rotational factor and a vibrational factor. A typical example is $\mu_{xy}\mathbf{P}_x\mathbf{P}_y$. Inserting this operator into the integral (1), we obtain

$$\int \psi_V{}^*\psi_R{}^*\mu_{xy}\mathbf{P}_x\mathbf{P}_y\psi_{R'}\psi_V\,d\tau = \int \psi_V{}^*\mu_{xy}\psi_V\,d\tau_V \int \psi_R{}^*\mathbf{P}_x\mathbf{P}_y\psi_{R'}\,d\tau_R$$

$$= \langle\mu_{xy}\rangle\,(RV\,|\mathbf{P}_x\mathbf{P}_y|\,R'V). \tag{2}$$

where the brackets $\langle\,\rangle$ denote the quantum mechanical average and is independent of the quantum numbers R.

Another typical example is of the form $\mathbf{h}_x\mathbf{P}_x$, which upon insertion into (1) gives

$$\int \psi_V{}^*\psi_R{}^*\mathbf{h}_x\mathbf{P}_x\psi_{R'}\psi_V\,d\tau = \int \psi_V{}^*\mathbf{h}_x\psi_V\,d\tau_V \int \psi_R{}^*\mathbf{P}_x\psi_{R'}\,d\tau_R$$

$$= \langle\mathbf{h}_x\rangle\,(R, V\,|\mathbf{P}_x|\,\mathbf{R}', V). \tag{3}$$

Thus the matrix equation can be written as

$$\mathbf{H}_0 + \lambda\mathbf{H}_1 = \mathbf{E}_V + \tfrac{1}{2}\sum_{gg'} \langle\mu_{gg'}\rangle(R, V\,|\mathbf{P}_g\mathbf{P}_{g'}|\,R', V) - \lambda\sum_{g} \langle\mathbf{h}_g\rangle(R, V\,|\mathbf{P}_g|\,R', V). \tag{4}$$

In (4), E_V is the vibrational energy, a matrix diagonal in R, and whose diagonal elements are all E_V. This energy derives from the purely vibrational terms in **H**.

The terms in λ^2 can next be considered. These terms are given by 3a4. The elements $(R\,|\mathbf{H}_2|\,R')$ are now a sum of terms as can be seen from 3a4. A typical form for the first factor in the numerator of 3a4 would be

$$\int \psi_V{}^*\psi_R{}^*(\tfrac{1}{2}\mu_{xx}\mathbf{P}_x{}^2 - \mathbf{h}_x\mathbf{P}_x)\psi_{R''}\psi_{V''}\,d\tau = \tfrac{1}{2}(R, V\,|\mu_{xx}\mathbf{P}_x{}^2|\,R'', V'') - (R, V\,|\mathbf{h}_x\mathbf{P}_x|\,R'', V''). \tag{5}$$

The second such factor might well be

$$\int \psi_{V''}{}^*\psi_{R''}{}^*(\tfrac{1}{2}\mu_{yy}\mathbf{P}_y{}^2 - \mathbf{h}_y\mathbf{P}_y)\psi_{R'}\psi_V\,d\tau = \tfrac{1}{2}(R'', V''|\mu_{yy}\mathbf{P}_y{}^2|\,R', V) - (R'', V''\,|\mathbf{h}_y\mathbf{P}_y|\,R', V). \tag{6}$$

The contribution of these factors to $\lambda^2 E_2$ is, according to 3a4,

$$\begin{aligned}\sum_{R''V''}{}' \{&\tfrac{1}{4}(R, V\,|\mu_{xx}\mathbf{P}_x{}^2|\,R'', V'')(R'', V''|\,\mu_{yy}\mathbf{P}_y{}^2|\,R', V)\\ &- \tfrac{1}{2}(R, V\,|\mu_{xx}\mathbf{P}_x{}^2|\,R'', V'')(R'', V''\,|\mathbf{h}_y\mathbf{P}_y|\,R', V)\\ &- \tfrac{1}{2}(R, V\,|\mathbf{h}_x\mathbf{P}_x|\,R'', V'')(R'', V''\,|\mu_{yy}\mathbf{P}_y{}^2|\,R', V)\\ &+ (R, V\,|\mathbf{h}_x\mathbf{P}_x|\,R'', V'')(R'', V''|\,\mathbf{h}_y\mathbf{P}_y|\,R', V)\}/h\nu_{V,V''}.\end{aligned} \tag{7}$$

As is shown in (3) and (4) each of these matrix elements factors into a vibrational part and a rotation part. For instance,

$$(R, V\,|\mu_{xx}\mathbf{P}_x{}^2|\,R'', V'') = (V\,|\mu_{xx}|\,V'')(R\,|\mathbf{P}_x{}^2|\,R''). \tag{8}$$

With this in mind, the second term of (7), for example, can be written as follows:

$$\begin{aligned}&-\tfrac{1}{2}\sum_{R''V''}{}' (R, V\,|\mu_{xx}\mathbf{P}_x{}^2|\,R'', V'')(R'', V''\,|\mathbf{h}_y\mathbf{P}_y|\,R', V)/h\nu_{V,V''}\\ &= -\tfrac{1}{2}\sum_{R''}{}'(R\,|\mathbf{P}_x{}^2|\,R'')(R''\,|\mathbf{P}_y|\,R') \times \sum_{V''}{}' (V\,|\mu_{xx}|\,V'')(V''\,|\mathbf{h}_y|\,V)/h\nu_{V,V''}\\ &= -\tfrac{1}{2}(R\,|\mathbf{P}_x{}^2\mathbf{P}_y|\,R')\sum_{V''}{}' (V\,|\mu_{xx}|\,V'')(V''\,|\mathbf{h}_y|\,V)/h\nu_{V,V''}.\end{aligned} \tag{9}$$

Similarly, the other terms can be reduced to the following expressions:

$$\begin{aligned}&\tfrac{1}{4}(R\,|\mathbf{P}_x{}^2\mathbf{P}_y{}^2|\,R')\sum_{V''}{}' (V\,|\mu_{xx}|\,V'')(V\,|\mu_{yy}|\,V)/h\nu_{V'V''}\\ &- \tfrac{1}{2}(R\,|\mathbf{P}_x\mathbf{P}_y{}^2|\,R')\sum_{V''}{}' (V\,|\mathbf{h}_x|\,V'')(V''\,|\mu_{yy}|\,V)/h\nu_{V,V''}\\ &+ (R\,|\mathbf{P}_x\mathbf{P}_y|\,R')\sum_{V''}{}' (V\,|\mathbf{h}_x|\,V'')(V''\,|\mathbf{h}_y|\,V)/h\nu_{V,V''}.\end{aligned} \tag{10}$$

Each of these terms reduces to a constant independent of R multiplied by a rotational matrix element.

It is thus seen that the matrix **H** can be written as a polynomial in the matrices $\mathbf{P}_x$, $\mathbf{P}_y$, $\mathbf{P}_z$, with constant coefficients formed from integrals involving the vibrational wave functions.

By considering asymmetric rotors, where accidental vibrational degeneracy is excluded, it can be shown that the linear terms vanish identically. If there is an accidental degeneracy, this treatment is not sufficient in any event. From (4) it is seen that the coefficients of $\mathbf{P}_x$, $\mathbf{P}_y$, and $\mathbf{P}_z$ are $-\langle\mathbf{h}_x\rangle$, $-\langle\mathbf{h}_y\rangle$, and $-\langle\mathbf{h}_z\rangle$, respectively. **H** and $\mathbf{P}_g$ are Hermitian matrices; hence the matrix formed from the elements $(V|\mathbf{h}_g|V')$ is also Hermitian. Now, $\mathbf{h}_g$ is a pure imaginary operator as is easily seen from 3a1. (Each operator $\mathbf{p}_g$ contains $-i\hbar$; all other terms are real.) Furthermore, for a nondegenerate, asymmetric rotor, ψ_V is real. Thus the diagonal term

$$(V|\mathbf{h}_g{}^*|V) = -(V|\mathbf{h}_g|V) = (V|\mathbf{h}_g|V) = 0. \tag{11}$$

The cubic terms in $\mathbf{P}_x$, $\mathbf{P}_y$, and $\mathbf{P}_z$ can also be eliminated. It can be shown that the coefficients of $\mathbf{P}_x{}^3$, $\mathbf{P}_y{}^3$, and $\mathbf{P}_z{}^3$ vanish, for example,

$$\frac{\frac{1}{2}\mathbf{P}_x{}^3 \sum_{V''}{}' (V|\mu_{xx}|V'')(V''|\mathbf{h}_x|V) + (V|\mathbf{h}_x|V'')(V''|\mu_{xx}|V)}{h\nu_{V,V''}}. \tag{12}$$

Since $\mathbf{h}_x$ is a pure imaginary operator, $(V|\mathbf{h}_x|V'') = -(V''|\mathbf{h}_x|V)$. However, μ_{xx} is a real operator; hence $(V|\mu_{xx}|V'') = (V''|\mu_{xx}|V)$; hence the coefficient of $\mathbf{P}_x{}^3$ vanishes.

The other cubic terms all occur in pairs of the type $\mathbf{P}_x{}^2\mathbf{P}_y - \mathbf{P}_y\mathbf{P}_x{}^2$ with some coefficient. These may be treated in the following manner by adding and subtracting $\mathbf{P}_x\mathbf{P}_y\mathbf{P}_x$:

$$\mathbf{P}_x{}^2\mathbf{P}_y - \mathbf{P}_y\mathbf{P}_x{}^2 = \mathbf{P}_x(\mathbf{P}_x\mathbf{P}_y - \mathbf{P}_y\mathbf{P}_x) + (\mathbf{P}_x\mathbf{P}_y - \mathbf{P}_y\mathbf{P}_x)\mathbf{P}_x = -i\hbar(\mathbf{P}_x\mathbf{P}_z + \mathbf{P}_z\mathbf{P}_x). \tag{13}$$

The last reduction results from the commutation rules 2b1. Thus either the coefficients of the cubic terms vanish or the cubic terms reduce to quadratic terms.

The conclusion is that as a result of the perturbation treatment, the rotational matrix $\mathbf{H} = \mathbf{H}_0 + \lambda\mathbf{H}_1 + \lambda^2\mathbf{H}_2$ which corresponds to a single vibrational quantum state, may be written as a polynomial in $\mathbf{P}_x$, $\mathbf{P}_y$, and $\mathbf{P}_z$; that is,

$$\mathbf{H} = \mathbf{E}_V + \tfrac{1}{2}\sum \alpha_{gg'}\mathbf{P}_g\mathbf{P}_{g'} + \tfrac{1}{4}\sum_{gg'jj'} \tau_{gg'jj'}\mathbf{P}_g\mathbf{P}_{g'}\mathbf{P}_j\mathbf{P}_{j'}. \tag{14}$$

The coefficients $\alpha_{gg'}$ and $\tau_{gg'jj'}$ depend on the vibrational quantum state to which **E** corresponds. If the vibrational wave functions are known with sufficient accuracy, these coefficients can be computed, and since the matrices of $\mathbf{P}_x$, $\mathbf{P}_y$, and $\mathbf{P}_z$ are known, the secular equation can, in principle, be set up to include quartic terms in $\mathbf{P}_x$, $\mathbf{P}_y$, and $\mathbf{P}_z$. The solution to this problem would include the change in moment of inertia with vibrational state, the coupling of vibration and rotation, and the effects of centrifugal distortion.

It is unfortunate that the use of the harmonic-oscillator approximation to ψ_V introduces an error that may be as large as the perturbation effects that have been considered. However, by making certain approximations, the secular equation may be reduced to the form 2g1, in which the rigid moments of inertia are replaced by three empirical parameters (effective inertial constants) which are to be determined from the experimental data.

3c THE FACTORED FORM OF THE SECULAR EQUATION

To obtain the form of the secular equation analogous to 2g1, it is necessary to neglect the quartic terms in 3b14. By analogy to diatomic molecules, these terms correspond, at least in part, to centrifugal distortion effects. Experimentally, these effects can be observed, especially for the higher rotational states; therefore they are not negligible and must be considered in any complete treatment of the energies. They are, however, small for low rotational states.

An approximate method for the classification of the terms in **H** is by the powers of the vibrational frequency. Thus the coefficients of the quadratic terms are of the order ν^0 plus corrections of the order ν^{-1}, whereas the coefficients of the quartic terms are of the order ν^{-2}. On this basis the quartic terms are omitted for the present.

If no other vibrational state perturbs the state under consideration, the **H** matrix becomes, on omission of the quartic terms,

$$\mathbf{H} = \mathbf{E}_v + \tfrac{1}{2}\{A_v\mathbf{P}_z^{\,2} + B_v\mathbf{P}_x^{\,2} + C_v\mathbf{P}_y^{\,2}\} + \delta\{\mathbf{P}_x\mathbf{P}_y + \mathbf{P}_y\mathbf{P}_x\} + \varepsilon\{\mathbf{P}_y\mathbf{P}_z + \mathbf{P}_z\mathbf{P}_y\} + \eta(\mathbf{P}_z\mathbf{P}_x + \mathbf{P}_x\mathbf{P}_z). \tag{1}$$

By an orthogonal transformation (1) may be reduced to a principal axes system, the transformation coefficients being numbers. If a molecule has orthorhombic symmetry, it can be shown from symmetry considerations that the cross-terms in (1) vanish. For molecules of lower symmetry, since the coefficients of $\mathbf{P}_g$ depend on the vibrational quantum number, the transformation and hence the orientation of the principal axes system vary slightly with vibrational state.

Upon reduction to the principal axis system, we obtain

$$\mathbf{H}' = \mathbf{E}_v + \frac{1}{2}\left\{\frac{\mathbf{L}_x^{\,2}}{(I_x)_v} + \frac{\mathbf{L}_y^{\,2}}{(I_y)_v} + \frac{\mathbf{L}_z^{\,2}}{(I_z)_v}\right\} \tag{2}$$

subject to

$$\mathbf{L}_x^{\,2} + \mathbf{L}_y^{\,2} + \mathbf{L}_z^{\,2} = \mathbf{P}_x^{\,2} + \mathbf{P}_y^{\,2} + \mathbf{P}_z^{\,2}.$$

Furthermore, the commutation rules for $\mathbf{L}_x$, $\mathbf{L}_y$, and $\mathbf{L}_z$ are the same as those for $\mathbf{P}_x$, $\mathbf{P}_y$, and $\mathbf{P}_z$, 2b1. It can thus be seen that the problem is formally identical with that of the rigid rotor discussed in Chapter 2, except that I_x, I_y, and I_z, the rigid moments of inertia, must be replaced by $(I_x)_v$, $(I_y)_v$, and $(I_z)_v$, the effective moments of inertia. The energy levels can consequently be calculated just as outlined in the preceding chapter. In the future the notation used will be effective moments of inertia and $\mathbf{P}_g$ for angular momentum, bearing in mind the preceding results.

For orthorhombic molecules the expression for A_v is

$$\begin{aligned} A_v = {} & \tfrac{1}{2}(V\,|\mu_{gg}|\,V) + \sum_{V'}{}' (V\,|\mathbf{h}_g|\,V')(V'\,|\mathbf{h}_g|\,V)/h\nu_{V,V'} \\ & + i\hbar \sum_{V'}{}' \{(V\,|\mu_{gg'}|\,V')(V'\,|\mathbf{h}_j|\,V) \\ & - (V\,|\mu_{gj}|\,V')(V'\,|\mathbf{h}_{g'}|\,V)\}/h\nu_{V,V'}. \end{aligned} \tag{3}$$

From (3) it is readily seen that even in the simplest case $(I_x)_v$, etc., are not simply related to either the rigid rotor or the instantaneous moments of inertia. It is consequently necessary to be careful in giving geometrical interpretations to these quantities for excited vibrational states. Not even for the ground vibrational state do the equilibrium or rigid rotor moments of inertia equal the effective moments.

Several investigators have shown that the dependence of the moments of inertia on vibrational quantum number for the general case of the asymmetric rotor may be expressed as

$$(I_g)_V = (I_g)_e + \sum_i \alpha_{gi}(v_i + \tfrac{1}{2}) + \dots, \qquad g = x, y, z, \tag{4}$$

in which $(I_g)_e$ is the equilibrium moment of inertia, the v_i are the vibrational quantum numbers which characterize the vibrational state V, and $(I_g)_V$ is the moment of inertia in the vibrational state V. The α_{gi}'s are constants. Explicit expressions have been derived for many types of molecules [37, 54, 102–111].

In general, the α_{gi} are functions of the constants of the nonquadratic terms in the potential energy function and have some contribution from the interaction terms in the kinetic energy. If there is some local interaction such as a Fermi resonance or a Coriolis interaction, (4) is not adequate to express the observed moments, and additional corrections must be introduced.

For linear molecules and spherical and symmetric rotors it is customary to write the expression analogous to (4) in terms of the inertial constants, $B = h/8\pi^2 I_A$, etc.

$$B_V = B_e + \sum_i \alpha_i \left(v_i + \frac{d_i}{2}\right) + \dots, \tag{5}$$

in which d_i is the degeneracy of the ith vibrational state. In general, the linear terms in $(v_i + \frac{1}{2})$ in (4) and (5) are sufficient to represent the observed moments of inertia; however, for some molecules it has been necessary to include terms that are quadratic in the vibrational quantum numbers.

3d CENTRIFUGAL DISTORTION

The increased resolution which is now available in the infrared region necessitates a more complete treatment of the energy levels. Since the bonds between atoms are not rigid, the interatomic distances will vary with the speed of rotation, giving rise to a centrifugal distortion.

A physical picture of centrifugal distortion can be obtained by considering a diatomic molecule classically. The atoms may be considered as hard spheres joined by a rather rigid spring that obeys Hooke's Law, that is, harmonic forces. If the molecule is considered to rotate about an axis, then at equilibrium the centrifugal force equals the centripetal force. Letting $\mu = m_1 m_2/(m_1 + m_2)$ be the reduced mass of the molecule at equilibrium

$$k(r - r_0) = \mu r \omega^2 = \frac{P^2}{\mu r^3}, \tag{1}$$

where k is the force constant, r_0 the interatomic distance of the stationary molecule, ω the angular velocity, and P the angular momentum.

The energy H of the system is

$$H = \frac{P^2}{2\mu r^2} + \tfrac{1}{2}k(r - r_0)^2 \tag{2}$$

Combining (1) and (2), and making use of the expansion $r^2 = r_0^2[1 + 2(r - r_0)/r_0 + \ldots]$, we obtain

$$H = \frac{P^2}{2\mu r_0^2} - \frac{P^4}{2\mu^2 k r_0^6} + 0(P^6), \tag{3}$$

where only the first terms in the expansion of r^2 have been retained. The first term is the kinetic energy of a rigid rotor, and the second term is the contribution due to centrifugal forces. The centrifugal distortion of a polyatomic molecule has also been treated classically [35, 73, 123] and gives an Hamiltonian of the same form. From (3) it is seen that the main contribution from centrifugal distortion arises from the P^4 term in the kinetic energy expression. Therefore the Hamiltonian resulting from the perturbation problem may be used as the appropriate one for the centrifugal distortion problem.

3e QUANTUM MECHANICAL TREATMENT OF CENTRIFUGAL DISTORTION

Starting from 3b14 Wilson [124] has obtained the secular equation for orthorhombic molecules. He has also discussed methods of evaluating the $\tau_{gg'jj'}$ which appear as coefficients of the quartic terms [123, 124].

From 3b10 it is readily seen that the τ's are given by

$$\tau_{gg'jj'} = \frac{\sum'(V\,|\mu_{gg'}|\,V'')(V''\,|\mu_{jj'}|\,V)}{h\nu_{V,V''}} \tag{1}$$

where prime indicates that the state V is excluded from the sum. Methods of evaluating these coefficients are given in Appendix VII. In order to calculate these quantities from first principles a knowledge of geometrical data concerning the molecule, the normal modes of vibration (or force constants), and the fundamental vibrational frequencies are necessary.

For orthorhombic molecules, only τ_{xxxx}, τ_{yyyy}, τ_{zzzz}, τ_{xxyy}, τ_{yyzz}, τ_{zzxx}, τ_{xyxy}, τ_{yzyz}, τ_{zxzx}, and related coefficients in which jj' have been exchanged with gg' are nonvanishing. For nonlinear triatomic molecules such as H_2O or H_2S a greater simplification is possible, since for these molecules only τ_{xxxx}, τ_{yyyy}, τ_{zzzz}, $\tau_{xxyy} = \tau_{yyxx}$, $\tau_{yyzz} = \tau_{zzyy}$, $\tau_{zzxx} = \tau_{xxzz}$, and $\tau_{xzxz} = \tau_{zxzx} = \tau_{xzzx} = \tau_{zxxz}$ are different from zero.

The matrix elements of $\mathbf{P}_x$, $\mathbf{P}_y$, and $\mathbf{P}_z$ were derived in Chapter 2. By matrix multiplication we can obtain $\mathbf{P}_x^2$, $\mathbf{P}_y^2$, and $\mathbf{P}_z^2$, and from these we can finally obtain the elements of the matrices of the quartic terms of the type $\mathbf{P}_g\mathbf{P}_{g'}\mathbf{P}_j\mathbf{P}_{j'}$. The elements for water-type molecules are given by Wilson. The secular equation has nonvanishing elements of K, K; $K, K \pm 2$; and $K, K \pm 4$. Wilson has shown that by the introduction of a Wang-type transformation, the secular equation separates into four factors just as for the rigid rotor. Explicit matrix elements for this treatment are not given here because the perturbation treatment described in the following section is considerably more convenient. However, this treatment has been successfully applied to a photographic infrared band [34] of H_2S.

3f APPROXIMATE TREATMENT OF THE EFFECT OF CENTRIFUGAL DISTORTION

In this section a general quantum-mechanical approach to the centrifugal distortion problem based on first-order perturbation theory is outlined, which leads to an explicit correction term to the rigid rotor energies. This has the advantage of making use of the vast amount of rigid rotor information already available and is also somewhat easier to apply.

As a starting point, 3b14 is used after it has been transformed to a principal axis system. Only the rotational submatrices are considered since these differ only in the term E_v, which varies from state to state but does not affect the rotational energy levels. Thus

$$\mathbf{H} = \mathbf{H}_0' + \mathbf{H}_1' = A_v'\mathbf{P}_z^2 + B_v'\mathbf{P}_x^2 + C_v'\mathbf{P}_y^2 + \tfrac{1}{4}\sum_{gg'jj'} \tau_{gg'jj'}\mathbf{P}_g\mathbf{P}_{g'}\mathbf{P}_j\mathbf{P}_{j'}, \qquad (1)$$

in which

$$\mathbf{H}_0' = A_v'\mathbf{P}_z^2 + B_v'\mathbf{P}_x^2 + C_v'\mathbf{P}_y^2 \qquad (2)$$

and

$$\mathbf{H}_1' = \tfrac{1}{4}\sum_{gg'jj'} \tau_{gg'jj'}\mathbf{P}_g\mathbf{P}_{g'}\mathbf{P}_j\mathbf{P}_{j'}, \qquad (3)$$

where the A_v', B_v', C_v' are proportional to the reciprocal of the effective principal moments of inertia for the vibrational state under consideration. $\mathbf{H}_1'$ represents the centrifugal distortion term, and it is assumed that the effect of $\mathbf{H}_1'$ is small, so it may be treated by first-order perturbation theory. The solution for the eigenvalues of $\mathbf{H}_0'$, the zeroth-order approximation was discussed in Chapter 2.

As pointed out earlier, for many problems of interest $\mathbf{H}_1'$ simplifies through the vanishing of some of the constants $\tau_{gg'jj'}$. The orthorhombic case was mentioned earlier. For nonorthorhombic molecules other nonvanishing coefficients appear. These coefficients multiply operators that introduce $(K, K \pm 1)$ and $(K, K \pm 3)$ elements into the Hamiltonian when set up in a symmetric rotor basis. These terms alter the rigid rotor energy in the second order only; thus they do not need to be considered here. To show that these terms do not contribute in the first order, one has only to consider the effect of the Wang transformation, 2m1 on the energy matrix. If the energy matrix is that of a rigid rotor, the result is 2m1. This factoring fails [89] if the energy matrix is that of a general nonrigid asymmetric rotor. Then, the transformed energy matrix factors into four submatrices along the diagonal with elements $(K \mid K)$, $(K \mid K \pm 2)$ and $(K \mid K \pm 4)$, these factors corresponding to the four factors of the rigid rotor case. The only elements outside these factors are the $(K \mid K \pm 1)$ and $(K \mid K \pm 3)$ elements. Upon transformation of the energy matrix to the rigid asymmetric basis, no $(K \mid K \pm 1)$ or $(K \mid K \pm 3)$ enter the diagonal; thus in the first order they will not affect the energy even in the symmetric rotor limits.

3g THE DISTORTION TERM

The commutation rules 2b1 can be used to bring about some simplification. Terms of the $gg'gg'$ type can be reduced with the result that the coefficients of the other terms are changed and new terms in $\mathbf{P}_g^2$ are introduced. These new terms can be absorbed into $\mathbf{H}^0$ with a new definition of the A_v', etc. Thus we find

$$\mathbf{H} = \mathbf{H}_0 + \mathbf{H}_1 \tag{1}$$

$$\mathbf{H}_0 = A_v\mathbf{P}_z^2 + B_v\mathbf{P}_x^2 + C_v\mathbf{P}_y^2 \tag{2}$$

$$\mathbf{H}_1 = \tfrac{1}{4}\sum_{gg'}{}' \tau'_{ggg'g'}\mathbf{P}_g^2\mathbf{P}_{g'}^2. \tag{3}$$

The relations between the old and new coefficients are readily obtained and are given in Table 3g2.

Table 3g1—Relations Used to Reduce 3g1 to 3g4

$$\mathbf{P}^2 = \mathbf{P}_x^2 + \mathbf{P}_y^2 + \mathbf{P}_z^2$$

$$\mathbf{P}_x^2\mathbf{P}^2 + \mathbf{P}^2\mathbf{P}_x^2 = 2J(J+1)\mathbf{P}_x^2 = 2\mathbf{P}_x^4 + (\mathbf{P}_x^2\mathbf{P}_y^2 + \mathbf{P}_y^2\mathbf{P}_x^2) + (\mathbf{P}_z^2\mathbf{P}_x^2 + \mathbf{P}_x^2\mathbf{P}_z^2)$$

$$\mathbf{P}_y^2\mathbf{P}^2 + \mathbf{P}^2\mathbf{P}_y^2 = 2J(J+1)\mathbf{P}_y^2 = 2\mathbf{P}_y^4 + (\mathbf{P}_y^2\mathbf{P}_x^2 + \mathbf{P}_x^2\mathbf{P}_y^2) + (\mathbf{P}_z^2\mathbf{P}_y^2 + \mathbf{P}_y^2\mathbf{P}_z^2)$$

$$\mathbf{H}_0^2 = A^2\mathbf{P}_z^4 + B^2\mathbf{P}_x^4 + C^2\mathbf{P}_y^4 + AB(\mathbf{P}_z^2\mathbf{P}_x^2 + \mathbf{P}_x^2\mathbf{P}_z^2) + BC(\mathbf{P}_x^2\mathbf{P}_y^2 + \mathbf{P}_y^2\mathbf{P}_x^2) + CA(\mathbf{P}_y^2\mathbf{P}_z^2 + \mathbf{P}_z^2\mathbf{P}_y^2)$$

$$\frac{\partial \mathbf{H}_0^2}{\partial A} = 2(A - C)\mathbf{P}_z^4 + (B - C)(\mathbf{P}_x^2\mathbf{P}_z^2 + \mathbf{P}_z^2\mathbf{P}_x^2) + 2CJ(J+1)\mathbf{P}_z^2$$

$$\frac{\partial \mathbf{H}_0^2}{\partial A} = 2(A - B)\mathbf{P}_z^4 + (C - B)(\mathbf{P}_y^2\mathbf{P}_z^2 + \mathbf{P}_z^2\mathbf{P}_y^2) + 2BJ(J+1)\mathbf{P}_z^2$$

$$\langle \mathbf{P}_z^2 \rangle = \left\langle \frac{\partial \mathbf{H}_0}{\partial A} \right\rangle = \frac{\partial E_0}{\partial A}$$

Through considerable operator manipulation which is not repeated here, Kivelson and Wilson [69] arrive at the following expression for the energy of a nonrigid rotor to the first order.

$$E = E_0 + A_1E_0^2 + A_2E_0J(J+1) + A_3J^2(J+1)^2 + A_4J(J+1)\langle \mathbf{P}_z^2 \rangle + A_5\langle \mathbf{P}_z^4 \rangle + A_6E_0\langle \mathbf{P}_z^2 \rangle. \tag{4}$$

The relations necessary for this reduction are given in Table 3g1.

The values of the A_j in terms of the original τ's are given in Table 3g2. The A_j are independent of the rotational quantum numbers. E_0 is the rigid rotor energy which can be found by the methods of Chapter 2. The evaluation of $\langle \mathbf{P}_z^2 \rangle$ and $\langle \mathbf{P}_z^4 \rangle$ is considered next.

If good structural parameters and force constants are available, the A_j can be calculated although the calculation is quite tedious.

It should be noted that although there are nine distortion constants, only six linear combinations of them enter into the result (4). Usually it is probably desirable to evaluate these six A's from the experimental data. One would start by calculating the rigid rotor parameters from levels of low J, since the correction will be least for them. These parameters can be used to calculate E_0 for the higher rotational levels. The difference between the rigid energies and the observed energies is the distortion correction. The A's can then be evaluated by fitting the levels of high rotational quantum number. To obtain a higher degree of accuracy, a new set of parameters can be determined from the low J levels by using the A's found from the high J levels. The fitting process can then be repeated. Such an iterative process can be repeated until the desired accuracy is obtained. Sum rules given in Appendix V can be used effectively when proper assignments can be made.

The differences between the A_v and A_v' introduce an ambiguity into the definition of the moment of inertia besides that arising from the vibrational effects described in 3c. This ambiguity is transmitted to the structural parameters. If the classical analogy

Table 3g2—Centrifugal Distortion Coefficients

$$\tau'_{zzzz} = \tau_{zzzz}\hbar^4 \qquad \tau'_{xxyy} = (\tau_{xxyy} + 2\tau_{xyxy})\hbar^4$$
$$\tau'_{xxxx} = \tau_{xxxx}\hbar^4 \qquad \tau'_{yyzz} = (\tau_{yyzz} + 2\tau_{yzyz})\hbar^4$$
$$\tau'_{yyyy} = \tau_{yyyy}\hbar^4 \qquad \tau'_{zzxx} = (\tau_{zzxx} + 2\tau_{zxzx})\hbar^4$$

$$A = A' + (3\tau_{xyxy} - 2\tau_{zxzx} - 2\tau_{yzyz})\hbar^4/4$$
$$B = B' + (3\tau_{yzyz} - 2\tau_{xyxy} - 2\tau_{zxzx})\hbar^4/4$$
$$C = C' + (3\tau_{zxzx} - 2\tau_{xyxy} - 2\tau_{yzyz})\hbar^4/4$$

$$A_1 = 16R_6/(B - C)$$
$$A_2 = -[16R_6(B + C)/(B - C)^2 + 4\delta_J/(B - C)]$$
$$A_3 = -D_J + 2R_6 + 16R_6BC/(B - C)^2 + 2\delta_J(B + C)/(B - C)$$
$$A_4 = -[D_{JK} - 2\delta_J\sigma - 16R_6(A^2 - BC)/(B - C)^2 + 4R_6\sigma^2 + 4R_5(C + B)/(B - C)]$$
$$A_5 = -[D_K + 4R_5\sigma + 2R_6 - 4R_6\sigma^2]$$
$$A_6 = (8R_5 - 16R_6\sigma)/(B - C)$$

$$\sigma = (2A - B - C)/(B - C)$$
$$D_J = -(\tfrac{1}{32})(3\tau_{xxxx} + 3\tau_{yyyy} + 2\tau_{xxyy} + 4\tau_{xyxy})\hbar^4$$
$$D_K = D_J - \tfrac{1}{4}(\tau_{zzzz} - \tau_{zzxx} - \tau_{yyzz} - 2\tau_{xzxz} - 2\tau_{yzyz})\hbar^4$$
$$D_{JK} = -D_J - D_K - \tfrac{1}{4}\tau_{zzzz}\hbar^4$$
$$R_5 = -(\tfrac{1}{32})[\tau_{xxxx} - \tau_{yyyy} - 2(\tau_{xxzz} + 2\tau_{xzxz}) + 2(\tau_{yyzz} + 2\tau_{yzyz})]\hbar^4$$
$$R_6 = +(\tfrac{1}{64})[\tau_{xxxx} + \tau_{yyyy} - 2(\tau_{xxyy} + 2\tau_{xyxy})]\hbar^4$$
$$\delta_J = -\tfrac{1}{16}(\tau_{xxxx} - \tau_{yyyy})\hbar^4$$

is to be preserved of a rigid rotor with a centrifugal distortion correction, the quantities A_v', B_v', and C_v' should be used to define the moment of inertia; however, it is A_v, B_v, and C_v that are determined from the experimental data.

3h THE EVALUATION OF $\langle \mathbf{P}_z^2 \rangle$

The evaluation of $\langle \mathbf{P}_z^2 \rangle$ can be carried out in a straightforward way [69]. The evaluation depends on the identification that is made between the a, b, c inertial axes and the x, y, z molecule-fixed axes.

For present purposes, assume that the c-axis is identified with the z-axis (III representation). Then

$$\mathbf{H}_0 = A\mathbf{P}_a^2 + B\mathbf{P}_b^2 + C\mathbf{P}_c^2. \tag{1}$$

Let this problem be perturbed by a small change in C, say δC, such that the perturbation is given by the operator,

$$\mathbf{H}' = \delta C \mathbf{P}_c^2. \qquad (2)$$

Then $E(A, B, C)$ is an eigenvalue of the unperturbed problem, (1), and E' is the first-order energy correction due to (2), $E' = \langle \delta C \mathbf{P}_c^2 \rangle$, where the brackets denote the average value over the unperturbed eigenstate corresponding to $E(A, B, C)$.

Now $E(A, B, C + \delta C)$ is the corresponding eigenvalue of the operator

$$\mathbf{H} = A\mathbf{P}_a^2 + B\mathbf{P}_b^2 + (C + \delta C)\mathbf{P}_c^2. \qquad (3)$$

If ε is defined such that $E(A, B, C + \delta C) = E(a, b, c) + E' + \varepsilon$, then

$$\frac{\partial E(A, B, C)}{\partial C} = \lim_{\delta C \to 0} \frac{E' + \varepsilon}{\delta C}. \qquad (4)$$

ε is of the order of δC^2, $\Delta[\delta E(A, B, C)]/\delta C = \langle \mathbf{P}_c^2 \rangle$. Thus $\langle \mathbf{P}_z^2 \rangle$ is given by the derivative of the energy with respect to the inertial constant, with the particular value depending on the representation I, II, or III that is chosen. For convenience, these expressions are derived here making use of 2j6 and 2j9 or of 2k3 and 2k4 with results

$$\alpha = \frac{\partial E(A, B, C)}{\partial A} = \tfrac{1}{2}J(J + 1) + \tfrac{1}{2}E(\kappa) - \frac{\kappa + 1}{2}\frac{\partial E(\kappa)}{\partial \kappa}$$

$$= E(b) - b\frac{\partial E(b)}{\partial b}. \qquad (5)$$

$$\beta = \frac{\partial E(A, B, C)}{\partial B} = \frac{\partial E(\kappa)}{\partial \kappa}$$

$$= \tfrac{1}{2}J(J + 1) - \tfrac{1}{2}E(b) - \frac{1 - b}{2}\frac{\partial E(b)}{\partial b}. \qquad (6)$$

$$\gamma = \frac{\partial E(A, B, C)}{\partial C} = \tfrac{1}{2}J(J + 1) - \tfrac{1}{2}E(\kappa) - \frac{1 - \kappa}{2}\frac{\partial E(\kappa)}{\partial \kappa}$$

$$= \tfrac{1}{2}J(J + 1) - \tfrac{1}{2}E(b) + \frac{1 + b}{2}\frac{\partial E(b)}{\partial b}. \qquad (7)$$

The values of $\partial E/\partial \kappa$ may be obtained to sufficient accuracy from the tables of $E(\kappa)$ in Appendix III. The ΔE should be taken between two values of κ, which bracket the desired value of $E(\kappa)$. Since tables of $E(b)$ are not readily available, it is necessary to calculate $E(b)$ for two values of b, which bracket the desired value. For this reason the κ form of the derivatives is more convenient.

3i EVALUATION OF $\langle \mathbf{P}_z^4 \rangle$

Consider the Hamiltonian

$$\mathbf{H}^* = A\mathbf{P}_a^2 + B\mathbf{P}_b^2 + C\mathbf{P}_c^2 + \tfrac{1}{2}(A - C)q\mathbf{\Pi} = \mathbf{H}^0 + \tfrac{1}{2}(A - C)q\mathbf{\Pi}. \qquad (1)$$

All the old symbols have their familiar meaning, whereas q is independent of the rotational quantum numbers and $\mathbf{\Pi}$ is an operator diagonal in the limiting symmetric rotor basis. $\mathbf{H}^*$ can conveniently be rewritten as

$$\mathbf{H}^* = \tfrac{1}{2}(A + C)J(J + 1) + \tfrac{1}{2}(A - C)\boldsymbol{\lambda}'(\kappa, q) + \tfrac{1}{2}(A - C)FJ(J + 1), \tag{2}$$

in which F is the constant defined in Table 2j1 and $\boldsymbol{\lambda}'(\kappa, q) = \boldsymbol{\lambda}(\kappa) + q\mathbf{\Pi}$. Now in a manner similar to that used in 3h the average value of $\mathbf{\Pi}$ may be written as

$$\langle \mathbf{\Pi} \rangle = \left(\frac{2}{A - C}\right)\left(\frac{\partial E''}{\partial q}\right)_{q=0} = \left[\frac{\partial \boldsymbol{\lambda}'(\kappa, q)}{\partial q}\right]_{q=0}, \tag{3}$$

where E'' and λ' are the eigenvalues of $\mathbf{H}''$ and $\boldsymbol{\lambda}'$ respectively. Using a notation analogous to 2o10 an expression for λ' may be written down as a continued fraction; that is,

$$\begin{aligned}\lambda' = k_m &+ q\Pi_m \\ &- \cfrac{b_m}{k_{m-1} - \lambda' + q\Pi_{m-1} - \cfrac{b_{m-1}}{k_{m-2} - \lambda' + q\Pi_{m-2} - \cdots}} \\ &- \cfrac{b_{m+1}}{k_{m+1} - \lambda' + q\Pi_{m+1} - \cfrac{b_{m+2}}{k_{m+2} - \lambda' + q\Pi_{m+2} - \cdots}}\end{aligned} \tag{4}$$

In (4), Π_m represents the mth element of $\mathbf{\Pi}$ in the limiting symmetric basis. If the following abbreviations are introduced,

$$\begin{aligned} R_n &= \frac{b_{n+1}}{\left[k_n - \lambda - \left(\dfrac{b_n}{k_{n-1} - \lambda - \cdots}\right)\right]^2} \\ R_n' &= \frac{b_n}{\left[k_n - \lambda - \left(\dfrac{b_{n+1}}{k_{n+1} - \lambda - \cdots}\right)\right]^2} \end{aligned} \tag{5}$$

and (4) differentiated with respect to q, remembering (3), we find

$$\begin{aligned}\left(\frac{\partial \lambda'}{\partial q}\right)_{q=0} &= \langle \mathbf{\Pi} \rangle \\ &= \Pi_m + R_{m-1}\{(\Pi_{m-1} - \langle \mathbf{\Pi} \rangle) + R_{m-2}[(\Pi_{m-2} - \langle \mathbf{\Pi} \rangle) + R_{m-3}(\ldots)]\} \\ &\quad + R'_{m+1}\{(\Pi_{m+1} - \langle \mathbf{\Pi} \rangle) + R'_{m+2}[(\Pi_{m+2} - \langle \mathbf{\Pi} \rangle) + \ldots]\}.\end{aligned} \tag{6}$$

This series should behave similarly to the continued fraction (4) especially with regard to convergence. Rearranging (6) gives

$$\begin{aligned}\langle \mathbf{\Pi} \rangle = &\{1 + R_{m-1}[1 + R_{m-2}(1 + \ldots)] + R'_{m+1}[1 + R'_{m+2}(1 + \ldots)]\}^{-1} \\ &\times \{\Pi_m + R_{m-1}[\Pi_{m-1} + R_{m-2}(\ldots)] + R'_{m+1}[\Pi_{m+1} + R'_{m+2}(\ldots)]\}\end{aligned} \tag{7}$$

Except when $E(\kappa)/J(J + 1)$ is nearly equal to κ, only a few terms of the series should be needed, especially since $\langle \mathbf{\Pi} \rangle$ enters in a correction term to the rigid rotor energy and thus need not be known as accurately as the latter. It should be pointed out that the evaluation of R_n or R_n' requires only one arithmetic step when the rigid rotor energy is calculated by the continued fraction method.

If K^2 is substituted for Π_m and $(K \pm 2n)^2$ for Π_{m+n} and choosing the appropriate submatrix, this procedure yields $\langle \mathbf{P}_z^{\,2} \rangle$. If K^4 is substituted for Π_m and $(K \pm 2n)^4$ for Π_{m+n}, $\langle \mathbf{P}_z^{\,4} \rangle$ is obtained.

For certain cases of slight asymmetry, (7) can be expanded in terms of an asymmetry parameter. For a slightly asymmetric prolate rotor, $\delta = (\kappa + 1)/2 = (B - C)/(A - C) = -2b/(1 + b)$ is a convenient parameter. The expansions give

$$\langle \mathbf{P}_z^{\,2} \rangle = K^2 + \left[\frac{f(J, K-1)}{16(1-K)} + \frac{f(J, K+1)}{16(1+K)}\right]\delta^2 + 0(\delta^3) + \dots \tag{8}$$

$$\langle \mathbf{P}_z^{\,4} \rangle = K^4 + 2\left[\frac{(K^2 - 2K + 2)f(J, K-1)}{16(1-K)} + \frac{(K^2 + 2K + 2)f(J, K+1)}{16(1+K)}\right]\delta^2 + 0(\delta^3) + \dots, \tag{9}$$

where K is the quantum number of the limiting prolate symmetric rotor, and $f(J, n)$ is defined in 2j12. For cases of not too great asymmetry this result should be very satisfactory since the accuracy required for the correction term is not too great.

For very slight asymmetry rougher approximations suffice.

$$\langle \mathbf{P}_z^{\,4} \rangle \simeq \langle \mathbf{P}_z^{\,2} \rangle^2 \simeq K^4. \tag{10}$$

In the region where $E/J(J + 1)$ is very different from κ, K_{-1} is large, the asymmetry slight, a better approximation is

$$\langle \mathbf{P}_z^{\,4} \rangle \simeq \langle \mathbf{P}_z^{\,2} \rangle^2 \simeq \left(\frac{\partial E_0}{\partial A}\right)^2. \tag{11}$$

If this approximation is used in 3g4, the semiclassical relation proposed by Lawrence and Strandberg [73] and Cross [35] results.

With these methods of evaluating $\langle \mathbf{P}_z^{\,2} \rangle$ and $\langle \mathbf{P}_z^{\,4} \rangle$, the expression for the energy, 3g4 may be used either by calculating the distortion constants from force-constant data or they may be obtained empirically by fitting the rotational energy levels.

3j LINEAR, SPHERICAL, AND SYMMETRIC ROTORS

In the symmetric rotor limit R_5, R_6, and δ_J all vanish and $B_v = C_v$. (For definitions see Tables 3g1 and 3g2.) Making use of the equations used to derive 3g4, it can be shown that the factors multiplying R_5, R_6, and δ_J are finite. Since in the symmetric rotor limit $\langle \mathbf{P}_z^{\,4} \rangle = \langle \mathbf{P}_z^{\,2} \rangle^2 = K^4$, 3g4 reduces to the familiar result for a symmetric rotor [113].

$$E = E_0 - D_J J^2(J + 1)^2 - D_{JK} J(J + 1)K^2 - D_K K^4, \tag{1}$$

in which E_0 is given by either 2h2 or 2h3. The expression for linear and spherical rotors turns out to be simply

$$E = E_0 - DJ^2(J + 1)^2, \tag{2}$$

with E_0 given by 2i1. Such a result is quite plausible since the energy levels of both types of molecules are independent of K.

4

PERTURBATIONS TO THE RIGID ROTOR ENERGY LEVELS

4a HAMILTONIAN

In the preceding chapters the rigid rotor energy levels, the effect of vibrational state on the inertial constants, and corrections to the energy levels arising from centrifugal distortion effects have been discussed in considerable detail. This chapter gives a detailed discussion of some of the energy level perturbations that arise from the interaction of vibration with rotation. For this purpose it is necessary to investigate a form of the Hamiltonian which is more general than the rigid rotor Hamiltonian but at the same time is somewhat simpler than the general Hamiltonian. The form chosen for this discussion is 1f2.

$$\mathbf{H} = \left[\frac{(\mathbf{P}_x - \mathbf{p}_x)^2}{2I_x} + \frac{(\mathbf{P}_y - \mathbf{p}_y)^2}{2I_y} + \frac{(\mathbf{P}_z - \mathbf{p}_z)^2}{2I_z}\right] + \tfrac{1}{2}\sum_k \mathbf{p}_k{}^2 + \mathbf{V}. \qquad (1f2)$$

In this Hamiltonian there are two main sources from which perturbations can arise. One of these sources is the cross-terms between $\mathbf{P}_g$ and $\mathbf{p}_g$, and the other is the anharmonic terms in V.

Equation 1f2 may be expanded in a manner similar to that used in expanding 3a1. Remembering that the $\mathbf{P}_g$ commute with the $\mathbf{p}_g$, we obtain the result

$$\mathbf{H} = \frac{1}{2}\left(\frac{\mathbf{P}_x{}^2}{I_x} + \frac{\mathbf{P}_y{}^2}{I_y} + \frac{\mathbf{P}_z{}^2}{I_z}\right) - \left(\frac{\mathbf{p}_x\mathbf{P}_x}{I_x} + \frac{\mathbf{p}_y\mathbf{P}_y}{I_y} + \frac{\mathbf{p}_z\mathbf{P}_z}{I_z}\right)$$
$$+ \frac{1}{2}\left(\frac{\mathbf{p}_x{}^2}{I_x} + \frac{\mathbf{p}_y{}^2}{I_y} + \frac{\mathbf{p}_z{}^2}{I_z}\right) + \tfrac{1}{2}\sum_k \mathbf{p}_k^2 + \mathbf{V}. \qquad (1)$$

The first term in (1) represents the rigid rotor Hamiltonian which was discussed fully in Chapter 2. The last two terms are the vibrational Hamiltonian. The third term also depends on the vibrational coordinates only and cannot give rise to more than an additive term for each of the rotational-vibrational levels. It is therefore of no interest to this discussion except for overtone and combination bands with multiple components. The second term in the expanded Hamiltonian represents the interaction between vibration and rotation. It is this term that will be the major concern of this chapter. That this term does represent the interaction between vibration and rotation is easily seen from 1d5 in which the $\mathbf{p}_g$ are defined as

$$\mathbf{p}_x = \sum_k \mathfrak{X}_k \mathbf{p}_k, \qquad \mathbf{p}_y = \sum_k \mathfrak{Y}_k \mathbf{p}_k, \qquad \mathbf{p}_z = \sum_k \mathfrak{Z}_k \mathbf{p}_k,$$

with the $\mathfrak{X}$, $\mathfrak{Y}$, $\mathfrak{Z}$ defined by 1c13; that is,

$$-\mathfrak{X}_k = \sum_{\alpha l} \frac{\partial(\zeta_\alpha, \eta_\alpha)}{\partial(Q_k, Q_l)} Q_l = \sum_{\alpha l} \frac{\partial(Q_k, Q_l)}{\partial(\zeta_\alpha, \eta_\alpha)} Q_l = \sum_{\alpha l} (n_{\alpha k} m_{\alpha l} - m_{\alpha k} n_{\alpha l}) Q_l = -\sum_l \zeta_{kl}{}^x Q_l, \quad \text{etc.},$$

The $l_{\alpha k}$, $m_{\alpha k}$, $n_{\alpha k}$ are the coefficients in the transformation from mass-weighted Cartesian coordinates to normal coordinates. Since this transformation is orthogonal,

$$\frac{\partial Q_k}{\partial \xi_\alpha} = l_{\alpha k} = \frac{\partial \xi_\alpha}{\partial Q_k}, \quad \text{etc.}$$

The $\mathbf{p}_g$ may be written as, for example,

$$\begin{aligned} \mathbf{p}_x &= -\sum_k \sum_{\alpha l} (n_{\alpha k} m_{\alpha l} - m_{\alpha k} n_{\alpha l}) Q_l \mathbf{p}_k \\ &= i\hbar \sum_{\alpha, k, l} (n_{\alpha k} m_{\alpha l} - m_{\alpha k} n_{\alpha l}) Q_l \frac{\partial}{\partial Q_k} = -i\hbar \sum_{kl} \zeta_{kl}{}^x Q_l \frac{\partial}{\partial Q_k}, \end{aligned} \tag{2}$$

showing clearly that the $\mathbf{p}_g$ depend solely on the vibrational coordinates. The $\mathbf{P}_g$, of course, depend only on the rotational coordinates. The perturbations arising from the cross-terms between $\mathbf{P}_g$ and $\mathbf{p}_g$ have been called, in general, Coriolis interactions.

4b LINEAR MOLECULES

An intuitive feeling for the Coriolis-type interaction can best be obtained by considering the effect classically. If the motion of a body relative to a uniformly rotating coordinate system is considered, two accelerations act on the body aside from the acceleration due to external forces: the centrifugal and Coriolis accelerations. These accelerations may be considered as the result of two virtual forces. The effect of the centrifugal force was discussed in Chapter 3. A discussion of the Coriolis force follows.

The Coriolis force [6] acting on a body is given classically by

$$F_{\text{Cor}} = 2m_\alpha v_\alpha \omega \sin \phi = 2m(\boldsymbol{\omega} \times \mathbf{v}_\alpha), \tag{1}$$

where m_α is the mass of the body, v_α its velocity relative to the rotating axis system, and ϕ is the angle between the axis of rotation and the direction of v_α. It can now be seen that the Coriolis force exists only for $v_\alpha \neq 0$. It is directed normal to a plane which includes both the axis of rotation and the vector $\mathbf{v}_\alpha$.

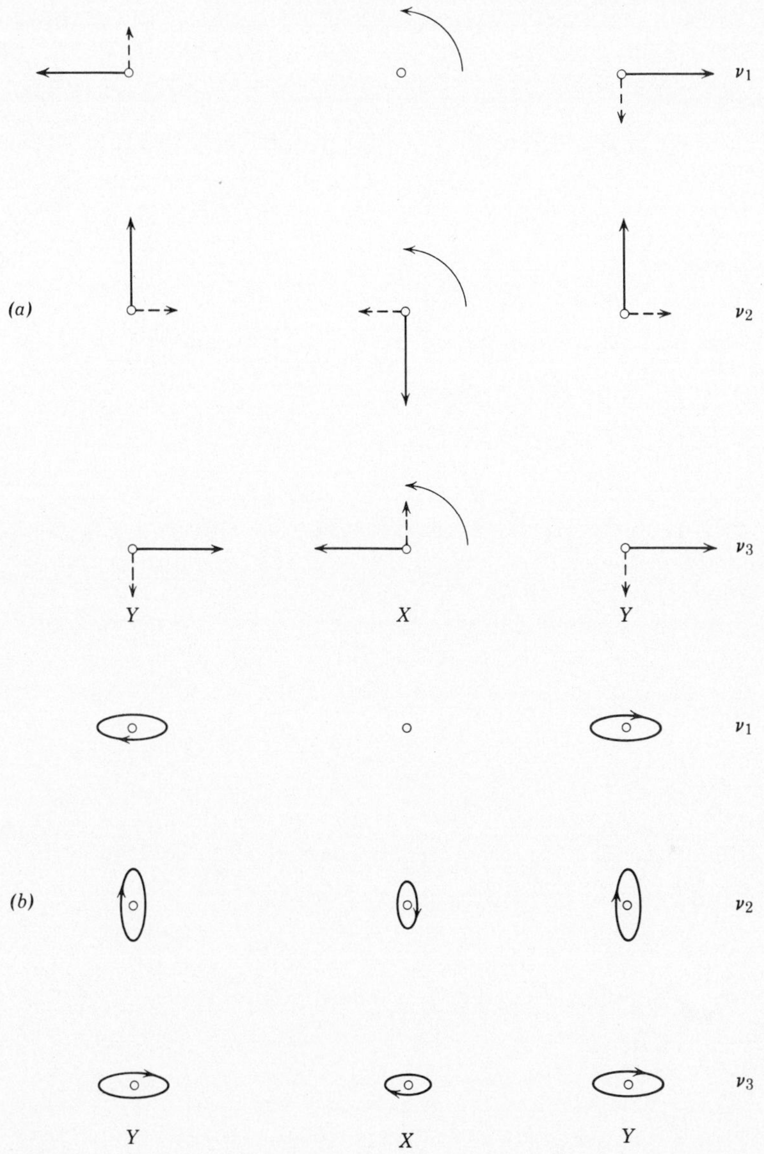

Fig. 4b1 Coriolis forces in a linear symmetrical XY_2 molecule. (*a*) The normal modes of vibration are shown schematically by solid straight arrows. The curved arrows indicate the assumed direction of rotation, whereas the Coriolis forces on each atom are indicated by the dashed arrows. (*b*) Resultant elliptical motions of the atoms. (From Herzberg, *Infrared and Raman Spectra*, copyright 1945, D. van Nostrand Company, Princeton, New Jersey.)

The effect in molecules may be seen by considering a linear-symmetric XY_2 molecule. The normal modes of vibration are shown schematically in Fig. 4b1*a*. The curved arrows indicate the assumed direction of rotation, and the Coriolis forces on each atom are indicated by the small arrows. There are now two ways that the physical result can be visualized. If ν_2 is considered, it is seen that the Coriolis forces tend to excite ν_3; however, since the v_α are those appropriate to ν_2, ν_3 is excited at the frequency of ν_2. Similarly, the Coriolis forces acting on ν_3 tend to excite ν_2 at the frequency of ν_3. Thus, if ν_2 and ν_3 are nearly the same frequency, the excitation of one of these frequencies would strongly excite the other. For ν_1, the Coriolis forces do not tend to excite another vibration but rather produce a rotation, and will hence have an effect on the total angular momentum. In general, this is not considered Coriolis interaction since the coupling is with rotation rather than vibration. In most molecules of this type the frequencies of ν_2 and ν_3 are sufficiently different so that this interaction is weak.

Another way to visualize this interaction is shown in Fig. 4b1*b*. As long as there is a Coriolis force on the atom as well as a vibrational force, the resulting force is such that the atoms no longer vibrate in a line about their equilibrium positions but execute ellipses about their equilibrium positions. As the interaction becomes smaller either because ν_2 and ν_3 have widely different frequencies or because the velocity is small, the ellipses will finally flatten out, degenerating to a straight line when the Coriolis force is zero.

A general rule, due to Jahn [62], has been formulated which enables one to predict when two states can interact through a Coriolis coupling. Two vibrational states will couple through a Coriolis interaction if the direct product of the symmetry species of the two vibrational states contains a rotational species. Linear molecules can only belong to two symmetry point groups, $C_{\infty v}$ or $D_{\infty h}$. The linear symmetric molecules belong to the point group $D_{\infty h}$. For the XY_2 molecule previously considered ν_1 is of species Σ_g^+, ν_2 is of species Π_u and ν_3 is of species Σ_u^+. The direct products are $\Sigma_g^+ \times \Pi_u = \Pi_u$, and $\Sigma_g^+ \times \Sigma_u^+ = \Sigma_u^+$. Since the rotations in $D_{\infty h}$ have the species Σ_g^- and Π_g, the rules predict that there will be no rotational-vibrational interaction of the Coriolis type between ν_1 and either ν_2 or ν_3. However, the product $\Sigma_u^+ \times \Pi_u = \Pi_g$, which is one of the rotational species; hence a Coriolis interaction would be expected between ν_2 and ν_3. Thus Jahn's rule confirms the result deduced from classical considerations.

For a nonsymmetric triatomic molecule XYZ, the point group is $C_{\infty v}$ and the vibrations ν_1, ν_2, ν_3 have the symmetry species Σ^+, Π, Σ^+ respectively. The rotational species are Σ^- and Π. By Jahn's rule, one would predict that both ν_1 and ν_3 could interact with ν_2, but ν_1 and ν_3 would not be expected to interact with each other.

For linear molecules the rotational constant in a given vibrational state may be expressed as

$$B_V = B_e + \sum_{i=1}^{3N-5} \alpha_i\left(v_i + \frac{d_i}{2}\right) + \ldots,$$

where d_i is the degeneracy of the ith normal mode, v_i the quantum number of the ith normal mode in the vibrational state V, and α_i is a constant representing the change in rotational constant per quantum of vibrational excitation of the ith normal mode. In general, there can be three contributions to the α_i. Even for an harmonic-oscillator, α_i would not be zero because $\langle 1/r^2\rangle \neq 1/r_e^{\,2}$; a second contribution to the α_i arises from the anharmonicity of the vibrations; that is, a quadratic potential function does not adequately describe a real molecule, and finally there is a contribution to α_i due to Coriolis perturbations. Detailed formulas for each of these contributions have been given by Nielsen [86, 89] but are not repeated here.

4c *l*-TYPE DOUBLING

Since the bending modes of a linear polyatomic molecule may be considered to take place at the same frequency in either of two perpendicular planes, both including the axis of the molecule, they are degenerate. This degeneracy, when coupled with rotation, gives rise to the phenomenon known as *l*-type doubling. Let the z-axis be the molecular axis, and assume the molecule is not rotating. The bending vibration may then take place in either the xz or yz plane. These are two degenerate frequencies which up to

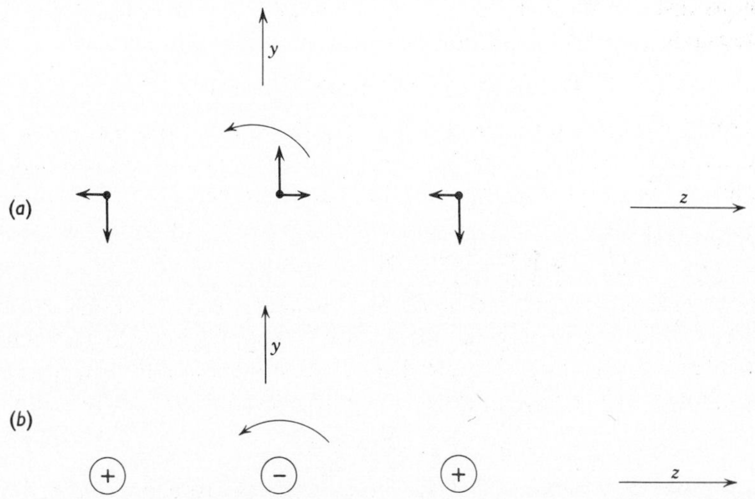

Fig. 4c1 The degenerate bending vibration may take place either in the yz plane as in (*a*) or in the xz plane as in (*b*). The curved arrow indicates that the rotation is in the yz plane. It is easily seen that the moment of inertia about the axis of rotation is slightly different in the two cases.

When the bending vibration is excited perpendicular to the axis of rotation, ν_3 is excited by a Coriolis force shown by the horizontal arrows. However, when the bending vibration is excited parallel to the axis of rotation the Coriolis force is zero.

now have been considered to have the same rotational constant B_V. Suppose now that the molecule rotates in the yz plane. It can be seen from Fig. 4c1 that the result of a bend in the yz plane is not quite the same as that in xz plane. The angle ϕ in 4b1 is $\pi/2$ for a bend in the yz plane, whereas it is zero for a bend in the xz plane; hence the effective moment of inertia about the axis of rotation differs slightly in the two cases. Thus when the bending vibration is excited perpendicularly to the angular momentum, ν_3 is excited by a Coriolis force, but when the bending vibration is excited parallel to the angular momentum, the Coriolis force is zero, This vibrational-rotational interaction slightly splits the degenerate levels, giving rise to the so-called *l*-type doubling.

A more detailed and accurate treatment of *l*-type doubling requires the use of quantum mechanics. The procedure is only outlined here; the more complicated cases of the symmetric and asymmetric rotors are treated in detail later. If the vibrational problem is studied in Cartesian coordinates [7], the solution is an oscillator along the y-axis and an oscillator along the x-axis with the allowed energies of $(v_y + \frac{1}{2})h\nu$ and $(v_x + \frac{1}{2})h\nu$, the total energy being the sum of the two parts $(v_y + v_x + 1)h\nu = (v + 1)h\nu$. This method of solution shows the degeneracy readily since the value of $v \geqslant 1$ can be achieved by more than one combination of values of v_x and v_y; yet all combinations lead to the same energy.

Suppose that the x and y oscillations are combined in the proper phase so that the oscillator moves in a circle or ellipse about the molecular axis. The problem may then be solved using cylindrical coordinates [7] which will specify the position of the oscillator by its distance from the origin r and an angle θ between r and one of the axes, say the x-axis. The wave function is then given by

$$\psi_{v,l} = N_{v,l}\rho^{|l|} \exp\left(-\tfrac{1}{2}\rho^2 + il\theta\right)F^{|l|}_{\frac{1}{2}(v+|l|)}(\rho^2). \tag{2}$$

$N_{v,l}$ is a normalizing constant, $\rho = 2\pi(\nu/h)^{1/2}(Q_x^{\,2} + Q_y^{\,2})^{1/2} = 2\pi(m\nu/h)^{1/2}r$, m is the mass of the oscillator and $F^{|l|}_{\frac{1}{2}(v+|l|)}(\rho^2)$ is the associated Laguerre polynomial. The energy is again given by $(v + 1)h\nu$, whereas the angular momentum $l\hbar$ can only assume the values

$$v\hbar, \qquad (v - 2)\hbar, \ldots, -v\hbar.$$

This angular momentum in the degenerate modes affects the vibrational-rotational energies, the result being an energy expression very much like that for a symmetric rotor in which l replaces K. For the example under consideration, the linear triatomic molecule, which has only one degenerate mode of vibration, the wave function, is

$$\psi_{v,l,J} = \psi_{v,l}R_{J,l}(\theta, \phi), \tag{3}$$

in which $\psi_{v,l}$ is given above and $R_{J,l}(\theta, \phi)$ is the symmetric rotor wave function discussed in Chapters 2 and 5. The obvious condition is $J \geqslant |l|$ since the total angular momentum cannot be less than the angular momentum about the molecular axis. The energy levels for this case are given by

$$E = \sum_i h\nu_i\left(v_i + \frac{d_i}{2}\right) + \sum_i g_{ii}l_i^{\,2} + B_v[J(J + 1) - l^2] + D_v[J(J + 1) - l^2]^2, \tag{4}$$

in which g_{ii} is a small constant characteristic of a given degenerate vibrational state. This expression for the energy still does not show the l-type doubling since l enters the energy only as the square, leaving the twofold degeneracy for $\pm l$. The l-type doubling arises when this degeneracy is removed. The problem is formally similar to that of the slightly asymmetric rotor which is described in detail in Chapter 2.

A rather complicated treatment of the energy level splitting has been carried out by H. H. Nielsen [89]. The actual derivation is not repeated here, merely the result for $l = 1$ being quoted since one would expect this case to give the largest effect (see discussion of approximate methods for calculating rigid rotor energy levels in Appendix IV). Thus, for $l = 1$

$$\Delta E = \frac{B_e^{\,2}}{\omega_i}\left[1 + 4\sum_k \zeta_{ik}\left(\frac{\omega_i^{\,2}}{\omega_k^{\,2} - \omega_i^{\,2}}\right)\right](v_i + 1)J(J + 1), \tag{5}$$

in which v_i is the quantum number of the degenerate vibration, ω_i, ω_k the molecular frequencies of the interacting levels, and the ζ_{ik} are the Coriolis parameters of the molecule which depend on the potential function of the molecule, the molecular geometry, and the masses. Nielsen has shown that, in general, this splitting is proportional to $B_e(B_e/\omega_i)^l$, and hence decreases very rapidly as l increases. One-half of that portion of (5) which has no quantum number dependence is generally denoted as q, so that the energy level splitting is

$$\Delta\nu = 2q(v_i + 1)J(J + 1). \tag{6}$$

4d THE SYMMETRIC ROTOR

The Coriolis interaction in symmetric rotors has been discussed by many authors. Some of these discussions can be found in the references [24, 39, 89, 115]. The development used here is essentially that due to Boyd and Longuet-Higgins [24]. This problem is discussed from a strictly quantum-mechanical standpoint since an intuitive feeling for the problem should have been obtained from the discussion for linear molecules. The vibrational-rotational interaction in symmetric rotors is considered as a perturbation to the rigid rotor harmonic-oscillator problem. The zero-order Hamiltonian is

$$\mathbf{H}^0 = \mathbf{H}_R{}^0 + \mathbf{H}_V{}^0 \tag{1}$$

in which, using a Type I representation,

$$\mathbf{H}_R{}^0 = \frac{\mathbf{P}_x{}^2 + \mathbf{P}_y{}^2}{2I_b} + \frac{\mathbf{P}_z{}^2}{2I_a}, \tag{2}$$

and $\mathbf{H}_V{}^0$ is the harmonic-oscillator Hamiltonian. The energy levels of the rigid symmetric rotor were deduced and discussed in Chapter 2.

To take into account the vibrational-rotational interaction a more exact form of the Hamiltonian than (1) must be employed. Thus the form used is the one usually chosen

$$\mathbf{H} = \mathbf{H}^0 + \mathbf{H}_c, \tag{3}$$

in which $\mathbf{H}^0$ is given by (2) and

$$\mathbf{H}_c = \frac{-(\mathbf{P}_x\mathbf{p}_x + \mathbf{P}_y\mathbf{p}_y)}{I_b} - \frac{\mathbf{P}_z\mathbf{p}_z}{I_a}. \tag{4}$$

The first-order correction to the energy levels is found by integrating $\mathbf{H}_c$ over the unperturbed wave functions, that is,

$$E = E^0 + E_c, \tag{5}$$

where

$$E_c = \int \psi_V{}^*\psi_R{}^*H_c\psi_V\psi_R \, d\tau. \tag{6}$$

In (6) the approximation has been made that vibrational-rotational wave functions of the product type may be used. Remembering that the $\mathbf{p}_g$ are functions only of the vibrational coordinates, (6) may be rewritten in the form

$$E_c = -\left[\frac{1}{I_b}\int(\psi_R{}^*\mathbf{P}_x\psi_R\psi_V{}^*\mathbf{p}_x\psi_V \, d\tau_V \, d\tau_R + \psi_R{}^*\mathbf{P}_y\psi_R\psi_V{}^*\mathbf{p}_y\psi_V \, d\tau_V \, d\tau_R) + \frac{1}{I_a}\int \psi_R{}^*\mathbf{P}_z\psi_R\psi_V{}^*\mathbf{p}_z\psi_V \, d\tau_V \, d\tau_R\right]. \tag{7}$$

Since $\mathbf{P}_x$ and $\mathbf{P}_y$ are nondiagonal in the representation which has been used, the first integral in the bracket vanishes. Thus it is necessary to consider only the contribution from the last term in the bracket. Integrating over the rotational coordinates leads to a rotational dependence

$$\int \psi_R{}^*\mathbf{P}_z\psi_R\,d\tau = K\hbar. \tag{8}$$

Before finding the vibrational dependence of (7) an explicit form for $\mathbf{p}_z$ must be found. From the definition of $\mathbf{p}_z$, 1d5, it is found

$$\begin{aligned} \mathbf{p}_z &= \sum_k \mathfrak{Z}_k\mathbf{p}_k = \sum_k \sum_{\alpha l} \frac{\partial(Q_k, Q_l)}{\partial(\xi_\alpha, \eta_\alpha)} Q_l\mathbf{p}_k \\ &= -i\hbar \sum_{l<k}\sum \left(Q_l\frac{\partial}{\partial Q_k} - Q_k\frac{\partial}{\partial Q_l}\right)\sum_\alpha \frac{\partial(Q_k, Q_l)}{\partial(\xi_\alpha, \eta_\alpha)}. \end{aligned} \tag{9}$$

If Q_k and Q_l are two nondegenerate normal coordinates, the contribution from the vibration part of the integral from the operator $-i\hbar[Q_l(\partial/\partial Q_k) - Q_k(\partial/\partial Q_l)$ vanishes when harmonic-oscillator wave functions are assumed. Hence the first-order Coriolis interaction between two nondegenerate vibrations of a symmetric rotor vanishes. It might be added here that this is true for all molecules if the vibrations are nondegenerate.

Now, if X_r and Y_r are the two real normal coordinates of a degenerate vibration, the energy correction due to the operator $-i\hbar[X_r(\partial/\partial Y_r) - Y_r(\partial/\partial X_r)]$ does not vanish. The vibrational part of the integral becomes

$$\int \psi_V{}^*\mathbf{p}_z\psi_V\,d\tau = \sum_r\sum_\alpha \frac{\partial(X_r, Y_r)}{\partial(\xi_\alpha, \eta_\alpha)}\int \psi_V{}^*\mathbf{p}_r\psi_V\,d\tau = \zeta_r\int \psi_V{}^*\mathbf{p}_r\psi_V\,d\tau, \tag{10}$$

where $\mathbf{p}_r = -i\hbar[X_r(\partial/\partial Y_r) - Y_r(\partial/\partial X_r)]$, and the summation is taken only over the degenerate modes (X_r, Y_r). ζ_r is quite analogous to the previously defined ζ's.

Just as in the degenerate modes of a linear molecule, two quantum numbers are required to specify a doubly degenerate state. The first is the vibrational quantum number v_r and the second, denoted by l_r, is the quantum number associated with the angular momentum $\mathbf{p}_r$. For the ground vibrational state both v_r and l_r are zero; thus the wave function is

$$\psi_{v_r l_r}(X_r, Y_r) = \psi_{0,0}(X_r, Y_r) = e^{-\frac{1}{2}\alpha_r(X_r{}^2+Y_r{}^2)}, \tag{11}$$

in which $\alpha_r = 4\pi^2\nu_r/h$. The state in which the degenerate mode (X_r, Y_r) is singly excited is doubly degenerate with the explicit wave functions

$$\begin{aligned} \psi_{1,1} &= \sqrt{\alpha_r}e^{-\frac{1}{2}\alpha_r(X_r{}^2+Y_r{}^2)}(X_r + iY_r) \\ \psi_{1,-1} &= \sqrt{\alpha_r}e^{-\frac{1}{2}\alpha_r(X_r{}^2+Y_r{}^2)}(X_r - iY_r). \end{aligned} \tag{12}$$

For the states described by the wave functions (12) the integral in (10) has the value $\pm\zeta_r\hbar$. This result combined with the result $K\hbar$ for the rotational contribution and the negative sign of $\mathbf{H}_c$ gives the energy correction E_c as

$$E_c = \mp\frac{\zeta_r\hbar^2}{I_a}K = \mp 2\zeta_r AK. \tag{13}$$

It is now necessary to determine when each of the signs in (13) occurs. For an optical transition from the ground state to the state $\psi_{1,1}$ to occur, $(X_r + iY_r)$ must transform in the same way as the electric moment operator $(M_x + iM_y)$ or $(M_x - iM_y)$. If $(X_r + iY_r)$ transforms as the former, the K selection rule is $\Delta K = +1$, and if it transforms as the latter, it is $\Delta K = -1$. If we choose the signs of X_r and Y_r so that (X_r, Y_r) transforms as (M_x, M_y), the transition from the ground state to $\psi_{1,1}$ occurs with $\Delta K = +1$. Such a choice is always possible since the sign of a normal coordinate is indeterminate. With this choice the energy correction becomes

$$E_c = -2\zeta_r AK, \qquad \Delta K = +1, \qquad K = -J, \qquad -J+1, \ldots, J-1, J. \tag{14}$$

Similarly, transitions from the ground state to the state $\psi_{1,-1}$ occur with $\Delta K = -1$, and the energy correction is

$$E_c = 2\zeta_r AK, \qquad \Delta K = -1, \qquad K = -J, \qquad -J+1, \ldots, J-1, J. \tag{15}$$

With the choice (14) the rotational energy levels of a symmetric rotor in a degenerate vibrational state, to this order of approximation, are given by

$$F(J, K) = BJ(J+1) + (A-B)K^2 - 2\zeta_r AK, \qquad \Delta K = +1. \tag{16}$$

Alternatively, (16) may be written

$$F(J, K) = BJ(J+1) + (A-B)K^2 \mp 2\zeta_r AK, \tag{17}$$

where the upper sign applies when $\Delta K = +1$ and the lower sign applies when $\Delta K = -1$.

4e THE ζ-SUM RULE

It can be seen from the definition of ζ that the ζ-value for a given vibration depends on the masses of the atoms, the geometry of the molecule, and the vibrational potential function. Usually, the vibrational potential function is not known sufficiently well to enable making the calculation of ζ-values with any degree of precision. However, several authors [6, 24, 39, 77, 80, 115] have shown that, for harmonic oscillations, the sum of the ζ-values of the vibrations in a given degenerate symmetry species is independent of the force system chosen to represent the molecule. The proof of this sum rule may be accomplished in several ways. The method outlined here is due to Boyd and Longuet-Higgins. Although it is not as elegant as some other proofs, it seems to have the advantage of including all the steps explicitly.

Since a treatment of molecular vibrations normally starts with symmetry coordinates, this is a logical starting place for this proof. For a complete discussion of symmetry coordinates, see Wilson, Decius, and Cross [9].

Consider a particular doubly degenerate representation of the molecular point group. Let $\begin{pmatrix} R_{11} & R_{12} \\ R_{21} & R_{22} \end{pmatrix}$ be the transformation matrix representing the symmetry operation R, and let (S_j, T_j), $j = 1, 2, \ldots$ be any set of orthonormal symmetry coordinates such that

$$\begin{bmatrix} RS_j \\ RT_j \end{bmatrix} = \begin{bmatrix} R_{11} & R_{12} \\ R_{21} & R_{22} \end{bmatrix} \begin{bmatrix} S_j \\ T_j \end{bmatrix} \quad \text{for all } j. \tag{1}$$

The term orthonormal means that (S_j, T_j) are orthonormal combinations of the mass-weighted Cartesian coordinates.

The normal coordinates (X_r, Y_r) form such a set if the phases are chosen so that

$$\begin{bmatrix} RX_r \\ RY_r \end{bmatrix} = \begin{bmatrix} R_{11} & R_{12} \\ R_{21} & R_{22} \end{bmatrix} \begin{bmatrix} X_r \\ Y_r \end{bmatrix}. \tag{2}$$

It is not necessary to require that (S_j, T_j) be orthogonal to the translations and rotations of the molecule, although the number of symmetry coordinates in a given symmetry species equals the number of vibrations, rotations, and translations in that symmetry species.

From (1) and (2) we find

$$X_r = \sum_j L_{rj} S_j \qquad Y_r = \sum_j L_{rj} T_j, \tag{3}$$

where the L_{rj} are real orthogonal matrices. Hence the ζ's are given by

$$\zeta_r = \sum_{j'} \sum_j L_{rj} L_{rj'} \sum_\alpha \frac{\partial(S_j, T_{j'})}{\partial(\xi_\alpha, \eta_\alpha)}, \tag{4}$$

which in view of the orthogonal character of the L_{rj} leads to

$$\sum_r \zeta_r = \sum_j \sum_\alpha \frac{\partial(S_j, T_j)}{\partial(\xi_\alpha, \eta_\alpha)}. \tag{5}$$

Since the symmetry coordinates were chosen without regard to the molecular potential function, the right-hand side of (5), and consequently the left-hand side, is independent of the molecular potential function. The sum in (5) includes any translational and rotational coordinates that belong to this symmetry species, so that in order to obtain the correct expression for the ζ-sum rule these should be subtracted out; thus

$$\Sigma_r' \zeta_r = \sum_j \sum_\alpha \frac{\partial(S_j, T_j)}{\partial(\xi_\alpha, \eta_\alpha)} - \zeta_{\text{rot}} - \zeta_{\text{trans}}, \tag{6}$$

the prime on the summation indicating that the sum is taken over only the vibrational modes of the symmetry species. ζ_{rot} and ζ_{trans} are defined as

$$\sum_\alpha \frac{\partial(R_x, R_y)}{\partial(\xi_\alpha, \eta_\alpha)} \quad \text{and} \quad \sum_\alpha \frac{\partial(T_x, T_y)}{\partial(\xi_\alpha, \eta_\alpha)}, \quad \text{respectively.} \tag{7}$$

These terms occur in the ζ-sum for a particular symmetry species only if the corresponding motions belong to that symmetry species. The contributions can be evaluated quite generally. The translational normal coordinates (T_x, T_y) are

$$T_x = \frac{1}{M^{1/2}} \sum_\alpha m_\alpha x_\alpha, \qquad T_y = \frac{1}{M^{1/2}} \sum_\alpha m_\alpha y_\alpha, \tag{8}$$

where M is the mass of the molecule. The value of ζ_{trans} then becomes

$$\zeta_{\text{trans}} = \sum_\alpha \frac{\partial(T_x, T_y)}{\partial(\xi_\alpha, \eta_\alpha)} = \sum_\alpha \frac{1}{m_\alpha} \frac{\partial(T_x, T_y)}{\partial(x_\alpha, y_\alpha)} = \sum_\alpha \frac{1}{m_\alpha} \begin{vmatrix} \dfrac{m_\alpha}{M^{1/2}} & 0 \\ 0 & \dfrac{m_\alpha}{M^{1/2}} \end{vmatrix} = 1. \tag{9}$$

The rotational normal coordinates (R_x, R_y) are

$$R_x = \sum_\alpha \frac{m_\alpha(Y_\alpha z_\alpha - Z_\alpha y_\alpha)}{I_b^{1/2}}, \qquad R_y = \sum_\alpha \frac{m_\alpha(Z_\alpha x_\alpha - X_\alpha z_\alpha)}{I_b^{1/2}}, \tag{10}$$

where X_α, Y_α, Z_α are the Cartesian coordinates of the αth atom relative to the center of mass of the molecule. If the molecule has neither a center of symmetry nor a symmetry plane perpendicular to the z-axis (rotor axis), (R_x, R_y) transform like (M_x, M_y) under the group operations; hence

$$\zeta_{\text{rot}} = \sum_\alpha \frac{1}{m_\alpha} \frac{\partial(R_x, R_y)}{\partial(x_\alpha, y_\alpha)} = \sum_\alpha \frac{1}{m_\alpha} \begin{vmatrix} 0 & \dfrac{-m_\alpha Z_\alpha}{I_b^{1/2}} \\ \dfrac{m_\alpha Z_\alpha}{I_b^{1/2}} & 0 \end{vmatrix} = \sum_\alpha \frac{m_\alpha Z_\alpha^2}{I_b}, \tag{11}$$

or, as is usually written,

$$\zeta_{\text{rot}} = 1 - \frac{I_a}{2I_b} = 1 - \frac{B}{2A}. \tag{12}$$

When the molecule possesses a center of symmetry or a horizontal plane of symmetry, (R_x, R_y) as defined in (9) belongs to a different representation from (M_x, M_y); thus (R_x, R_y) belongs to a symmetry species not active in the infrared. It is still useful to define Coriolis coefficients in these cases since the modes in these species may be active in combination bands. If the molecule has an n-fold axis of symmetry, and if the phase of (X_r, Y_r) is chosen so that

$$\begin{vmatrix} C_n X_r \\ C_n Y_r \end{vmatrix} = \begin{vmatrix} \cos\theta & \sin\theta \\ -\sin\theta & \cos\theta \end{vmatrix} \begin{vmatrix} X_r \\ Y_r \end{vmatrix}, \tag{13}$$

where C_n rotates X toward Y and $0 < \theta < \pi$, the Coriolis coefficient ζ_r is defined

$$\zeta_r = \sum_\alpha \frac{\partial(X_r, Y_r)}{\partial(\xi_\alpha, \eta_\alpha)}. \tag{14}$$

The relative signs of (R_x, R_y) are such that $\theta = 2\pi/n$, thus $0 < \theta < \pi$; therefore regardless of whether or not (R_x, R_y) belong to the active species, ζ_{rot} is given by (12).

In order that the method of calculating ζ-sums be understood, a specific case is worked out in detail and then the general case covered.

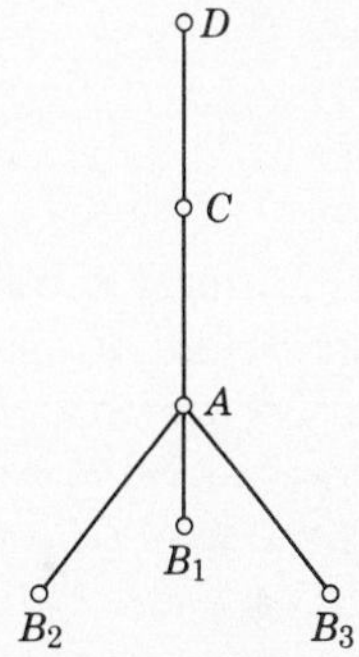

Fig. 4e1 The molecule AB_3CD of point group C_{3v}. ACD is the symmetry axis.

For illustrative purposes, consider the molecule AB_3CD of the point group C_{3v}. Such a molecule might be CH_3CN. The arrangement of the atoms in this molecule is shown in Fig. 4e1.

A set of symmetry coordinates which forms a complete orthonormal set for the motions of species E is

$$\begin{aligned}
S_1 &= 3^{-1/2}(\xi_1 + \xi_2 + \xi_3) & T_1 &= 3^{-1/2}(\eta_1 + \eta_2 + \eta_3) \\
S_2 &= -12^{-1/2}(2\xi_1 - \xi_2 - \xi_3) - 2^{-1}(\eta_3 - \eta_2) & T_2 &= 2^{-1}(\xi_2 - \xi_3) \\
S_3 &= 2^{-1/2}(-\zeta_2 + \zeta_3) & &\quad + 12^{-1/2}(2\eta_1 - \eta_2 - \eta_3) \\
S_4 &= \xi_4, & T_3 &= 6^{-1/2}(2\zeta_1 - \zeta_2 - \zeta_3) \\
&\text{etc.} & T_4 &= \eta_4
\end{aligned} \tag{15}$$

The motions corresponding to the first three pair of these symmetry coordinates are shown in Fig. 4e2. The relative signs of each pair have been chosen so that for a counterclockwise rotation they transform as

$$\begin{vmatrix} C_3S_j \\ C_3T_j \end{vmatrix} = \begin{vmatrix} -\frac{1}{2} & \sqrt{\frac{3}{4}} \\ -\sqrt{\frac{3}{4}} & -\frac{1}{2} \end{vmatrix} \begin{vmatrix} S_j \\ T_j \end{vmatrix}, \tag{16}$$

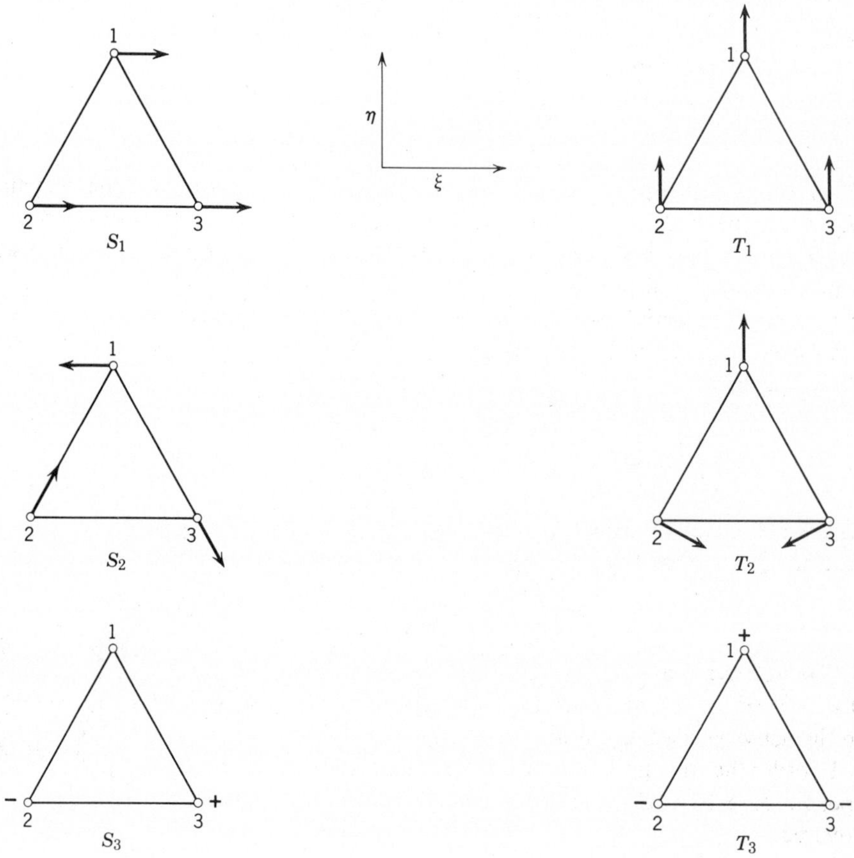

Fig. 4e2 Illustrative symmetry coordinates involving B atom displacements in AB_3CD molecule (C_{3v} symmetry).

which is (13) with $\theta = 120°$ and is the same as (M_x, M_y). The ζ-sum is now given by (7)

$$\sum_r{}' \zeta_r = \sum_j \sum_\alpha \frac{\partial(S_j, T_j)}{\partial(\xi_\alpha, \eta_\alpha)} - 1 - \left(1 - \frac{B}{2A}\right), \tag{17}$$

The Jacobians may be evaluated directly from (15); thus

$$\begin{aligned} &\sum_\alpha \frac{\partial(S_1, T_1)}{\partial(\xi_\alpha, \eta_\alpha)} = 1 \qquad \sum_\alpha \frac{\partial(S_2, T_2)}{\partial(\xi_\alpha, \eta_\alpha)} = -1 \\ &\sum_\alpha \frac{\partial(S_3, T_3)}{\partial(\xi_\alpha, \eta_\alpha)} = 0 \qquad \sum \frac{\partial(S_4, T_4)}{\partial(\xi_\alpha, \eta_\alpha)} = 1, \text{ etc.} \end{aligned} \tag{18}$$

The contributions to the sum of (S_1, T_1) and (S_2, T_2) cancel each other, and the contribution from (S_3, T_3) is zero. Each atom on the rotor axis contributes $+1$ to the sum. For a C_{3v} molecule the zeta sum is given quite generally as

$$\sum_r \zeta_r = \text{number of atoms on the symmetry axis} - 2 + \frac{B}{2A}\,. \tag{19}$$

This expression is valid for any molecule with the point group C_{3v} since the net contribution of any set of three equivalent atoms must be zero.

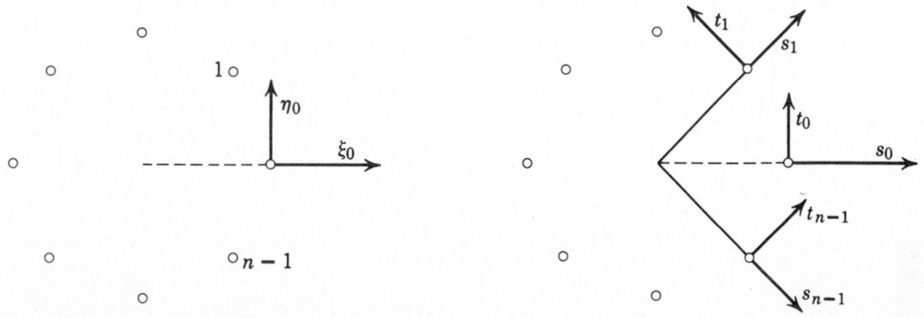

Fig. 4e3 The coordinate system for the case of an n-fold axis of rotation. The points represent the n-equivalent atoms in the plane.

For n atoms symmetrically placed about the symmetry axis, point group C_n or C_{nv}, it can be shown that the contribution of these atoms to the ζ-sum is likewise zero for each degenerate species. This case is shown schematically in Fig. 4e3. The atoms are numbered counterclockwise from 0 to $n - 1$, and the x-axis is taken along the radius vector of the atom numbered zero. The mass-weighted radial and tangential displacements of the αth atom are defined as

$$\begin{bmatrix} s_\alpha \\ t_\alpha \end{bmatrix} = \begin{bmatrix} \cos\alpha\theta & \sin\alpha\theta \\ -\sin\alpha\theta & \cos\alpha\theta \end{bmatrix}\begin{bmatrix} \xi_\alpha \\ \eta_\alpha \end{bmatrix} \qquad \begin{matrix} \alpha = 0, 1, 2, \ldots, n-1 \\ \theta = \dfrac{2\pi}{n} \end{matrix} \tag{20}$$

A complete set of orthonormal symmetry coordinates is now constructed from the s_α and t_α for the motions in the x, y plane

$$\begin{bmatrix} S_j \\ T_j \end{bmatrix} = n^{-1/2} \sum_\alpha \begin{bmatrix} \cos j\alpha\theta & -\sin j\alpha\theta \\ \sin j\alpha\theta & \cos j\alpha\theta \end{bmatrix}\begin{bmatrix} s_\alpha \\ t_\alpha \end{bmatrix} \qquad j = 0,1,2,\ldots n-1 \tag{21}$$

The transformation matrix for (S_j, T_j) under a counterclockwise rotation C_n is

$$\begin{aligned} \begin{bmatrix} C_n S_j \\ C_n T_j \end{bmatrix} &= n^{-1/2} \sum_\alpha \begin{bmatrix} \cos j\alpha\theta & -\sin j\alpha\theta \\ \sin j\alpha\theta & \cos j\alpha\theta \end{bmatrix}\begin{bmatrix} s_{\alpha+1} \\ t_{\alpha+1} \end{bmatrix} \\ &= n^{-1/2} \sum_\alpha \begin{bmatrix} \cos j(\alpha-1)\theta & -\sin j(\alpha-1)\theta \\ \sin j(\alpha-1)\theta & \cos j(\alpha-1)\theta \end{bmatrix}\begin{bmatrix} s_\alpha \\ t_\alpha \end{bmatrix} \\ &= \begin{bmatrix} \cos j\theta & \sin j\theta \\ -\sin j\theta & \cos j\theta \end{bmatrix}\begin{bmatrix} S_j \\ T_j \end{bmatrix} \end{aligned} \tag{22}$$

The transformation matrix for $(S_{n-j}, -T_{n-j})$ is therefore

$$\begin{bmatrix} \cos(n-j)\theta & -\sin(n-j)\theta \\ \sin(n-j)\theta & \cos(n-j)\theta \end{bmatrix} = \begin{bmatrix} \cos j\theta & \sin j\theta \\ -\sin j\theta & \cos j\theta \end{bmatrix} \tag{23}$$

Thus $(S_{n-j}, -T_{n-j})$ belong to the same representation as (S_j, T_j) which is doubly degenerate unless $j = 0$ or $j = n/2$, n even. The contribution of (S_j, T_j) to the ζ-sum of this doubly degenerate representation E_k is

$$\sum_\alpha \frac{\partial(S_j, T_j)}{\partial(\xi_\alpha, \eta_\alpha)} = \sum_\alpha \frac{\partial(S_j, T_j)}{\partial(s_\alpha, t_\alpha)} \cdot \frac{\partial(s_\alpha, t_\alpha)}{\partial(\xi_\alpha, \eta_\alpha)}$$
$$= n^{-1} \sum_\alpha \begin{vmatrix} \cos j\alpha\theta & -\sin j\alpha\theta \\ \sin j\alpha\theta & \cos j\alpha\theta \end{vmatrix} \begin{vmatrix} \cos \alpha\theta & \sin \alpha\theta \\ -\sin \alpha\theta & \cos \alpha\theta \end{vmatrix} = 1 \tag{24}$$

Similarly,

$$\sum_\alpha \frac{\partial(S_{n-j}, -T_{n-j})}{\partial(\xi_\alpha, \eta_\alpha)} = -\frac{\partial(S_{n-j}, T_{n-j})}{\partial(\xi_\alpha, \eta_\alpha)} = -1. \tag{25}$$

It is thus seen that (S_j, T_j) and $(S_{n-j}, -T_{n-j})$ make equal and opposite contributions to the ζ-sum of E_k, so the net contribution of the off-axis atoms is zero. For the representation E_1 which contains (T_x, T_y) and (R_x, R_y) and the perpendicular motions of the off-axis atoms, the ζ-sum will be the same as the special case of C_{3v} given explicitly by (19). For the other degenerate species, E_2, E_3, etc., which only occur for $n > 4$ the ζ-sum is zero since the axial atoms do not move and the contributions of the off-axis atoms vanish.

An extension of these arguments enables the inclusion of groups which contain centers of symmetry and horizontal planes of symmetry.

The following are the three different cases:

1. The molecule has only one degenerate symmetry class. Here the ζ-sum is given by (19), which applies also to the group V_d, in which S_4 has the same role as C_4 in C_{4v}.

2. The molecule has a center of symmetry. It is now necessary to distinguish between E_g, which contains the perpendicular rotations, and E_u which contains the perpendicular translations. Each equivalent pair of axial atoms will contribute +1 to the ζ-sum in each species, but the central atom, if present, contributes only to the sum for E_u. The sums in this case become

$$E_g\colon \sum_j{}' \zeta_j = (\text{number of pairs of axial atoms}) - 1 + \frac{B}{2A} \tag{26}$$

$$E_u\colon \sum_j{}' \zeta_j = (\text{number of pairs of axial atoms} + \text{central atom if present}) - 1. \tag{27}$$

3. The molecule possesses a horizontal plane of symmetry but no center of symmetry. The degenerate species are E' and E'', the former being symmetric and the latter antisymmetric with respect to reflection in the plane. The sums in this case are the same as in Case 2, E' corresponding to E_u, and E'' corresponding to E_g. For molecules with $n > 4$ these formulas hold only for the E_1, E_{1g}, E_{1u}, E_1' and E_1''. The representations with subscripts greater than one have ζ-sums of zero as demonstrated for C_n.

The actual evaluation of the individual ζ's requires a knowledge of the masses, molecular geometry, and potential function and may be evaluated from their definition 4d10 or 4e4 once the appropriate transformations are known. The whole problem of zetas and their evaluation have been elegantly discussed [78, 79]. This treatment is given in Appendix VIII.

4f OVERTONE AND COMBINATION BANDS

The wave function for any excited state of the degenerate vibration (X_r, Y_r) is written as

$$\psi_{v_r,l_r}(X_r, Y_r), \tag{1}$$

where v_r is the vibrational quantum number and $l_r = v_r, v_r - 2, \ldots, -v_r$ is the quantum number of $\mathbf{p}_r$. In order for transitions from the ground state ψ_{00} to the state ψ_{v_r,l_r} to be optically active,

$$\int \psi_{00}{}^*(M_x - iM_y)\psi_{v_r,l_r}\, d\tau \neq 0$$

or

$$\int \psi_{00}{}^*(M_x + iM_y)\psi_{v_r,l_r}\, d\tau \neq 0. \tag{2}$$

ψ_{00} is real and belongs to the totally symmetric species. ψ_{v_r,l_r} and $\psi_{v_r,-l_r}$ are complex conjugates and form the basis for one of the doubly degenerate representations. A necessary condition for the integrals in (2) to be nonvanishing is that $(\psi_{v_r,l_r}, \psi_{v_r,-l_r})$ transform like $[(M_x + iM_y), (M_x - iM_y)]$ under all operations of the molecular point group. The only way overtone bands may appear is through the presence of either electrical or mechanical anharmonicity. Usually both types of anharmonicity are present, but for present purposes it is simpler to assume that the occurrence of overtone bands is due to electrical anharmonicity since with this assumption the wave functions remain in their unperturbed form. Such an assumption has no effect on the validity of the selection rules.

For those groups containing neither a horizontal plane of reflection nor a center of inversion, the symmetry species of a degenerate pair of wave functions may be determined from their behavior under the rotation C_n alone (for D_{2d}, C_n should read S_4). If there is a center of inversion, the behavior of the degenerate pair under i must be known, and if a horizontal plane is present without a center of inversion their behavior under σ_h must be known.

From 4d12 it can be verified that $\psi_{1,1}$ and $\psi_{1,-1}$ transform like $X_r + iY_r$ and $X_r - iY_r$ since $X_r{}^2 + Y_r{}^2$ is invariant under the group operations. In general, $(\psi_{v_r,l_r}, \psi_{v_r,-l_r})$ will transform like $[(X + iY_r)^{l_r}, (X_r - iY_r)^{l_r}]$. If the relative signs of (X_r, Y_r) are chosen correctly, it follows from 4e3 that

$$\begin{bmatrix} C_n(X_r + iY_r) \\ C_n(X_r - iY_r) \end{bmatrix} = \begin{bmatrix} e^{-i\theta_r} & 0 \\ 0 & e^{i\theta_r} \end{bmatrix} \begin{bmatrix} X_r + iY_r \\ X_r - iY_r \end{bmatrix} \quad \text{with } 0 < \theta_r < \pi \tag{3}$$

Thus

$$\begin{bmatrix} C_n\psi_{v_r,l_r} \\ C_n\psi_{v_r,-l_r} \end{bmatrix} = \begin{bmatrix} e^{-il_r\theta_r} & 0 \\ 0 & e^{il_r\theta_r} \end{bmatrix} \begin{bmatrix} \psi_{v_r,l_r} \\ \psi^*_{v_r,l_r} \end{bmatrix}. \tag{4}$$

If (4) is compared to the transformation equation for $M_x \pm iM_y$,

$$\begin{bmatrix} C_n(M_x + iM_y) \\ C_n(M_x - iM_y) \end{bmatrix} = \begin{bmatrix} e^{-2\pi i/n} & 0 \\ 0 & e^{2\pi i/n} \end{bmatrix} \begin{bmatrix} M_x + iM_y \\ M_x - iM_y \end{bmatrix}. \tag{5}$$

It is seen that the conditions for (2) not to vanish are

$$l_r\theta_r = \frac{2\pi}{n} + 2\pi m \qquad m = 0, \pm 1, \pm 2, +, \ldots. \tag{6}$$

For molecules whose point group contains i or σ_h this condition must be modified by the rules on g, u, or $'$, $''$. ψ_{v_r,l_r} is optically active only when $(X_r + iY_r)^{v_r}$ transforms under i or σ_h in the same way as does $(M_x + iM_y)$. The effective Coriolis coefficient for the state ψ_{v_r,l_r} is

$$\begin{aligned} \zeta &= \frac{1}{\hbar} \iint \psi^*_{v_r,l_r} \mathbf{p}_z \psi_{v_r,l_r}\, dX_r\, dY_r \\ &= \frac{1}{\hbar} \sum_\alpha \frac{\partial(X_r, Y_r)}{\partial(\xi_\alpha, \eta_\alpha)} \iint \psi^*_{v_r,l_r} \mathbf{p}_r \psi_{v_r,l_r}\, dX_r\, dY_r \\ &= \frac{1}{\hbar} \zeta_r l_r \hbar = l_r \zeta_r. \end{aligned} \tag{7}$$

The result may be applied to the molecule C_2H_6 which belongs to the point group D_{3d}. The molecule has a center of inversion; hence, from the g, u selection rule, the only active overtones are the odd overtones of the E_u vibrations.

From (6) with $\theta_r = 2\pi/3$

$$l_r = -2, 1, 4 \tag{8}$$

The state $v_r = 3$ has the components $l_r = 3, 1, -1, -3$, of which only the component with $l_r = 1$ is optically accessible for $\Delta K = +1$ and $\zeta_{\text{eff}} = \zeta_r$.

This same information can be obtained from purely group theoretical considerations. Using the methods outlined in [9], Chapter 7, Section 3, or Appendix XIV, the symmetry species of the components of the $v = 3$ wave functions can be determined. For $v = 3$, $|l| = 1$ the species is E_u, which is the same species as M_x and M_y, hence is optically accessible from the ground state with $\Delta K = +1$. For $v = 3$, $|l| = 3$ the species are A_{1u} and A_{2u}. A transition from the ground state to A_{1u} is forbidden. A_{2u}, however, is the species of M_z; thus the transition from the ground state to the A_{2u} level is allowed, but in this instance $\Delta K = 0$.

For a molecule with D_{2d} symmetry $\theta_r = \pi/2$ for S_4. The condition on perpendicular overtones becomes, from (8),

$$l_r = -3, 1, 5. \tag{9}$$

Again only odd perpendicular overtones are allowed since odd l values restrict v to odd values. For $v = 3$, only the components with $l_r = -3, 1$ are active for $\Delta K = +1$, with effective Coriolis coefficients of $-3\zeta_r$ and ζ_r.

Treating the D_{2d} from symmetry considerations, we find the $v = 3$ level of the double degenerate level has two doubly degenerate components, both of which should be optically active.

The procedure to be used for combination bands can be deduced by an extension of the arguments used for overtone bands. The excited state wave function now takes the form

$$\Pi_r^1\, \psi_{v_r,l_r}(X_r,\, Y_r)\Pi_s^2\, \psi_s(Q_s), \tag{10}$$

where the first product is over the degenerate modes and the second over the nondegenerate modes. The condition analogous to (6) becomes

$$\Sigma_r^1\, l_r\theta_r = \frac{2\pi}{n} + 2\pi m. \tag{11}$$

Again the g, u, or $'$, $''$ rule must be satisfied; that is, $\Pi^1(X_r + i\,Y_r)^{v_r}$ must transform under i or σ_h in the same way as does $M_x + iM_y$.

If any of the nondegenerate modes are present in the excited state, (11) must be modified to

$$\Sigma^1\, l_r\theta_r + \Sigma^2\, v_s\theta_s = \frac{2\pi}{n} + 2\pi m \tag{12}$$

and $\Pi_r^1\,(X_r + i\,Y_r)^{v_r}\Pi_s^2 Q_s{}^{v_s}$ must transform like $M_x + iM_y$ under i or σ_h. θ_s is either zero or π, depending on whether Q_s is symmetric or antisymmetric for the operation C_n. If $n = 3, 5, \ldots, \theta_s$ is necessarily zero, but if $n = 4, 6, \ldots, \theta_s$ may be either zero or π. (For D_{2d}, $\theta_s = 0$ for A_1, A_2, and $\theta_s = \pi$ for B_1, B_2). The effective Coriolis coefficient is given by

$$\zeta = \frac{1}{\hbar}\int \psi_v{}^*\mathbf{p}_z\psi_v\, d\tau \tag{13}$$

with ψ_v given by (10). Since only the degenerate modes contribute to the integral, the result is

$$\zeta_{\text{eff}} = \sum_r l_r\zeta_r. \tag{14}$$

These rules may be applied to the state of a molecule with the point group D_{3h}, which has one quantum of excitation E' and two quanta of excitation E''.

From (12) we find

$$l'\,\frac{2\pi}{3} + l''\,\frac{2\pi}{3} = \frac{2\pi}{3} + 2m\pi, \tag{15}$$

which is satisfied by

$$l' + l'' = -2, 1, 4. \tag{16}$$

There are six possible combinations of l' and l'' possible in this state,

$$\begin{matrix} (1, 2) & (1, 0) & (1, -2) \\ (-1, 2) & (-1, 0) & (-1, -2). \end{matrix} \tag{17}$$

Of these possibilities only (1, 0) and (−1, 2) satisfy (15); therefore this vibrational state gives rise to two active perpendicular bands with $\zeta = \zeta'$ and $\zeta = -\zeta' + 2\zeta''$.

Using the group theoretical approach it is found [9] that the components of this state have the species $A_1{}' + A_2{}' + 2E'$. The component $A_1{}'$ cannot combine with the ground state optically, whereas $A_2{}'$ can combine with the ground state only with $\Delta K = 0$. The $2E'$ components can, of course, combine with the ground state with $\Delta K = +1$ confirming the other analysis.

There is a further consideration for overtone and combination bands with more than one active component. These components are split further through the term $\mathbf{p}_z^2/2I_z$. For a symmetric rotor the purely vibrational energy is given by

$$H_v = H_v^0 + H_v', \tag{18}$$

in which H^0 is the harmonic-oscillator Hamiltonian and H' is given by

$$H' = \frac{\mathbf{p}_z^2}{2I_z}. \tag{19}$$

For an overtone band the contribution from H' is

$$E_v' = l_r^2\zeta_r^2 A,$$

and this correction applies to the vibrational energy of levels giving rise to parallel as well as perpendicular bands. The correction from this term for combination levels in somewhat more complicated if two or more degenerate levels are simultaneously excited. For the combination band just considered, the correction would be

$$E' = (l_1^2\zeta_1^2 + l_2^2\zeta_2^2 + 2l_1l_2\zeta_1\zeta_2)A.$$

4g HIGHER ORDER CORIOLIS EFFECTS IN SYMMETRIC ROTORS

In 4d it was shown that unless (Q_k, Q_l) are the two components of a degenerate vibration, the first-order effect due to rotational-vibrational interaction vanishes as long as the vibrational forces are harmonic. However, if two nondegenerate vibrational levels occur closely together and satisfy the Jahn criterion, it is possible to have a Coriolis interaction between two levels which are not the components of a degenerate pair. The interaction here is a second-order perturbation, the actual solution of which is a special case of the asymmetric rotor problem which is described later. As yet no pair of levels interacting in this way have been studied in detail. The theoretical treatment has been given by Nielsen [89], to which the reader is referred for details.

Another higher order effect arising from the interaction of vibration and rotation was first predicted by Wilson [121] and subsequently worked out in detail by Nielsen [89]. The symmetries of the rotational levels of a symmetric rotor are discussed in detail in Chapter 5, but the results are anticipated here to give some background for this interaction.

Consider a doubly degenerate vibrational level of a symmetric rotor of the point group C_{3v}. As is shown in Chapter 5, the symmetry of the rotational wave functions is A for $K = 0$, $A + A$ for $K = 3p$, and E for $K = 3p \pm 1$. The symmetry of the vibrational wave function is E and the symmetry of $\psi_v\psi_R$ is found by taking the direct product of ψ_v and ψ_R. The result of this is shown schematically in Fig. 4g1 in which it is assumed that ζ is positive. For $K = 0$, $\psi_v\psi_R$ is, of course, E, and for $K = 3p$ $\psi_v\psi_R$ is still E. However, for $K = 3p \pm 1$, the sublevels have the species $A + A + E$. There can be no further splitting of the E levels for any perturbation with the symmetry of the molecule, whereas the coinciding A levels may be split by an appropriate perturbation. Such a splitting is analogous to the l-type doubling in a linear molecule.

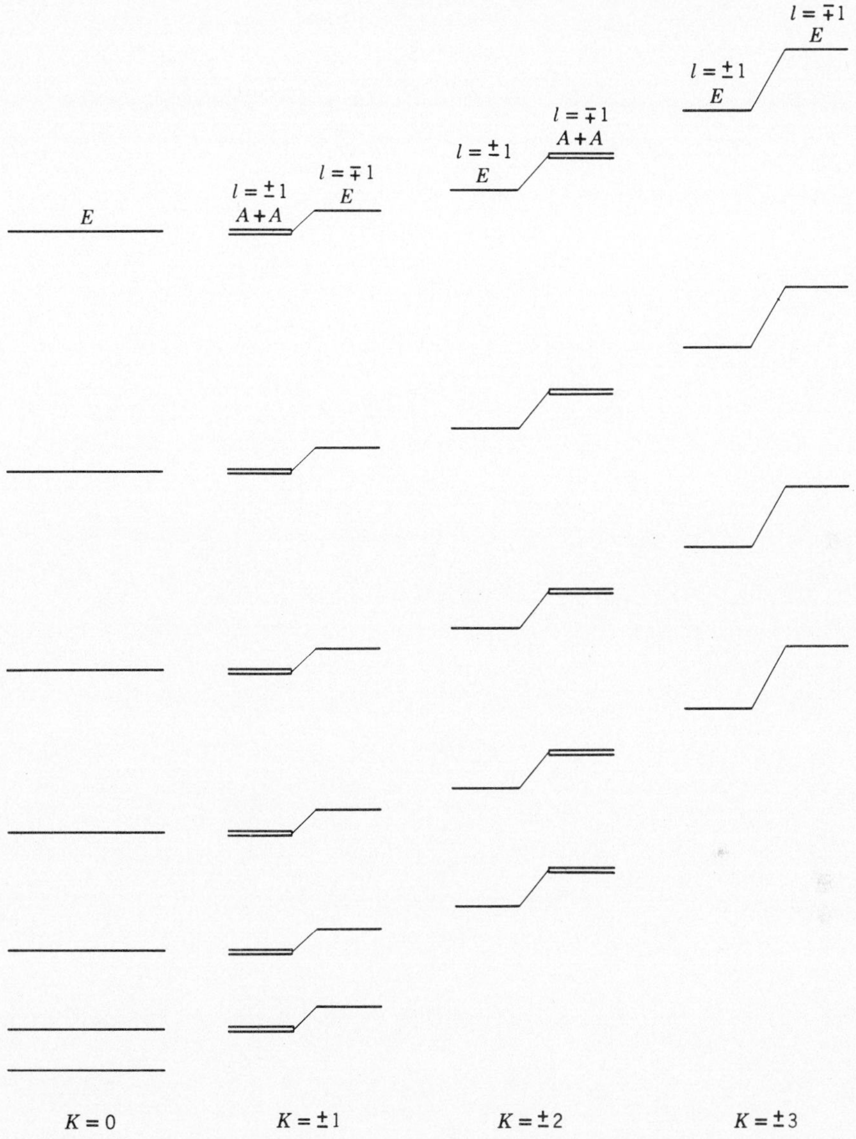

Fig. 4g1 Vibrational-rotational levels of a doubly degenerate vibrational state of a symmetric rotor of the point group C_{3v}. For this figure ζ has been assumed positive. The instances of two coincident A levels, which can be further split, are doubly drawn.

This interaction has been worked out [89] to the second order of approximation and the interaction matrix element has been found to be

$$(v, l, K\,|\mathbf{H}'|\,v, l \pm 2, K \pm 2)$$
$$= q\{[J(J+1) - K(K+1)][J(J+1) - (K+1)(K \pm 2)]\,[(v \mp l)(v \pm l + 2)]\}^{1/2}. \quad (1)$$

Close examination of (1) shows that the levels interact pairwise, and to this approximation only the $A + A$ levels for $K = 1$ are actually split. Figure 4g2 shows the

K, l	4, 1	−4, −1	4, −1	−4, 1	3, 1	−3, −1	3, −1	−3, 1	2, 1	−2, −1	2, −1	−2, 1	1, 1	−1, −1	1, −1	−1, 1	0, 1	0, −1
4, 1	E_8										$q(v+1)4\sqrt{7}$							
−4, −1		E_8										$4\sqrt{7}q(v+1)$						
4, −1			E_7															
−4, 1				E_7														
3, 1					E_6										$6\sqrt{7}q(v+1)$			
−3, −1						E_6										$6\sqrt{7}q(v+1)$		
3, −1							E_5											
−3, 1								E_5										
2, 1									E_4									$q(v+1)6\sqrt{10}$
−2, −1										E_4							$q(v+1)6\sqrt{10}$	
2, −1	$4\sqrt{7}q(v+1)$										E_3							
−2, 1		$4\sqrt{7}q(v+1)$										E_3						
1, 1													E_2	$20q(v+1)$				
−1, −1													$20q(v+1)$	E_2				
1, −1					$q(v+1)6\sqrt{7}$										E_1			
−1, 1						$q(v+1)6\sqrt{7}$										E_1		
0, 1										$q(v+1)6\sqrt{10}$							E_0	
0, −1									$q(v+1)6\sqrt{10}$									E_0

Factors

	±4, ±1	±2, ∓1
±4, ±1	E_8	$q(v+1)4(7)^{1/2}$
±2, ∓1	$q(v+1)4(7)^{1/2}$	E_3

	±4, ∓1
±4, ∓1	E_7

	±3, ±1	±1, ±1
±3, ±1	E_6	$q(v+1)6(7)^{1/2}$
±1, ∓1	$q(v+1)6(7)^{1/2}$	E_1

	±3, ∓1
±3, ∓1	E_5

	±2, ±1	0, ∓1
±2, ±1	E_4	$q(v+1)6(10)^{1/2}$
0, ∓1	$q(v+1)6(10)^{1/2}$	E_0

	1, 1	−1, −1
1, 1	E_2	$20q(v+1)$
−1, −1	$20q(v+1)$	E_2

Fig. 4g2 Factors of the secular equation for a degenerate vibrational level of a symmetric rotor belonging to the point group C_{3v}. ($J = 4$, $|l| = 1$).

factors of the secular equation. It should be pointed out that for $K = 1$ the problem is formally identical with the linear case since (1) reduces to

$$(v, 1, 1 \,|\mathbf{H}'|\, v, -1, -1) = qJ(J+1)(v+1).$$

Only recently has this splitting been observed in the infrared region. An example is discussed in Chapter 7.

4h SPHERICAL ROTOR

The energy levels of a nondegenerate state of a rigid spherical rotor are given quite adequately by 2i2. The most common type of spherical rotor is the XY_4 molecule in which the X-atom is tetrahedrally surrounded by the four Y-atoms. These molecules belong to the point group T_d. The vibrational energy levels have the symmetries A_1, E, and F_2, of which only transitions to the F_2 levels are normally optically active. The F_2 levels, being triply degenerate, are subject to a Coriolis perturbation in the same manner that the doubly degenerate states of the symmetric rotor molecules were.

If (X_r, Y_r, Z_r) are the three components of an F_2 level, a set of acceptable wave functions can be written as

$$Z_r, \qquad (X_r + iY_r), \qquad (X_r - iY_r). \tag{1}$$

Carrying out the treatment as for the symmetric rotor, we find that the energy corrections give rise to three sets of levels

$$\begin{aligned} F^+ &= B_vJ(J+1) + 2B_v\zeta_r(J+1) \\ F^0 &= B_vJ(J+1) \\ F^- &= B_vJ(J+1) - 2B_v\zeta_rJ. \end{aligned} \tag{2}$$

These same formulas also hold for the F_1 states although they will be active in the infrared only in certain overtone and combination bands. The selection rule for this type band is given here since it is simply stated. For a transition from the ground state to an F_2 state if $\Delta J = +1$, transitions occur to the F^- levels, $\Delta J = 0$ transitions to the F^0 levels are allowed, and for $\Delta J = -1$, transitions to the F^+ levels are allowed.

As for the symmetric rotor if two vibrational states satisfy the Jahn criterion, and lie closely enough together, there is a second order Coriolis interaction which further splits the rotational energy levels. This effect has been treated in great detail by Jahn [61]. Jahn's work has been further extended by Hecht [53].

The details of Jahn's or Hecht's work are much too voluminous to be included here; hence the reader is referred to the original literature.

4i ASYMMETRIC ROTOR

In an asymmetric rotor there are, in general, no degenerate vibrational modes. There are, however, some near degeneracies which are generally referred to as accidental degeneracies. The lack of degenerate modes eliminates the type of Coriolis interaction such as occurs in the highly symmetric molecules. However, if two vibrational modes satisfy the Jahn criterion and are sufficiently close in frequency, there is a perturbation of the vibrational-rotational levels. Starting from the Hamiltonian given in 4a1, Wilson [122] has worked out the matrix elements for this problem. Since only levels of the same J in the two vibrational states couple, one has a $2(2J + 1)$ degree secular equation to solve. There is a certain amount of factoring which can be achieved, although the diagonalization of the energy matrix is still a formidable chore. A part of the infinite secular equation is shown in Fig. 4i1. The doubly shaded blocks, representing groups of elements diagonal in V, but having all values of R and R', are very similar to the rigid rotor secular equation. These elements come from the purely rotational and purely vibrational part of $\mathbf{H}$. The elements of the singly shaded blocks arise from the coupling terms $-\mathbf{P}_x\mathbf{p}_x/I_x$, $-\mathbf{P}_y\mathbf{p}_y/I_y$, and $-\mathbf{P}_z\mathbf{p}_z/I_z$. As has been pointed out these coupling terms do not contribute to the elements diagonal in V for non-degenerate states. The coupling terms may be regarded as a perturbation whose first order effect vanishes. Ordinarily, the second-order effect will be quite small unless the two states V and V' have energies E_V and $E_{V'}$ very close to one another. Then the second-order perturbation theory is inadequate.

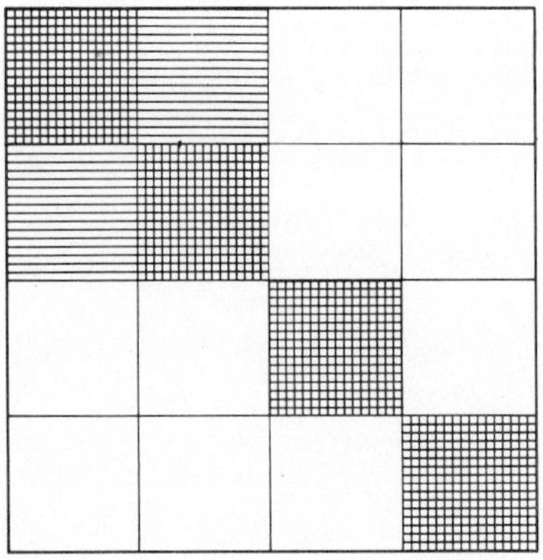

Fig. 4i1 Part of infinite secular equation for H. The division is into vibrational blocks. The doubly shaded blocks are diagonal in V, the singly shaded areas connect two vibrational states of nearly the same energy and represent the interaction.

Wilson has treated this effect by retaining the elements connecting the adjacent states V and V' while neglecting all other terms nondiagonal in V. In Fig. 4i1 the singly shaded blocks which connect two states of nearly the same energy are retained, the nonshaded blocks are neglected, for they represent the coupling between states of widely different energy. The energy levels of the two interacting states are determined by diagonalizing the double-sized block in the upper left-hand corner of Fig. 4i1. In evaluating the matrix elements the usual assumption is made about the wave function, that it is a product of a purely rotational and a purely vibrational part, $\psi_{RV} = \psi_R\psi_V$. With this assumption, the matrix elements of the $\mathbf{P}_g$ given in Chapter 2 and the information given earlier in this chapter the matrix elements can be evaluated. A Type III representation is chosen. Since the energy is independent of M, the matrix elements for each J,J block are given below. For simplicity the J and M indices are dropped.

$$(K, V|\mathbf{H}|K, V) = K^2 + \sigma_V - \sigma$$

$$(K, V'|\mathbf{H}|K, V') = K^2 + \sigma_{V'} - \sigma$$

$$(K, V|\mathbf{H}|K \pm 2, V) = (K, V'|\mathbf{H}|K \pm 2, V')$$

$$= -\tfrac{1}{2}b^*\{[J^2 - (K \pm 1)^2][(J+1)^2 - (K \pm 1)^2]\}^{1/2} \tag{1}$$

$$(K, V|\mathbf{H}|KV') = -(K, V'|\mathbf{H}|KV) = iG_zK$$

$$(K, V|\mathbf{H}|K \pm 1, V') = -(K, V'|\mathbf{H}|K \pm 1, V)$$

$$= \tfrac{1}{2}(iG_y \mp G_x)[(J \mp K)(J \pm K + 1)]^{1/2}.$$

Thus the matrix of **H** for each J has $2(2J + 1)$ rows and columns labelled by K and V, leading to a secular equation of degree $2(2J + 1)$ in σ.

The symbols have the meaning

$$a = b^* \left(\frac{A + B}{A - B}\right) \qquad c = \frac{A - B}{2b^*}$$

$$\sigma = \frac{E}{c} - J(J + 1)a \qquad \sigma_V = \frac{E_V}{c}$$

$$G_x = \frac{4A}{c} \sum_{\alpha kl} \frac{\partial(Q_k, Q_l)}{\partial(\eta_\alpha, \zeta_\alpha)} I_{kl}, \tag{2}$$

$$G_y = \frac{4B}{c} \sum_{\alpha kl} \frac{\partial(Q_k, Q_l)}{\partial(\zeta_\alpha, \xi_\alpha)} I_{kl}, \qquad I_{kl} = \int \psi_{V'}{}^* Q_k \left(\frac{\partial}{\partial Q_l}\right) \psi_V \, d\tau_V,$$

$$G_z = \frac{4C}{c} \sum_{\alpha kl} \frac{\partial(Q_k, Q_l)}{\partial(\xi_\alpha, \eta_\alpha)} I_{kl}.$$

The form of I_{kl} restricts ψ_V to a function of the vibrational coordinates alone but not necessarily to harmonic-oscillator functions. However, if the harmonic-oscillator approximation is used, ψ_V and $\psi_{V'}$ can be written down immediately and I_{kl} evaluated with the result

$$I_{kl} = \frac{1}{2} \left[(V_k + 1) V_l \left(\frac{\nu_l^0}{\nu_k^0}\right) \right]^{1/2}. \tag{3}$$

Many asymmetric rotor molecules possess orthorhombic symmetry (point groups V, V_h, and C_{2v}). For this group of molecules some simplification of the secular equation is possible. For this symmetry the $\mathbf{p}_x$, $\mathbf{p}_y$, $\mathbf{p}_z$, each has a distinctive irreducible representation which is, in turn, different from the complete symmetry of the molecule. Also, for example, G_x vanishes unless $\psi_{V'}\psi_V$ has the same symmetry properties as $\mathbf{p}_x$. Therefore for a given pair of states V, V', at most one of the quantities G_x, G_y, or G_z, will be nonvanishing. If it should occur that $\psi_{V'}\psi_V$ has a symmetry different from that of each of $\mathbf{p}_x$, $\mathbf{p}_y$, $\mathbf{p}_z$, this type of interaction will vanish completely.

In each of the three possible cases, $G_x \neq 0$, $G_y \neq 0$, $G_z \neq 0$, the secular equation can be factored into four factors. This factoring is brought about by a Wang-type transformation, the transformation matrix being diagonal in V and V', and the VV and $V'V'$ blocks being identical to the transformation matrix given in 2m1.

It can be seen from the treatment that certain other conditions must be satisfied before the two states V and V' will perturb each other strongly from this type of interaction, even when their energies are approximately the same. The first of these is the Jahn condition, that $\psi_V\psi_V{}'$ have the same symmetry properties as one of the angular momenta $\mathbf{p}_x$, $\mathbf{p}_y$, or $\mathbf{p}_z$. The second is that I_{kl} shall not be too small for some pair of normal coordinates k and l. Since ψ_V is approximately given by harmonic-oscillator functions, this requires that the two states V and V' differ in two and only two of their vibrational quantum numbers v_k and that these quantum numbers change by one unit from V to V'. Since $E_V \sim E_{V'}$ this practically limits the state to one in which $v_k{}' = v_k + 1$ and $v_l{}' = v_l - 1$ for some pair of normal modes k and l. Finally, there is the restriction that the coefficient of I_{kl}, must not be too small. This coefficient depends on the nature of the normal modes of vibration k and l, so it is both a geometrical and dynamical question whether or not a given pair of normal modes will have a large coefficient.

Before starting an analysis, it is well to investigate if these conditions are fulfilled. If so, then the expanded secular equation should be used. This is true regardless of whether or not the perturbing level is optically active. The use of (1) requires either a knowledge of the exact normal modes of vibration, or the use of one or more additional empirical parameters G_x, etc. With the obvious modification this secular equation may be used for a similar situation of two nearly equal nondegenerate states in a symmetric rotor.

4j FERMI RESONANCE

The next interaction which is taken up has as its primary source the neglected terms in the potential energy function rather than neglected terms in kinetic energy portion of the Hamiltonian. The zero-order solution of the molecular vibrational problem is the harmonic-oscillator approximation. This solution is corrected to take into account the effect of anharmonic terms in the potential function by perturbation calculations. The first result of considering the anharmonic terms is that overtone frequencies are not exactly integral multiples of the fundamental frequencies and that combination frequencies are not exactly equal to the sums of the fundamental frequencies making up the combination. A second effect which often arises when it is necessary to use successive approximations is the so-called resonance effect. Resonance becomes important when the approximate theory predicts two or more energy levels with very nearly the same energy. The more nearly these levels approach degeneracy the greater the resonance effect. The result of these resonance effects is that the approximate levels appear to repel each other; thus the observed separation of the levels is greater than the separation calculated from the approximate theory.

The matrix elements for this perturbation are given in the usual manner by

$$(V_i\,|\mathbf{H}'|\,V_j') = \int \psi_i^{0*}\mathbf{H}'\psi_j^{0}\,d\tau, \tag{1}$$

in which ψ_i^0 and ψ_j^0 are the approximate wave functions of the resonating vibrational states, and $\mathbf{H}'$ is the perturbation operator. In general, $\mathbf{H}'$ consists of the cubic quartic and higher terms in the molecular potential function. Strictly, $\mathbf{H}'$ also includes any terms from the kinetic energy part of the Hamiltonian which have been neglected in the zero-order approximation, but the contributions of these terms to this interaction are generally negligibly small. $\mathbf{H}'$ must have the full symmetry of the molecule, that is, be totally symmetric. Therefore, for the matrix element defined in (1) to be nonvanishing, ψ_i^0 and ψ_j^0 must have the same symmetry in order for this interaction to exist.

If the resonating levels have very nearly the same energy, the effect can be calculated from first-order perturbation theory. The secular determinant is for only two resonating levels

$$\begin{vmatrix} E_i^0 - \lambda & H_{ij} \\ H_{ji} & E_j^0 - \lambda \end{vmatrix} = 0, \tag{2}$$

the solutions of which may be written as

$$\lambda = \frac{(E_i^0 + E_j^0)}{2} \pm \tfrac{1}{2}[4\,|H_{ij}|^2 + (E_i^0 - E_j^0)^2]^{1/2}$$
$$= \bar{E}_{ij} \pm \tfrac{1}{2}\,[4\,|H_{ij}|^2 + \delta_0^2]^{1/2}. \tag{3}$$

The eigenfunctions of the resulting states are, of course, linear combinations of the zero-order eigenfunctions.

$$\psi_i' = a\psi_i^0 - b\psi_j^0 \qquad a^2 + b^2 = 1 \qquad \psi_j' = b\psi_i^0 + a\psi_j^0. \tag{4}$$

If $\delta = [4\,|H_{ij}'|^2 + \delta_0^2]^{1/2}$ is the observed level separation,

$$a = \left[\frac{\delta + \delta_0}{2\delta}\right]^{1/2} \qquad b = \left[\frac{\delta - \delta_0}{2\delta}\right]^{1/2}. \tag{5}$$

If more than two vibrational levels of the same symmetry have nearly the same energy, the secular equation (2) becomes more complicated. For three interacting levels the secular equation is cubic, and for four interacting levels the secular equation is a quartic.

Table 4j1—Vibrational Quantum Number Dependence of the Matrix Elements for the More Common Cases of Fermi Resonance

$\langle v_1, v_2^l, v_3\,|\mathbf{H}'|\,v_1 - 1, v_2^l + 2, v_3\rangle = v_1^{1/2}[(v_2 + 2)^2 - l^2]^{1/2}$	CO_2
$\langle v_1, v_2, v_3\,|\mathbf{H}'|\,v_1 - 2, v_2, v_3 + 2\rangle = [v_1(v_1 - 1)(v_3 + 1)(v_3 + 2)]^{1/2}$	H_2O
$\langle v_1, v_2^l, v_3\,|\mathbf{H}'|\,v_1 + 3, v_2^l, v_3 - 2\rangle = [v_3(v_3 - 1)(v_1 + 1)(v_1 + 2)(v_1 + 3)]^{1/2}$	HCN

The resonance-type interaction was first recognized by Fermi [44] in the vibrational spectrum of CO_2. In this spectrum $2\nu_2 \sim \nu_1$; hence the levels in CO_2 which have the quantum numbers (V_1, V_2^l, V_3) and $(V_1 - 1, V_2^l + 2, V_3)$ all have nearly the same energy and hence can be expected to show a resonance interaction. For most linear triatomic molecules this resonance interaction is present.

A similar resonance was recognized by Darling and Dennison [37] in the water vapor spectrum. The two stretching fundamentals of H_2O are nearly the same frequency but are of different symmetry species, hence cannot interact in the first order. However, the first overtones of these fundamentals are still nearly equal in energy and do have the correct symmetry. Thus in the H_2O spectrum levels characterized by the quantum numbers (V_1, V_2, V_3) and $(V_1 - 2, V_2, V_3 + 2)$ have nearly the same energy and the same symmetry, hence they can and do interact strongly through a resonance interaction. This second-order resonance interaction has also been observed in the spectra of H_2S [14], C_2D_2 [17], and C_2H_2 [20]. A resonance between levels characterized by the quantum numbers (V_1, V_2, V_3) and $(V_1 + 3, V_2, V_3 - 2)$ has been observed in the spectrum of HCN [18]. The vibrational quantum number dependence of the interaction matrix element for these three cases is given in Table 4j1. Not only does this interaction affect the purely vibrational frequencies but it also modifies the rotational-vibrational energies. Since the moments of inertia have a dependence on the normal coordinates, they are affected by any interaction which modifies these coordinates. It was shown in Chapter 3 that in general this variation of moment of inertia with vibrational state can be accounted for by an effective moment of inertia for each vibrational state. The effect of the resonance interaction can be best seen by recognizing that the effective inertial constant in a vibrational state is given by

$$\langle B \rangle = \int \psi_i^* B(Q) \psi_i \, d\tau. \tag{6}$$

For an unperturbed state the ψ_i is, of course, ψ_i^0, whereas for the resonance perturbed states the ψ_i's are given by (4). The result is a new effective inertial constant for each of the resonating states. Thus one would not expect any drastic irregularities in the rotational fine structure of a vibrational band due to this perturbation; rather one would expect the effect to be apparent only in the effective inertial constant.

For two interacting levels of a linear molecule the effective moments are found to be

$$\begin{aligned} B_i &= a^2 B_i^0 + b^2 B_j^0 \\ B_j &= b^2 B_i^0 + a^2 B_j^0 \end{aligned} \tag{7}$$

in which B_i^0 and B_j^0 are the constants which would be observed if there were no resonance present. A similar expression was found to hold for the Darling and Dennison resonance. For their more general case, the expression (7) was found to hold for each of the moments of inertia of the asymmetric rotor.

4k INVERSION DOUBLING

If certain types of symmetry operations are carried out on a molecule, the energy of the system remains unchanged. One might expect that the wave function also would not be changed. If the rotational wave equation for a symmetric top is written down, it takes the form

$$(\mathbf{H} - E)\psi = 0. \tag{1}$$

If $\mathbf{H}$ is written in terms of Cartesian coordinates, it may be seen that inversion through the center of mass leaves $\mathbf{H}$ unaffected. Such an operation consists of replacing each of the Cartesian coordinates by its negative, and since only even powers of the coordinates appear in $\mathbf{H}$, the operation cannot effect the Hamiltonian.

Consider now the effect of the transformation of inversion on (1). $\mathbf{H}$ does not change, the energy does not change, and the new solution of (1), ψ', must correspond to the same energy. Assuming ψ' to be nondegenerate, the most that ψ_0 and ψ' can differ by is a constant factor, say a

$$\psi' = a\psi_0. \tag{2}$$

When the inversion transformation is again applied, a new solution ψ'' is obtained which must be equal to ψ_0 since the second transformation merely reverses the first, but at the same time ψ'' must equal $a\psi'$. Thus

$$\psi'' = \psi_0 = a\psi' = a^2\psi_0, \tag{3}$$

and $a = \pm 1$. This says that if an inversion through the center of mass is carried out, the wave function for a nondegenerate state is either unchanged (symmetric) or at most changes sign (antisymmetric) for the inversion operation. This argument is quite general and applies to the wave function of any molecule.

It is now necessary to investigate in detail what happens for a symmetric rotor molecule. For this purpose, consider a planar XY_3 molecule as shown in Fig. 4k1.

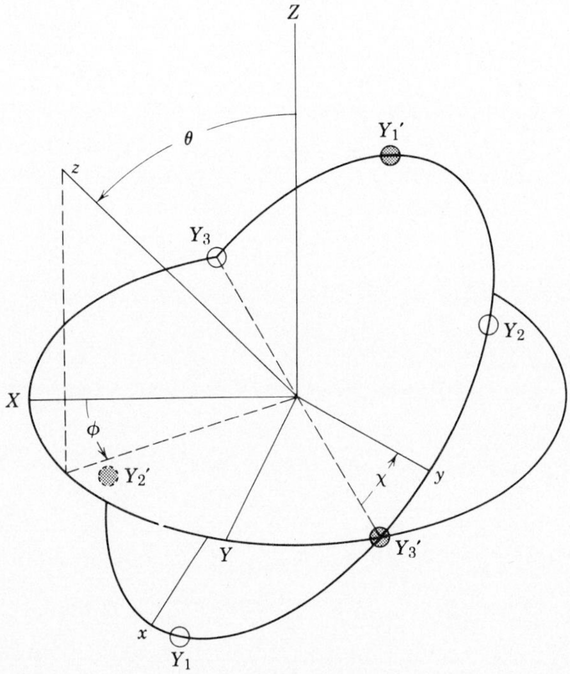

Fig. 4k1 The X atom is at the origin. The 3 Y atoms, before inversion, are shown as open circles in the xy plane. The inversion through the center of mass (Y atoms now shown as solid circles) does not change the direction of the molecular axis z, the positive direction of which is defined as the direction of advance of right-handed screw rotated in direction of Y_1 to Y_2 to Y_3. Hence the angles θ and ϕ do not change and $\chi' = \chi + \pi$.

If an inversion is carried out through the center of mass, θ and ϕ remain unchanged while $\chi' = \chi + \pi$. Since χ enters the symmetric rotor wave function only through the exponential $e^{iK\chi}$, the result of the inversion of this wave function is merely

$$\psi_{J,K,M}(\theta, \varphi, \chi) \xrightarrow{\text{inversion}} e^{iK\pi}\psi_{J,K,M}(\theta, \varphi, \chi) = (-1)^K \psi_{J,K,M}(\theta, \phi, \chi). \qquad (4)$$

Consequently, the rotational wave function is either plus or minus, depending on whether K is even or odd.

The vibrational wave function can be treated in a similar manner. The vibration which is considered is that in which the X molecule oscillates through the Y_3 plane, the ν_2 vibration. If z represents the distance of the X atom above or below the Y_3 plane, the harmonic-oscillator wave function for this motion is

$$\psi_V(z) = N_V e^{-\frac{1}{2}Q^2} H_V(Q), \tag{5}$$

where

$$Q = 2\pi \left[\frac{m\nu_0}{h}\right]^{\frac{1}{2}} z.$$

The $H_V(Q)$ are the Hermite polynomials, and $H_V(Q)$ involves only the even or odd powers of z, depending on whether V is even or odd. The lowest wave function, containing H_0, corresponds to the ground state, with H_1, H_2 . . . , etc. corresponding to the higher overtones. Thus it can be seen that the result of an inversion through the center of mass on ψ_V is

$$\psi_V \xrightarrow{\text{inversion}} (-1)^V \psi_V \tag{6}$$

The total wave function of the molecule to a high degree of approximation may be written as a product

$$\psi_T = \psi_e \psi_V \psi_R \psi_S, \tag{7}$$

in which ψ_e is the electronic wave function and ψ_S the nuclear spin wave function. Consideration of ψ_S is deferred until the next chapter.

The behavior of $\psi_e \psi_V \psi_R$ under inversion depends on the behavior of each of the three parts. Since the electronic wave functions for almost all stable polyatomic molecules in the ground state are symmetric, they need not be considered further. From (4) and (6) it is easily seen that for V even the states with K even are plus and with K odd are minus, whereas the converse is true when V is odd.

There are a large number of molecules such as CH_3X, and pyramidal XY_3 which have the same symmetry properties with respect to inversion as planar XY_3.

The pyramidal XY_3 can be thought of as a planar molecule whose potential function is not harmonic, but such that the X-atom spends most of its time out of the Y_3 plane. As long as the X-atom vibrates with an harmonic potential (quadratic), there is a potential minimum when the X-atom is in the plane of the Y_3-atoms; this potential gives rise to a set of equally spaced energy levels just as any harmonic oscillator has. A potential function which would cause the X-atom to spend most of its time out of the Y_3 plane is one with a maximum at $z = 0$ when X is in the plane, and two minima at some distance z above and below the plane. If this central maxima were very high, it would lead to two sets of equally spaced energy levels, one on each side of the maximum, and the corresponding levels on each side would be of essentially the same energy. It is true there would be some quantum-mechanical tunneling which would allow the particle to oscillate from one side of the maximum to the other, but for a very high barrier this period of vibration would be very long, even as long as years. However, as the central maximum becomes lower, the oscillator levels that were previously degenerate split into pairs as shown in Fig. 4k2*b*. Whether the central maximum is low or high the symmetry properties of the wave functions are the same.

Let ψ_0 and ψ_1 be wave functions for a pair of states that are split by tunneling of the X-atom through the plane of the three Y atoms. If $h\nu_1 - h\nu_0 = h\nu_i$, the wave function for the system including time may be written

$$\psi = 2^{-\frac{1}{2}}(\psi_0 + \psi_1 e^{-2\pi i \nu_i t})\, e^{-2\pi i \nu_0 t}, \tag{8}$$

in which ν_0 is the frequency corresponding to ψ_0. At time $t = 0$, $\psi = 2^{-1/2}(\psi_0 + \psi_1)$ is the wave function corresponding to the X-atom being on the negative side of the Y_3 plane. After a time $\frac{1}{2}\nu_i$, ψ becomes $\psi = 2^{-1/2}(\psi_0 - \psi_1)e^{-\pi i\nu_0 t}$ so that the X-atom has moved over to the positive side of this plane. Thus the X-atom oscillates between positions on each side of the Y_3 plane with a frequency ν_i (for the complete cycle). This frequency ν_i depends on the barrier height. This splitting is smallest for the

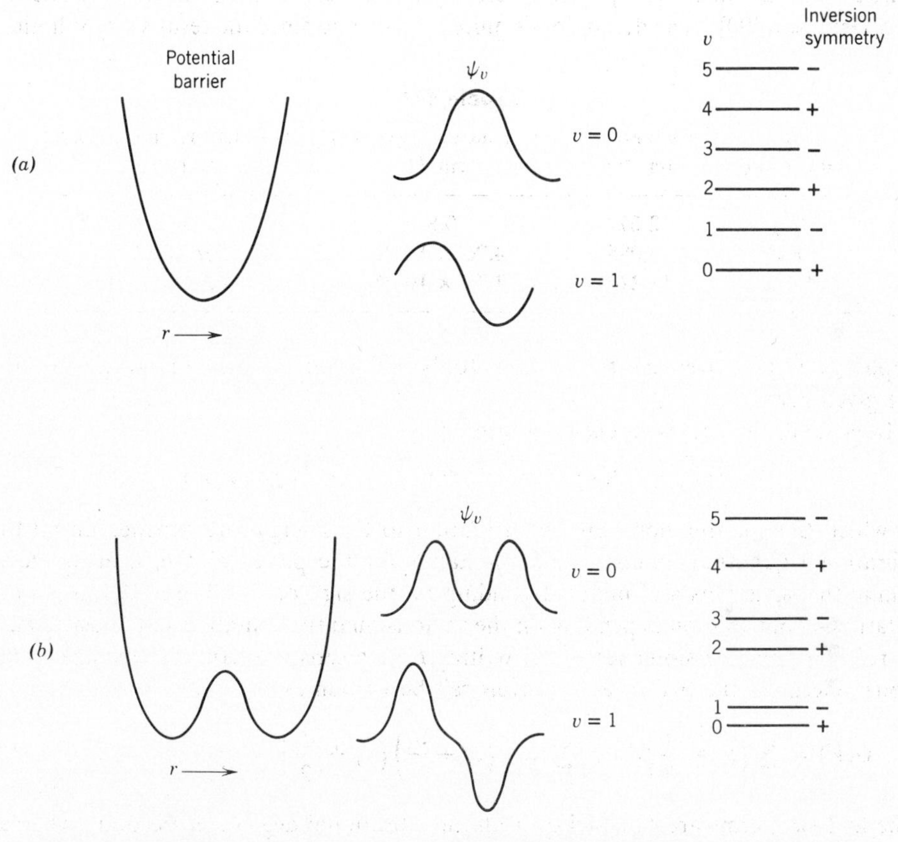

Fig. 4k2 Vibrational energy levels and wave functions for the inversion of a symmetric top. In (*a*) the potential minimum is at the planar configuration; (*b*) indicates the effects of a central maximum.

lowest levels, increasing as the levels get closer to the top of the barrier. For sufficiently high levels, that is, those with energy greater than the barrier, the effect of the barrier vanishes leaving a set of vibrational levels such as those corresponding to the out-of-plane oscillation of a planar XY_3 molecule.

For the NH_3 molecule the barrier height is such that the inversion frequency in the lowest level is about 0.8 cm^{-1}. Up to this time an inversion splitting has not been observed in any other symmetric top molecule even with extremely high resolution available in the microwave region.

Costain and Sutherland [30] have estimated the ground state splitting for PH_3 and

AsH_3 by using a potential function found to be suitable for NH_3. The results of their calculations are summarized in Table 4k1. As is readily seen from their numbers the splittings are such that they are not likely to be seen in the infrared.

In instances where the inversion splitting is large enough to be observed, which is only in the NH_3 spectrum so far, there are, of course, interactions between the vibrational and rotational motions. The problem has been investigated in detail by Sheng, Barker, and Dennison [108]. More recently it has been considered again by Nielsen and Dennison [90]. The derivation is quite lengthy and since the results are of limited

Table 4kl

MOLECULE	BARRIER (cm^{-1})	ν_i (GROUND STATE) (cm^{-1})	ν_i (FIRST EXCITED STATE) (cm^{-1})
NH_3	2,077	0.8	36.6
PH_3	6,085	4.76×10^{-4}	2.38×10^{-4}
AsH_3	11,220	3.7×10^{-18}	6.0×10^{-16}

application in infrared spectral analysis (only NH_3) only the more important results are given here.

In general, the energy levels are written

$$T(V, p, J, K) = G(V) + F(V, J, K) \pm \tfrac{1}{2}E_i(C, J, K), \tag{9}$$

in which $G(V)$ is the vibrational contribution to the energy and depends on all the vibrational quantum numbers v_i, l_i. p stands for the parity of the inversion level taking the signs plus and minus depending on the sigh of E_i. The $F(V, J, K)$ is the rotational contribution depending on the various quantum numbers in the usual way. E_i refers to the inversion energy. By writing E_i in a separate term, the average vibrational energy of the two inversion levels can be written as

$$G(V) = \sum \left(v_i + \frac{d_i}{2}\right)\omega_i + \sum_{i,j\geqslant 1} X_{ij}\left(v_i + \frac{d_i}{2}\right)\left(v_j + \frac{d_j}{2}\right) + \sum_{i,j\geqslant i} g_{ij}l_il_j + \dots. \tag{10}$$

The first two terms are summed over all the vibrational degrees of freedom, whereas the last term is summed over only the degenerate modes. The rotational contribution to the energy is given by

$$\begin{aligned} F(V, J, K) = {} & B_vJ(J+1) + (A_v - B_v)K^2 - D_{vJ}J^2(J+1)^2 \\ & - D_{vJK}J(J+1)K^2 - D_{vK}K^4 \mp 2A_v\zeta_vK \pm \dots, \end{aligned} \tag{11}$$

which is the usual expression for the rotational energy. The last term applies only with a degenerate state. The energy of inversion is given by

$$\begin{aligned} E_i(V, J, K) = {} & E_V{}^0 \exp\,[b_VJ(J+1) + (a_V - b_V)K^2 + d_{VJ}J^2(J+1)^2 \\ & + d_{VJK}J(J+1)K^2 + d_{VK}K^4 + \dots \pm \tfrac{1}{2}f_n(A_1A_2)], \end{aligned} \tag{12}$$

where $E_V{}^0$ is the splitting in the rotationless level, and the exponential represents the interaction of vibration and rotation. The term $f_n(A_1A_2)$ is important for a nearly degenerate pair formed by states with over-all symmetry A_1 and A_2. The form of

$f_n(A_1A_2)$ has been given by Nielsen and Dennison [90]. For $K = 3$ in a vibrational state of A symmetry

$$f_3(A_1A_2) = B_3(J-2)J(J+1)(J+2)(J+3)(J-1) \tag{13}$$

For $K = 2$ in one K-component of vibrational states with E symmetry

$$f_2(A_1A_2) = B_2(J-1)J(J+1)(J+2), \tag{14}$$

and for $K = 1$ in the other component of vibrational states of E symmetry

$$f_1(A_1A_2) = B_1J(J+1). \tag{15}$$

4*l* INTERNAL ROTATION

Another example of multiple potential minima occurs when there is the possibility of restricted rotation of one part of the molecule with respect to the other. An example of such a molecule would be X_2Y_6, in which one XY_3 group can rotate with respect to the other. For X_2Y_6 there are, regardless of the equilibrium configuration chosen, three identical configurations for which $V(\alpha) = V(\alpha \pm 2\pi/3)$, where α is the angle of relative rotation of the two XY_3 groups. Examples of this type of molecule are C_2H_6, C_2F_6, etc. Another case which might be considered is that of the planar X_2Y_4 molecule (for instance, C_2H_4, C_2F_4) for which the potential function has the property $V(\alpha) = V(\alpha + \pi)$. These cases with some of the energy levels are given in Fig. 4*l*1.

In the preceding examples, in which the nuclei have the same mass, the energy levels in the harmonic approximation are the same, and there is a threefold and two-fold degeneracy, respectively. If the potential deviates from strictly harmonic, the degeneracy is, at least, partially removed; that is, there is some tunneling. The levels below the potential maxima give rise to torsional oscillations if tunneling is neglected. Unlike inversion the levels above the potential maxima do not give oscillations, but turn into levels of a rotor corresponding to free rotation.

For the double minima problem the levels are split into two sublevels as shown in Fig. 4*l*1*b*, with the splitting rapidly increasing with vibrational quantum number. In the triple minima case, the levels are again split into two sublevels, one of which is doubly degenerate. The degenerate level is alternately the higher and the lower level as shown in Fig. 4*l*1*a*. The splittings given here hold only for $K = 0$, for there is a strong splitting dependence on K which is discussed in the following paragraph. The potential function generally assumed is of the form

$$V = \frac{V_0}{2}(1 - \cos n\alpha), \tag{1}$$

where V_0 is the barrier or maximum height, and n the number of minima.

The energy levels for internal rotation have been discussed for several different models by numerous authors [29, 33, 60, 94, 126], with the complexity of treatment corresponding to the complexity of the model. The model chosen for discussion here is the one treated by Wilson, Lin, and Lide [126]. Their development of the problem is followed very closely since their treatment has the advantage of being readily related to the general method used up to now. The model is useful for molecules such as CH_3BF_2 or nitromethane, and demonstrates the principles and techniques which must be used to treat any internal rotation problem. For other treatments of the problem and of other models the reader is referred to the original papers [27, 28, 68, 71, 75].

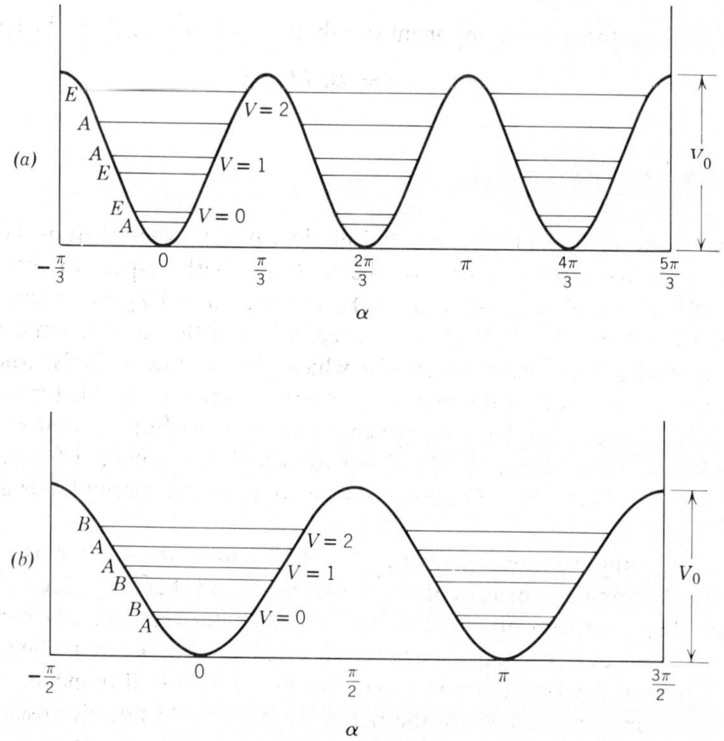

Fig. 4.1 (*a*) Potential function for the case of three identical minima such as X_2Y_6. The usual cosine form of the potential has been assumed, $V = (V_0/2)(1 - \cos 3\alpha)$. (*b*) Potential function for two identical minima such as for the planar X_2Y_4 molecule, $V = (V_0/2)(1 - \cos 2\alpha)$. (From Herzberg, *Infrared and Raman Spectra*, copyright 1945, D. van Nostrand Company, Princeton, New Jersey.)

The model considered is that of two connected rigid bodies. One of the rigid bodies (a symmetric top) has two equal principal moments of inertia about axes perpendicular to a principal axis of the whole molecule. This model has four degrees of rotational freedom, three for over-all rotation and one for rotation of the rotor about its unique axis. The Cartesian axes x, y, z are rigidly attached to the framework part of the molecule and coincide with the principal axis of inertia. The z-axis is regarded as the symmetry axis of the rotor. The Eulerian angles θ, ϕ, χ of x, y, z relative to the space-fixed axes describe the over-all orientation of the molecule, whereas the angle α gives the relative orientation of the framework and rotor.

The kinetic energy may be written as

$$2T = I_x\omega_x^2 + I_y\omega_y^2 + I_z\omega_z^2 + I_\alpha\dot{\alpha}^2 + 2I_\alpha\dot{\alpha}\omega_z. \tag{2}$$

I_α is the moment of inertia of the rotor about its symmetry axis; the other symbols have their usual meaning. The Hamiltonian is obtained by the methods of Chapter 1, using the following momenta definition:

$$\mathbf{p} = \frac{\partial T}{\partial \dot{\alpha}} \qquad \mathbf{P}_x = \frac{\partial T}{\partial \omega_x}, \quad \text{etc.,} \tag{3}$$

which leads to the Hamiltonian in a Type III representation

$$\mathbf{H} = A\mathbf{P}_x^{\,2} + B\mathbf{P}_y^{\,2} + C\mathbf{P}_z^{\,2} + F\mathbf{p}^2 - 2C\mathbf{p}\mathbf{P}_z + \mathbf{V}(\alpha) \tag{4}$$

with $\mathbf{V}(\alpha)$ given by (1), A and B the usual inertial constants, and

$$C = \frac{\hbar^2}{2(I_z - I_\alpha)} \qquad F = \frac{\hbar^2 I_z}{2[I_\alpha(I_z - I_\alpha)]} \tag{5}$$

In this model the coefficients are all constants. C involves the moment of inertia of the framework alone, and F contains the reduced moment of the two parts of the molecule.

The quantities $\mathbf{P}_x$, $\mathbf{P}_y$, and $\mathbf{P}_z$ were defined by (3), but by using the basic definition of angular momentum it can be shown that they are equal to the components of the total angular momentum of the molecule including contributions arising from internal rotation. Similarly it can be shown that $\mathbf{p}$ is the total contribution of the top atoms to the z component of the angular momentum including both the internal and over-all motions.

To simplify later equations we may write

$$\mathbf{H}' = \frac{\mathbf{H} - D(\mathbf{P}_x^{\,2} + \mathbf{P}_y^{\,2} + \mathbf{P}_z^{\,2})}{C - D}, \tag{6}$$

where

$$D = \tfrac{1}{2}(A + B).$$

$\mathbf{H}'$ is now rewritten

$$\mathbf{H}' = b^*(\mathbf{P}_x^{\,2} - \mathbf{P}_y^{\,2}) + \mathbf{P}_z^{\,2} - d\mathbf{p}\mathbf{P}_z + f\mathbf{p}^2 + \mathbf{V}'(\alpha) \tag{7}$$

with

$$b^* = \frac{1}{2}\frac{(A - B)}{C - D} \qquad f = \frac{F}{C - D}$$

$$d = \frac{2C}{C - D} \qquad \mathbf{V}' = \frac{V}{C - D}.$$

If Λ is an eigenvalue of $\mathbf{H}'$, the energy is given by

$$E = J(J + 1)D + (C - D)\Lambda.$$

Since $\mathbf{H}'$ and $\mathbf{H}$ are both diagonal in J and M, only one JM block need be considered at a time.

4m SYMMETRY

As discussed in Chapter 2, the asymmetric rotor wave functions are invariant to the operations of the Four group. In this the cross-term $\mathbf{p}\mathbf{P}_z$ is not invariant unless $\mathbf{p}$ changes sign when $\mathbf{P}_z$ does. If $V(\alpha)$ is an even function of α, the operations of the

Four-group whose effect is given in Table 4m1 leave **H** invariant and will again belong to the Four-group. Thus every solution of the wave equation belongs to one of the species A, B_x, B_y, B_z of this group just as for the asymmetric rotor, but here $V(\alpha)$ must be included. It is also true that the energy matrix can be factored into at least four factors by using the symmetry.

Sometimes $V(\alpha)$ possesses additional symmetry properties, as for example in a molecule such as nitromethane. Here $V(\alpha)$ is presumably invariant under $\alpha \to \alpha + 2\pi k/6$, $k = 0, 1, 2, \ldots$. Likewise $\alpha \to 2\pi k/6 - \alpha$ would leave $V(\alpha)$ invariant.

Table 4m1—Effect of Four-Group Symmetry Operations [126]

E	C_x	C_y	C_z
$P_i \to P_i$	$P_x \to P_x$	$P_y \to P_y$	$P_z \to P_z$
	$P_{y,z} \to -P_{y,z}$	$P_{x,z} \to -P_{x,z}$	$P_{x,y} \to -P_{x,y}$
$p \to p$	$p \to -p$	$p \to -p$	$p \to p$
$(\alpha \to \alpha)$	$(\alpha \to -\alpha)$	$(\alpha \to -\alpha)$	$(\alpha \to \alpha)$
$(\theta \to \theta)$	$(\theta \to \pi - \theta)$	$(\theta \to \pi - \theta)$	$(\theta \to \theta)$
$(\phi \to \phi)$	$(\phi \to \pi + \phi)$	$(\phi \to \pi + \phi)$	$(\phi \to \phi)$
$(\chi \to \chi)$	$(\chi \to 2\pi - \chi)$	$(\chi \to \pi - \chi)$	$(\chi \to \pi + \chi)$

Table 4m2—Character Table for Case s = 3 [126]

CLASS	E	$2C_3$	$3C_xC_3$	C_z	$2C_zC_3$	$3C_yC_2$
A	1	1	1	1	1	1
B_z	1	1	−1	1	1	−1
E_1	2	−1	0	2	−1	0
B_x	1	1	1	−1	−1	−1
B_y	1	1	−1	−1	−1	1
E_2	2	−1	0	−2	1	0

Table 4m3—Character Table for Case s = 6 [126]

	E	$2C_3$	$3C_xC_3$	C_z	$2C_zC_3$	$3C_yC_3$	C_6^3	$2C_3C_6^3$	$3C_xC_3C_6^3$	$C_zC_6^3$	$2C_zC_3C_6^3$	$2C_yC_3C_6^3$
A_e	1	1	1	1	1	1	1	1	1	1	1	1
B_{ze}	1	1	−1	1	1	−1	1	1	−1	1	1	−1
E_{1e}	2	−1	0	2	−1	0	2	−1	0	2	−1	0
B_{xe}	1	1	1	−1	−1	−1	1	1	1	−1	−1	−1
B_{ye}	1	1	−1	−1	−1	1	1	1	−1	−1	−1	0
E_{2e}	2	−1	0	−2	1	0	2	−1	0	−2	1	0
A_o	1	1	1	1	1	1	−1	−1	−1	−1	−1	−1
B_{zo}	1	1	−1	1	1	−1	−1	−1	+1	−1	−1	+1
E_{1o}	2	−1	0	2	−1	0	−2	1	0	−2	1	0
B_{zo}	1	1	1	−1	−1	−1	−1	−1	−1	1	1	1
B_{yo}	1	1	−1	−1	−1	1	−1	−1	1	1	1	−1
E_{2o}	2	−1	0	−2	1	0	−2	1	0	2	−1	0

These operations are isomorphous with the point group C_{3v}, but this group is not used in the same way as when group theory is applied to vibrations.

The symmetry of $V(\alpha)$ cannot be applied to $\mathbf{H}$ indiscriminately, for $\mathbf{p}$ occurs in a cross-term. The "rotations" $\alpha \rightarrow \alpha + 2\pi k/s$ do leave $\mathbf{H}$ invariant as do the "reflections" $\alpha \rightarrow (2\pi k/s) - \alpha$, if $\mathbf{P}_z \rightarrow -\mathbf{P}_z$ at the same time. The operations of the group V and the s internal rotations $\alpha \rightarrow \alpha + 2\pi k/s$ generate a group of $4s$ operations. For $s = 3$, the group is isomorphous with the group C_{6v}, the characters table being that in Table 4m2. The one-dimensional species are labeled to correspond to the species for the subgroup V. The degenerate species E_1 would become $A + B_z$, and E_2 would become $B_x + B_y$ in the subgroup.

The case $s = 6$ can be constructed from Table 4m2. The new group has the additional operation $C_6^3(\alpha \rightarrow \alpha + \pi)$ which commutes with all the operations of C_{6v}; hence the new group has twice as many classes and twice as many species as those of $s = 3$. The result is given in Table 4m3.

These higher symmetries permit further factoring of the energy matrix if the basis functions are chosen to have symmetries in accord with the various species.

4n THE ENERGY MATRIX

If the asymmetry b^* is zero and there were no barrier, the reduced Hamiltonian $\mathbf{H}'$ is

$$\mathbf{P}_z^2 - d\mathbf{p}\mathbf{P}_z + f\mathbf{p}^2, \tag{1}$$

which leads to a diagonal matrix with the basis functions

$$S_{J,K,M}(\theta, \phi)e^{iK\chi}e^{im\alpha}, \tag{2}$$

where S is the θ, ϕ part of the symmetric rotor and

$$K = 0, \pm 1, \pm 2, \ldots, \pm J \qquad m = 0, \pm 1, \pm 2, \ldots, \infty. \tag{3}$$

These wave functions may be used to set up a matrix for the general form of $\mathbf{H}'$, 416. The asymmetry term b^* gives rise to off-diagonal terms the same as for the rigid rotor. The barrier potential usually has the form

$$V' = \tfrac{1}{2}V_0'(1 - \cos s\alpha) = \tfrac{1}{2}V_0' - \tfrac{1}{4}V_0'(e^{is\alpha} + e^{-is\alpha}), \tag{4}$$

where s is the number of equivalent minima and $(C - D)V_0' = V_0$ is the barrier height. The nonvanishing elements are

$$\begin{aligned}(K, m\,|\mathbf{H}|\,K, m) &= (K, m\,|\mathbf{H}' - \tfrac{1}{2}V_0'|\,K, m) = K^2 - dmK + fm^2 \\ (K, m\,|\mathbf{H}|\,K \pm 2, m) &= -\tfrac{1}{2}b^*\{[J^2 - (K \pm 1)^2][(J + 1)^2 - (K \pm 1)^2]\}^{1/2} \\ (K, m\,|\mathbf{H}|\,K, m \pm s) &= -\tfrac{1}{4}V_0'.\end{aligned} \tag{5}$$

The constant $\frac{1}{2}V_0'$ has been incorporated so that the eigenvalues of $\mathbf{H}$ are related to those of $\mathbf{H}'$ through the relation

$$\Lambda' = \Lambda + \tfrac{1}{2}V_0'.$$

If $V_0' \neq 0$, this form of $\mathbf{H}$ corresponds to an infinite secular equation. There are, however, no elements connecting even and odd K values, so that the secular equation factors into one for even and one for odd K values. This is part of the factoring into symmetry species.

4o FREE INTERNAL ROTATION

If the barrier height $V_0' = 0$, the energy matrix becomes diagonal in m, and the matrix factors into one block for each value of m in addition to the even odd factors. For a given set of JMm values the secular equation has the elements

$$
\begin{aligned}
(K|\mathbf{H}|K) &= K^2 - dmK - \lambda \\
(K|\mathbf{H}|K \pm 2) &= -\tfrac{1}{2}b^*\{[J^2 - (K \pm 1)^2][(J+1)^2 - (K \pm 1)^2]\}^{1/2}.
\end{aligned} \tag{1}
$$

The energy is related to the roots λ by the relation

$$E = DJ(J+1) + Fm^2 + (C - D)\lambda. \tag{2}$$

This secular equation is similar to that given for the rigid rotor in Chapter 2, except for the added term $-dmK$ on the diagonal. This term spoils the additional factoring

$$
\begin{vmatrix}
16 + 4md - \lambda & 6\sqrt{3}b & 0 & 0 & 0 \\
6\sqrt{3}b & 4 + 2md - \lambda & (210)^{1/2}b & 0 & 0 \\
0 & (210)^{1/2}b & -\lambda & (210)^{1/2}b & 0 \\
0 & 0 & (210)^{1/2}b & 4 - 2md - \lambda & 6\sqrt{3}b \\
0 & 0 & 0 & 6\sqrt{3}b & 16 - 4md - \lambda
\end{vmatrix} = 0.
$$

Fig. 4o1 The secular equation for a free internal rotor with $J = 5$ and even.

possible in the rigid case. The situation is shown for K even $J = 5$ in Fig. 4o1. For each nonvanishing value of $|m|$, there will be two identical equations; hence all levels except $m = 0$ are doubly degenerate. With $m = 0$, the matrix takes the form of the ordinary rigid rotor matrix.

The energy levels of free rotation must conform to the symmetry restrictions given in 4m. If $s = 3$ or 6, and m is not a multiple of 3, the symmetry is one of the degenerate species. If m is a multiple of 3, the levels are of species A or B, which would split into nondegenerate components if the barrier were sufficiently high. When K is even, the species may be A, B_z or E_1; if K is odd, B_x, B_y or E_2. For $s = 6$, the species are further divided into even (e) or odd (o), depending on whether m is even or odd.

For small values of b^* the energy levels may be expanded in powers of b^*, which to the second order gives

$$
\begin{aligned}
\lambda = K^2 - m\,dK &+ \frac{b^{*2}}{8}\{[J^2 - (K-1)^2][(J+1)^2 - (K-1)^2]/(2K - md - 2)\} \\
&- \frac{b^{*2}}{8}\{[J^2 - (K+1)^2][(J+1)^2 - (K+1)^2]/(2K - md + 2)\ldots.
\end{aligned} \tag{3}
$$

For large m this reduces to

$$\lambda = K^2 - m\,dK. \tag{4}$$

4p LOW BARRIER

When the barrier height V_0 is small but not negligible compared to F, a solution can be obtained using the Van Vleck-Jordahl perturbation technique. The reduced

Hamiltonian **H** is split into an unperturbed diagonal matrix $\mathbf{H}^0$ with the elements

$$(K, m\,|\mathbf{H}^0|\,Km) = K^2 - dmK + fm^2 \tag{5}$$

and a perturbation matrix with the elements

$$\begin{aligned}(K, m\,|\mathbf{H}_p|\,K \pm 2, m) &= -\tfrac{1}{2}b^*\{[J^2 - (K \pm 1)^2][(J+1)^2 - (K \pm 1)^2]\}^{1/2}\\ (K, m\,|\mathbf{H}_p|\,K, m \pm s) &= -\tfrac{1}{4}V_0'.\end{aligned} \tag{6}$$

The result of the transformation after neglecting all off-diagonal terms in m of order $V_0'^2/16$ becomes

$$\begin{aligned}(K\,|\mathbf{H}|\,K) &= K^2 - dmK\\ &\quad + \frac{1}{s}\left(\frac{V_0'^2}{16}\right)\{(1/[Kd - (2m+s)f]) - (1/[Kd - (2m-s)f])\} - \lambda\\ (K\,|\mathbf{H}|\,K \pm 2) &= \tfrac{1}{2}b^*\{[J^2 - (K \pm 1)^2][(J+1)^2 - (K \pm 1)^2]\}^{1/2}.\end{aligned} \tag{7}$$

The secular equation is similar in its properties to the free rotation case. If $m = \pm\frac{1}{2}s$, the approach is not valid when s is even because a near degeneracy occurs; hence the unperturbed states with $m = \pm\frac{1}{2}s$ must be considered. The nonvanishing elements then become

$$\begin{aligned}\left(K, \frac{s}{2}\,|\mathbf{H}|\,K, \frac{s}{2}\right) &= K^2 - \tfrac{1}{2}\,dsK + \frac{1}{s}\left[\left(\frac{V_0'}{4}\right)^2\!/(Kd - 2sf)\right] - \lambda\\ \left(K, -\frac{s}{2}\,|\mathbf{H}|\,K, -\frac{s}{2}\right) &= K^2 + \tfrac{1}{2}\,dsK - \frac{1}{s}\left[\left(\frac{V_0'}{4}\right)^2\!/(Kd + 2sf)\right] - \lambda\\ \left(K, \frac{s}{2}\,|\mathbf{H}|\,K, -\frac{s}{2}\right) &= -\tfrac{1}{4}V_0'\\ \left(K, \frac{s}{2}\,|\mathbf{H}|\,K \pm 2, \frac{s}{2}\right) &= \tfrac{1}{2}b^*\{[J^2 - (K \pm 1)^2][(J+1)^2 - (K \pm 1)^2]\}^{1/2}\\ &= \left(K, -\frac{s}{2}\,|\mathbf{H}|\,K \pm 2, -\frac{s}{2}\right).\end{aligned} \tag{8}$$

Because of the symmetry $(K, s/2\,|\mathbf{H}|\,K', \pm s/2) = (-K, -s/2\,|\mathbf{H}|\,-K;\ \mp s/2)$ this secular equation can be factored into two factors similar to the Wang equation. This leads to a splitting of the degeneracy.

When m is not a multiple of $s/2$, the levels are inherently doubly degenerate for all barriers. If m is a multiple of $s/2$, the degeneracy is split at high enough barriers. In the present approximation the splitting apears only for $m = s/2$. The higher the multiple of $s/2$ the higher the order of approximation required to show the splitting.

The quantum number m is no longer a good quantum number if there is a barrier; however, if the barrier is low, m can still be used to index the levels.

4q HIGH BARRIERS

When the barrier is high enough so that the torsional levels of a given symmetry are widely separated compared to the rotational levels, another perturbation treatment becomes possible. If the asymmetry is small then the asymmetry and the coupling terms are treated as perturbations by the Van Vleck-Jordahl transformation. Once again a rotational secular equation is obtained. The unperturbed Hamiltonian is

$$\mathbf{H}_0 = D\mathbf{P}^2 + (C - D)\mathbf{P}_z^{\,2} + F\mathbf{p}^2 + \mathbf{V}(\alpha), \tag{1}$$

and the perturbation operator is

$$\mathbf{H}_1 = \tfrac{1}{2}(A - B)(\mathbf{P}_x^{\,2} - \mathbf{P}_y^{\,2}) - 2C\mathbf{p}\mathbf{P}_z. \tag{2}$$

$\mathbf{H}_0$ is diagonal in the representation for which the basis functions are

$$S_{J,K,M}(\theta, \phi)\, e^{iK\chi} U_{v\kappa}(\alpha), \tag{3}$$

in which the $U_{v\kappa}$ are the eigenfunctions of

$$(F\mathbf{p}^2 + \mathbf{V}_\alpha)U_{v\kappa} = E_{v\kappa}U_{v\kappa}. \tag{4}$$

The torsional states are described by the quantum numbers v_κ, where v is the principal quantum number of the vibrational state and κ is a degeneracy index.

For the cosine potential function used previously, the nondegenerate eigenvalues of the torsional equation (4), which is related to the Mathieu equation, can be obtained from published Tables* by using the definitions

$$E_v = \tfrac{1}{4}s^2Fbe_r(S) \quad \text{or} \quad \tfrac{1}{4}s^2Fbo_r(S) \tag{5}$$

with

$$S = \left(\frac{4}{s^2}\right)\left(\frac{V_0}{F}\right). \tag{6}$$

The eigenvalues be_r and bo_r are given in tables as functions of the parameter S. The values of the quantum number $v = 0, 1, 2, 3, \ldots$ are identified respectively with the eigenvalues be_0, bo_1; be_1, bo_2; be_3, etc. Some of the degenerate levels have been tabulated by Kilb [55]. The treatment of the high barrier case has been extended by several authors to higher degrees of approximation and more complicated molecules [29, 55, 76]. Figure 4m1 shows the correlation of energy levels corresponding to the various limiting cases. Because of space limitations and limited applicability to the infrared, this work is not covered here. Resolution in the infrared, although greatly improved, is still not sufficient to allow one to see the splittings due to internal rotation in many molecules. For these treatments the reader is referred to the original literature.

With appropriate simplifications, 4o2 can be reduced to the case of two freely rotating symmetric rotors which presumably is the situation in dimethylacetylene, [82], an over-all symmetric rotor in which both rotating groups are symmetric rotors. This problem was first studied by Howard [57] and later by Pryce, who is quoted by Mills and Thompson [82]. The result for a nondegenerate mode of vibration is

$$E = BJ(J + 1) - B(K_1 + K_2)^2 + 2AK_1^{\,2} + 2AK_2^{\,2}, \tag{7}$$

with K_1 and K_2 being the angular momentum associated with each rotating part of the molecule in units of $\hbar$. If the vibrational state in question is degenerate, the Coriolis interaction must be taken into account and Mills and Thompson give the result

$$E = B(J)(J + 1) - B(K_1 + K_2)^2 + 2AK_1^{\,2} + 2AK_2^{\,2} - 4\zeta A(K_1l_1 + K_2l_2) + 2\zeta^2A(l_1^{\,2} + l_2^{\,2}). \tag{8}$$

* *Tables Relating to Mathieu Functions*, Columbia University Press, New York, 1951.

5

LINE STRENGTHS AND SELECTION RULES

5a INTRODUCTION

The successful analysis of vibrational-rotational band spectra depends on using all the information available in the data. In the preceding chapters the energy levels have been discussed, from which it is possible to calculate the frequencies of the transitions. All possible energy differences between the energy levels represent a staggering number. Fortunately, not all of these energy differences represent permitted energy changes. The permitted transitions are governed by selection rules that not only limit the number of allowed transitions but also bring considerable order into the spectrum. Thus, before being able to use the knowledge concerning the energy levels, the selection rules must be known.

Another kind of information contained in the experimental data is the transition intensity. The transition intensity is as important to consider in analytical work as the transition frequency. In this chapter the selection rules and line strengths are considered, thus completing the information needed to undertake the analysis of vibrational-rotational bands.

The discussion starts with a consideration of the symmetry properties of the vibrational-rotational states since these symmetry properties have a direct bearing on the selection rules and the nuclear spin degeneracies which contribute to the intensities of the spectral transitions.

Symmetries of the rotational states of the asymmetric rotor were discussed in Chapter 2.

5b SYMMETRY PROPERTIES AND STATISTICAL WEIGHTS

Linear molecules

The total eigenfunction in the simplest approximation may be written as product of the electronic, vibrational, rotational, and nuclear spin parts:

$$\psi = \psi_e \psi_V \psi_R \psi_S. \tag{1}$$

For the present the effect of the ψ_S is neglected and attention focused on the rest of the wave function. The rotational eigenfunctions of a linear polyatomic molecule are negative or positive, depending on whether the total eigenfunction changes sign or remains unchanged when all the particles are reflected through the center of mass. If the electronic and vibrational wave functions ψ_e and ψ_V remain unchanged by all the symmetry operations of the molecular point group, that is, are totally symmetric, the symmetry characteristic of the wave function depends only on ψ_R. Here the rotational wave functions with even J are positive, those with odd J are negative.

If the linear molecule also has a center of symmetry, that is, belongs to the point group $D_{\infty h}$, the wave functions have an additional symmetry property, symmetric or antisymmetric with respect to an interchange of the identical nuclei. The wave function $\psi = \psi_e \psi_V \psi_R$ must either remain unchanged or change sign only when a reflection of the nuclei is carried out through the center of the molecule (see Fig. 4k1). If the wave function remains unchanged, the corresponding energy levels are called symmetric in the nuclei; if the wave function changes sign, the energy levels are antisymmetric in the nuclei. In the ground vibrational state of symmetric electronic states (in $D_{\infty h}$ molecules, a Σ_g^+ electronic state) the positive rotational states are symmetric and the negative states are antisymmetric.

Linear molecules without a center of symmetry, point group $C_{\infty v}$, even if they have identical nuclei, do not interchange identical nuclei when an inversion through the center of mass is carried out; therefore the symmetric-antisymmetric property of the alternate levels is no longer present. For such a molecule each rotational state is doubly degenerate because of two equivalent configurations that are separated by a barrier which is generally so high that negligible tunneling results from one configuration to the other. These two forms cannot be transformed into one another by a simple rotation of the whole molecule. One of the wave functions of this degenerate pair is symmetric, and the other is antisymmetric in the nuclei. If the barrier should be low enough to allow appreciable tunneling, the degenerate pair would split into a symmetric and antisymmetric level of slightly different energies.

For a linear molecule of the point group $C_{\infty v}$, the statistical weight of a rotational level in a totally symmetric electronic state and in the absence of an external field is merely the spatial degeneracy caused by the $2J + 1$ values M can assume. An additional degeneracy arises from the nuclear spins. This degeneracy is the same for all levels and hence for spectroscopic purposes can be ignored; this degeneracy, however, is important in thermodynamic applications. The degeneracy is given by

$$(2I_1 + 1)(2I_2 + 2)\ldots,$$

where I_1 is the spin of nucleus 1, etc.

If a molecule belongs to the point group $D_{\infty h}$, alternate rotational levels have different statistical weights arising from nuclear statistics. If there is only one nucleus with nonzero spin on each side of the center of symmetry, the problem is in every respect similar to that of homopolar diatomic molecules. For instance, $C_2^{12}H_2$ has the same type of degeneracy as H_2 since C^{12} has a zero spin. As is well known, there are three symmetric and one antisymmetric spin states for spin one-half. Since protons obey Fermi-Dirac statistics, the total eigenfunction must be antisymmetric. Assuming ψ_e and ψ_V to be totally symmetric, as is found for practically all stable molecules in the ground state, the antisymmetric spin states must be paired with the symmetric rotational states, and the symmetric spin states must be paired with the antisymmetric rotational states. Thus the antisymmetric rotational states are triply degenerate as a result of the spin statistics.

If the molecule were C_2D_2, the ratio of symmetric spin states to antisymmetric spin states is 6 : 3. However, the deuteron with integral mass number obeys Bose-Einstein statistics; hence the total eigenfunction must be symmetric. Thus the proper pairings show that the symmetric rotational states are sixfold degenerate, whereas the antisymmetric rotational states are threefold degenerate. Of course, if all spins are zero, with the possible exception of the nucleus at the center of symmetry, the antisymmetrical levels will be absent. Examples are $C^{12}O_2^{16}$, $C^{12}S_2^{32}$, $C_3^{12}O_2^{16}$.

To a high degree of approximation, intercombinations between symmetric and antisymmetric rotational levels are forbidden for any type of radiation and for collisions. Hence there are ortho and para species of linear symmetric polyatomic molecules, although no one has yet been able to separate the two species.

In general, if only one pair of nuclei of a $D_{\infty h}$ molecule have $I \neq 0$, the ratio of the symmetric to antisymmetric rotational states is

$$\frac{I}{I+1} \quad \text{for Fermi-Dirac statistics}$$

and

$$\frac{I+1}{I} \quad \text{for Bose-Einstein statistics}$$

For symmetric linear molecules with more than one pair of identical nuclei the weight factors of the symmetrical and antisymmetrical rotational levels are obtained by an extension of the method outlined here. This method was first used by Placzek and Teller [91]. Assuming that all nuclear spins are uncoupled, for a linear molecule $X(YZA)_2$ the number of nuclear spin configurations of the nuclei on one side of the center is $(2I_Y + 1)(2I_Z + 1)(2I_A + 1)$. The total number of spin configurations therefore is the square of this quantity, provided there is no contribution due to X. There are $(2I_Y + 1)(2I_Z + 1)(2I_A + 1)$ configurations for which a reflection leaves the configuration unchanged; these spin functions are symmetric with respect to a simultaneous interchange of all identical nuclei. The other spin configurations all occur in pairs, one of which is symmetric and the other antisymmetric. Thus there are

$$\tfrac{1}{2}[(2I_Y + 1)^2(2I_Z + 1)^2(2I_A + 1)^2 - (2I_Y + 1)(2I_Z + 1)(2I_A + 1)]$$

antisymmetric spin functions and

$$\tfrac{1}{2}[(2I_Y + 1)^2(2I_Z + 1)^2(2I_A + 1)^2 + (2I_Y + 1)(2I_Z + 1)(2I_A + 1)]$$

symmetric spin functions. The total eigenfunction including nuclear spin can be only totally symmetric or totally antisymmetric, depending on whether the resultant statistics is Bose-Einstein or Fermi-Dirac. If there is an even number of nuclei in the group,

YZA, following Fermi statistics, the resultant statistics is Bose; the resultant statistics are Fermi if the group has an odd number of nuclei obeying Fermi statistics. If all nuclei follow Bose statistics, the resultant statistics are Bose. The antisymmetric and symmetric spin states are combined with the rotational states just as in the simpler case.

Symmetric rotors

The rotational levels of a symmetric rotor have the positive or negative property, just as the linear molecule, depending on whether the total eigenfunction changes sign or remains unchanged upon reflection of all particles at the center of mass (see Fig. 4k1). If the molecule is nonplanar, the reflection of the nuclei at the center of mass results in a configuration that cannot be obtained by rotations alone. The result is two modifications, rather like optical isomers, which are separated by a potential maximum that is usually very high. Since the only difference in the two forms is the method of description, about which the molecule can know nothing, each form of the molecule has the same principal moments of inertia, and hence the same rotational energy levels, the difference arising solely from the positive and negative characteristics of the wave functions. Consequently, each of the energy levels of a nonplanar symmetric rotor is doubly degenerate. If the potential maximum separating the two forms is not exceedingly high, a slight splitting of this degeneracy occurs, and the wave functions contain equal contributions of the left- and right-handed species. It has been shown that one of these levels has the positive character and one has the negative character. Since whenever there is a positive level there is an almost coinciding negative level, this distinction for nonplanar symmetric rotor molecules is academic unless the splitting itself is sufficient to be observed.

For a planar symmetric rotor, the right- and left-handed species do not occur, because the inversion of the nuclei can be replaced by a suitable rotation. This behavior, which always corresponds to an oblate symmetric rotor, results in one set of energy levels. The even K states are positive and the odd K states are negative for a totally symmetric vibrational-electronic state. It should be pointed out that in the nonplanar case for $K \neq 0$ there are four sublevels for each K due to the K degeneracy and the positive negative degeneracy. In the planar case, $K \neq 0$, there is only the K degeneracy.

For a molecule which is accidentally a symmetric rotor, the only symmetry property of the rotational levels is the positive or negative character; that is, either the eigenfunction changes sign or does not change sign upon reflection through the center of mass.

If the molecule is a symmetric rotor because of the presence of a threefold or higher axis of symmetry, one must consider other symmetry properties of the rotational levels since certain of the rotations are symmetry operations of the molecular point group. All those symmetry operations of the point group which are equivalent to rotations form a subgroup called the rotational subgroup. For instance, the rotational subgroup of D_{3h} or D_{3d} is D_3 and for T_d is T. As for the wave functions of the vibrational levels, the rotational wave functions belong to one of the symmetry species of the rotational subgroup. Thus the wave functions of the rotational levels of a molecule with the rotational subgroup D_3 are of the species A_1, A_2, or E, and for the subgroup T the possible species are A, E, or F.

The symmetry species of the rotational levels is readily found by considering the effect of the symmetry operations on the rotational wave function. The rotational wave function for a symmetric rotor is given by 211,

$$\psi_{JKM}(\theta, \phi, \chi) = \Theta(\theta)e^{iK\chi}e^{iM\phi}. \tag{211}$$

To illustrate the process, consider a molecule of the type CH_3X which, because of the threefold axis of rotation, is a symmetric rotor of the molecular point group C_{3v}. The character table for C_{3v} is given in Table 5b1.

Table 5b1—Character Table for C_{3v}

C_{3v}	E	$2C_3$	$3\sigma_v$
A_1	+1	+1	+1
A_2	+1	+1	−1
E	+2	−1	0

The rotational subgroup of C_{3v} is C_3; that is, C_{3v} without the three reflection planes. The character table is shown in Table 5b2.

Table 5b2—Character Table for C_3

C_3	E	$2C_3$
A	1	1
E	2	−1

The problem becomes that of finding the behavior of the symmetric rotor eigenfunctions under the two operations of the group E, the identity operation, and C_3, the rotation of the molecule by $2\pi/3$ about the rotor axis. Since the identity operation consists of leaving the eigenfunction alone, it is easily shown that

$$\psi_{J,K,M}(\theta, \phi, \chi) \xrightarrow{E} \psi_{J,K,M}(\theta, \phi, \chi). \tag{1}$$

Thus, for $K = 0$, $\chi_E' = 1$; for $|K| \neq 0$ since the levels are doubly degenerate, $\chi_E' = 2$. A rotation by $2\pi j/n$ about the symmetry axis changes only the Eulerian angle χ which becomes $\chi + 2\pi j/n$. Hence the transformation becomes

$$\psi_{J,K,M}(\theta, \phi, \chi) \xrightarrow{C_n^j} \psi_{J,K,M}\left(\theta, \phi, \chi + \frac{2\pi j}{n}\right) = e^{2\pi jKi/n}\psi_{J,K,M}(\theta, \phi, \chi) \tag{2}$$

since χ enters the wave function only through the factor $e^{iK\chi}$. For a given J and M, $K \neq 0$, there are two functions $\psi_{J,K,M}$ and $\psi_{J,-K,M}$; thus the character of the rotation $C_n^{\,j}$ is

$$\chi_C' = e^{2\pi ijK/n} + e^{-2\pi ijK/n} = 2\cos\left(\frac{2\pi jK}{n}\right) \qquad \text{if } K \neq 0. \tag{3}$$

If $K = 0$, the character is +1.

For the special case $j/n = \frac{1}{3}$ the character becomes $2\cos 2\pi K/3$. It is now possible to determine the characters for the representation of C_3 by the $\psi_{J,K,M}(\theta, \phi, \chi)$. The results are tabulated in Table 5b3.

The symmetries of the rotational levels are now found by decomposing the characters in the standard manner [9]. Such a process gives the result shown in Table 5b4. Thus the symmetry species of the rotational levels are readily found.

Table 5b3—Characters for the Representation of C_3 Formed by ψ_{JKM}
$(p = 0, 1, 2, \cdots)$

	χ_E'	χ_{C_3}'
$K = 0$	1	1
$\|K\| = 3p$	2	2
$\|K\| = 3p \pm 1$	2	−1

The procedure is the same regardless of the rotational subgroup. Consider the rotational subgroup D_3 with the character table given in Table 5b5.

For this group one must consider the effect on the symmetric rotor wave function of a twofold rotation about an axis perpendicular to the symmetry axis and making an angle α with the x' axis, which is the origin of the angle χ measuring the rotation of the axes about the symmetry axis. Such a rotation corresponds to changing θ

Table 5b4—Symmetry of the Rotational Functions of a Molecule with Rotational Symmetry C_3

$K = 0$	A
$\|K\| = 3p$	$2A$
$\|K\| = 3p \pm 1$	E

into $\pi - \theta$, ϕ into $\pi + \phi$, and χ into $\pi - \chi + 2\alpha$. These changes will become clearer upon consulting Fig. 5b1.

Operating on $\psi_{J,K,M}(\theta, \phi, \chi)$, we find

$$\psi_{JKM}(\theta, \phi, \chi) \xrightarrow{C_2} \psi_{J,K,M}(\pi - \theta, \pi + \phi, \pi - \chi + 2\alpha)$$
$$= (-1)^{(J+n+M)} e^{iK(\pi+2\alpha)} \psi_{J,-K,M}(\theta, \phi, \chi),$$

where n is the larger of $|K|$ and $|M|$. Therefore if $K \neq 0$, $\chi_{C_2}' = 0$; if $K = 0$, $\chi_{C_2}' = (-1)^J$.

Table 5b5—Character Table of D_3

D_3	E	$2C_3$	$3C_2$
A_1	1	1	1
A_2	1	1	−1
E	2	−1	0

It is now possible to make Table 5b6 for D_3 similar to Table 5b3 for C_3. Thus the rotational levels have the symmetries given in Table 5b7 in the subgroup D_3.

These same general considerations can be applied to any polyatomic molecule, whether a symmetric, spherical, or asymmetric rotor, in order to find the symmetries. The asymmetric rotor was considered in Chapter 2 since the rotational symmetries of the levels were of great importance in solving the secular equation.

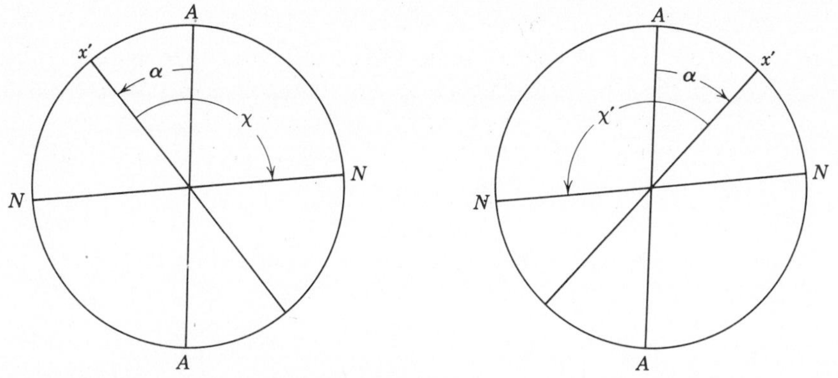

Fig. 5b1. The $x'y'$ plane of a symmetrical rotor showing the effect on the angle χ of a twofold rotation about an axis AA in the $x'y'$ plane. NN is the nodal line. χ', the angle after rotation, is equal to $\pi - \chi + 2\alpha$.

It remains now to consider the effect of nuclear spin on the degeneracies of the total wave function. Consider again the molecule of the CH_3X type. The number of spin states is given by $(2I_H + 1)^3$. Since $I_H = \frac{1}{2}$, there are eight different spin states which

Table 5b6—Characters for the Representation of D_3 Formed by ψ_{JKM}

	χ_E'	χ_{C_3}'	χ_{C_2}'
$K = 0$ J even	1	1	1
$K = 0$ J odd	1	1	−1
$\|K\| = 3p$	2	2	0
$\|K\| = 3p \pm 1$	2	−1	0

can be written down. If the nuclear wave function for $I = +\frac{1}{2}$ is represented by plus and for $I = -\frac{1}{2}$ by minus, the wave functions can be represented as in Fig. 5b2.

We must now ascertain what happens as the identical nuclei are interchanged. Clearly, functions a and h are symmetric to the interchange of any two nuclei since

Table 5b7—Symmetry of the Rotational Function of a Molecule with Rotational Symmetry D_3

$K = 0$ J even	A_1
$K = 0$ J odd	A_2
$\|K\| = 3p$	$A_1 + A_2$
$\|K\| = 3p \pm 1$	E

they do not change. For function b, if 2 and 3 are interchanged, the function is symmetric, but if 1 and 2 are interchanged, the function b is transformed into c, and if 1 and 3 are interchanged, function b is transformed into d. These functions are then equivalent as are the set e, f, and g. Acceptable functions, symmetric or antisymmetric, can be formed by taking linear combinations of these sets. To obtain the characters for the representation formed by the spin functions, we can return to the functions represented in Fig. 5b2, and make use of the fact that if a given permutation transforms one spin function into a different function, the corresponding row of the transformation matrix has a zero diagonal element; if it transforms a function into itself, the diagonal element is unity. The character for a given permutation is therefore equal to

	1	2	3
a	+	+	+
b	−	+	+
c	+	−	+
d	+	+	−
e	+	−	−
f	−	+	−
g	−	−	+
h	−	−	−

Fig. 5b2. Nuclear spin states for CH_3X.

the number of spin functions unchanged by the given permutation. Thus the identity permutation leaves all functions unchanged, hence has a character of 8. The permutation (12)(23) (which is equivalent to a threefold rotation) leaves only a and h unchanged. The same is true of the permutation corresponding to C_3^2 (13)(23).

Thus in C_3 the characters for the representation formed by the spin states are $\chi_E' = 8$, $\chi_{C_3}' = 2$, $\chi_{C_3^2}' = 2$, so that the ψ_S have the symmetry $4A + 2E$. Here the

Table 5b8—Weights of the Vibrational-Rotational Levels of CH_3X

$I_X = 0$

WAVE FUNCTION	SYMMETRY			SYMMETRY		
$\psi_e\psi_V$	A			E		
ψ_S	$4A$	$2E$	TOTAL	$4A$	$2E$	TOTAL
$K = 0$	4	0	4	0	4	4
$\|K\| = 3p$	8	0	8	0	8	8
$\|K\| = 3p \pm 1$	0	4	4	8	4	12

total wave function must have the symmetry A. The combination of the rotational functions with the spin functions can be accomplished by using the multiplication properties of C_3 which are: $A \times A = A$, $A \times E = E$, $E \times E = 2A + E$. The results of this combination are given in Table 5b8.

If X has a spin, the weight given in Table 5b8 must be multiplied by $(2I_X + 1)$. If $I_X = I_{Cl} = \frac{3}{2}$, as for methylchloride, the weights would be multiplied by 4, if $X = D$, $I_D = 1$, and the above weights would be multiplied by 3.

If we consider CD_3X there are $(3)^3$ or 27 spin functions which can be set down in a manner similar to that in which the CH_3X functions were set down. If X has no spin,

Table 5b9—Weights of the Vibrational-Rotational Levels of CD_3X

$I_X = 0$

WAVE FUNCTION	SYMMETRY			SYMMETRY		
$\psi_e \psi_V$	A			E		
ψ_S	$11A$	$8E$	TOTAL	$11A$	$8E$	TOTAL
$K = 0$	11	0	11	0	16	16
$\lvert K \rvert = 3p$	22	0	22	0	32	32
$\lvert K \rvert = 3p \pm 1$	0	16	16	22	16	38

then ψ_S has the symmetry $11A + 8E$, from which the weights of the vibrational-rotational levels may be deduced. The result is given in Table 5b9.

Again, if X has a nonzero spin the weights in 5b9 must be multiplied by $(2I_X + 1)$. This final factor for the nonzero spin of X is not important in calculating relative intensities since it is the same for all rotational-vibrational levels. It is important, however, in thermodynamic calculations.

This same procedure can be used to determine the weights of the rotational levels for any symmetric or asymmetric rotor. The weights for some of the more common molecules are given in Table 5b10.

Table 5b10—Some Ground State Statistical Weight Factors

MOLECULES	STATISTICAL WEIGHT FACTOR: SYMMETRIC ROTATIONAL LEVELS	STATISTICAL WEIGHT FACTOR: ANTISYMMETRIC ROTATIONAL LEVELS
NO_2, SO_2, CS_2, CO_2	1	0
H_2O, H_2CO, H_2S, F_2O	1	3
D_2O, D_2S, D_2CO	6	3
$C_2H_2D_2$, $C_2F_2D_2$, D_2CCH_2	15	21
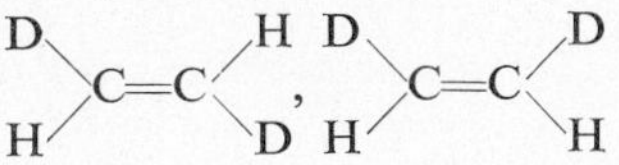	15	21
$C_2H_2Cl_2$, 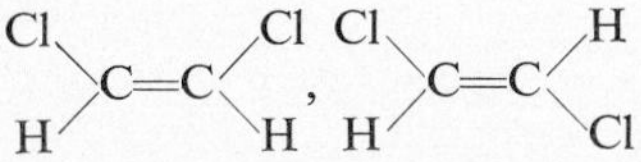	78	66
CD_2Cl_2, 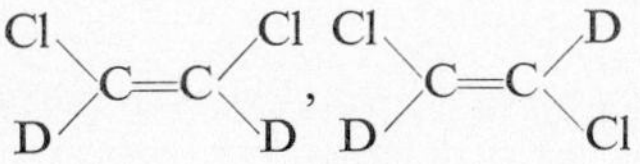	153	171

5c FORMULATION OF SELECTION RULE AND LINE STRENGTH PROBLEM

Once again the problem will be formulated in terms of the rigid rotor harmonic-oscillator model described in Chapter 1, where it is assumed that the rotational and vibrational motions are completely independent. As for the energy levels, perturbations arising from the inadequacy of this model are considered after the simple problem has been solved.

To these approximations the Hamiltonian function is written as

$$\mathbf{H} = \mathbf{H}_V + \mathbf{H}_R. \tag{1}$$

The wave equation is separable into one part dependent only on the vibrational coordinates and the other dependent on the rotational coordinates. The solutions have the general properties of separable coordinate problems. The total energy is the sum of the energy of the two parts

$$E = E_V + E_R. \tag{2}$$

The wave function is the product of ψ_V, depending only on the vibrational coordinates, and ψ_R, depending only on the rotational coordinates

$$\psi = \psi_V \psi_R. \tag{3}$$

Both the functions ψ_R, where R represents all the rotational quantum numbers, and ψ_V, where V represents the aggregate of vibrational quantum numbers, form a complete orthogonal set and each function ψ_R and ψ_V is considered to be separately normalized. To find the selection rules and line strengths, the matrix elements representing the electric moment must be considered. The two sets of axes, x, y, z fixed in the molecule and X, Y, Z fixed in space, enter into the description of the problem. The electric moment relative to one of the moving axes, say x, is

$$M_x = (M_x)_0 + \sum_k \left(\frac{\partial M_x}{\partial Q_k}\right)_{Q_k=0} Q_k + \dots, \tag{4}$$

where $(M_x)_0$ is the permanent electric moment in the x-direction, and Q_k is the kth normal coordinate of the vibrational problem. The component of the electric moment along a space-fixed axis, say the X-axis, becomes

$$\mathbf{M}_X = \mathbf{M}_x \cos xX + \mathbf{M}_y \cos yX + \mathbf{M}_z \cos zX, \tag{5}$$

where the direction cosines between the two coordinate systems are functions of the rotational coordinates alone.

In general, (5) can be written

$$\mathbf{M}_F = \sum_g \mathbf{\Phi}_{Fg} \mathbf{M}_g \qquad F = X, Y, Z; \quad g = x, y, z, \tag{6}$$

the $\mathbf{\Phi}_{Fg}$ being the direction cosines between the two sets of coordinate axes. The matrix elements of the electric moment are

$$(V', R' \,|\mathbf{M}_F|\, V'', R'') = \int \psi_{V'}{}^* \psi_{R'}{}^* \mathbf{M}_F \psi_{V''} \psi_{R''} \, d\tau \tag{7}$$

Substituting (4) and (5) into (7) enables the separation of (7) into two independent parts:

$$(V', R' |\mathbf{M}_F| V', R'') = \sum_g (M_g)_0 \int \psi_{R'}{}^* \mathbf{\Phi}_{Fg} \psi_{R''} \, d\tau$$
$$= \sum_g (M_g)_0 (V'R' |\mathbf{\Phi}_{Fg}| V'R'') \tag{8}$$

and

$$(V', R' |\mathbf{M}_F| V'', R'') = \sum_g \int \psi_{V'}{}^* \sum_k \frac{\partial M_g}{\partial Q_k} Q_k \psi_{V''} \int \psi_{R'}{}^* \mathbf{\Phi}_{Fg} \psi_{R''}{}^* \, d\tau$$
$$= \sum_g (V' |\mathbf{M}_g| V'')(R' |\mathbf{\Phi}_{Fg}| R''). \tag{9}$$

The $(R' |\Phi_{Fg}| R'')$ can be thought of as the matrix element along the F-direction corresponding to a unit electric moment along the g-direction, associated with the change in rotational quantum numbers from R' to R''. It is independent of the normal coordinates of the system. $(V' |\mathbf{M}_g| V')$ is the matrix element of the electric moment along g associated with a change in vibrational quantum numbers from V' to V''. For an harmonic-oscillator it is readily shown that $(V' |\mathbf{M}_g| V'')$ is nonzero only when the individual quantum numbers change in the manner

$$v_1, v_2 \ldots v_k \ldots v_n \rightarrow v_1, v_2 \ldots v_k \pm 1 \ldots v_n.$$

In this instance, its explicit value may be calculated in terms of the $\partial M_g/\partial Q_k$.

Equation (8) gives the expression for the electric moment matrix elements for pure rotational spectra, that is, transitions between rotational levels in the same vibrational state. These transitions depend on the presence of a permanent electric moment in the molecule under investigation.

The vibrational-rotational bands arise from simultaneous changes in the vibrational and rotational quantum numbers. The electric moment matrix elements are given by (9). In general, the calculation of the $(V' |\mathbf{M}_g| V'')$ is quite difficult and requires more information than is usually available, especially since the harmonic-oscillator approximation is never quite adequate. For the most part, in band analysis, absolute intensities are not necessary, only the relative intensities of the vibrational-rotational transitions are. Thus it is not necessary to know the exact values of the $(V' |\mathbf{M}_g| V'')$, merely that the quantity is nonvanishing. In cases of interest this is always true, for otherwise no band would be observed. Hence we can calculate the line strengths which are defined as $\sum_{Fg} |(R' |\mathbf{\Phi}_{Fg}| R'')|^2$, and, by combining these appropriately with weight factors, the relative intensities of vibrational-rotational transitions may be calculated.

The selection rules are determined once we know which of the matrix elements $(R' |\mathbf{\Phi}_{Fg}| R'')$ are nonvanishing. As will be seen in the following discussion these rules are such that they may be stated quite generally.

5d SELECTION RULES FOR A SYMMETRIC ROTOR

Since the components of angular momentum transform in the same way as do coordinates under rotation, we may write

$$\mathscr{P}_F = \sum_g \mathbf{\Phi}_{Fg} \mathbf{P}_g \tag{1}$$

and

$$\mathbf{P}_g = \sum_F \mathbf{\Phi}_{Fg}\mathscr{P}_F. \tag{2}$$

The commutation rules involving the direction cosines are as follows [36]:

$$\begin{aligned} &[\mathbf{\Phi}_{Fg}, \mathbf{\Phi}_{F'g'}] = 0 && [\mathscr{P}_F, \mathbf{\Phi}_{Fg}] = [\mathbf{P}_g, \mathbf{\Phi}_{Fg}] = 0 \\ &[\mathscr{P}_X, \mathbf{\Phi}_{Yg}] = -[\mathscr{P}_Y, \mathbf{\Phi}_{Xg}] = i\hbar\mathbf{\Phi}_{Zg}, && [\mathbf{P}_x, \mathbf{\Phi}_{Fy}] = -[\mathbf{P}_y, \mathbf{\Phi}_{Fx}] = -i\hbar\mathbf{\Phi}_{Fz} \\ &[\mathscr{P}_Y, \mathbf{\Phi}_{Zg}] = -[\mathscr{P}_Z, \mathbf{\Phi}_{Yg}] = i\hbar\mathbf{\Phi}_{Xg}, && [\mathbf{P}_y, \mathbf{\Phi}_{Fz}] = -[\mathbf{P}_z, \mathbf{\Phi}_{Fy}] = -i\hbar\mathbf{\Phi}_{Fx} \\ &[\mathscr{P}_Z, \mathbf{\Phi}_{Xg}] = -[\mathscr{P}_X, \mathbf{\Phi}_{Zg}] = i\hbar\mathbf{\Phi}_{Yg}, && [\mathbf{P}_z, \mathbf{\Phi}_{Fx}] = -[\mathbf{P}_x, \mathbf{\Phi}_{Fz}] = -i\hbar\mathbf{\Phi}_{Fy} \end{aligned} \tag{3}$$

The relations (3) together with the known matrix elements of $\mathscr{P}_F$ and $\mathbf{P}_g$ deduced in Chapter 2 are now used to determine the nonvanishing elements of the matrices of the direction cosines $\mathbf{\Phi}_{Fg}$. For convenience the representation used is that which diagonalizes $\mathbf{P}^2$, $\mathbf{P}_z$, and $\mathscr{P}_Z$; this representation leads directly to the rotational contribution to the intensities of the symmetric rotor, for this is the representation in which the symmetric rotor energy is diagonal.

It is convenient to consider first which elements of the $\mathbf{\Phi}_{Fg}$ have nonvanishing elements in the J, K, M representation, thereby obtaining the selection rules for the symmetric rotor.

The selection rules for J may be obtained by considering the second commutator of $\mathbf{P}^2$ with the $\mathbf{\Phi}_{Fg}$. The procedure used is illustrated here by deriving the first commutator.

Consider the relation

$$\begin{aligned} \mathbf{P}^2\mathbf{\Phi}_{Fx} - \mathbf{\Phi}_{Fx}\mathbf{P}^2 &= (\mathbf{P}_x^2 + \mathbf{P}_y^2 + \mathbf{P}_z^2)\mathbf{\Phi}_{Fx} - \mathbf{\Phi}_{Fx}(\mathbf{P}_x^2 + \mathbf{P}_y^2 + \mathbf{P}_z^2) \\ &= (\mathbf{P}_y^2 + \mathbf{P}_z^2)\mathbf{\Phi}_{Fx} - \mathbf{\Phi}_{Fx}(\mathbf{P}_y^2 + \mathbf{P}_z^2), \end{aligned} \tag{4}$$

the terms with exclusively x subscripts vanishing because of the commutation relations. Then

$$\begin{aligned} \mathbf{P}_y^2\mathbf{\Phi}_{Fx} - \mathbf{\Phi}_{Fx}\mathbf{P}_y^2 &= \mathbf{P}_y(\mathbf{\Phi}_{Fx}\mathbf{P}_y + i\hbar\mathbf{\Phi}_{Fz}) - (\mathbf{P}_y\mathbf{\Phi}_{Fx} - i\hbar\mathbf{\Phi}_{Fz})\mathbf{P}_y \\ &= i\hbar(\mathbf{P}_y\mathbf{\Phi}_{Fz} + \mathbf{\Phi}_{Fz}\mathbf{P}_y), \end{aligned} \tag{5}$$

and finally

$$\begin{aligned} \mathbf{P}_z^2\mathbf{\Phi}_{Fx} - \mathbf{\Phi}_{Fx}\mathbf{P}_z^2 &= \mathbf{P}_z(\mathbf{\Phi}_{Fx}\mathbf{P}_z - i\mathbf{\Phi}\hbar_{Fy}) - (\mathbf{P}_z\mathbf{\Phi}_{Fx} + i\hbar\mathbf{\Phi}_{Fy})\mathbf{P}_z \\ &= -i\hbar(\mathbf{P}_z\mathbf{\Phi}_{Fy} + \mathbf{\Phi}_{Fy}\mathbf{P}_z). \end{aligned} \tag{6}$$

Substituting (5) and (6) into (4) gives

$$\mathbf{P}^2\mathbf{\Phi}_{Fx} - \mathbf{\Phi}_{Fx}\mathbf{P}^2 = i\hbar[(\mathbf{\Phi}_{Fz}\mathbf{P}_y - \mathbf{\Phi}_{Fy}\mathbf{P}_z) + (\mathbf{P}_y\mathbf{\Phi}_{Fz} - \mathbf{P}_z\mathbf{\Phi}_{Fy})]. \tag{7}$$

A similar treatment of $\mathbf{P}^2\mathbf{\Phi}_{Fy} - \mathbf{\Phi}_{Fy}\mathbf{P}^2$ and $\mathbf{P}^2\mathbf{\Phi}_{Fz} - \mathbf{\Phi}_{Fz}\mathbf{P}^2$ leads to the general relation

$$\mathbf{P}^2\mathbf{\Phi} - \mathbf{\Phi}\mathbf{P}^2 = i\hbar(\mathbf{P} \times \mathbf{\Phi} - \mathbf{\Phi} \times \mathbf{P}), \tag{8}$$

in which the direction cosines are written as a vector to which they are equivalent. The expanded second commutator $[\mathbf{P}^2, [\mathbf{P}^2, \mathbf{\Phi}]]$ is given by

$$\mathbf{P}^4\mathbf{\Phi} - 2\mathbf{P}^2\mathbf{\Phi}^2\mathbf{P}^2 + \mathbf{\Phi}\mathbf{P}^4, \tag{9}$$

which by repeated applications of (3) can be shown to be

$$\mathbf{P}^4\mathbf{\Phi} - 2\mathbf{P}^2\mathbf{\Phi}\mathbf{P}^2 + \mathbf{\Phi}\mathbf{P}^4 = 2\hbar^2(\mathbf{P}^2\mathbf{\Phi} + \mathbf{\Phi}\mathbf{P}^2) - 4\hbar^2\mathbf{P}(\mathbf{P} \cdot \mathbf{\Phi}). \tag{10}$$

If (10) is now converted to a matrix element relation, we find

$$(J'K'M'\,|\mathbf{P}^4\mathbf{\Phi} - 2\mathbf{P}^2\mathbf{\Phi}\mathbf{P}^2 + \mathbf{\Phi}\mathbf{P}^4|\,J, K, M)$$
$$= 2\hbar^2(J', K', M'\,|\mathbf{P}^2\mathbf{\Phi} + \mathbf{\Phi}\mathbf{P}^2|\,J, K, M) - 4\hbar^2(J', K', M'\,|\mathbf{P}(\mathbf{P}\cdot\mathbf{\Phi})|\,J, K, M), \quad (11)$$

and therefore

$$\hbar^4[J'^2(J'+1)^2 - 2J'(J'+1)J(J+1) + J^2(J+1)^2](J', K', M'\,|\mathbf{\Phi}|\,J, K, M)$$
$$= 2\hbar^4[J'(J'+1) + J(J+1)](J', K', M'\,|\mathbf{\Phi}|\,J, K, M) \quad (12)$$
$$- 4\hbar^2(J', K', M'\,|\mathbf{P}(\mathbf{P}\cdot\mathbf{\Phi})|\,J, K, M).$$

If J' and J are equal, then (12) can be reduced to

$$(J, K', M'\,|\mathbf{\Phi}|\,J, K, M) = \frac{(J, K', M'\,|\mathbf{P}(\mathbf{P}\cdot\mathbf{\Phi})|\,J, K, M)}{J(J+1)\hbar^2}. \quad (13)$$

It is already known that $\mathbf{P}_z$ is diagonal in J and K. Similarly, $\mathbf{P}\cdot\mathbf{\Phi}$ is readily shown to be diagonal in J and K, since $\mathbf{P}\cdot\mathbf{\Phi}$ is a scalar operator that is invariant under operations generated by $\mathbf{P}_z$ and $\mathbf{P}^2$. This invariance can be expressed by the relation $[\mathbf{P}_z, (\mathbf{P}\cdot\mathbf{\Phi})] = 0 = [\mathbf{P}^2, (\mathbf{P}\cdot\mathbf{\Phi})]$, which in turn leads to the relationship between the matrix elements

$$(J', K', M'\,|\mathbf{P}_z(\mathbf{P}\cdot\mathbf{\Phi}) - (\mathbf{P}\cdot\mathbf{\Phi})\mathbf{P}_z|\,J, K, M)$$
$$= [(J', K', M'\,|\mathbf{P}_z|\,J', K', M')\,(J', K', M'\,|\mathbf{P}\cdot\mathbf{\Phi}|\,J, K, M)$$
$$- (J, K, M\,|\mathbf{P}_z|\,J, K, M)\,(J', K', M'\,|\mathbf{P}\cdot\mathbf{\Phi}|\,J, K, M)] \quad (14)$$
$$= (K' - K)\,(J', K', M'\,|\mathbf{P}\cdot\mathbf{\Phi}|\,J, K, M) = 0.$$

Therefore $(J', K', M'\,|\mathbf{P}\cdot\mathbf{\Phi}|\,J, K, M) = 0$ unless $K' = K$.

If $\mathbf{P}^2$ is substituted for $\mathbf{P}_z$ in (14) it can also be shown that $J' = J$ for a nonvanishing matrix element. It can further be shown that $\mathbf{P}\cdot\mathbf{\Phi}$ is independent of K. This proof is left for the reader.

Thus (13) may be simplified to

$$(J, K', M'\,|\mathbf{\Phi}|\,J, K, M) = \frac{(J, M'\,|\mathbf{P}\cdot\mathbf{\Phi}|\,J, M)}{J(J+1)\hbar^2}(J, K'\,|\mathbf{P}|\,J, K), \quad (15)$$

which leads to a nonvanishing matrix element only if $J' - J = 0$.

Returning to (12), the assumption is made that $J' \neq J$. By (14), this makes the last term on the right of (12) zero. The bracket on the left of (12) is factored to give

$$[J'(J'+1) - J(J+1)]^2 = [(J' - J)^2(J' + J + 1)^2], \quad (16)$$

whereas

$$2[J'(J'+1) + J(J+1)] = (J' + J + 1)^2 + (J' - J)^2 - 1. \quad (17)$$

Therefore

$$[(J' - J)^2(J' + J + 1)^2 - (J' + J + 1)^2 - (J' - J)^2 + 1]$$
$$(J', K', M'\,|\mathbf{\Phi}|\,J, K, M) = 0 \quad (18)$$

or

$$[(J' + J + 1)^2 - 1][(J' - J)^2 - 1](J', K', M'\,|\mathbf{\Phi}|\,J, K, M) = 0, \qquad J' \neq J. \quad (19)$$

In order to have a nonvanishing matrix element one of the brackets must vanish. The first bracket cannot because $J' \neq J$ and $J', J \geqslant 0$. The second vanishes only when $J' - J = \pm 1$, which completes the derivation of the selection rules for J,

$$J = 0, \pm 1 \tag{20}$$

To determine the selection rules on M it will prove convenient to consider not $\mathbf{\Phi}_{Xg}$ and $\mathbf{\Phi}_{Yg}$ but the quantity $\boldsymbol{\phi} = \mathbf{\Phi}_{Yg} - i\mathbf{\Phi}_{Xg}$ from which the matrices of $\mathbf{\Phi}_{Yg}$ and $\mathbf{\Phi}_{Xg}$ are readily obtained. From the matrix of $\boldsymbol{\phi}$ the selection rules on M are readily obtained.

$$[\mathscr{P}_Z, \boldsymbol{\phi}] = [\mathscr{P}_Z, \mathbf{\Phi}_{Yg} - i\mathbf{\Phi}_{Xg}] = [\mathscr{P}_Z, \mathbf{\Phi}_{Yg}] - i[\mathscr{P}_Z, \mathbf{\Phi}_{Xg}] = -i\hbar\mathbf{\Phi}_{Xg} + \hbar\mathbf{\Phi}_{Yg} \tag{21}$$

or

$$[\mathscr{P}_Z\boldsymbol{\phi} - \boldsymbol{\phi}\mathscr{P}_Z] = \hbar\boldsymbol{\phi}. \tag{22}$$

By taking the $J, K, M;\ J', K', M'$ matrix component of (22),

$$\begin{aligned} M\hbar(J, K, M\,|\boldsymbol{\phi}|\,J', K', M') - (J, K, M\,|\boldsymbol{\phi}|\,J', K', M')M'\hbar \\ = \hbar(J, K, M\,|\boldsymbol{\phi}|\,J', K', M') \end{aligned} \tag{23}$$

or

$$(M - M' - 1)(J, K, M\,|\boldsymbol{\phi}|\,J', K', M') = 0. \tag{24}$$

Thus it is readily seen that

$$(J, K, M\,|\boldsymbol{\phi}|\,J', K', M') = 0 \qquad \text{unless } M' = M - 1. \tag{25}$$

Repeating this process using $\boldsymbol{\phi}^* = \mathbf{\Phi}_{Yg} + i\mathbf{\Phi}_{Xg}$ we find that

$$(J, K, M\,|\boldsymbol{\phi}|\,J', K', M') = 0 \qquad \text{unless } M' = M + 1. \tag{26}$$

Hence the only nonvanishing matrix elements of $\mathbf{\Phi}_{Yg}$ and $\mathbf{\Phi}_{Xg}$ are those for which

$$M' = M \pm 1. \tag{27}$$

Since $\mathbf{\Phi}_{Zg}$ commutes with $\mathscr{P}_Z$, the only nonvanishing components of $\mathbf{\Phi}_{Zg}$ are those for which $M' = M$. Therefore the complete selection rules for M are

$$\Delta M = 0, \pm 1. \tag{28}$$

These same arguments may be repeated using $\mathbf{P}_z$ and $(\mathbf{\Phi}_{Fy} - i\mathbf{\Phi}_{Fx})$ together with the commutation rules (3) to show that the selection rules for K are

$$\Delta K = 0, \pm 1. \tag{29}$$

By collecting the complete results for the symmetric rotor selection rules in the general case, $(V'\,|\mathbf{M}_g|\,V'')$ gives

$$\Delta J = 0, \pm 1, \qquad \Delta K = 0, \pm 1, \qquad \Delta M = 0, \pm 1. \tag{30}$$

In applying these results to the symmetric rotor it is usually possible to make use of a further restriction imposed by the vanishing of one or more of the $(V'\,|\mathbf{M}_g|\,V'')$. The symmetric rotor has been defined as having two of its principal moments of inertia equal. Consider the case $A = B \neq C$ in the representation that diagonalizes P_z. The unique inertial axis here, the z-axis, is designated as the symmetry axis. The equality of two or more of the principal moments of inertia almost invariably arises from the symmetry properties of the molecule, and an n-fold, $n \geqslant 3$, proper or improper axis of symmetry coincides with the unique axis of the molecule. The molecular symmetry that makes the molecule a necessarily symmetric rotor also imposes certain restrictions on the $(V'\,|\mathbf{M}_g|\,V'')$, which are summarized with their corresponding selection rules as follows.

Case I. Forbidden bands

$$(V'\,|\mathbf{M}_x|\,V'') = (V'\,|\mathbf{M}_y|\,V'') = (V'\,|\mathbf{M}_z|\,V'') = 0. \tag{31}$$

The $V' \rightarrow V''$ transition is forbidden.

Case II. Parallel-type bands

$$(V'\,|\mathbf{M}_x|\,V'') = (V'\,|\mathbf{M}_y|\,V'') = 0 \qquad (V'\,|\mathbf{M}_z|\,V'') \neq 0. \tag{32}$$

Because the electric moment oscillates along the z-axis only, the selection rules for rotation depend on the elements of Φ_{Fz}, which are (assuming field-free molecules in all orientations)

$$\begin{array}{lll} K \neq 0 & \Delta J = 0, \pm 1 & \Delta K = 0, \\ K = 0 & \Delta J = \pm 1 & \Delta K = 0. \end{array} \tag{33}$$

Case III. Perpendicular-type bands

$$(V'\,|\mathbf{M}_x|\,V'') = (V'\,|\mathbf{M}_y|\,V'') \neq 0 \qquad (V'\,|\mathbf{M}_z|\,V'') = 0. \tag{34}$$

The electric moment oscillates perpendicularly to the symmetry axis, hence the selection rules depend on $\mathbf{\Phi}_{Fy}$ and $\mathbf{\Phi}_{Fx}$, and are

$$\Delta J = 0, \pm 1, \qquad \Delta K = \pm 1. \tag{35}$$

Case IV. General case

$$(V'\,|\mathbf{M}_g|\,V'') \neq 0.$$

For certain transitions involving simultaneous changes in more than one vibrational quantum number, or not originating in the ground vibrational state, it is possible that all the $(V'\,|\mathbf{M}_g|\,V'')$ have nonvanishing values; the selection rules are given in (33) and (35). Such bands have the appearance of superimposed parallel and perpendicular bands (see 7d).

5e DEPENDENCE OF THE $\mathbf{\Phi}_{Fg}$ ON *J*, *K*, AND *M*

It is convenient to use the operators $\mathbf{J} = [\mathscr{P}_Y - i\mathscr{P}_X]$ and $\boldsymbol{\phi} = [\mathbf{\Phi}_{Yg} - i\mathbf{\Phi}_{Xg}]$ to determine the dependence of the $\mathbf{\Phi}_{Fg}$ on M. These two operators commute as is easily shown.

$$\begin{aligned} [\mathbf{J}, \boldsymbol{\phi}] &= [\mathscr{P}_Y - i\mathscr{P}_X, \mathbf{\Phi}_{Yg} - i\mathbf{\Phi}_{Xg}] \\ &= [\mathscr{P}_Y, \mathbf{\Phi}_{Yg}] - i[\mathscr{P}_Y, \mathbf{\Phi}_{Xg}] - i[\mathscr{P}_X, \mathbf{\Phi}_{Yg}] - [\mathscr{P}_X, \mathbf{\Phi}_{Xg}], \end{aligned} \tag{1}$$

which reduces by use of the commutation rules 5d3 to

$$[\mathbf{J},\boldsymbol{\phi}] = 0 - \hbar\boldsymbol{\Phi}_{Z_g} + \hbar\boldsymbol{\Phi}_{Z_g} + 0 = 0. \tag{2}$$

The matrix elements of **J** were derived in Chapter 2 and were found to have the nonvanishing elements

$$(J, K, M-1\,|\mathbf{J}|\,J, K, M) = [(J+M)(J-M+1)]^{1/2}\hbar \qquad J \geqslant M \geqslant -J+1. \tag{3}$$

Setting down the most general $M-1$, $M+1$ matrix component which satisfies the selection rules for the equation

$$\mathbf{J}\boldsymbol{\phi} = \boldsymbol{\phi}\mathbf{J}, \tag{4}$$

we obtain

$$\begin{aligned}&(J, K, M-1\,|\mathbf{J}|\,J, K, M)(J, K, M\,|\boldsymbol{\phi}|\,J', K', M+1)\\ &= (J, K, M-1\,|\boldsymbol{\phi}|\,J', K', M)(J', K', M\,|\mathbf{J}|\,J', K', M+1) \qquad J'-J = 0, \pm 1.\end{aligned} \tag{5}$$

From (3)

$$\begin{aligned}&[(J+M)(J-M+1)]^{1/2}(J, K, M\,|\boldsymbol{\phi}|\,J', K', M+1)\\ &= (J, K, M-1\,|\boldsymbol{\phi}|\,J', K', M)\,[(J'+M+1)(J'-M)]^{1/2}.\end{aligned} \tag{6}$$

It is now convenient to consider separately the three cases $J'-J = 0, +1, -1$. If $J'-J = 0$, (6) becomes

$$\frac{(J, K, M\,|\boldsymbol{\phi}|\,J, K', M+1)}{[(J+M+1)(J-M)]^{1/2}} = \frac{(J, K, M-1\,|\boldsymbol{\phi}|\,J, K', M)}{[(J-M+1)(J+M)]^{1/2}}. \tag{7}$$

Since this relation holds for any value of M, this ratio must be independent of M. This ratio is designated as $(J, K\,|\boldsymbol{\phi}|\,J, K')$. Thus the dependence of the matrix elements of $\boldsymbol{\phi}$ on M is found to be

$$\begin{aligned}&(J, K, M\,|\boldsymbol{\phi}|\,J, K', M+1)\\ &= [(J+M+1)(J-M)]^{1/2}(J, K\,|\boldsymbol{\phi}|\,J, K').\end{aligned} \tag{8}$$

For $J' = J-1$, (6) may be written

$$\begin{aligned}&[(J+M)(J-M+1)]^{1/2}(J, K, M\,|\boldsymbol{\phi}|\,J-1, K', M+1)\\ &= (J, K, M-1\,|\boldsymbol{\phi}|\,J-1, K', M)\,[(J+M)(J-M-1)]^{1/2},\end{aligned} \tag{9}$$

which upon multiplication by $[(J-M)/(J+M)]^{1/2}$ is rewritten as

$$\begin{aligned}\frac{(J, K, M\,|\boldsymbol{\phi}|\,J-1, K', M+1)}{[(J-M-1)(J-M)]^{1/2}} &= \frac{(J, K, M-1\,|\boldsymbol{\phi}|\,J-1, K', M)}{[(J-M)(J-M+1)]^{1/2}}\\ &= (J, K\,|\boldsymbol{\phi}|\,J-1, K').\end{aligned} \tag{10}$$

Again each ratio is independent of M and is set equal to a constant independent of M, leading to

$$\begin{aligned}&(J, K, M\,|\boldsymbol{\phi}|\,J-1, K', M+1)\\ &= [(J-M)(J-M-1)]^{1/2}(J, K\,|\boldsymbol{\phi}|\,J-1, K').\end{aligned} \tag{11}$$

For $J' = J+1$, (6) becomes

$$\begin{aligned}&[(J+M)(J-M+1)]^{1/2}(J, K, M\,|\boldsymbol{\phi}|\,J+1, K', M+1)\\ &= (J, K, M-1\,|\boldsymbol{\phi}|\,J+1, K', M)\,[(J+M+2)(J-M+1)]^{1/2}.\end{aligned} \tag{12}$$

After multiplication of (12) by $[(J + M + 1)/(J - M + 1)]^{1/2}$ and rewriting, we obtain

$$\frac{(J, K, M\,|\boldsymbol{\phi}|\, J + 1, K', M + 1)}{[(J + M + 2)(J + M + 1)]^{1/2}} = \frac{(J, K, M - 1\,|\boldsymbol{\phi}|\, J + 1, K', M)}{[(J + M + 1)(J + M)]^{1/2}}$$
$$= -(J, K\,|\boldsymbol{\phi}|\, J + 1, K'), \tag{13}$$

leading to

$$(J, K, M\,|\boldsymbol{\phi}|\, J + 1, K', M + 1)$$
$$= -[(J + M + 1)(J + M + 2)]^{1/2}(J, K\,|\boldsymbol{\phi}|\, J + 1, K'). \tag{14}$$

The dependence of $\boldsymbol{\Phi}_{Zg}$ on M is found by making use of $\mathbf{J}^* = \mathscr{P}_Y + i\mathscr{P}_X$. Starting with the relation

$$[\mathbf{J}^*, \boldsymbol{\phi}] = [\mathbf{J}, \boldsymbol{\phi}] + 2i[\mathscr{P}_X, \boldsymbol{\Phi}_{Yg} - i\boldsymbol{\Phi}_{Xg}] = -2\hbar\boldsymbol{\Phi}_{Zg}, \tag{15}$$

which has been reduced through the use of the commutation rules 5d3. The only nonvanishing elements of $\mathbf{J}^*$ are

$$(J, K, M\,|\mathbf{J}^*|\, J, K, M - 1) = \hbar[(J + M)(J - M + 1)]^{1/2}$$
$$(J \geqslant M \geqslant -J + 1). \tag{16}$$

Since the matrices of $\mathbf{J}^*$ and $\boldsymbol{\phi}$ are known, the matrix of $\boldsymbol{\Phi}_{Zg}$ can be determined directly from (15). From 5d the selection rule $M' = M$ is known. Now for $J' = J$,

$$-2\hbar(J, K, M\,|\boldsymbol{\Phi}_{Zg}|\, J, K', M) \tag{17}$$
$$= (J, K|\boldsymbol{\phi}|\, J, K')[(J + M)(J - M + 1)]^{1/2}[(J - M + 1)(J + M)]^{1/2}\hbar$$
$$- (J, K|\boldsymbol{\phi}|\, J, K')[(J - M)(J + M + 1)]^{1/2}[(J + M + 1)(J - M)]^{1/2}\hbar$$
$$= 2M\hbar(J, K|\boldsymbol{\phi}|\, J, K'),$$

or

$$-(J, K, M\,|\boldsymbol{\Phi}_{Zg}|\, J, K', M) = M(J, K, |\boldsymbol{\phi}|\, J, K').$$

Since $\boldsymbol{\Phi}_{Zg}$ is real, $(J, K|\boldsymbol{\phi}|\, J, K') = (J, K'\,|\boldsymbol{\phi}|\, J, K)^*$.

For $J' = J - 1$, from (15) it follows that

$$-2\hbar(J, K, M\,|\boldsymbol{\Phi}_{Zg}|\, J - 1, K', M)$$
$$= (J, K, M\,|\mathbf{J}^*|\, J, K, M - 1)(J, K, M - 1\,|\boldsymbol{\phi}|\, J - 1, K', M) \tag{18}$$
$$- (J, K, M\,|\boldsymbol{\phi}|\, J - 1, K', M + 1)(J - 1, K', M + 1\,|\mathbf{J}^*|\, J - 1, K', M)$$
$$= 2[J^2 - M^2]^{1/2}(J, K|\boldsymbol{\phi}|\, J - 1, K')\hbar,$$

or

$$(J, K, M\,|\boldsymbol{\Phi}_{Zg}|\, J - 1, K', M) = [J^2 - M^2]^{1/2}(J, K|\boldsymbol{\phi}|\, J - 1, K'). \tag{19}$$

For $J' = J + 1$, we find in an analogous manner that

$$(J, K, M\,|\boldsymbol{\Phi}_{Zg}|\, J + 1, K', M) = [(J + 1)^2 - M^2]^{1/2}(J, K\,|\boldsymbol{\phi}|\, J + 1, K'). \tag{20}$$

Since $\boldsymbol{\Phi}_{Zg}$ is real, it can be seen from equation (19) that

$$(J, K|\boldsymbol{\phi}|\, J - 1, K') = (J - 1, K'|\boldsymbol{\phi}|\, J, K)^*. \tag{21}$$

Thus the matrix $(J, K|\boldsymbol{\phi}|\, J', K')$ as defined is Hermitian. This fact enables us to obtain the matrix $\boldsymbol{\phi}^*$. For example, from (11)

$$(J - 1, K', M + 1\,|\boldsymbol{\phi}^*|\, J, K, M) = (J, K, M|\boldsymbol{\phi}|\, J - 1, K', M + 1)^*$$
$$= (J - 1, K'|\boldsymbol{\phi}|\, J, K)[(J - M)(J - M - 1)]^{1/2} \tag{22}$$

The other components of $\boldsymbol{\phi}^*$ are found in a similar manner.

Table 5e1—Values of the Elements of All the Direction Cosine Matrices, Separated into the Three Factors [36]

MATRIX ELEMENT FACTOR	J'		
	$J+1$	J	$J-1$
$(J\|\mathbf{\Phi}_{Fg}\|J')$	$\{4(J+1)[(2J+1)(2J+3)]^{\frac{1}{2}}\}^{-1}$	$[4J(J+1)]^{-1}$	$[4J(4J^2-1^{\frac{1}{2}}]^{-1}$
$(J, K\|\mathbf{\Phi}_{Fz}\|J', K)$	$2[(J+1)^2-K^2]^{\frac{1}{2}}$	$2K$	$-2[J^2-K^2]^{\frac{1}{2}}$
$(J, K\|\mathbf{\Phi}_{Fy}\|J', K\pm 1)$ $= \pm i(J, K\|\mathbf{\Phi}_{Fx}\|J', K\pm 1)$	$\mp[(J\pm K+1)(J\pm K+2)]^{\frac{1}{2}}$	$[(J\mp K)(J\pm K+1)]^{\frac{1}{2}}$	$\pm[(J\mp K)(J\mp K-1)]^{\frac{1}{2}}$
$(J, M\|\mathbf{\Phi}_{Zg}\|J', M)$	$2[(J+1)^{-2}-M^2]^{\frac{1}{2}}$	$2M$	$-2[J^2-M^2]^{\frac{1}{2}}$
$J, M\|\mathbf{\Phi}_{Yg}\|J', M\pm 1)$ $= \mp i(J, M\|\mathbf{\Phi}_{Xg}\|J', M\pm 1)$	$\mp[(J\pm M+1)(J\pm M+2)]^{\frac{1}{2}}$	$[(J\mp M)(J\pm M+1)]^{\frac{1}{2}}$	$\pm[(J\mp M)(J\mp M-1)]^{\frac{1}{2}}$

In a manner entirely analogous to that used to obtain the dependence of the $\mathbf{\Phi}_{Fg}$ matrices on M, we can deduce the dependence of the $\mathbf{\Phi}_{Fg}$ on K. It is necessary to use the elements of the $\mathbf{P}_g$ for this derivation rather than the $\mathscr{P}_F$. Since this is true, the long procedure is not repeated, but only the results stated. Each matrix element $(J, K, M|\mathbf{\Phi}_{Fg}|J', K', M')$ can be written in the form $(J, K, M|\mathbf{\Phi}_{Fg}|J'K'M') = (J, M|\mathbf{\Phi}_{Fg}|J', M')(J, K|\mathbf{\Phi}_{Fg}|J, K')(J|\mathbf{\Phi}_{Fg}|J')$, where

$$
\begin{aligned}
&(J, K|\mathbf{\Phi}_{Fz}|J+1, K) = [(J+K+1)(J-K+1)]^{1/2} \\
&(J, K|\mathbf{\Phi}_{Fz}|J, K) = K \\
&(J, K|\mathbf{\Phi}_{Fz}|J-1, K) = -(J^2-K^2)^{1/2}.
\end{aligned} \tag{23}
$$

$$
\begin{aligned}
&\quad (J, K|\mathbf{\Phi}_{Fy}|J+1, K\pm 1) = \pm i(J, K|\mathbf{\Phi}_{Fx}|J+1, K\pm 1) \\
&= \mp\tfrac{1}{2}[(J\pm K+1)(J\pm K+2)]^{1/2} \\
&\quad (J, K|\mathbf{\Phi}_{Fy}|J, K\pm 1) = \pm i(J, K|\mathbf{\Phi}_{Fx}|J, K\pm 1) \\
&= \tfrac{1}{2}[(J\mp K)(J\pm K+1)]^{1/2} \\
&\quad (J, K|\mathbf{\Phi}_{Fy}|J-1, K\pm 1) = \pm i(J, K|\mathbf{\Phi}_{Fx}|J-1, K\pm 1) \\
&= \mp\tfrac{1}{2}[(J\mp K)(J\mp K-1)]^{1/2}.
\end{aligned} \tag{24}
$$

Now that the dependence of the matrix elements of the $\mathbf{\Phi}_{Fg}$ on K and M is known, it is possible to deduce their dependence on J. Using 5d1 and 5d3 together with the known dependence on K and M, the evaluation is straightforward but somewhat tedious; thus the actual evaluation is left as an exercise for the reader. The results are

$$
\begin{aligned}
&(J|\mathbf{\Phi}_{Fg}|J+1) = \{4(J+1)[(2J+1)(2J+3)]^{1/2}\}^{-1} \\
&(J|\mathbf{\Phi}_{Fg}|J) = [4J(J+1)]^{-1} \\
&(J|\mathbf{\Phi}_{Fg}|J-1) = [4J(4J^2-1)^{1/2}]^{-1}
\end{aligned} \tag{25}
$$

This completes the determination of the direction-cosine matrix elements. The complete results are summarized in Table 5e1.

5f SYMMETRIC ROTOR LINE STRENGTHS

The direction cosine matrix elements may now be used to obtain expressions for the line strengths of the symmetric rotor. These matrix elements are directly applicable since they were derived in the representation for which the symmetric rotor wave functions are the basis functions. Returning to 5c9 the sum $\sum_g |(R'|\Phi_{Fg}|R'')|^2 = |\sum_g (R'|\Phi_{Fg}|R'')|^2$ is the function to be evaluated. The factor $|(V'|\mathbf{M}_g|V'')|^2$ is considered to be a nonzero constant for the vibrational-rotational transition under consideration. In the absence of an external field (electric or magnetic) these terms may be summed over the Zeeman components.

In order to include all the degenerate components contributing to a given transition $J_K \rightarrow J_K'$, the direction cosine elements $(J, K, M|\Phi_{Fg}|J', K', M')$ are squared and summed over M, M' and F. In the absence of external fields X, Y, and Z are equivalent, and the summation of the squared elements over F may be accomplished by multiplying the squared elements for any given F by 3. Thus,

$$\sum_{FMM'} |(J, K, M|\Phi_{Fg}|J', K', M')|^2$$
$$= 3\,|(J|\Phi_{Zg}|J')|^2 \cdot |(J, K|\Phi_{Zg}|J', K')|^2 \cdot \sum_{MM'} |(J, M|\Phi_{Zg}|J', M')|^2, \quad (1)$$

in which Z on the right-hand side could be replaced by X or Y. This quantity is called the line strength of the transition with component of the electric moment M_g by analogy with the term used in atomic spectra [1]. The results of this process for a parallel type band are as follows:

(a) $|(J, K|\Phi_{Zz}|J+1, K)|^2$

$$= \frac{(2-\delta_{K,0})(J+K+1)(J-K+1)}{(J+1)^2[(2J+1)(2J+3)]} 3 \sum_{M=-J}^{J} (J+M+1)(J-M+1)$$

$$= \frac{(2-\delta_{K,0})(J+K+1)(J-K+1)}{(J+1)^2[(2J+1)(2J+3)]} [(2J+1)(J+1)(2J+3)]$$

$$= \frac{(2-\delta_{K,0})(J+K+1)(J-K+1)}{J+1}$$

(b) $|(J, K|\Phi_{Zz}|J-1, K)|^2$ (2)

$$= \frac{(2-\delta_{K,0})(J^2-K^2)}{J^2(4J^2-1)} 3 \sum_{-J}^{J} (J^2-M^2)$$

$$= \frac{(2-\delta_{K,0})(J^2-K^2)}{J^2(4J^2-1)} [J(2J+1)(2J-1)]$$

$$= \frac{(2-\delta_{K,0})(J^2-K^2)}{J}$$

(c) $|(J, K|\Phi_{Zz}|J, K)|^2$

$$= \frac{(2-\delta_{K,0})K^2}{J^2(J+1)^2} \left(3 \sum_{-J}^{J} M^2\right)$$

$$= \frac{(2-\delta_{K,0})K^2}{J^2(J+1)^2} [J(J+1)(2J+1)] = \frac{(2-\delta_{K,0})(2J+1)K^2}{J(J+1)},$$

in which J and K always refer to the lower state and δ is the Kronecker delta. It should be remembered that (2) does not correspond exactly to (1) because the values given in (2) contain the necessary factor for the K degeneracy.

For perpendicular bands the expressions for the line strengths are calculated in an analogous manner. Here it is necessary to take the sum of $\Phi_{Zy}{}^2$ and $\Phi_{Zx}{}^2$ to get the complete expression. The results, again including the K degeneracy factor, are:

$$(a)\quad |(J, K\,|\boldsymbol{\Phi}_{Zy}|\, J+1, K\pm 1)|^2 + |(J, K\,|\boldsymbol{\Phi}_{Zx}|\, J+1, K\pm 1)|^2 = \frac{(J\pm K+1)(J\pm K+2)}{J+1}$$

$$(b)\quad |(J, K\,|\boldsymbol{\Phi}_{Zy}|\, J, K\pm 1)|^2 + |(J, K\,|\boldsymbol{\Phi}_{Zx}|\, J, K\pm 1)|^2 = \frac{(2J+1)(J\mp K)(J\pm K+1)}{J(J+1)} \qquad (3)$$

$$(c)\quad |(J, K\,|\boldsymbol{\Phi}_{Zy}|\, J-1, K\pm 1)|^2 + |(J, K\,|\boldsymbol{\Phi}_{Zx}|\, J-1, K\pm 1)|^2 = \frac{(J\mp K)(J\mp K-1)}{J}$$

5g LINEAR AND SPHERICAL ROTOR LINE STRENGTHS

The line strengths for a linear molecule are readily deduced from the expressions for the symmetric rotor by replacing K by l. There are then three general types of bands for which the transitions can be distinguished. If $l = 0$, and $\Delta l = 0$, $(\Sigma - \Sigma)$ transition, the line strengths are given by 5f2 with $K = 0$, and the band is similar to that observed for a diatomic molecule since it is readily seen that $\Delta J = \pm 1$.

A second type band occurs when $\Delta l = 1$, $(\Sigma - \Pi, \Pi - \Delta$, etc., transition), in which case the line strengths are given by the appropriately substituted expressions of 5f3. In this instance, $\Delta J = 0, \pm 1$.

The third type band occurs when $\Delta l = 0$, $l \neq 0$ $(\Pi - \Pi, \Delta - \Delta$, etc., transition). The line strengths are given by 5f2, again $\Delta J = 0, \pm 1$.

The line strengths of the rigid spherical rotor can easily be deduced from 5f1. It would be necessary to sum over g and K as well as F and M. However, the rigid model breaks down for a spherical rotor as was indicated earlier. Recent work shows that even the first-order Coriolis interaction discussed in Chapter 4 is not sufficient for optically active fundamentals of spherical rotors if the moment of inertia is sufficiently small as for methane. For a treatment of this molecule the reader is referred to the work of Jahn [61].

For heavier spherical rotors such as $SiCl_4$, GeD_4, CBr_4, the higher order splitting probably cannot be observed, and the rigid line strengths should be a fairly good approximation. By carrying out the additional summations previously outlined, the following expressions are obtained:

$J \rightarrow J+1$	$(2J+1)(2J+3)$
$J \rightarrow J$	$(2J+1)^2$
$J \rightarrow J-1$	$4J^2 - 1$

5h ASYMMETRIC ROTOR LINE STRENGTHS

As pointed out in Chapter 2, the symmetric rotor basis functions employed to deduce the matrix elements of the direction cosines belong to the internal rotation group D_∞, whereas the asymmetric rotor wave functions belong to the Four group **V**. In order to calculate the asymmetric-rotor line strengths and to correlate them properly with the components of the degenerate pairs to which they converge in the limiting symmetric rotor cases, we must transform to a set of symmetric rotor basis functions, the Wang functions, which belong to the Four group. The transformation X, defined by 2m1 applied to the direction-cosine matrices of the symmetric rotor in the D_∞ representation, gives

$$\Phi_{Fg}{}^{S} = X'\Phi_{Fg}X. \tag{1}$$

These elements of the $\Phi_{Fg}{}^{S}$ yield the intensities of the limiting symmetric rotor transitions in an unusual form in that the line strengths of the two component transitions connecting two doubly degenerate pairs of energy levels are given in terms of a species classification of energy levels which applies over the entire range of asymmetry, including both the oblate and prolate limiting cases.

The asymmetric rotor wave functions may be expressed as a linear combination of Wang functions of the same symmetry. Hence the direction-cosine matrices for the asymmetric rotor, $\Phi_{Fg}{}^{A}$ may be calculated from the Φ_{Fg} given in Table 5e1 by the following series of operations:

$$\Phi_{Fg}{}^{A} = T_1'\Phi_{Fg}{}^{S}T_2 = T_1'X'\Phi_{Fg}XT_2, \tag{2}$$

where T_1 and T_2 are the transformation matrices for the lower and upper states, respectively. For the present, just as was done for the symmetric rotor, we shall consider only cases where the asymmetry is approximately the same in the two states so that $T_1 \sim T_2 \sim T$.

The transformation matrix $\mathbf{T}$ is diagonal with respect to J. For each J it is split into four submatrices, one for each of the four species of levels. The submatrices given by 2p7 may be calculated by the procedure described in detail in Chapter 2.

In the absence of external fields that remove the space degeneracy, X and T are both diagonal in J and M, and the factors $(J\,|\Phi_{Fg}|\,J')$ and $(J, M\,|\Phi_{Fg}|\,J', M)$ are invariant under transformation by XT. Thus in making numerical computations only the factor $(J, K\,|\Phi_{Fg}|\,J', K')$ need be transformed. The line strengths are calculated from an equation similar to 5f1.

$$\sum_{F,M,M'} \left|(J, \tau, M|\mathbf{\Phi}_{Fg}|\,J', \tau', M')\right|^2$$
$$= 3\left|(J\,|\mathbf{\Phi}_{Zg}|\,J')\right|^2 \cdot \left|(J, \tau\,|\mathbf{\Phi}_{Zg}|\,J'\tau')\right|^2 \cdot \sum_{MM'} \left|(J, M\,|\mathbf{\Phi}_{Zg}|\,J', M')\right|^2. \tag{3}$$

The amount of calculation is minimized by transforming $(J, K\,|\mathbf{\Phi}_{Zz}|\,J', K')$ and $(J, K\,|\mathbf{\Phi}_{Zy} + i\mathbf{\Phi}_{Zx}|\,J', K')$, the elements of $(J, \tau\,|\mathbf{\Phi}_{Zy}{}^{A}|\,J', \tau')$ and $(J, \tau\,|\mathbf{\Phi}_{Zx}{}^{A}|\,J', \tau')$ being obtained from the latter transformation by inspection.

The orthogonal properties of the direction-cosine matrices aid in calculating the line strengths since they result in several kinds of stability under unitary transformations [1]. The following sum rules may be used to detect any errors in calculation:

$$\sum_{J'=J-1}^{J} \sum_{\tau'} |(J, \tau | \Phi_{Zg}{}^{A} | J', \tau')|^2 = (2J)^2$$

$$\sum_{J'=J}^{J+1} \sum_{\tau'} |(J, \tau | \Phi_{Fg}{}^{A} | J', \tau')|^2 = [2(J+1)]^2$$

$$\sum_{F, J' \tau' M' M} |(J, \tau, M | \Phi_{Fg}{}^{A} | J', \tau', M')|^2 = 2J + 1 \tag{4}$$

$$\sum_{F, g, \tau', M', M} |(J, \tau, M | \Phi_{Fg}{}^{A} | J', \tau', M')|^2 = 2J' + 1$$

$$\sum_{F, \tau, \tau', M', M} |(J, \tau, M | \Phi_{Fg}{}^{A} | J', \tau', M')|^2 = \tfrac{1}{3}(2J+1)(2J'+1).$$

5i ASYMMETRIC ROTOR SELECTION RULES

Since the transformation matrices are diagonal with respect to J, the selection rules for J are the same for an asymmetric rotor as they are for the symmetric rotor, that is, $\Delta J = 0, +1, -1$ corresponding to Q, R, P-branches, respectively. The selection rules for K_{-1} and K_1 can be obtained from group theoretical considerations. The components of the change in electric moment along the molecular axes a, b, and c belong respectively to the representations B_a, B_b, and B_c of the Four group. The product of the characters of the representations of the initial and final wave functions and of the moment vector must be +1 for all group operators. As a result, if the representation of one of them is A, the other two must belong to the same representation. If no representation is A, then all representations must be different. Table 5i1 gives the permitted changes in representation for each component of the electric moment. These changes can be interpreted in terms of the symmetric rotor rules applied to each K index separately. A relaxation in the K selection rules removes the restriction

Table 5i1

Direction of the electric moment permitting transitions between states belonging to the representations of the Four group [36].

	REPRESENTATION	A *ee*	B_a *eo*	B_b *oo*	B_c *oe*
A	*ee*	—	*a*	*b*	*c*
B_a	*eo*	*a*	—	*c*	*b*
B_b	*oo*	*b*	*c*	—	*a*
B_c	*oe*	*c*	*b*	*a*	—

$\Delta K = 0, \pm 1$ in the parallel and perpendicular directions of the electric moment, respectively. ΔK is now restricted merely to even or odd changes. However, as is seen presently, the $\Delta K = 0, \pm 1$ transitions are still those of greatest significance for the asymmetric case.

Let **e** and **o** be operators representing even and odd changes in K. The combined operators **eo**, **oo**, and **oe**, which are to be applied to the double suffix, belong to the representations of the Four group, $eo(B_a)$, $oo(B_b)$, and $oe(B_c)$, respectively, and hence are directly related to the components of the electric moment a, b, c, which also belong to these respective representations. The fourth operator **ee** belongs to the representation **ee** (or A), but no branches of this type occur.

Table 5i2

Allowed changes in representation, labeled by the KK notation, for the three components of the electric moment, which show the selection rules in terms of parity changes in the K's [36].

INITIAL REPRESENTATION	FINAL REPRESENTATION FOR MOMENT PARALLEL TO		
	a (LEAST)	b (MIDDLE)	c (GREATEST)
ee	*eo*	*oo*	*oe*
eo	*ee*	*oe*	*oo*
oo	*oe*	*ee*	*eo*
oe	*oo*	*eo*	*ee*
Parity change is in	K_1	$K_{-1}K_1$	K_{-1}

The results for the asymmetric rotor may be summarized as follows and also in Table 5i2.

For a change in electric moment parallel to the axis of least moment of inertia a (A-type band), the parity of the K_{-1} index does not change.

For a change in electric moment parallel to the axis of greatest moment of inertia c (C-type band), the parity of the K_1 index does not change.

For a change in electric moment parallel to the axis of intermediate moment of inertia b (B-type band), the parity of both indices change. These rules are independent of J or ΔJ values.

In the general case, **e** can stand for $\Delta K = 0, \pm 2, \pm 4, \ldots$, and **o** can stand for $\Delta K = \pm 1, \pm 3, \ldots$. However, not all numerical combinations of ΔK_{-1} and ΔK_1 are possible because the sum $K_{-1} + K_1$ for any level is equal to J for even levels* and $J + 1$ for odd levels. The permitted values of $\Delta(K_{-1} + K_1) = \Delta K_{-1} + \Delta K_1$ are given in Table 5i3.

A convenient notation for the designation of branches within a band is needed. The Q, R, and P notation can be retained for $\Delta J = 0, +1, -1$. The change in K_{-1} and K can be denoted by q and r appended as subscripts; these will define the "sub-branches." As an example, consider $R_{0,1}$, which would indicate $\Delta J = +1, \Delta K_{-1} = 0$, $\Delta K_1 = +1$, or $P_{\bar{2},1}$ which would denote $\Delta J = -1, \Delta K_{-1} = -2, \Delta K_1 = 1$.

*For even levels $J + K_{-1} + K_1$ is even, whereas for odd levels this same quantity is odd.

Table 5i3—Permitted Changes in $\Delta(K_{-1} + K_1) = \Delta K_{-1} + \Delta K_1$

This table can also be used to find the parity of the levels from which the various transitions can arise. For example, if $\Delta J = 0$, $\Delta K_{-1} = 1$, and $\Delta K_1 = -1$, the sum $\Delta(K_{-1} + K_1)$ is 0, and the transitions can arise from even and odd levels to give both ${}^{b,e}Q_{1,\bar{1}}$ and ${}^{b,o}Q_{1,\bar{1}}$ sub-branches. If $\Delta J = -1$, $\Delta K_{-1} = 1$, $\Delta K_1 = -3$, then the sum is -2, and such transitions arise only from odd levels to give only ${}^{b,o}P_{1,\bar{3}}$.

INITIAL PARITY OF $J + K_{-1} + K_1$	P	Q	R
Even	$-1, 0$	$0, 1$	$1, 2$
Odd	$-1, -2$	$0, -1$	$1, 0$

In the identification of a sub-branch, the direction of the electric moment as determined by the two ΔK's is indicated as a superscript as is the parity of the initial level, that is, ${}^{a,e}R_{0,1}$.

The distribution of sub-branches in the matrix of line strengths among all states is shown in Fig. 5i1, where the nonzero elements have been indicated by inserting values

	0_{00} e	1_{01} e	1_{11} o	1_{10} e	2_{02} e	2_{12} o	2_{11} e	2_{21} o	2_{20} e	3_{03} e	3_{13} o	3_{12} e	3_{22} o	3_{21} e	3_{31} o	3_{30} e
$0_{00}e$		01	11*	10†												
$1_{01}e$	$0\bar{1}$	—	10†	$1\bar{1}*$	01	11*	10†	—	$2\bar{1}$							
$1_{11}o$	$\bar{1}\bar{1}*$	$\bar{1}0\dagger$	—	$0\bar{1}$	$\bar{1}1*$	01	—	10†	$1\bar{1}*$							
$1_{10}e$	$\bar{1}0\dagger$	$\bar{1}1*$	01	—	$\bar{1}2\dagger$	—	01	11*	10†							
$2_{02}e$		$0\bar{1}$	$1\bar{1}*$	$1\bar{2}\dagger$	—	10†	$1\bar{1}*$	$2\bar{1}$	—	01	11*	10†	—	$2\bar{1}$	$3\bar{1}*$	$3\bar{2}\dagger$
$2_{12}o$		$\bar{1}\bar{1}*$	$0\bar{1}$	—	$\bar{1}0\dagger$	—	$0\bar{1}$	$1\bar{1}*$	$1\bar{2}\dagger$	$\bar{1}1*$	01	—	10†	$1\bar{1}*$	$2\bar{1}$	—
$2_{11}e$		$\bar{1}0\dagger$	—	$0\bar{1}$	$\bar{1}1*$	01	—	10†	$1\bar{1}*$	$\bar{1}2\dagger$	—	01	11*	10†	—	$2\bar{1}$
$2_{21}o$		—	$\bar{1}0\dagger$	$\bar{1}\bar{1}*$	$\bar{2}1$	$\bar{1}1*$	$\bar{1}0\dagger$	—	$0\bar{1}$	—	$\bar{1}2\dagger$	$\bar{1}1*$	01	—	10†	$1\bar{1}*$
$2_{20}e$		$\bar{2}1$	$\bar{1}1*$	$\bar{1}0\dagger$	—	$\bar{1}2\dagger$	$\bar{1}1*$	01	—	$\bar{2}3$	$\bar{1}3*$	$\bar{1}2\dagger$	—	01	11*	10†
$3_{03}e$					$0\bar{1}$	$1\bar{1}*$	$1\bar{2}\dagger$	—	$2\bar{3}$	—	10†	$1\bar{1}*$	$2\bar{1}$	—	$3\bar{2}\dagger$	$3\bar{3}*$
$3_{13}o$					$\bar{1}\bar{1}*$	$0\bar{1}$	—	$1\bar{2}\dagger$	$1\bar{3}*$	$\bar{1}0\dagger$	—	$0\bar{1}$	$1\bar{1}*$	$1\bar{2}\dagger$	—	$2\bar{3}$
$3_{12}e$					$\bar{1}0\dagger$	—	$0\bar{1}$	$1\bar{1}*$	$1\bar{2}\dagger$	$\bar{1}1*$	01	—	10†	$1\bar{1}*$	$2\bar{1}$	—
$3_{22}o$					—	$\bar{1}0\dagger$	$\bar{1}\bar{1}*$	$0\bar{1}$	—	$\bar{2}1$	$\bar{1}1*$	$\bar{1}0\dagger$	—	$0\bar{1}$	$1\bar{1}*$	$1\bar{2}\dagger$
$3_{21}e$					$\bar{2}1$	$\bar{1}1*$	$\bar{1}0\dagger$	—	$0\bar{1}$	—	$\bar{1}2\dagger$	$\bar{1}1*$	01	—	10†	$1\bar{1}*$
$3_{31}o$					$\bar{3}1*$	$\bar{2}1$	—	$\bar{1}0\dagger$	$\bar{1}\bar{1}*$	$\bar{3}2\dagger$	—	$\bar{2}1$	$\bar{1}1*$	$\bar{1}0\dagger$	—	$0\bar{1}$
$3_{30}e$					$\bar{3}2\dagger$	—	$\bar{2}1$	$\bar{1}1*$	$\bar{1}0\dagger$	$\bar{3}3*$	$\bar{2}3$	—	$\bar{1}2\dagger$	$\bar{1}1*$	01	—

Fig. 5i1 Distribution of sub-branches in the matrix of the line strengths [36]. Plain numerals are used for branches appearing with a component along *a*, an asterisk the component along *b*, and a dagger the component along *c*.

of ΔK_{-1} and ΔK_1. Plain numerals are used for branches appearing with a component along a, an asterisk for the component along b, and a dagger for the component along c. It can be seen that the transitions of any sub-branch appear only along diagonals of the blocks. Half the sub-branches arise from either all odd levels or all even levels. Examination of the numerical values of the others, which arise from both odd and even levels, shows that the strengths alternate in value. A smooth trend of numbers is obtained by separating the latter into e and o parts (arising from even and odd levels, respectively), each of which is called a sub-branch, for example, ${}^{b,e}Q_{\bar{1},1}$ and ${}^{b,o}Q_{\bar{1},1}$. The e and o parts then have the same number of transitions as the other sub-branches. The values of the line strengths decrease away from the principal diagonal as ΔK_{-1}, ΔK_1 increase; that is, as the departure from symmetric rotor selection rules increases.

The selection rules for the molecule with internal rotation are also determined through the procedure just outlined. Consider the case with $s = 3$, where s has been defined in 4n, and the electric moment change along the z-axis. Here M_z would have the symmetry B_z. Since the product of the characters of the representations of the initial and final wave functions and the electric moment vector must be $+1$ for all group operators, the selection rules are $A \leftrightarrow B_z$, $B_x \leftrightarrow B_y$, $E_1 \leftrightarrow E_1$, and $E_2 \leftrightarrow E_2$. If the electric moment vector has the symmetry B_x, the selection rules become $A \leftrightarrow B_x$, $B_y \leftrightarrow B_z$, and $E_1 \leftrightarrow E_2$. If $s = 6$, the same selection rules apply, with the additional condition that the e species go only to e species and the o species go only into o species must be imposed.

This method is quite general and may be used to determine the selection rules of any type molecules as well as the nonvanishing matrix elements of other operators than the electric moment operator. One needs know only the symmetry properties of the initial and final states and of the operator.

5j THE STRUCTURE OF THE SPECTRUM

The principal lines in the spectrum of the asymmetric rotor are those which occur in the symmetric rotor, if the doubly degenerate levels are first resolved into the Wang components which retain their identity as the rotor becomes asymmetric. The resolution of this degeneracy which splits the symmetric rotor energy levels into those of the asymmetric rotor is also responsible for a splitting of the p, q, and r sub-branches of the oblate or prolate rotor spectra into the sub-branches of the asymmetric rotor spectra.

The principal sub-branches of the asymmetric rotor are those for which both K's change by 0 or ± 1, and thus are allowed in both the prolate and oblate symmetric limits. These sub-branches are listed first in the table of line strengths found in Appendix IX and are indicated by heavy lines in Fig. 5j1.

The next most important sub-branches are those in which one of the K's changes by 0 or ± 1, whereas the absolute change in the other K is greater than one. These correspond to allowed transitions in one symmetric limit and to forbidden transitions of zero intensity in the other symmetric limit. These sub-branches, such as $P_{1\bar{3}}$, are listed second in Appendix IX and are shown by light lines in Fig. 5j1. It should be noted that in the Wang resolution, transitions of the symmetric rotor can have changes of

K greater than one in absolute value, although *K* is no longer a true quantum number.

Finally, some sub-branches of the asymmetric rotor which are forbidden in both symmetric limits are listed in the Appendix. These transitions not only have very low intensities even for the most asymmetric rotor but do not occur below certain values of *J*. Even with a favorable Boltzmann factor these sub-branches can normally

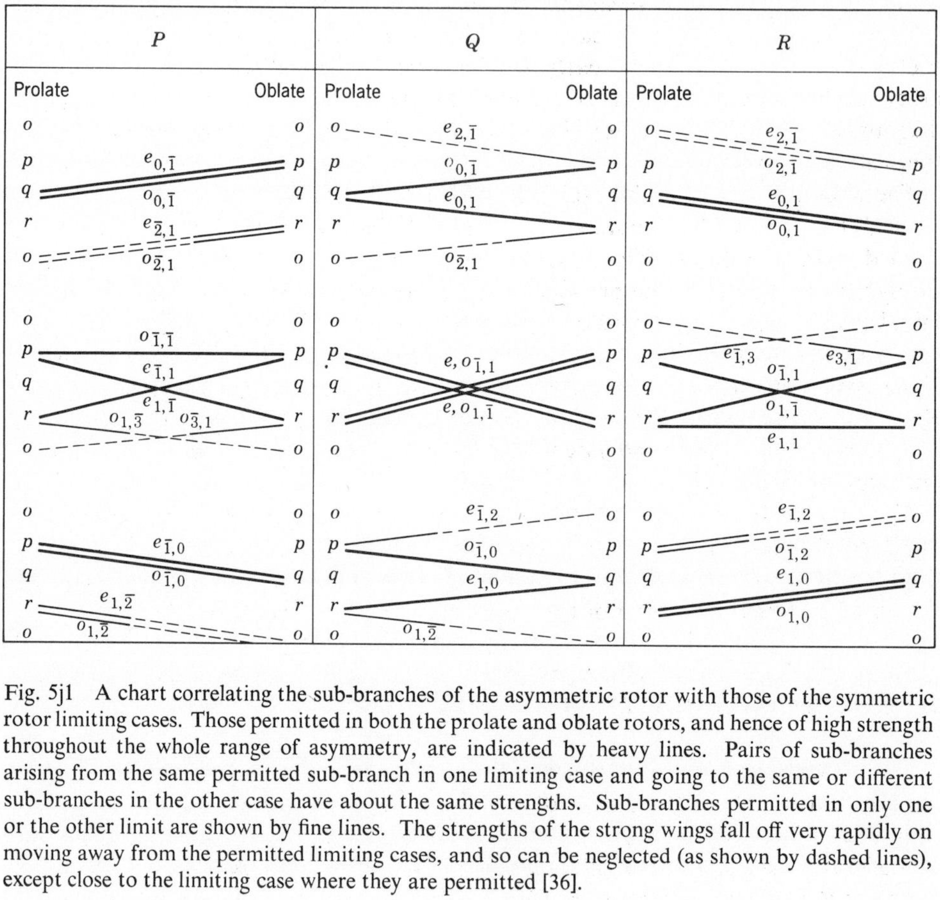

Fig. 5j1 A chart correlating the sub-branches of the asymmetric rotor with those of the symmetric rotor limiting cases. Those permitted in both the prolate and oblate rotors, and hence of high strength throughout the whole range of asymmetry, are indicated by heavy lines. Pairs of sub-branches arising from the same permitted sub-branch in one limiting case and going to the same or different sub-branches in the other case have about the same strengths. Sub-branches permitted in only one or the other limit are shown by fine lines. The strengths of the strong wings fall off very rapidly on moving away from the permitted limiting cases, and so can be neglected (as shown by dashed lines), except close to the limiting case where they are permitted [36].

be disregarded. Only the "first-order" forbidden sub-branches ($|\Delta\tau| = 5, 6, 7, 8$) are listed. These transitions appear first for $J \geqslant 3$.

5k CALCULATION OF RELATIVE INTENSITIES FROM LINE STRENGTHS

The intensity of a spectral line may, in principle, be evaluated by the application of the quantum theory of transition probabilities. The intensity of a transition $n'' \rightarrow n$ is

$$I_{n'',n'} = \frac{8\pi^3 \nu N g_{n''}(1 - e^{-h\nu/kT}) \exp(-E_{n''}/kT)}{3hc \sum_n g_{n''} \exp(-E_n/kT)} |(n''|\mathbf{M}|n')|^2, \qquad (1)$$

in which n stands for all the quantum numbers describing the state, $E_{n''}$ the energy of the lower state, $g_{n''}$ is its weight factor. N is the number of molecules per cubic centimeter, and ν is the frequency of the absorption line. The last term is, for present purposes, given by 5c7 squared and then summed over all F.

The denominator plus the constants in the numerator is a constant for any given molecule. Since the frequency from one end of a vibrational-rotational band to the other does not vary appreciably, $(1 - e^{-h\nu/kT})$ and ν are essentially constant for a given band and are so treated.

The $g_{n''}$ is the multiplicity of the lower state and includes only the nuclear spin degeneracies and degeneracies not arising from K or M. Exponential $(-E_{n''}/kT)$ varies from energy level to energy level and must be considered along with the nuclear spin statistics in computing relative intensities. As pointed out earlier the purely vibrational part of $(V', R' \, |\mathbf{M}| \, V'', R'')$ is not important in calculating relative intensities as long as it is nonvanishing, and the mere appearance of the band assures that this is true. The other part of the last term, that is, of $|(n'' \, |\mathbf{M}| \, n')|^2$, is the line strength, whose expressions were deduced earlier in the chapter. Thus the relative intensities in a vibrational-rotational band can be calculated to a high degree of approximation by combining the line strengths with the Boltzmann factor exp $(-E_{n''}/kT)$ and the appropriate nuclear spin statistics.

5*l* INTENSITY PERTURBATIONS

The perturbations of intensities have not received nearly as much attention as the corresponding effects on the energy levels. One might expect that the intensity effects might be more pronounced than the same effects on the energy since it is the square of the electric moment matrix elements which enter into intensity calculations. Intensity perturbations on rotational-vibrational transitions of polyatomic molecules have been examined by a few workers [10, 42, 85]. The general results of their work are discussed here and the reader is referred to the original papers for the details.

Nielsen [85] has considered the effect of vibrational-rotational interaction on two slightly anisotropic oscillators in a plane. In essence this is a perturbation similar to the Coriolis interaction for degenerate modes of a symmetric rotor. The results of his treatment show that the P-branch of the higher frequency band and the R-branch of the lower frequency band are enhanced at the expense of the other two branches.

Allen [10] has considered the problem when $T_1 \neq T_2$ in 5h2. There, a very definite and systematic effect on the sub-bands of the asymmetric rotor is found, although for normal infrared transitions the effect is small and can be ignored. However, in the spectra of molecular fragments there is the possibility of κ changing by as much as 0.5 between the ground and excited states; these effects then become important and have been observed in the spectrum of NH_2 [96]. A summary of the results of the treatment is given in Table 5*l*1.

Emerson and Eggers [42] have considered the effect of centrifugal distortion on the quantity $(V_1 \, |\mathbf{M}_g| \, V_2)$ which enters the expression for the intensity and is defined in the usual way in 5c9. Where the physical dimensions of the molecule change due to a centrifugal force acting on the molecule, the distortion of the molecule is dependent on the rotational energy and the way it is distributed about the three principal inertial axes. The changes in physical dimensions due to centrifugal distortion can be calculated classically [35] and have the effect of changing the reference origin of the normal coordinates of the vibrations. The normal coordinates of the distorted molecule can now be written as

$$Q_i^{R'} = Q_i^{R''} + \delta_i^{R'',R'}, \tag{1}$$

where $\delta_i^{R'',R'}$ is the change in Q_i due to the difference in the centrifugal deformation of the molecule in the R'' and R' rotational states. It should be noted that $\delta_i^{R'',R}$ has

Table 5II—Summary of Sub-bands Whose Intensity Is Enhanced

κ_g = ground state, κ_e = excited state

	CASE I $\kappa_g < \kappa_e$				CASE II $\kappa_e < \kappa_g$			
Stronger Sub-bands	$b, e_{R_{1,1}}$	$b, e_{P_{\bar{1},1}}$	$b, e_{Q_{1,\bar{1}}}$	$b, o_{Q_{1,\bar{1}}}$	$b, o_{R_{1,\bar{1}}}$	$b, o_{P_{\bar{1},1}}$	$b, e_{Q_{\bar{1},1}}$	$b, o_{Q_{\bar{1},1}}$
	$a, e_{P_{0,\bar{1}}}$	$a, o_{P_{0,\bar{1}}}$	$a, e_{Q_{0,\bar{1}}}$	$a, o_{Q_{\bar{2},1}}$	$a, e_{R_{0,\bar{1}}}$	$a, o_{R_{0,1}}$	$a, o_{Q_{0,\bar{1}}}$	$a, e_{Q_{2,\bar{1}}}$
	$c, o_{Q_{\bar{1},0}}$	$c, e_{Q_{\bar{1},2}}$	$c, e_{R_{1,0}}$	$c, o_{R_{1,0}}$	$c, e_{Q_{1,0}}$	$c, o_{Q_{1,\bar{2}}}$	$c, o_{P_{\bar{1},0}}$	$c, o_{P_{\bar{1},0}}$

the same magnitude for a pair of conjugate transitions but opposite signs. A pair of transitions a and b are conjugate when their rotational states are related as follows: $J_a'' = J_b'$; $J_a' = J_b''$; $\tau_a'' = \tau_b'$; and $\tau_a' = \tau_b''$. This correction has been applied to the symmetric fundamentals ν_1 and ν_2 of H_2S. For this molecule

$$(V_1 \,|M_b|\, V_1 + 1) = (V_1 + 1)^{1/2} \left[C_1 (M_0)_b \delta_1 + D_1 \frac{\partial M_b}{\partial Q_1} \right] \tag{2}$$

when all terms containing powers of δ greater than one are neglected. From (2) it is seen that the line of the conjugate pair which is enhanced depends on the relative signs of M_0 and $\partial M / \partial Q$. The result of this treatment shows that since δ_1 has a specific value for each allowed transition, the quantity $(V_1 \,|\mathbf{M}_b|\, V_1 + 1)$ must be evaluated separately for each rotational-vibrational transition.

6

ANALYSIS OF VIBRATIONAL-ROTATIONAL BANDS OF LINEAR MOLECULES

6a Σ–Σ TYPE BANDS

With the expressions for the energy levels, line strengths, and the selection rules it is now possible to proceed with the analysis of observed spectra. The rotational term values of the most general type of linear molecule are given by 4c4.

$$F(J) = B_v[J(J+1) - l^2] - D_v[J(J+1) - l^2]^2.$$

Σ states are those for which $l = 0$; hence a Σ–Σ type transition is characterized by $l = 0$, $\Delta l = 0$. Under these conditions 4c4 can be reduced so that the rotational part is given by 2i1, and the selection rules are $\Delta J = \pm 1$. The energy levels and allowed transitions are shown schematically on the left side of Fig. 6a1.

The frequencies of the rotational-vibrational transitions are given by the expression

$$\nu_m = \nu_0 + (B' + B'')m + (B' - B'' - D' + D'')m^2 - 2(D' + D'')m^3 - (D' - D'')m^4, \quad (1)$$

where m is a running index with $m = -J$ for $\Delta J = -1$ (P-branch) and $m = J + 1$ for $\Delta J = 1$ (R-branch). The J's refer always to the lower state, the primed constants to the excited state, and the double primed constants to the lower state. ν_0 is the band center or frequency of the vibrational transition. If $D' \approx D''$, an approximation that is justified in many cases, then (1) reduces to a cubic equation; that is,

$$\nu_m = \nu_0 + (B' + B'')m + (B' - B'')m^2 - 2(D' + D'')m^3. \quad (2)$$

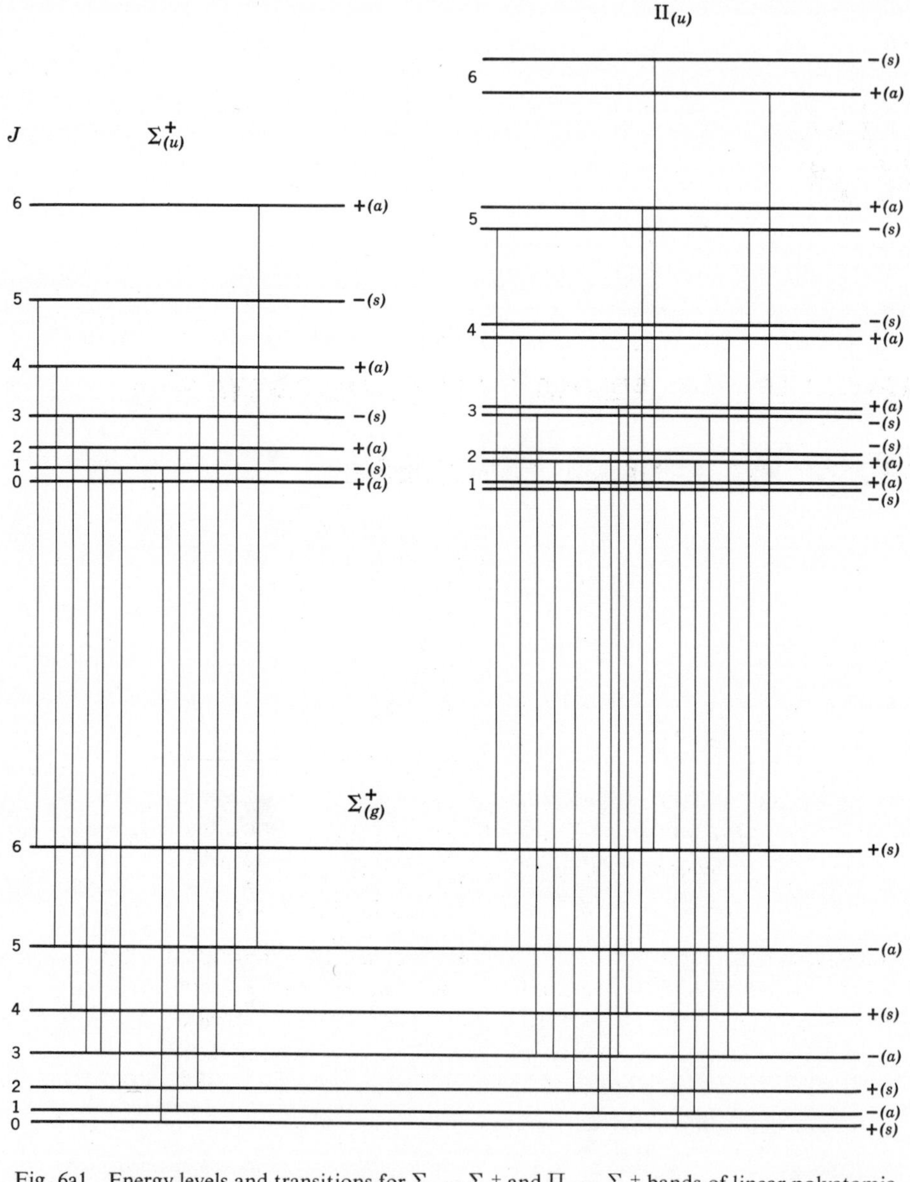

Fig. 6a1 Energy levels and transitions for $\Sigma_u - \Sigma_g^+$ and $\Pi_u - \Sigma_g^+$ bands of linear polyatomic molecules of symmetry $D_{\infty h}$. It is assumed that B' and B'' are equal. ($D' = D'' = 0$). The plus and minus signs refer to the symmetry with respect to inversion. The s and a indicate symmetric or antisymmetric, respectively, on exchange of all the symmetrically equivalent nuclei. If the molecular symmetry is $C_{\infty v}$ instead of $D_{\infty h}$, the g and u property does not exist, nor does the s and a symmetry apply.

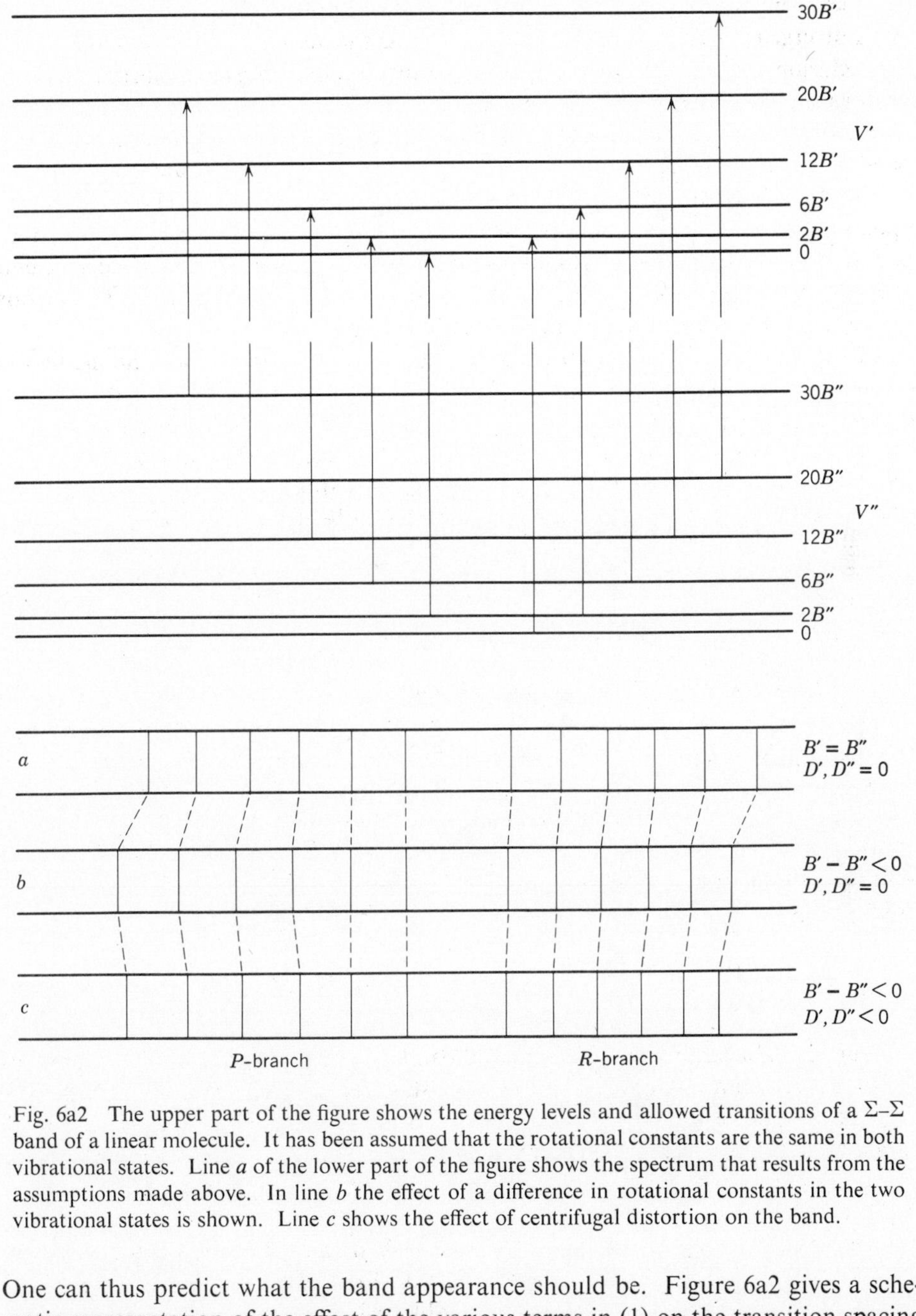

Fig. 6a2 The upper part of the figure shows the energy levels and allowed transitions of a Σ–Σ band of a linear molecule. It has been assumed that the rotational constants are the same in both vibrational states. Line *a* of the lower part of the figure shows the spectrum that results from the assumptions made above. In line *b* the effect of a difference in rotational constants in the two vibrational states is shown. Line *c* shows the effect of centrifugal distortion on the band.

One can thus predict what the band appearance should be. Figure 6a2 gives a schematic representation of the effect of the various terms in (1) on the transition spacing of the band. If $B' = B''$, that is, there is no change of moment of inertia with vibrational state, the rotational energy levels would be spaced in the two vibrational states as shown in the upper part of Fig. 6a2. The allowed transitions are also shown in this part of the figure. The spacing of the lines in the band would be that shown in line *a* of the figure. Now suppose $B' - B''$ is not zero. Since $B'' > B'$ in most cases, this coefficient is negative and leads to a convergence of the R-branch as J increases, and a divergence of the P-branch as J increases as shown in line b of the figure. With the inclusion of the centrifugal distortion terms one finds a general shrinkage of the band as shown in line c.

One way to analyze such a band is to look for the gap that locates the center of the band, assign ground state J values beginning with $J = 0$ in the R-branch and $J = 1$ in the P-branch, and set up a series of equations of the type (1) or (2), the choice depending upon the precision of the data. The set of equations we obtain in this way may then be solved by the method of least squares for the best estimates of ν_0 and the coefficients of m, m^2, m^3, and m^4. The values of the inertial constants and centrifugal distortion constants in the two states may then be deduced.

An alternate, and sometimes preferable, method of analysis is called the method of combinations and differences. If we locate a transition in the P-branch and the transition in the R-branch with the same value of J in the excited state, the energy difference between these two transitions is independent of the excited state and represents the energy difference between the two ground state levels appearing in the two transitions. The expression for this difference, $\Delta F_2''$, can readily be deduced to be

$$\Delta F_2'' = R(J-1) - P(J+1) = (4B'' - 6D'')(J+\tfrac{1}{2}) - 8D''(J+\tfrac{1}{2})^3. \quad (3a)$$

Similarly,

$$\Delta F_2' = R(J) - P(J) = (4B' - 6D')(J+\tfrac{1}{2}) - 8D'(J+\tfrac{1}{2})^3. \quad (3b)$$

A set of equations (3a) for the ground state and a set (3b) for the excited state can be set up. These two sets of equations can be solved by the method of least squares, and we obtain directly the inertial constants and centrifugal distortion constants.

Since all bands originating in the ground vibrational state should have the same $\Delta F_2''$ values, if one observes several bands originating in this state, the observed $\Delta F_2''$ values can be averaged before setting up the equations (3a). Such a procedure gives the most reliable values of the ground state constants. The band center may be determined from a set of equations

$$\frac{R(J-1) + P(J)}{2} = \nu_0 + (B' - B'' - D' + D'')\,J^2 - (D' - D'')J^4$$

Figure 6a3 is a reproduction of an HCN band [18] near 6519 cm^{-1}. This represents the transition from the ground state to the state $(v_1, v_2, v_3) = (0, 0, 2)$, a Σ–Σ type transition. There is another overlapping band near 6480 cm^{-1} that is discussed in 6d. It will be noted that there is the predicted gap at the center, the first line on the high frequency side of the gap being $R(0)$ and the first line on the low frequency side of the gap being $P(1)$. Here there is no question about the assignment of rotational quantum numbers. The frequencies of the lines in the Σ–Σ transition are listed together with the assignment of rotational quantum numbers in Table 6a1. The ΔF_2 values for both the ground and excited states are listed in Table 6a2. The constants obtained from a least squares analysis of the data using (3a), (3b), and (4) are listed in Table 6a3.

An improved value of $B_0 = B''$ and $D_0 = D''$ can be obtained by taking the average of the $\Delta F_2''$ values obtained from the analysis of all the bands originating in the ground vibrational state. Table 6a4 is a compilation of these $\Delta F_2''$ values obtained from several different HCN bands. The agreement between the $\Delta F_2''$ values from band to band

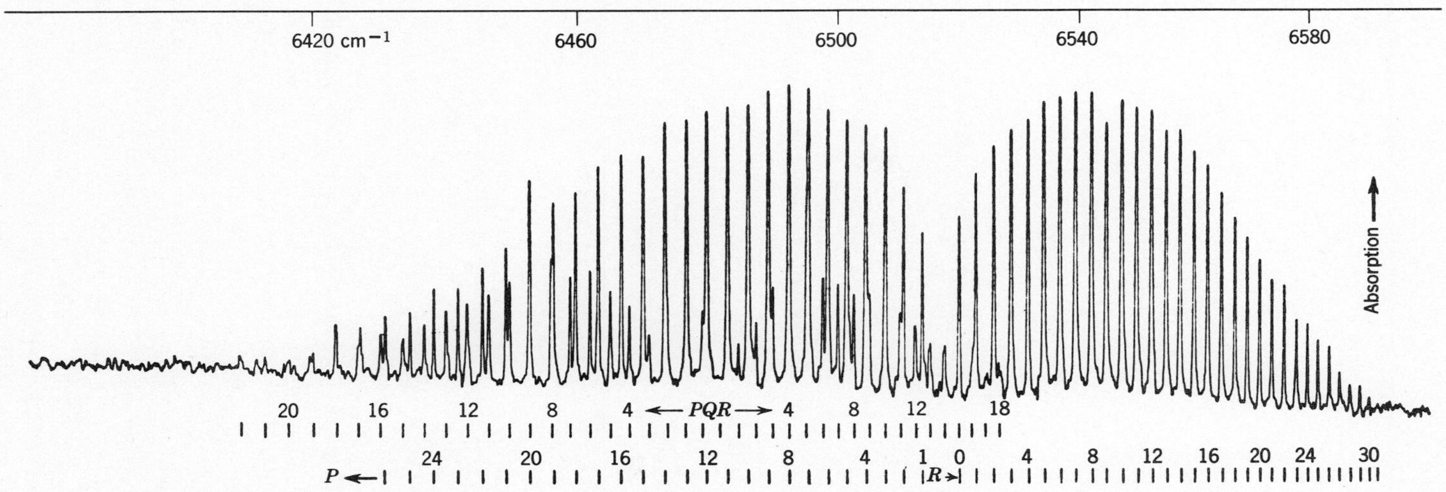

Fig. 6a3 The absorption of HCN near 6519 cm^{-1}. The main band, $2\nu_3$, is a Σ–Σ transition, whereas the hot band is the Π–Π transition $\nu_2^1 + 2\nu_3 - \nu_2^1$. It is to be noted that the weak Q-branch of the Π–Π band shows up clearly [18].

Table 6a1—Frequencies and Assignments for HCN Bands in Fig. 1 [18]

	$2v_3$ 6519.60 cm^{-1}		$2v_3 + v_2^1 - v_2^1$ 6480.84 cm^{-1}	
m	R	P	R	P
1	6522.53	6516.66		
2	6225.40	6513.61	6486.60	
3	6528.23	6510.61	6489.26	6471.60
4	6531.03	6507.55		6468.62
5	6533.75	6504.39		6465.52
6	6536.48	6501.26		6462.38
7	6539.12	6498.02	6500.57	6459.20
8	6541.72	6494.78	6503.01	
9	6544.31	6491.48	6505.51	
10	6546.82	6488.15		6449.20
11	6549.30	6484.79		6445.96
12	6551.75	6481.34		6442.53
13	6554.15	6477.88	6515.48	6439.05
14	6556.50	6474.44	6517.89	6435.57
15	6558.82	6470.88	6520.17	6432.08
16	6561.04	6467.28		6428.52
17	6563.33	6463.66		
18	6565.52	6459.98	6527.02	
19	6567.64	6456.23		6417.31
20	6569.76	6452.56		6413.68
21	6571.84	6448.80		6409.94
22	6573.86	6444.93		6406.16
23	6575.77	6441.05		6402.29
24	6577.70	6437.14		6398.36
25	6579.61	6433.17		6394.38
26	6581.44	6429.12		6390.47
27	6583.25	6425.05		
28	6584.99	6421.12		
29	6586.67	6417.11		
30	6588.35	6412.87		
31	6589.98	6408.65		
32	6591.56	6404.41		
33	6593.12	6400.09		
34	6594.60	6395.90		
35	6596.07	6391.53		
36	6597.48			

should be noted. The value of B_0 obtained by this process is also given in Table 6a3 for comparison purposes.

Another interesting comparison can be made between the constants obtained by the method of combination-differences and those obtained from using (1). In principle, the results from the two methods should be the same. The results given in Table 6a3 indicate this is essentially true. The difference $D' - D''$ is so small that unless

Table 6a2—ΔF_2 Values of 6519 cm^{-1} Band of HCN

J	$\Delta F_2''$	$\Delta F_2'$
1	8.92	8.74
2	14.79	14.62
3	20.68	20.42
4	26.64	26.20
5	32.49	32.09
6	38.46	37.86
7	44.34	43.70
8	50.24	49.53
9	56.16	55.34
10	62.03	61.15
11	67.96	66.96
12	73.87	72.81
13	79.71	78.62
14	85.62	84.38
15	91.54	90.16
16	97.38	96.05
17	103.35	101.86
18	109.29	107.66
19	115.08	113.53
20	120.96	119.28
21	126.91	125.06
22	132.81	130.84
23	138.63	136.65
24	144.53	142.47
25	150.49	148.27
26	156.39	154.13
27	162.13	159.94
28	167.88	165.55
29	173.80	171.34
30	179.70	177.11
31	185.57	182.91
32	191.47	188.71
33	197.27	194.51
34	203.07	200.17
35		205.95

Table 6a3—Rotational Constants for 6520 cm^{-1} HCN Band

	ΔF_2 METHOD (cm^{-1})	SERIES EXPANSION (3) (cm^{-1})
B'	1.47748	1.47746
B''	1.47791	1.47790
D'	2.80×10^{-6}	2.7×10^{-6}
D''	2.59×10^{-6}	2.7×10^{-6}
ν_0	6519.596	6519.600
Average $B_0 = 1.47791$		

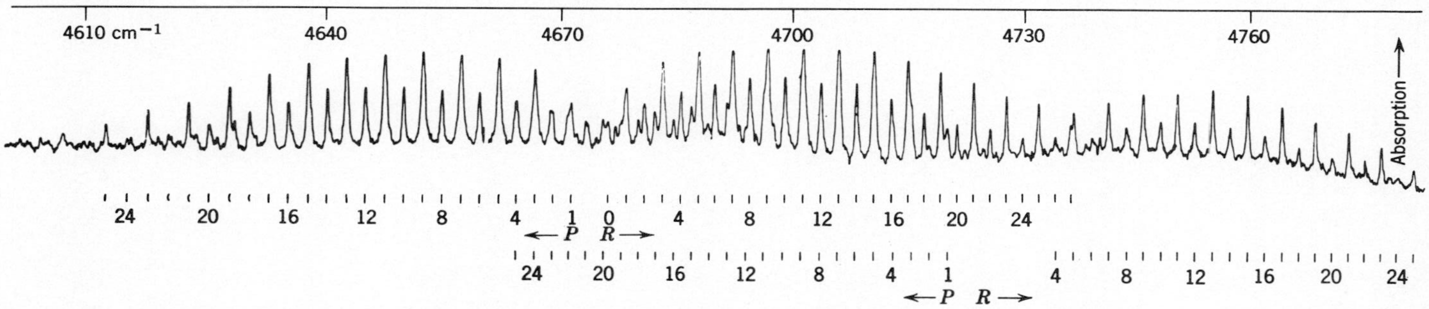

Fig. 6a4 Acetylene absorption bands at 4673.38 and 4722.72 cm^{-1}. The cell length was 10 meters and the gas pressure was 1 atmosphere [20].

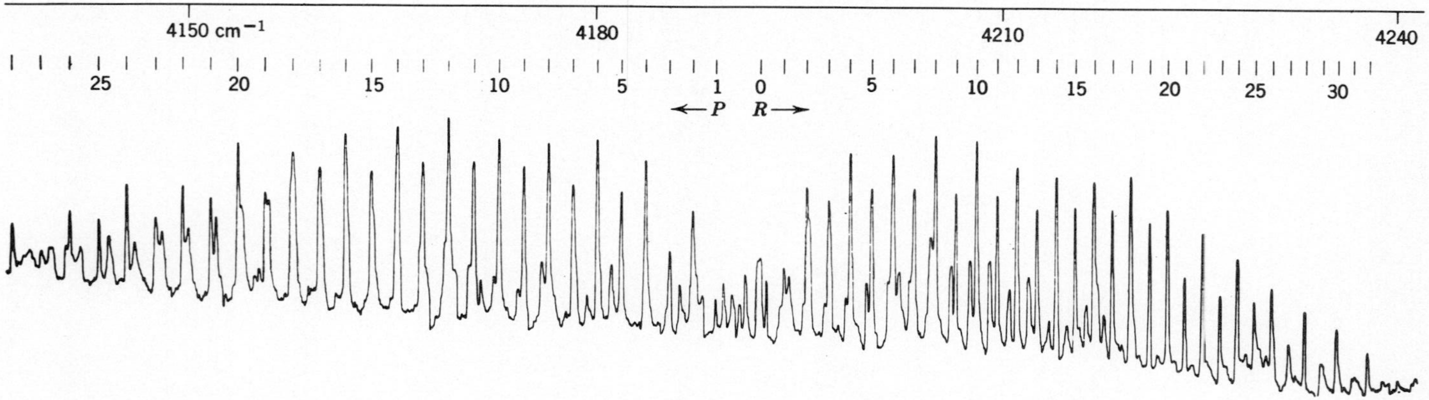

Fig. 6a5 Combination band $\nu_2 + \nu_3$ of C_2D_2. The pressure was 11 cm (Hg), and the cell length was 6 meters. Nonlabeled peaks belong to $\nu_2 + \nu_3 + \nu_4{}^1 - \nu_4{}^1$ at 4181.22 cm^{-1} and $\nu_2 + \nu_3 + \nu_5{}^1 - \nu_5{}^1$ at about 4187.0 cm^{-1} [17].

transitions of very high J are observed we might just as well use (2) for the determination of the constants.

If the linear molecule belongs to the point group $D_{\infty h}$ rather than $C_{\infty v}$, there is additional information contained in the spectrum that will help with the analysis, because the nuclear spin of the equivalent nuclei, if it is nonzero, will give rise to an alternation in intensity in the rotational fine structure. Figure 6a4 is a reproduction of an acetylene (C_2H_2) band [20], where the alternation of intensity is clearly discernable. Here protons obey Fermi statistics, hence the antisymmetric rotational levels of the ground state are three times as populous as the symmetric levels. Thus, it

Table 6a4—Comparison of $\Delta F_2''$ Values from Several HCN Bands

ν_0	4004	5393	6519	11674
1	8.89	8.91	8.92	8.87
2	14.78	14.78	14.79	14.79
3	20.71	20.70	20.68	20.72
4	26.61	26.60	26.64	26.62
5	32.48	32.47	32.49	32.54
6	38.45	38.42	38.46	38.45
7	44.33	44.31	44.34	44.38
8	50.25	50.23	50.24	50.27
9	56.13	56.12	56.16	56.16
10	62.06	62.06	62.03	62.07
11	67.96	67.98	67.96	67.98
12	73.85	73.85	73.87	73.84
13	79.75	79.76	79.71	79.74
14	85.68	85.67	85.62	85.65
15	91.54	91.57	91.54	91.56
16	97.35	97.43	97.38	97.45
17	103.31	103.16	103.35	103.34
18		109.24	109.29	109.21
19	115.11	115.12	115.08	115.10
20	121.00		120.96	121.00

is readily seen that the transitions originating in states with J odd will be three times as intense as those originating in states with J even The intensity alternation can be of considerable help in assigning quantum numbers. Figure 6a5 [17] shows a similar case for a band of C_2D_2. Here since deuterons obey Bose statistics and D has a spin of 1, the transitions originating in states with J even are twice as intense as those originating in states with J odd. If the symmetry of acetylene is destroyed as for C_2HD there will be no intensity alternation as can be seen in Fig. 6a6 [19]. It was the absence, of an intensity alternation which showed nitrous oxide to be NNO not NON [92].

6b Π–Σ BANDS

In a $\Pi - \Sigma$ band $l = 0$ in the ground state and $l = \pm 1$ in the excited state; that is, $l = 0$, $\Delta l = \pm 1$. Such a transition would arise from the bending fundamental of a linear triatomic molecule. The excited state term values are given by 4c4 with $|l| = 1$, whereas the ground state term values would have $l = 0$. The rotational selection rules for this type of band are $\Delta J = 0, \pm 1$, $J = 0 \not\leftrightarrow J = 0$, and $+ \leftrightarrow -$, $s \not\leftrightarrow a$. The energy levels and allowed transitions for a Π–Σ band are shown schematically on the right side of Fig. 6a1. It should be noted that for $\Delta J = \pm 1$, the transition from the ground state is always to the lower component of the one-doublet, whereas for $\Delta J = 0$ the transition is to the upper component. Thus the rotational constants for the excited state determined from the *P*- and *R*-branches differs from those determined

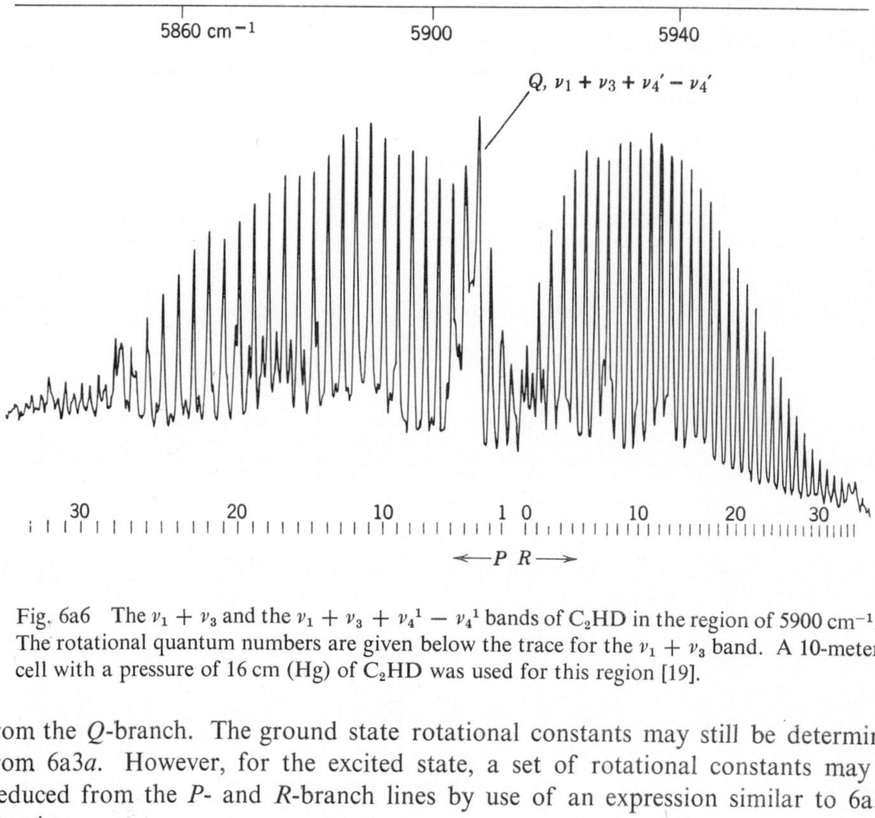

Fig. 6a6 The $\nu_1 + \nu_3$ and the $\nu_1 + \nu_3 + \nu_4{}^1 - \nu_4{}^1$ bands of C_2HD in the region of 5900 cm^{-1}. The rotational quantum numbers are given below the trace for the $\nu_1 + \nu_3$ band. A 10-meter cell with a pressure of 16 cm (Hg) of C_2HD was used for this region [19].

from the *Q*-branch. The ground state rotational constants may still be determined from 6a3*a*. However, for the excited state, a set of rotational constants may be deduced from the *P*- and *R*-branch lines by use of an expression similar to 6a3*a*; that is,

$$\Delta F_2^- = (4B^- + 2D^-)(J + \tfrac{1}{2}) - 8D^-(J + \tfrac{1}{2})^3. \tag{1}$$

A second set of constants for the excited state may be determined from the *Q*-branch lines by

$$\nu_{Q_J} = \nu_0 - (B^+ + D^+) + (B^+ - B'' + 2D^+)J(J+1) - (D^+ - D'')J^2(J+1)^2. \tag{2}$$

For a Π state 4*q* defined in 4c6 is merely the difference $B^+ - B^-$.

In only a few cases has the *Q*-branch been resolved sufficiently to enable a determination of the constants in the upper *l* state. Since $B^+ - B''$ is generally quite small, the *Q*-branch transitions tend to gather in a sharp peak. However, the *Q*-branch transitions have been resolved for the HCN bands [97] with excited states

$(V_2, V_2^l, V_3) = (0, 1^1, 1)$ at 4004 cm^{-1} and $(1, 1^1, 1)$ at 6083 cm^{-1} as well as the $(1, 1^1, 1)$ band of N_2O at 4062 cm^{-1} and the 4092 band of C_2H_2 [119]. Several bands of C_2H_2 in the photographic infrared have been observed under conditions which resolved the Q-branch transitions [45, 46, 47, 81].

Figure 6b1 is a trace of a Π–Σ band of HCN [18] near 2800 cm^{-1}. Here one can start the numbering of the transitions on the high frequency side of the collected Q-branch with $R(0)$. On the low frequency side of this band the first line is $P(2)$ not $P(1)$.

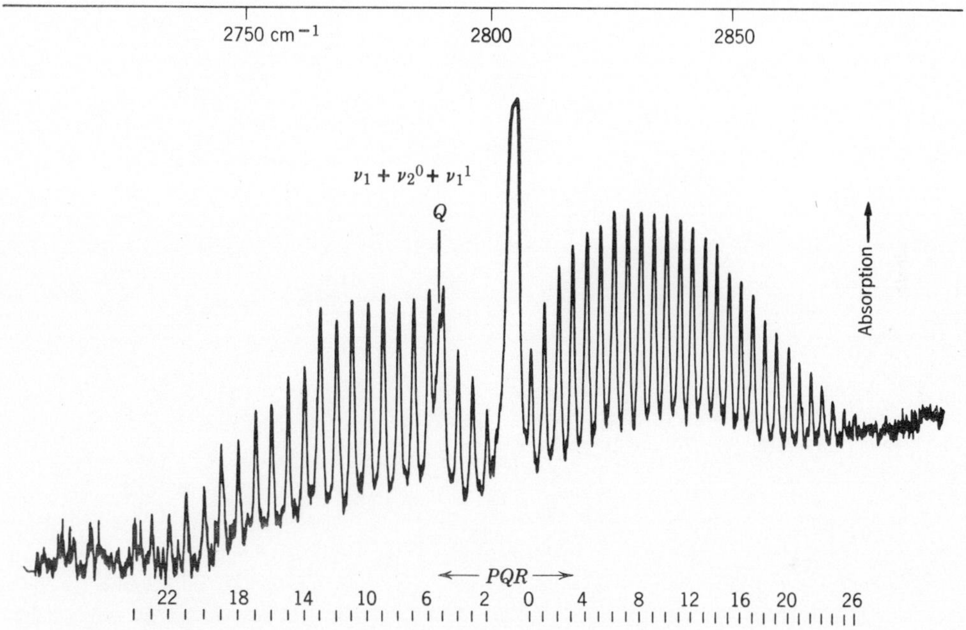

Fig. 6b1 The absorption of HCN near 2800 cm^{-1}. The absorption consists of the main band $\nu_1 + \nu_2^1 (\Pi \leftarrow \Sigma)$ and a weaker hot band $\nu_1 + 2\nu_2^\circ - \nu_2^1 (\Sigma \leftarrow \Pi)$. The numbering of the transitions is for the main band [18].

Since $J \geqslant l$, there is no level in the excited state with $J = 0$; thus there can be no $J = 1$ to $J = 0$ transition. The appearance of the Q-branch rules out determining any constants from it; hence we could only determine the excited state constants B^- and D^- from these data. Again, since HCN has $C_{\infty v}$ symmetry there is no intensity alternation in this band.

The $(0, 1^1, 1)$ band of HCN near 4004 cm^{-1} has an appearance very similar to that of the band in Fig. 6b1. However, the Q-branch of this band has been resolved [97] as has the $(1, 1^1, 1)$ HCN band near 6083 cm^{-1}. Figure 6b2 is a tracing of a portion of the Q-branch of the 4004 cm^{-1} band which shows the transitions from $J = 3$ up through $J = 12$. When such data are available it is possible to determine B^+ as well as B^- and hence q.

For a molecule with the point group $D_{\infty h}$ one would expect to find an intensity alternation similar to that found for Σ–Σ bands. Since the ground state in the

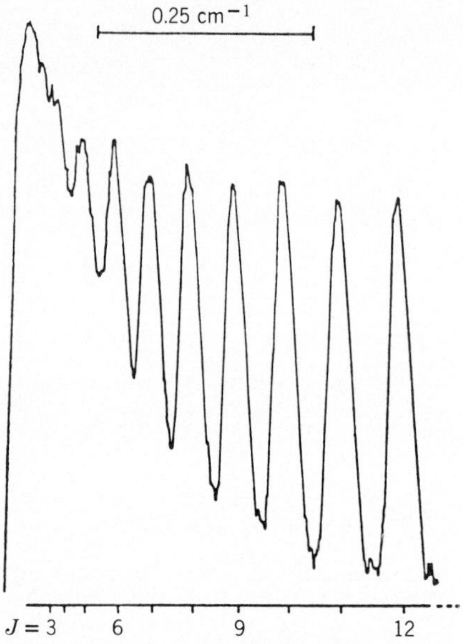

Fig. 6b2 Part of the Q-branch of the 01^11-000 band of HCN at 4004 cm^{-1}, obtained with an 8-meter absorption path and a pressure of a few hundredths of a millimeter of Hg [97].

Π–Σ band is the Σ state, the intensity patterns in the P- and R-branches would be the same as for a Σ–Σ band.

The expression for the P- and R-branch line positions analogous to 6a1 for a Π–Σ band is

$$\nu_m = \nu_0 - (B' + D') + (B' + B'' + 2D')m + (B' - B'' + D' + D'')m^2 - 2(D' + D'')m^3 - (D' - D'')m^4,$$

where the symbols all have their familiar meanings. It should be pointed out that here the constant which would be obtained from a least squares treatment is no longer the band center but $\nu_0 - (B' + D')$.

6c Σ–Π TRANSITIONS

Σ–Π transitions have many of the features of a Π–Σ transition. The selection rules are the same, and the term values of the lower state are given by 4c4 with $|l| = 1$. The excited state rotational term values are given by 4c4 with $l = 0$. Thus such a transition is characterized by $|l| = 1$, $\Delta l = \pm 1$.

The P- and R-branch transitions still involve only the lower of the sets of energy levels in the Π state, and the Q-branch only involves the upper levels of the Π state. The energy levels and allowed transitions of a Σ^+–Π transition are shown schematically on the left side of Fig. 6c1.

Such bands occur in the infrared spectra of linear molecules as difference bands, where the upper state is one of the stretching modes and the lower state is one of the bending modes. These bands can be important in determining infrared inactive vibrational frequencies, for optical transitions are often allowed between a level which does not combine with the ground state and one of the low lying bending fundamentals. The frequency of ν_1 in C_2D_2 was determined in this manner [17]. Similarly, these transitions fall many times in the region which can be investigated with photoconductive detectors and can be observed under high resolution enabling a good analysis. If the upper state also combines with the ground state, these two vibrational frequencies furnish a precise method of determining the lower frequency without having observed it directly.

These bands have the same general appearance as the Π–Σ bands; however, since the lower vibrational state is not the ground state, there must be a considerable population of the lower state in order to observe this type band. Thus these bands will be more easily observed at considerably elevated temperatures or when the Π state is rather low in energy. These bands at or near room temperature will, in general, be

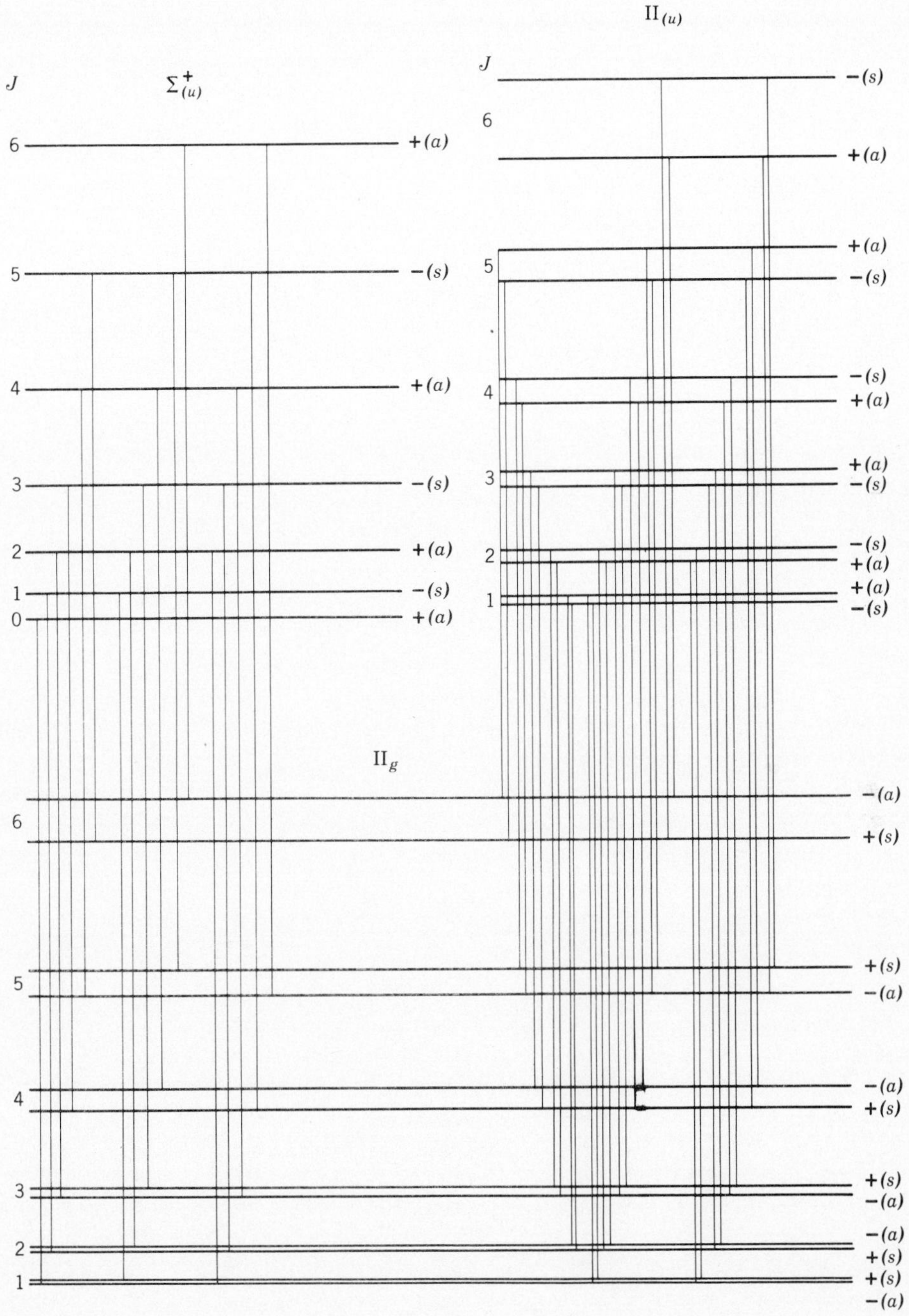

Fig. 6c1 A schematic representation of the energy levels and allowed transitions of a Σ–Π and a Π–Π transition of a linear molecule of $\mathbf{D}_{\infty h}$ symmetry is shown. If the molecule has $\mathbf{C}_{\infty v}$ symmetry, the *g*, *u* and *s*, *a* designations should be dropped.

much weaker than Π–Σ transitions for which the lower vibrational state is the ground state.

For molecules belonging to the point group $D_{\infty h}$ the intensity alternation in the *P*- and *R*-branches of a band with a Π_g lower state and a Σ_μ^+ upper state is the same as for a molecule with a Σ_g^+ ground state. For nuclei obeying Fermi statistics the transitions originating in odd *J* levels are stronger than those originating in even *J* levels, whereas the converse is true if the nuclei obey Bose statistics. However, when the lower state is Π_μ and the upper state is Σ_g^+, for nuclei which obey Fermi statistics the even *J* levels are strong, whereas the odd levels are strong if Bose statistics are followed.

In general, if $\Delta l = \pm 1$, the selection rules are the same as in the Σ–Π and Π–Σ cases. The bands have the *P, Q, R*-structure. Up to the present no bands have been resolved with $l > 1$ $\Delta l = \pm 1$. However, the same considerations would apply in considering these bands as apply in the cases already considered with $\Delta l = \pm 1$.

6d Π–Π TRANSITIONS

A Π–Π transition occurs when $|l| = 1$ and $\Delta l = 0$. The same general principles apply to any transition for which $l \neq 0$, $\Delta l = 0$ such as Δ–Δ, etc. For these bands the selection rule is again $\Delta J = 0, \pm 1$. These bands then have the *P, Q, R*-structure also. Here, however, only M_z does not equal zero, so the *Q*-branch is weak. Just as for a parallel band of a symmetric rotor, the *Q*-branch intensity decreases from the first line rather than going through an intensity maximum and then decreasing as for the Σ–Π, Π–Σ transitions just discussed. This can readily be seen from 5f2*c* by considering $l = 1$, $\Delta l = 0$, $\Delta J = 0$, and comparing this result to 5f3*b* where $l = 0$, $\Delta l = +1$, $\Delta J = 0$. In the spectra of linear molecules Π–Π bands generally appear accompanying a strong Σ–Σ transition. In Fig. 6a3 a second band can be clearly seen centered near 6480 cm^{-1}. The weak *Q*-branch appears on the side of the *P*(12) line of the main band and may be seen to be of the same order of intensity as the *P*- and *R*-branch transitions.

The energy levels and allowed transitions of a Π–Π transition are shown schematically on the right side of 6c1. It is noted that due to the *l*-type doubling of the two states each line is actually double. With the resolution available in Fig. 6a3 this doubling is not resolved until very high *J*. Also the Π–Π hot bands in Figs. 6a4 and 6a5 do not show the doubling of the lines. If the doubling is not resolved, we do not find any intensity alternation for molecules of the $D_{\infty h}$ point group. This is true because each line is really two transitions, one originating in a symmetric state and one originating in an antisymmetric state. However, if the splitting is resolved, then the components should show an intensity alternation which can be deduced in the usual manner.

Rank and his co-workers [98, 107] have resolved the *l*-type splitting in a number of bands of N_2O and HCN. Figure 6d1 is a portion of the $(3, 2^0, 0) - (0, 0, 0)$ band of N_2O overlapped by the Π–Π band $(3, 3^1, 0)$–$(0, 1^1, 0)$. The doublet splitting for several of the *R*-branch lines can readily be seen. The splitting formula is

$$\Delta\nu = 4(q' + q'' + \mu)m + 4(q' - q'')m^2 - 4\mu m^3, \tag{1}$$

where $\mu = D_d - D_c$. (The subscripts c and d refer to the two sublevels in each split π level). In this equation it is assumed that μ is constant for all states. By measuring the splitting for a number of P- and R-branch lines, it is possible to determine the q value for each state.

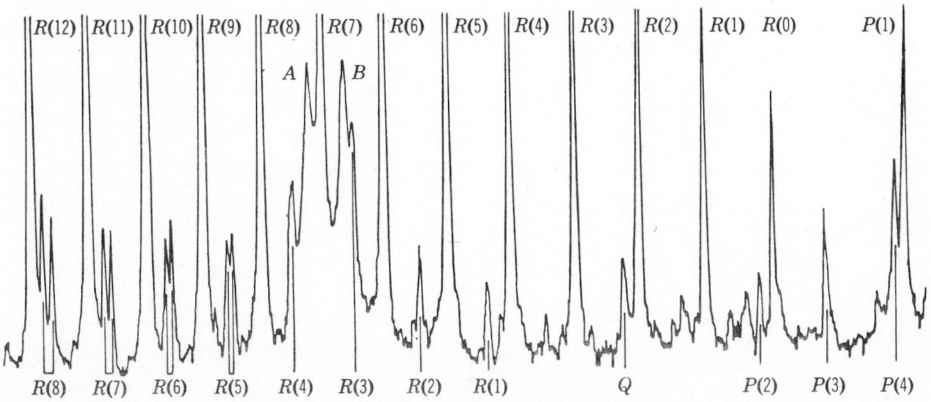

Fig. 6d1 Section of the N_2O absorption spectrum near 5030 cm^{-1}. The upper row of J numbers refers to the fine structure lines of the 32°0–000 cold band. The lower row refers to the fine structure doublets of the associated 33¹0–01¹0 hot band. The doublet structure is resolved for $J \geqslant 5$ in the R-branch. Lines A and B are extraneous to the N_2O sprectum and are probably due to atmospheric water. The other small unidentified lines are believed to come from isotope bands [107].

Guenther, Wiggins, and Rank [50] have investigated the interesting example of l-type doubling in CS_2. They chose the $(0, 1^1, 3)$–$(0, 1^1, 0)$ band near 4553 cm^{-1}, a Π_g–Π_u transition. This is a particularly interesting case since S^{32} has zero nuclear spin. In a linear symmetric triatomic molecule, if the identical nuclei have zero spin,

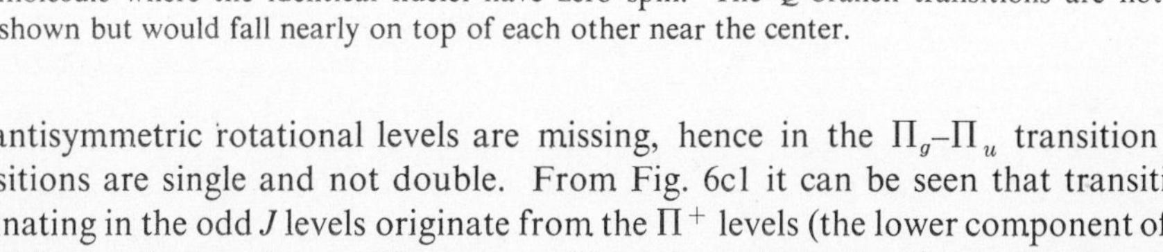

Fig. 6d2 P- and R-branch transitions in a Π_g–Π_u band of a linear symmetric triatomic molecule where the identical nuclei have zero spin. The Q-branch transitions are not shown but would fall nearly on top of each other near the center.

the antisymmetric rotational levels are missing, hence in the Π_g–Π_u transition the transitions are single and not double. From Fig. 6c1 it can be seen that transitions originating in the odd J levels originate from the Π^+ levels (the lower component of the l-doubling), whereas those with even J will originate from the Π^- levels (the upper component of the l-doubling). Thus every other line in the Π^+ and Π^- sub-bands will be missing as is shown in Fig. 6d2.

6e DETERMINING THE MOLECULAR PARAMETERS

When a band under analysis is overlapped by a second band, as is usually true in high resolution work, some of the measured lines cannot be uniquely assigned to any single transitions, for they may be a blend of a transition from each band so that the observed peak may not correspond uniquely to either transition. Even when the two transitions are resolved it may be that they are so close together that the positions of the two peaks are apparently shifted. Then the uncertain measurements should not be used to determine the molecular constants. However, it should be emphasized that there should be a real reason to omit the observation. Simply because omitting the line reduces the standard deviation is not a sufficient reason for omitting the measurement. Although a certain amount of judgment must be used in these case, it should be judgment with integrity.

When the inertial constants have been determined in a sufficient number of vibrational states, the constants in 3c5 may be determined,

$$B_v = B_e + \sum_i \alpha_i \left(v_i + \frac{d_i}{2}\right). \tag{3c5}$$

which enables one to determine the equilibrium value of the inertial constant B_e. This constant is related to the equilibrium value of the moment of inertia through the relation

$$B_e = \frac{h}{8\pi^2 c I_e};$$

thus the equilibrium value of the moment of inertia may be obtained. If the α_i cannot be evaluated, the common practice is to use B_0, the inertial constant of the ground vibrational state. The difference between r_e and r_0, the bond distance, is generally not too great.

If a linear symmetrical XY_2 molecule is under investigation, there is only one molecular parameter to determine, r_{XY}, hence knowing B_e (or B_0) is sufficient to determine it for

$$I = 2(m_Y r_{XY}^2). \tag{1}$$

However, if the molecule contains more than three atoms, there is more than one structural parameter to determine, and the one observable is not sufficient to determine the structure.

In these instances one resorts to determining B_e (or B_0) for a molecule in which a new isotope of one of the atoms has been substituted. Since the molecular electronic structure is presumably the same when the isotopic atom has been substituted, it is assumed that the molecular potential function is also the same. The assumption is undoubtedly good if the isotopic masses are not too different. The substitution of deuterium for hydrogen is the worst case since the mass changes 100% but even then the results are not too bad. The general expression for the moment of inertia of a linear triatomic molecule is

$$I = \frac{m_1 m_2 r_{12}^2 + m_1 m_3 r_{13}^2 + m_2 m_3 r_{23}^2}{M}, \tag{2}$$

where the r_{ij} are the distances between atoms i and j and M is the molecular weight of the molecule. For the more general case of a linear molecule with n atoms the moment of inertia is given by

$$I = \frac{\sum\sum_{i<j} m_i m_j r_{ij}^2}{M}. \tag{3}$$

Thus by determining B_e for a sufficient number of isotopically substituted molecules the complete structure of the molecule can be determined by the use of (1), (2), or (3).

When B_0's are used in the structural determination, especially when the deuterium isotope is concerned, caution should be exercised in interpreting the results. The quantity B_0 is proportional to $\overline{(1/r_0^2)}$ which in structural determinations is assumed to be the same as $1/(\bar{r}_0)^2$. This equality is not even true for an harmonic oscillator. A considerable body of evidence now points to the fact that $(r_0)_{HX}$ is not exactly the same as $(r_0)_{DX}$ [72]. This is related also to the interaction of vibration and rotation as shown by 3c3. For a more detailed discussion of molecular structure determinations by molecular spectroscopy see Costain [31].

As a further precaution, when reporting the results of an analysis always give the B values and not the moments of inertia. The B values are derived directly from the experimental data and are related to the moment of inertia through values of universal constants $\hbar$ and c. Since these constants change slightly with each new determination, I will also change with each new set of values for the constants. One should also give the source of the values of the constants used in any calculation.

7

ANALYSIS OF SYMMETRIC ROTOR SPECTRA

7a INTRODUCTION

The vibrational-rotational bands of a symmetric rotor are conveniently grouped into three types. If a molecule is a symmetric rotor due to symmetry, that is, the presence of a threefold or higher axis of rotation, the fundamental vibrational bands are of two types. The first type is observed when the change in the electric moment is parallel to the rotor axis, that is, $M_z \neq 0$, $M_x = M_y = 0$. Hence these bands are called parallel bands. In this instance the selection rules were previously found to be

$$K \neq 0 \qquad \Delta J = 0, \pm 1 \qquad \Delta K = 0$$

$$K = 0 \qquad \Delta J = \pm 1 \qquad \Delta K = 0. \tag{1}$$

In addition to these selection rules the usual restriction on plus$\leftrightarrow$ minus must be applied.

The second type of band, which occurs when the electric moment change is perpendicular to the rotor axis, is called a perpendicular band. Thus the conditions on the electric moment change are $M_z = 0$, $M_x = M_y \neq 0$. The selection rules in this case as found in Chapter 5 are

$$\Delta J = 0, \pm 1 \qquad \Delta K = \pm 1; \tag{2}$$

again the plus, minus considerations apply.

A third type of band arises only in certain overtone and combination bands or in the spectrum of an accidentally symmetric rotor. For these hybrid bands none of the electric moment components vanish, so such a band should have the characteristics of both a parallel and a perpendicular band.

Each of these band types is discussed and illustrated by real spectra in this chapter.

7b PARALLEL-TYPE BANDS

Because the structure of a parallel-type band is considerably simpler than that of the other types, it serves as a convenient starting place. The selection rules were given in 7a1. In order to be able to analyze parallel-type bands it is necessary first to understand the structure of the band.

The term values of a prolate symmetric rotor are given by

$$F_v(J, K) = B_vJ(J + 1) + (A_v - B_v)K_{-1}^2 - D_J^vJ^2(J + 1)^2 - D_{JK}^vJ(J + 1)K_{-1}^2 - D_K^vK_{-1}^4. \quad (1)$$

Similarly, the term values of an oblate symmetric rotor are given by

$$F_0(J, K) = B_vJ(J + 1) + (C_v - B_v)K_1^2 + \ldots \quad (2)$$

Since the general structure of parallel bands of both types of symmetric rotors is the same, only the prolate case is discussed in detail. The energy levels of a prolate symmetric rotor up through $K = 2$ are shown in Fig. 7b1 for two totally symmetric vibrational states V' and V''. Following the selection rules 7a1, the allowed transitions are also indicated. It is found that the transitions can be sorted by the K_{-1} values into sub-bands. It can be seen that for $K_{-1} = 0$, the sub-band has the appearance of a Σ–Σ type band of a linear molecule. If $K_{-1} > 0$, the sub-band has an appearance similar to that of a Π–Σ band of a linear molecule in which the l-type splitting of the Π state is zero. Since $K \leqslant J$, as K increases there are more and more lines missing from the beginning of the P-, Q-, and R-branches. Thus for the sub-band with $K = 1$, the first lines in the respective branches are P_2, Q_1, R_1; for $K = 2$ the first lines are P_3, Q_2, and R_2, etc. These sub-bands are shown in Fig. 7b2 in a schematic way, and in the lowest line they are superimposed so one can appreciate the over-all appearance of a parallel band. The energies and intensities are calculated using the inertial constants of the CH_3D molecule although no intensity alternation is shown.

From (1) an expression for the observed frequency of each line in the P- and R-branches may be written as

$$\begin{aligned}\nu_{m,K}^{P,R} &= \nu_0 + (B' + B'')m + (B' - B'' - D_J' + D_J'')m^2 \\ &\quad - 2(D_J' + D_J'')m^3 - (D_J' - D_J'')m^4 \\ &\quad + \{[(A' - B') - (A'' - B'')] - [D_{JK}' + D_{JK}'']m - [D_{JK}' - D_{JK}'']m^2\}K^2 \\ &\quad - (D_K' - D_K'')K^4, \end{aligned} \quad (3)$$

in which $m = J + 1$ for the R-branch and $-J$ for the P-branch.

The Q-branch transitions are given by

$$\begin{aligned}\nu_{J,K}^{Q} &= \nu_{\text{sub}} + (B' - B'')J(J + 1) - (D_J' - D_J'')J^2(J^2 + 1) \\ &\quad - (D_{JK}' - D_{JK}'')J(J + 1)K^2, \end{aligned} \quad (4)$$

in which

$$\nu_{\text{sub}} = \nu_0 + [(A' - B') - (A'' - B'')]K^2 - (D_K' - D_K'')K^4. \quad (5)$$

From (5) it is evident that the centers of the sub-bands are separated by the quantity $[(A' - B') - (A'' - B'')](2K + 1)$ if centrifugal stretching terms are neglected. This

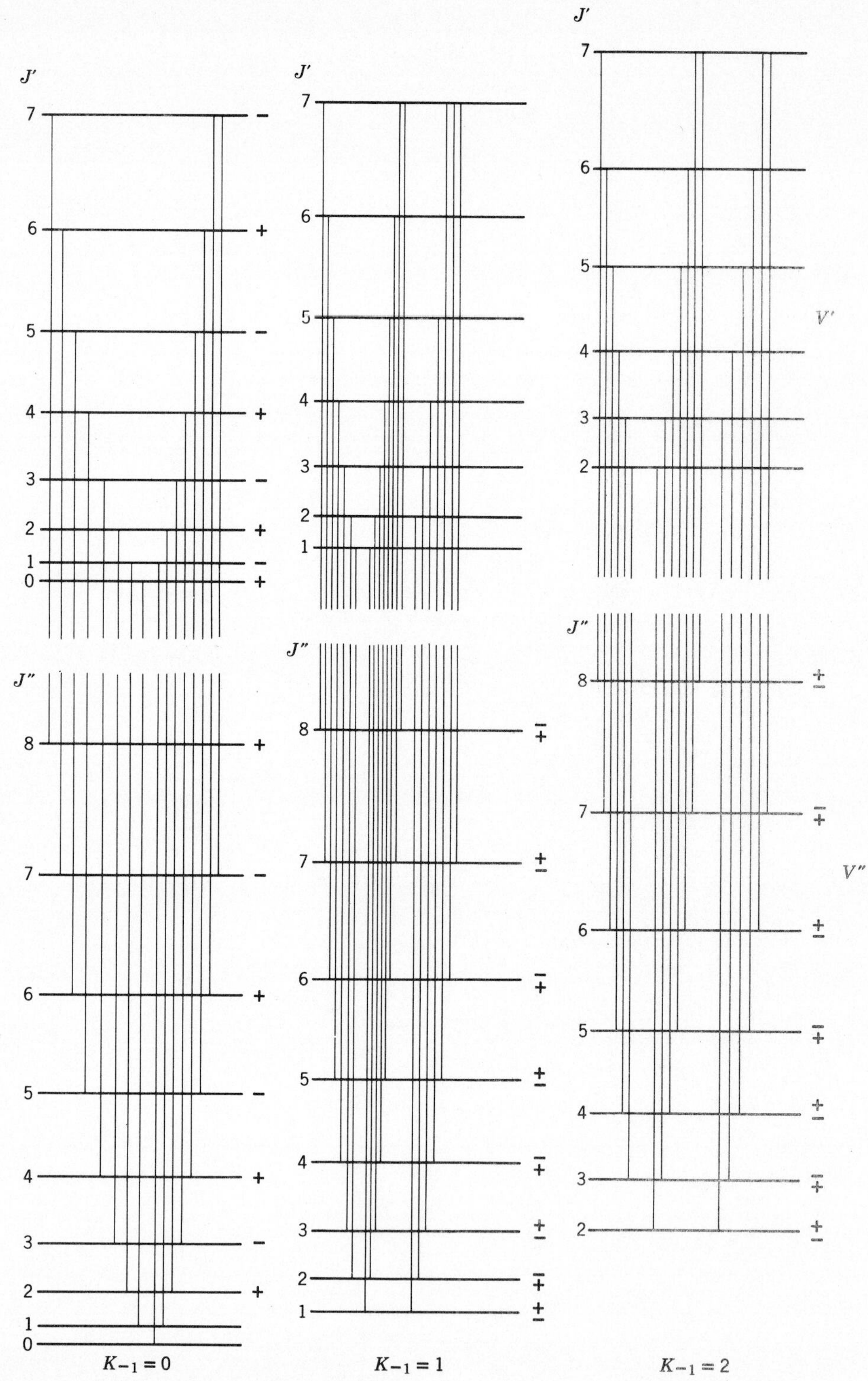

Fig. 7b1 The energy levels and allowed transitions for the first three sub-bands of a parallel band of a prolate symmetric rotor.

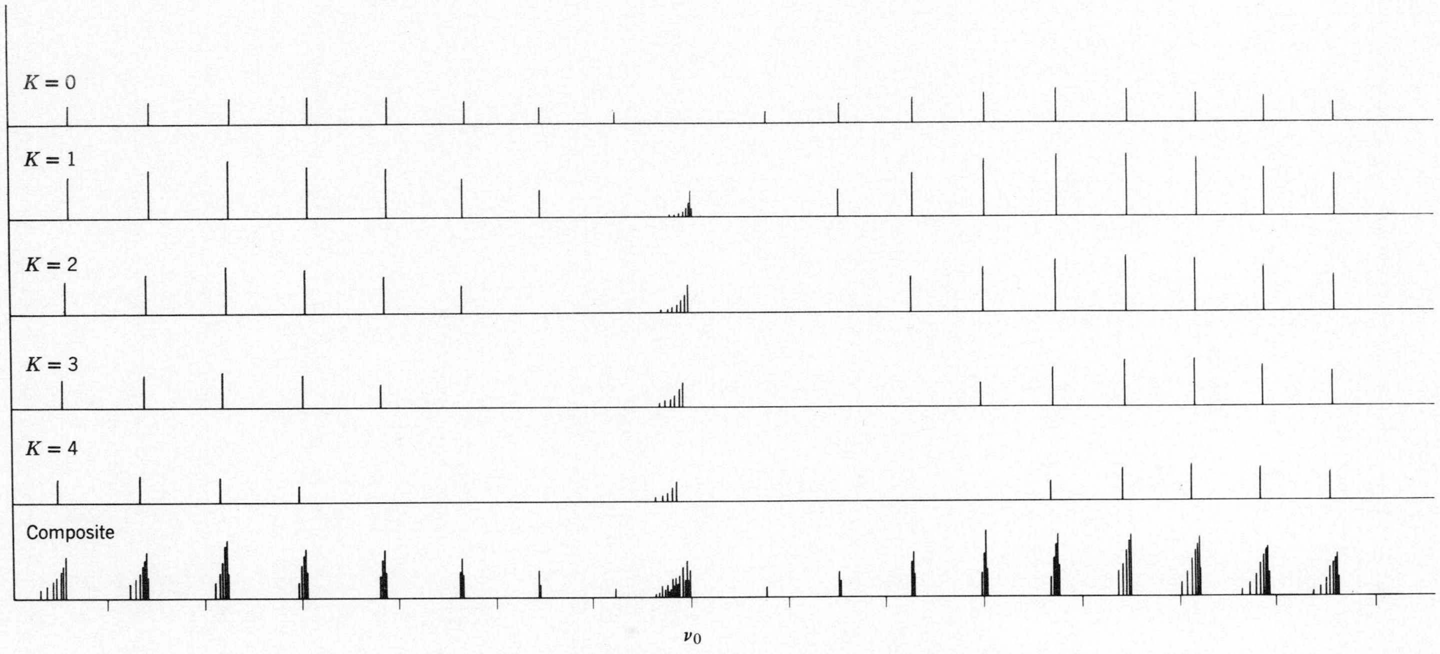

Fig. 7b2 The sub-band structure of a parallel-type band, calculated for 300°K, using the inertial constants of CH_3D. No nuclear spin statistics are included. In the lowest line the sub-bands are superimposed so one can appreciate the over-all appearance of the band. (From Herzberg, *Infrared and Raman Spectra*, copyright 1945, D. Van Nostrand Company, Inc., Princeton, New Jersey.)

is safe for low K since the difference $D_K' - D_K''$ is generally exceedingly small. This difference in sub-band centers can be seen in Fig. 7b2. It is for this reason that one sees, with sufficient resolution, the different K-components in the P- and R-branches. This phenomenon is called K-splitting and the splitting increases as K^2.

Since each of these sub-bands is analogous to a linear-molecule band, one can employ the same type of difference relations as in 6a3*a* and 6a3*b* for an analytical tool; thus

$$\Delta F_2''(J, K) = R(J-1, K) - P(J+1, K)$$
$$= (4B'' - 6D_J'' - 4D_{JK}''K^2)(J + \tfrac{1}{2}) - 8D_J''(J + \tfrac{1}{2})^3, \tag{6}$$

and the corresponding expression for the excited state

$$\Delta F_2'(J, K) = R(J, K) - P(J, K). \tag{7}$$

The band origin ν_0 can be obtained from the sum relation

$$\frac{R(J-1, K) + P(J, K)}{2} = \nu_0 + [(A' - B') - (A'' - B'')]K^2$$
$$- (D_K' - D_K'')K^4 + [(B' - B'')$$
$$- (D_J' - D_J'') + (D_{JK}' - D_{JK}'')K^2]J^2$$
$$- (D_J' - D_J'')J^2(J^2 + 1). \tag{8}$$

Now that the structure of a parallel band is understood the next step is to apply the principles to the analysis of an observed band.

The band chosen as an illustrative example is the parallel band of deuteromethane at 2200 cm^{-1}[16]. Deuteromethane is a prolate symmetric top, and the band at 2200 cm^{-1} is due to the normal coordinate which is essentially the stretching of the C—D bond. The observed absorption is shown in Fig. 7b3. It is readily seen that the absorption in this region has the features of a parallel band as shown in Fig. 7b2.

The first problem is the ever present one of assigning quantum numbers to the observed absorption peaks. One should start by assigning the quantum numbers J to the P- and R-branch transitions. One clue to the assignment of the J quantum numbers is the number of components observed due to K-splitting. If the transitions are designated by the J of the ground state, one can see from Fig. 7b1 that the transitions $R(J)$ should have $J + 1$ components, and the $P(J)$ transition will have J components. Since the quantity $[(A' - B') - (A'' - B'')]$ is generally quite small, the first few components will frequently not be resolved. Fortunately, when a molecule is a symmetric rotor because of symmetry, there is additional information to help with the assignments. In this case the nuclear spin statistics (see Chapter 5) are such that transitions for which K is a multiple of three are stronger than the others. The intensity alternation is readily seen in the figure. By assigning the stronger K components as 3, 6, etc., and remembering the total number of components allowed for a given $\Delta J = \pm 1$, it is possible to assign the J quantum numbers. Once these have been determined the assignment of the K quantum numbers follows readily, especially in view of the helpful intensity alternation. Once J and K have been assigned for a couple of P- and R-branch groups the sub-bands can readily be identified. By setting a pair of dividers to the distance between two successive transitions in the P- or R-branch of a sub-band it is then possible to step off this distance to higher or lower frequency and in this manner identify all the transitions of a particular sub-band.

This process should be repeated until all the transitions in all the sub-bands have been identified. When this is done for the CH_3D band under consideration, the sub-bands are identified as shown in Table 7b1. It will be noted that in the *R*-branch regions of the sub-bands there are several values of the quantum numbers for which no frequency

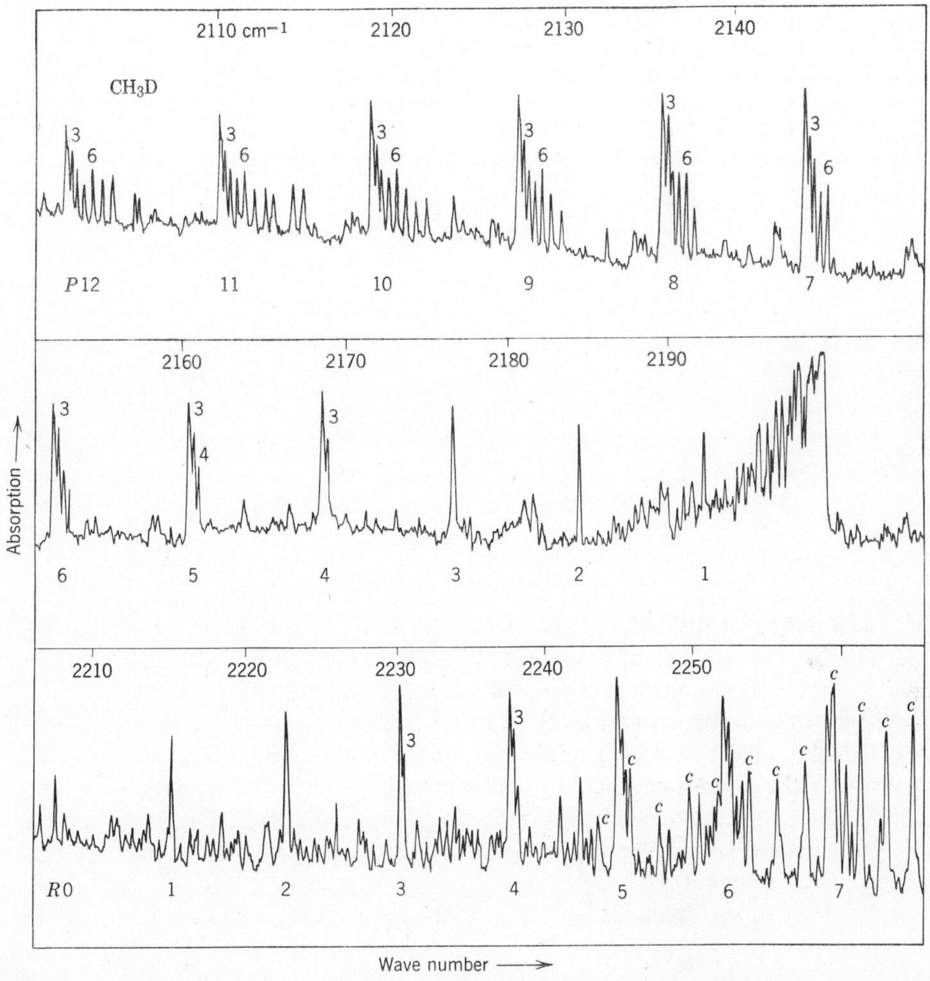

Fig. 7b3 The parallel band, ν_2, of CH_3D at 2200 cm^{-1}. The numbers under the spectrum are the ground state *J* values, whereas the numbers above the spectrum are the ground state *K* values. Lines marked *c* are due to atmospheric CO_2 [15].

has been listed. Unfortunately, the high frequency part of this band is obscured to some extent by atmospheric CO_2 absorption. The lines not listed could not be measured because they absorb at the same place as does CO_2.

After assigning the quantum numbers it remains to determine the derived constants from the observed frequencies. This can be done in a variety of ways which are described in some detail. The preferred method of determining the inertial parameters

Table 7bI—Sub-bands of the CH_3D at 2200 cm^{-1} (R-Branch Lines for Which

J''	$K = 0, 1$	$\Delta F_2''(J, 1)$	$K = 2$	$\Delta F_2''(J, 2)$	$K = 3$	$\Delta F_2''(J, 3)$	$K = 4$	$\Delta F_2''(J, 4)$
$\Delta J = +1$								
0	2207.70							
1	15.32	23.25						
2	22.82	38.79	2222.93					
3	30.23	54.32	30.35	54.30	2230.54			
4	37.56	69.79	37.66	69.80	37.86	69.77	2238.13	
5	44.78	85.30	44.89	85.27	45.09	85.27	45.34	85.26
6	51.94	100.76		100.76	52.23	100.74	52.48	100.71
7	58.97	116.26	59.05		59.31	116.20		116.17
8	65.90	131.68	66.02	131.64	66.20	131.68	66.47	
9	72.75	147.07	72.85	147.07	73.05	147.04	73.31	147.03
10	79.48	162.48	79.58	162.45	79.80	162.45	80.04	162.41
$\Delta J = -1$		177.86		177.81		177.83		177.77
1	2192.29							
2	84.45							
3	76.53		2176.63					
4	68.50		68.63		2168.84			
5	60.44		60.55		60.77		2161.04	
6	52.26		52.39		52.59		52.87	
7	44.02		44.13		44.35		44.63	
8	35.68		35.81		36.03		36.31	
9	27.29		27.41		27.63		27.92	
10	18.83		18.95		19.16		19.44	
11	10.27		10.40		10.60		10.90	
12	01.62		01.77		01.97		02.27	

and the centrifugal distortion constants is by means of the difference relations (6) and (7). The advantages of this method become apparent as it is explained. The first advantage is that one is working with energy differences of one vibrational state, and hence the number of constants to be determined simultaneously is smaller. If several bands of the molecule originating in the lowest vibrational state are to be analyzed at the same time, the $\Delta F_2''$ values should be the same as determined from each band. This serves as a check on the assignments in each of the bands. Also by using this method any small perturbations in the excited state will have no effect on the ground state constants.

If the $\Delta F_2''$ values are divided by $2J + 1$, the following is obtained:

$$\frac{\Delta F_2''(J, K)}{(2J + 1)} = (2B'' - 3D_J'' - 2D_{JK}''K^2) - 4D_J''(J + \tfrac{1}{2})^2. \qquad (9)$$

This, of course, is the equation of a straight line with a slope equal to $-4D_J''$ and an intercept of $2B'' - 3D_J'' - 2D_{JK}''K^2$. Thus if the $\Delta F_2''(J, K)$ values divided by $(2J + 1)$ are plotted against $(J + \frac{1}{2})^2$ for constant values of K, one should obtain a series of straight lines, each with the same slope but with slightly different intercepts. The difference between the different intercepts is then $-2D_{JK}''(K_2^2 - K_1^2)$. The graphical method of treating (9) is somewhat subjective and depends on the worker's judgment as to what is the best straight line through the points which have been plotted. However, with an experienced worker the graphical method is capable of giving very precise results.

No Frequencies Are Given Are Obscured by Atmosphere CO_2 Absorption)

$K = 5$	$\Delta F_2''(J, 5)$	$K = 6$	$\Delta F_2''(J, 6)$	$K = 7$	$\Delta F_2''(J, 7)$	$K = 8$	$\Delta F_2''(J, 8)$	$K = 9$	$K = 10$
2245.68									
52.80	100.70	2253.21							
59.84	116.13	60.25	116.11	2260.71					
	131.56	67.18	131.53	67.66	131.46				
		74.00	146.92	74.47	146.88	2274.99			
		80.75	162.30	81.21	162.24		2162.15		
			177.65		177.59				
2153.23									
44.98		2145.42							
36.67		37.10		2137.61					
28.28		28.72		29.25		2129.83			
19.80		20.26		20.78		21.38		2121.99	
11.26		11.70		12.23		12.84		13.30	2114.33
02.65		03.10		03.62		04.25			05.75

The $\Delta F_2''(J, K)$ values for the sub-bands of the CH_3D band under consideration are included in Table 7b1. In this band the sub-bands for $K = 0$ and $K = 1$ are not resolved so the $\Delta F_2''$ values for these sub-bands are not reliable enough to be used in an analysis. Since for every J value in the P- and R-branches the observed peak is a composite peak of the $K = 0$ and $K = 1$ transition, the measured wavelength of the peak does not correspond to the wavelength of either transition; thus the inclusion of these data will likely falsify the results. A graphical presentation of the ground state data given in Table 7b1 would appear much like that shown in Fig. 7b5.

A better way to evaluate these constants is to solve the equations (9) by the method of least squares for the best estimates of the constants B'', D_J'', and D_{JK}''. For this band, 35 equations of the type (9) were solved by the method of least squares in order to determine the constants. The constants determined for this band by the two methods of solution of the equations (9) are compared in Table 7b2. The comparison shows that the two methods are of comparable accuracy.

The excited state parameters are determined in an analogous manner using the $\Delta F_2'(J, K)$ values and either a graphical or a least squares method of solution. The band center ν_0 and the quantities $[(A' - B') - (A'' - B'')]$ and $(D_K' - D_K'')$ which govern the K-splitting are determined from the combination relation (8). Again a series of relations may be set up which can be solved by the method of least squares. Since the coefficients of J^2 and $J^2(J^2 + 1)$ have already been determined from the difference relations, their values may be used in the equations (8), leaving only the quantities ν_0, $[(A' - B') - (A'' - B'')]$ and $(D_K' - D_K'')$ to be determined. Alternatively, one can determine all five constants from this set of equations, the coefficients of J^2 and $J^2(J^2+1)$ serving as a check on the previous determinations from the difference equations. The complete results of the analysis of the CH_3D band are given in Table 7b3. These results are based on a least squares solution of all the pertinent equations.

Another and less desirable method of deriving the constants from the data involves the use of (3). In this case an equation of the type (3) can be written for each assigned

Table 7b2—The Constants of the 2200 cm^{-1} Band of CH_3D

	LEAST SQUARES (cm^{-1})	GRAPHICAL (cm^{-1})
$(2B'' - 3D_J'')$	7.760	7.760
$4D_J''$	−0.00020	−0.00020
$2D_{JK}''$	−0.00024	−0.00025

frequency. Again the unresolved peaks of the $K = 0$ and $K = 1$ sub-bands should not be used. The resulting set of equations is solved by the method of least squares. In this method one will obtain at least twice as many equations to solve simultaneously as in the previous method; but there are twice as many constants to be simultaneously determined. Thus there is no real gain if the same transitions are used for both methods. However, if the theory is correct, then one should obtain the same values of the constants by using either method. There is some advantage to using this second method for deriving the band constants if for some reason only a part of the band can

Table 7b3—The Rotational Constants of CH_3D from the 2200 cm^{-1} Band [16]

	GROUND (cm^{-1})	EXCITED (cm^{-1})
B	3.880	3.838
D_J	5×10^{-5}	5.5×10^{-5}
D_{JK}	1.2×10^{-4}	1.2×10^{-4}

$$\nu_0 = 2200.02 \text{ cm}^{-1}$$
$$[(A' - B') - (A'' - B'')] = 0.03963 \text{ cm}^{-1}$$
$$D_K' - D_K'' = 2 \times 10^{-5}$$

be observed. If only the P- or R-branch of a band is observed, this method is the only one which can be used if no previous knowledge of the molecular constants is available. There are many times when one branch of a band will be obscured either by an overlapping band of the same molecule or, if no vacuum instrument is available, by atmospheric absorption by H_2O or CO_2. In general, one should use this method only as a last resort.

When only one-half of the band is observed, another method of analysis can be resorted to if reliable values of the ground state $\Delta F_2''$ values are available from some other source such as the previous analysis of another band of the molecule. By the use of (6) the frequencies of the transitions in the unmeasured branch of the band can be calculated. For instance, if only the P-branch of a band can be measured, and if the quantum numbers can be assigned to the observed absorption peaks, then from (6)

$$R(J-1, K) = P(J+1, K) + \Delta F_2''(J, K). \qquad (10)$$

The measured values of the P-branch transitions and the calculated values of the R-branch can then be used in (7) to determine the constants for the excited state of the band. It should be emphasized that this method is not applicable unless one has prior knowledge of the ground state constants.

Before concluding the section on parallel-type bands it might be well to examine a band due to an oblate symmetric rotor. Considerable work has been done on the spectrum of CD_3H, which is readily seen to be an oblate symmetric rotor. In this limit the expressions for $\Delta F_2'(J, K)$ and $\Delta F_2''(J, K)$ are exactly the same as for a prolate symmetric rotor. Equations (3), (4), (5), and (8) are also applicable if A' and A'' are replaced by C' and C'', and if one realizes that the quantum number in this limit is K_1, not K_{-1}. Figure 7b4 is a microphotometer trace of a portion of the P-branch of a parallel band of CD_3H in the photographic infrared [23] centered at 11266.94 cm^{-1}. It can be seen by comparing Fig. 7b4 and Fig. 7b3 that the general characteristics of the bands are the same. Again, the sub-bands with K a multiple of three are more intense due to the nuclear spin statistics, and so are once again a help in assigning quantum numbers. In this band the $K = 0, 1, 2$ sub-bands were not resolved. The observed frequencies and the assignments are given in Table 7b4. This investigator chose to use the ΔF_2 method of analysis, but derived the constants by the graphical approach. Figure 7b5 shows his plot of $\Delta F_2''/(J + \frac{1}{2})$ versus $(J + \frac{1}{2})^2$ for each sub-band. From this plot he deduced the values of B'', D_J'', and D_{JK}'' given in Table 7b5.

To determine the other constants the investigator made use of (8). Assuming $(D_J' - D_J'')$ is negligible and plotting $R(J-1, K) + P(J, K)$ versus J^2 for constant values of K a series of lines of slightly different slopes are obtained. This difference in slope is due to the $(D_{JK}' - D_{JK}'')K^2$ dependence of the coefficient of J^2. Making allowance for this dependence, one can determine D_{JK}' as well as $(B' - B'')$. These plots are shown in Fig. 7b6. A plot of the intercepts in Fig. 7b6 against K^2 enables the determination of $[(C' - B') - (C'' - B'')]$ and ν_0. The constants determined in this manner are also given in Table 7b5.

A second parallel band of CD_3H has been observed at 5865.02 cm^{-1} [120]. The measurements on this extend through $J = 9$ in the R-branch and $J = 10$ in the P-branch. These workers chose to analyze their data by using a modification of (3). They assumed that the centrifugal distortion coefficients are the same in both vibrational states. Under these conditions (3) takes the form

$$\nu = \nu_0 + (B' - B'')m + (B' - B'')m^2 - 4D_Jm^3$$
$$+ [(C' - B') - (C'' - B'') - 2D_{JK}m]K^2 \qquad (11)$$

Rather than work with all the observed frequencies, these workers determine the position of the $K = 0$ transitions by plotting the component splitting versus K^2 and extrapolating back to the positions of the $K = 0$ transitions. Their method is shown in Fig. 7b7. The constants were determined from these extrapolated values for the $K = 0$ transitions using (11) which for these transitions has only the first four terms. Their

Table 7b4—Observed and Calculated Wave Numbers (ν_{vac}) of the Lines in the 8900 A Band of CD_3H [23]

		P(J, K)		R(J, K)	
J	K	OBS	CALC	OBS	CALC
0	0			11273.36	11273.37
1	0, 1			279.67	279.68
2	0, 1	11253.71	11253.71	285.88	285.85
3	0, 1, 2	246.93	246.94	292.00	291.92
4	0, 1, 2	239.98	240.00	297.84	279.82
	3	240.19	240.21	298.05	298.04
	4			298.26	298.24
5	0, 1, 2	233.04	232.94	303.56	303.59
	3	233.33	233.15	303.81	303.81
	4			304.03	304.01
6	0, 1, 2	225.77	225.75	309.19	309.22
	3	225.97	225.96	309.43	309.43
	4	226.18	226.16	309.63	309.64
	5	226.43	226.42	309.90	309.90
	6			310.24	310.23
7	0, 1, 2	218.41	218.45	314.69	314.71
	3	218.66	218.65	314.92	314.92
	4	218.86	218.85	315.12	315.13
	5	—	—	315.39	315.39
	6	219.40	219.42	315.70	315.71
8	0, 1, 2	210.99	211.01	320.09	320.05
	3	211.22	211.22	—	—
	4	211.41	211.42	320.48	320.47
	5	211.66	211.67	320.75	320.74
	6	212.00	211.98	321.04	321.06
	7	212.30	212.35	321.43	321.44
	8			321.82	321.88
9	0, 1, 2	203.44	203.46	325.25	325.26
	3	203.68	203.66	325.47	325.47
	4	203.87	203.86	325.70	325.67
	5	204.10	204.11	—	—
	6	204.42	204.42	326.24	326.26
	7			326.62	326.64
	8			327.06	327.08

Table 7b4 (continued)

J	K	$P(J, K)$ OBS	$P(J, K)$ CALC	$R(J, K)$ OBS	$R(J, K)$ CALC
10	0, 1, 2	11195.81	11195.78	11330.38	11330.31
	3	—	—	—	—
	4	196.19	196.18	330.70	330.73
	5	196.43	196.43	331.02	330.99
	6	196.75	196.74	331.30	331.31
	7	197.10	197.10	331.70	331.69
	8	197.54	197.52	332.16	332.13
	9	197.97	198.00	332.65	332.63
	10			333.16	333.18
11	0, 1, 2	188.00	187.98	335.31	335.21
	3	188.21	188.18	—	—
	4	—	—	—	—
	5	—	—	—	—
	6	188.94	188.93	—	—
	7	189.27	189.29	336.60	336.59
	8	189.72	189.71	337.07	337.03
	9			337.56	337.53
	10			338.14	338.08
12	0, 1, 2	180.11	180.05	340.14	339.97
	3	—	—	—	—
	4	180.45	180.45	—	—
	5	180.72	180.69	340.66	340.64
	6	180.98	181.00	341.07	340.96
	7	181.36	181.36	—	—
	8	—	—	—	—
	9	182.31	182.24	342.38	342.27
13	0, 1, 2	172.11	172.00	—	—

Table 7b5—Rotational Constants of $\mathbf{CD_3H}$ [23]

11266.94 cm^{-1} = $4\nu_1$ BAND

$B_0 = B''$	3.2787 cm^{-1}
B'	3.2146
$C' - C''$	−0.0350
ν_0	11266.94
D_J''	0.000046
D_J'	0.000052
$D_{JK''}$	-0.00003_5
$D_{JK'}$	-0.00003_0
α_1	0.0160

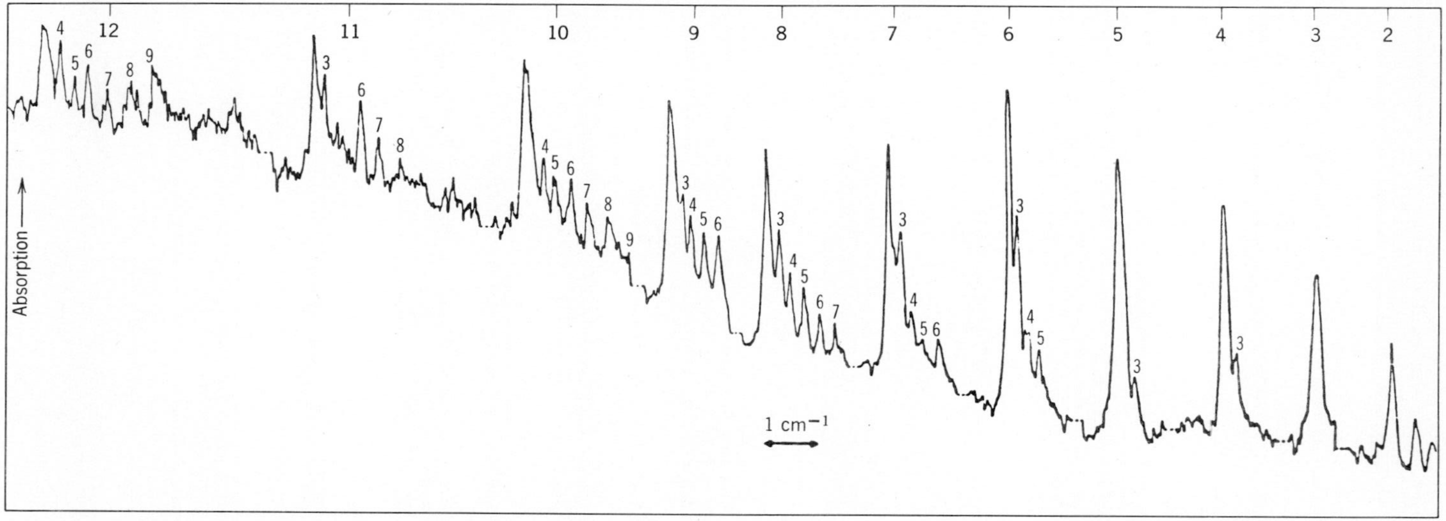

Fig. 7b4 Microphotometer trace of a portion of the *P*-branch of a parallel band of CD_3H in the photographic infrared centered at 11,266.94 cm^{-1}. To decrease the horizontal dimension, the portion of the record between the lines has been cut out. The *J* numbering is shown along the top and the *K* numbers are given near the associated peaks [23].

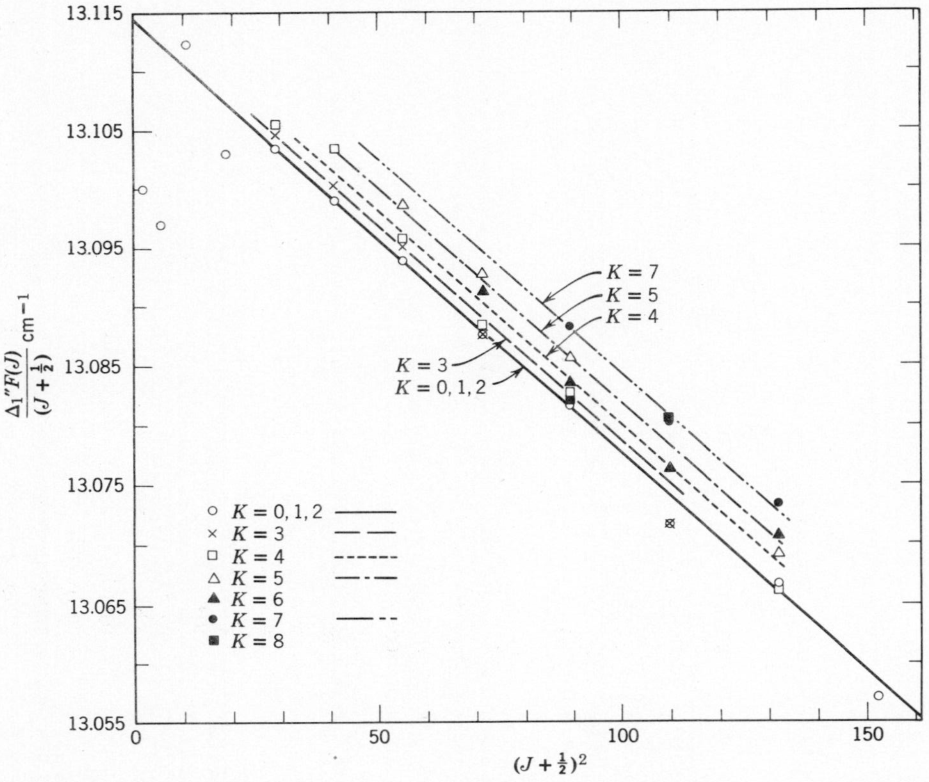

Fig. 7b5 Combination differences $\Delta F_2''(J, K)/(J + \frac{1}{2})$ for the ground state of CD_3H obtained from the data of the 11,266.94 cm^{-1} band [23].

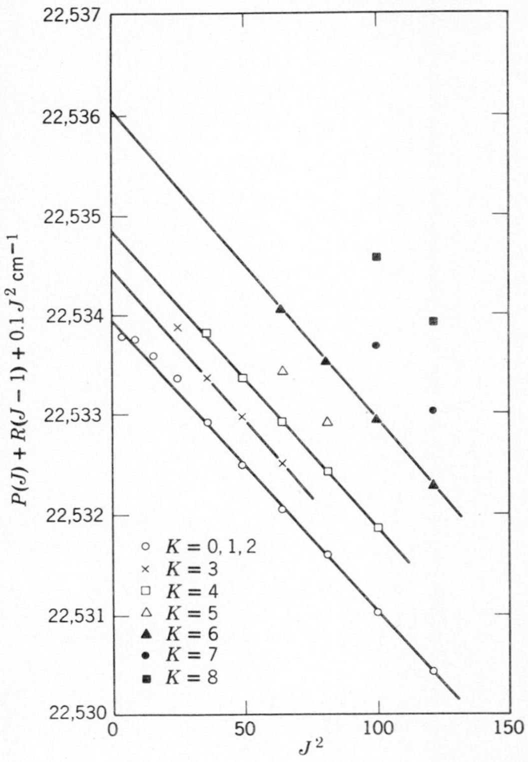

Fig. 7b6 Combination sums $R(J - 1, K) + P(J, K)$ for the 11,266.94 cm^{-1} band. The perturbed lines with $J = 3$ and 5 have been omitted [23].

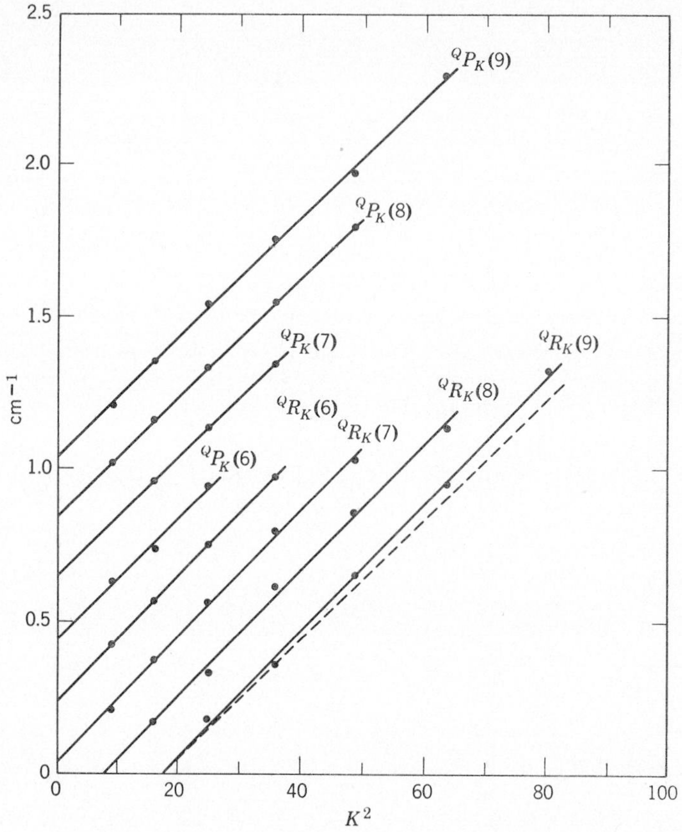

Fig. 7b7 A plot of the K fine structure component splitting versus K^2 for the CD_3H band at 5865 cm^{-1}. The dashed line is drawn parallel to the line labelled Q_{P_K} (9) [120].

data are given in Table 7b6. The $R(8)$ and $R(9)$ frequencies were not used in determining the constants because the excited states were perturbed in some manner not understood. The values of $[(C' - B') - (C'' - B'')]$ and D_{JK} were determined graphically as described previously.

Finally, a fundamental parallel band of CD_3H has been observed [100]. This band has a severe perturbation of the excited state, but it is still possible to identify many of the transitions and determine the $\Delta F_2''$ values. The ground state ΔF_2 values for all three bands are compared in Table 7b7. It is seen that the agreement from band to band is very satisfactory, especially in view of the wide differences in spectral region and the fact that the bands were measured by different workers in different laboratories. The ground state constants for the 2992.72 cm^{-1} band were determined by solving 43 equations of the type (6) by the method of least squares. Thus one band was analyzed using (6) and graphical solutions; another was analyzed by use of a modification of (3); and the last was analyzed by using (6) and a least squares reduction of the data. The results as given by the various authors for the ground state are given in Table 7b8. The agreement among the various workers is remarkable. The best set of ground state

Table 7b6—Vacuum Wave Numbers in the $2\nu_1$ Band of CHD_3 [120]

P-BRANCH			*R*-BRANCH		
J	*K*	*ν*	*J*	*K*	*ν*
2	0	5851.832	1	0	5877.899
3	0	45.118	2	0	84.206
4	0	38.349	3	0	90.476
5	0	31.504	4	0	96.646
6	0	24.578	5	0	5902.704
	3	24.764	6	0	8.737
	4	24.871		3	8.920
	5	25.076		4	9.062
7	0	17.580		5	9.248
	4	17.892		6	9.469
	5	18.069	7	0	14.631
	6	18.280		3	14.809
8	0	10.532		4	14.972
	3	10.704		5	15.158
	4	10.842		6	15.392
	5	11.017		7	15.626
	6	11.231	8	0	20.192
	7	11.480		4	20.517
9	0	3.394		5	20.683
	3	3.556		6	20.963
	4	3.701		7	21.205
	5	3.888		8	21.485
	6	4.106	9	0	25.921
	7	4.321		5	26.459
	8	4.645		6	26.636
10	0	5796.190		7	26.931
				8	27.226
				9	27.590

parameters would be obtained by averaging the three determinations of the $\Delta F_2''$ values and reducing them by the method of least squares. The analysis of the 2992 cm^{-1} band points out the advantage of using the $\Delta F_2''$ method of analysis. In this band the excited state levels were strongly perturbed and the use of (3) would have been impossible; however, since the ground state is not perturbed, the $\Delta F_2''$ values are valid in any event. In using the $\Delta F_2''$ method one always wants to be sure that the excited state levels are common to both the *P*- and *R*-branch, for if they are not, erroneous $\Delta F_2''$ values result.

For heavier molecules and also in special cases of the light molecules, the quantity $[(C' - B') - (C'' - B'')]$ is so small that no *K*-components in the *P*- and *R*-branches are observed. In these cases one can only assume that the observed absorption maxima in the *P*- and *R*-branches correspond to the same *K* value and hope that this is a low *K* value. One then applies (9), ignoring the *K* dependence of the first term. Figure 7b8 is an example of a parallel combination band of C_2D_6 for which the

Table 7b7—Ground State ΔF_2 Values from Three Different Bands of CD_3H

J	K	11266 cm^{-1} BAND [23]	5865 cm^{-1} BAND [120]	2992 cm^{-1} BAND [100]
1	0	19.65		19.67
2	0	32.74	32.781	32.81
	1			32.81
3	0	45.90	45.857	45.91
	1			45.95
	2			45.89
4	0	58.96	58.972	
	2			58.98
	3			59.00
5	0	72.07	72.068	
	2			72.08
	3	72.08		72.08
	4	72.08		72.12
6	0	85.15	85.124	
	2			85.18
	3	85.15		85.18
	4	85.17		85.19
	5			85.21
7	0	98.20	98.205	98.24
	1			98.22
	2			98.22
	3	98.21	98.216	98.22
	4	98.22	98.220	98.23
	5	98.24	98.231	98.23
	6	98.24	98.238	98.23
8	0	111.25	111.237	111.25
	1			111.25
	2			111.25
	3	111.24	111.25	111.25
	4	111.25	111.271	111.27
	5	111.29	111.270	111.27
	6	111.28	111.286	111.30
	7		111.305	111.31

Table 7b8—The Ground State Constants of CD_3H

	11266 cm^{-1} (GRAPHICAL)	5865 cm^{-1} (Eq. 3)	2992 cm^{-1} (L.S.)
B''	3.2787	3.2777	3.2790
D_J''	4.6×10^{-5}	3.9×10^{-5}	4.8×10^{-5}
D_{JK}	$-3._5 \times 10^{-5}$	-4×10^{-5}	$-3._5 \times 10^{-5}$

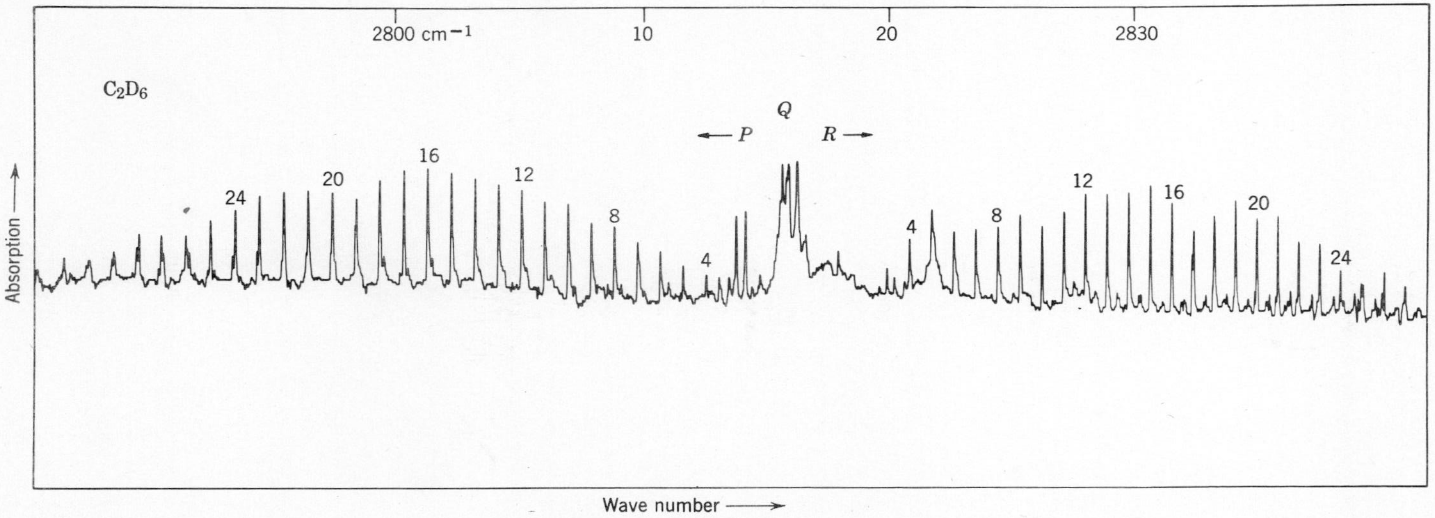

Fig. 7b8 A parallel combination band of C_2D_6 for which the quantity $[(A' - B') - (A'' - B'')]$ is very small, so that the K-components are not observed [15].

quantity $[(C' - B') - (C'' - B'')]$ is very small, and the K-components are not observed.

Thus far nothing has been said about the Q-branches as an aid in analysis. The reason is that the Q-branch region is generally of very little help in the analytical

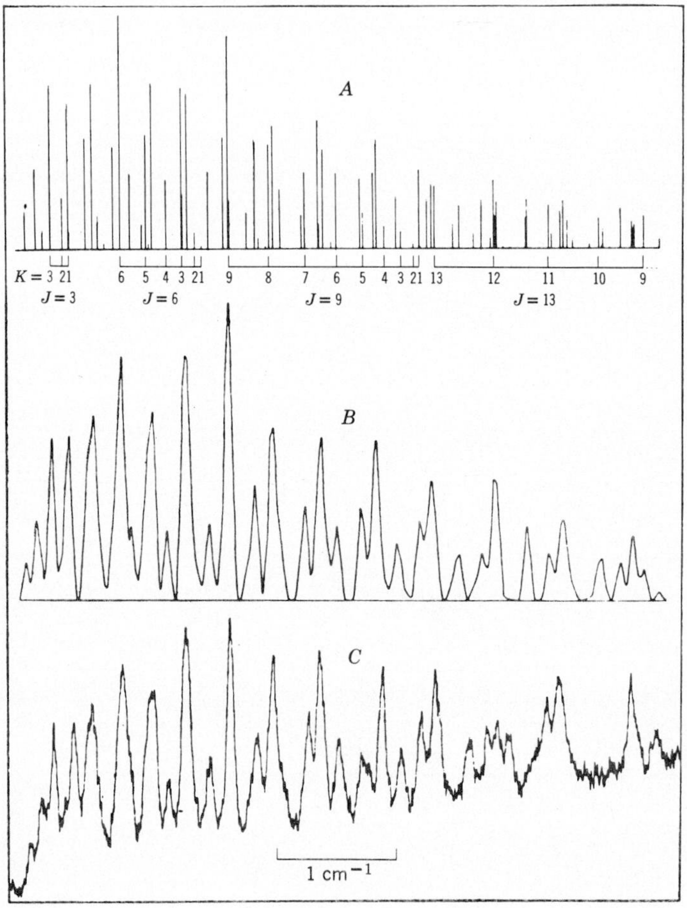

Fig. 7b9 The Q-branch of the CHD_3 band at 5865 cm^{-1}. (A) Positions and intensities of predicted components. (B) Synthetic branch assuming triangular line shape and half-width of 0.06 cm^{-1}. (C) Reproduction of spectrometer trace of the Q-branch [120].

process. Figure 7b9 [120] shows the problem involved. The many possible transitions are not separated very much on the energy scale, and absorption peaks do not correspond uniquely to any transition. Likewise no new information is to be obtained from the Q-branch region that cannot be obtained from the P- and R-branches if the K-components are resolved. However, Fig. 7b9 is a convincing demonstration of the validity of the theory. Even if one is not inclined to accept their slit function, which is as reasonable as any, the general features are certainly present in the top line of the figure.

In general, the analysis of a parallel-type band is a straightforward process and, although there are more parameters to be determined, the techniques are essentially those used in the analysis of the spectra of linear molecules. However, complications can arise, and the examples discussed were included to show how to cope with some of these situations.

7c PERPENDICULAR-TYPE BANDS

The selection rules for a perpendicular band are given in 7a2, and the energy levels for a nondegenerate state are again those given in 7b1. The selection rules give rise to more transitions than in the parallel case; hence the structure of the over-all band is more complex.

For each value of K, $\Delta K = \pm 1$; for each K and ΔK, ΔJ can be 0 ± 1. Thus for $K = 0$ there is a sub-band with $\Delta K = 1$, $\Delta J = \pm 1, 0$. This sub-band then has a P, Q, R structure and is identical to the sub-band due to $K = 0$, $\Delta K = -1$, and $\Delta J = \pm 1, 0$. For $K = 1$, $\Delta K = +1$ a similar sub-band is possible, as is also true with $K = 1$, $\Delta K = -1$.

Consider only the $\Delta J = 0$ transitions. Then one finds for the frequencies of the Q-branch lines, neglecting centrifugal distortion, and assuming both states to be nondegenerate

$$\nu^Q = \nu_0 + (A' - B') \pm 2(A' - B')K + [(A' - B') - (A'' - B'')]K^2 + (B' - B'')J(J + 1). \quad (1)$$

The sub-band centers correspond to the case when $J = 0$, thus

$$\nu_0^{Q_K} = \nu_0 + (A' - B') \pm 2(A' - B')K + [(A' - B') - (A'' - B'')]K^2. \quad (2)$$

It should be noticed in (2) that even when $K = 0$ the center of the sub-band $\nu_0^{Q_0}$ is not equal to ν_0 the frequency of the vibrational transition. Thus a perpendicular band is made up of a series of sub-bands with P, Q, R structure, with the Q-branches being separated by approximately $2(A' - B')$. This sub-band structure is shown in Fig. 7c1, and the lowest line shows the composite band. As can be seen from this figure the main features of this type band are a series of Q-branches, and if $B' - B''$ is small, as is usually so, these Q-branches do not have their structure resolved. If the molecule is fairly heavy, the P- and R-branch transitions are swamped by the Q-branches and in general are not resolved.

The intensities shown for the P- and R-branch transitions of the sub-bands were calculated from the intensity formula developed in Chapter 5, together with the Boltzmann factor using calculations from CH_3D inertial parameters. Again no nuclear spin statistics were included. It is to be noted that the stronger branch in the sub-bands is the one for which $\Delta J = \Delta K$; that is, for $\Delta K = -1$ sub-bands the P-branch is strong, whereas for $\Delta K = +1$ sub-bands the R-branch is strong.

Consider that the same K-component of the Q branches* can be observed; then from (1)

$$^RQ_K - {}^PQ_K = 4(A' - B')K,$$

and similarly (3)

$$^RQ_{(K-1)} - {}^PQ_{(K+1)} = 4(A'' - B'')K.$$

* The notation ^RQ_K stands for the set of Q-branch transitions where $K + 1 \leftarrow K$. Similarly, ^PQ_K denotes the Q-branch transitions with $K - 1 \leftarrow K$.

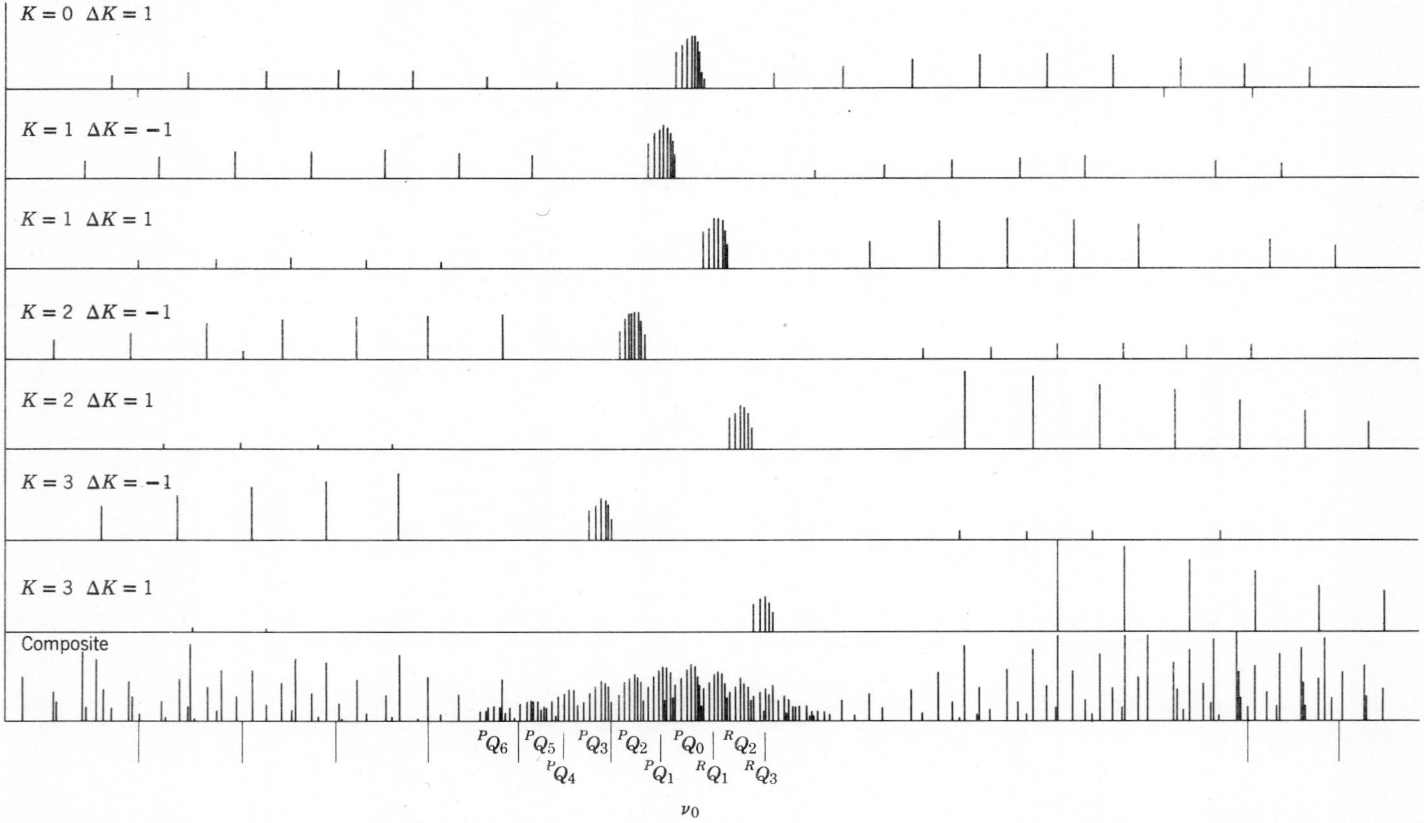

Fig. 7c1 The sub-band structure of a perpendicular band calculated using the inertial constants of CH_3D. No nuclear spin statistics are included. In the lowest line the sub-bands are superimposed so that one can appreciate the over-all appearance of the band. (From Herzberg, *Infrared and Raman Spectra*, copyright 1945, D. Van Nostrand Company, Inc., Princeton, New Jersey.)

Thus, if we can observe a perpendicular transition between two nondegenerate states, the analysis is quite straightforward, and all the other constants would be determined using techniques discussed under parallel bands. Unfortunately, such a transition is found only in an accidentally symmetric rotor, and as yet none has been found.

If a molecule is a symmetric rotor due to its symmetry (a threefold or higher symmetry axis), then perpendicular transitions originating in the ground state will always have a degenerate excited state, and the simple analytical procedure as given is not sufficient. The energy levels of a degenerate state of a symmetric rotor were discussed in Chapter 4, where it was found that the degeneracy was removed by the interaction of rotation and vibration. The degenerate state energy levels are given by 4d17.

$$F(J, K) = BJ(J+1) + (A-B)K^2 \mp 2\zeta AK + \text{centrifugal distortion terms}, \quad K = 0, 1, 2 \ldots. \tag{4}$$

The upper sign applies when $\Delta K = +1$ and the lower sign when $\Delta K = -1$. The expression for the Q-branch frequencies when a degenerate upper state is present becomes

$$\nu^Q = \nu_0 + [A'(1-2\zeta) - B'] \pm 2[A'(1-\zeta) - B']K + [(A'-B') - (A''-B'')]K^2 + (B'-B'')J(J+1), \tag{5}$$

and the sub-band centers are given by

$$\nu_0^{Q_K} = \nu_0 + [A'(1-2\zeta) - B'] \pm 2[A'(1-\zeta) - B']K + [(A'-B') - (A''-B'')]K^2. \tag{6}$$

The band structure is not altered by this interaction; the result is a new effective A' constant and hence a different separation for the Q-branch series. Also since ζ can vary considerably from state to state, we find drastically different Q-branch separations from band to band in the same molecule. The selection rules are such that the $\Delta K = -1$ transitions do not have the same excited state as the $\Delta K = +1$ transitions even when K in the excited state is the same. The transitions are shown in Fig. 7c2, in which it is clearly shown that the $\Delta K = +1$ and $\Delta K = -1$ transitions have totally different excited states.

The P- and R-branch transitions of the sub-bands of a perpendicular band are given by a formula very similar to that given before for linear or parallel type bands.

$$\nu^{P,R} = \nu_0^{Q_K} + (B' + B'')m + (B' - B'')m^2 \tag{7}$$

as long as centrifugal distortion terms are neglected.

Because of the splitting of the energy levels of the degenerate state, the simple difference relations (3) must be modified to become

$$\begin{aligned} {}^RQ_{(K-1)} - {}^PQ_{(K+1)} &= 4(A'' - A'\zeta - B'')K \\ {}^RQ_K - {}^PQ_K &= 4(A' - A'\zeta - B')K. \end{aligned} \tag{8}$$

In principle, the ζ's can be calculated from the force constants and molecular geometry if such information is available. Hence a value of A'' can be obtained from (7), (8), and a knowledge of the ζ's. In practice the information necessary to calculate

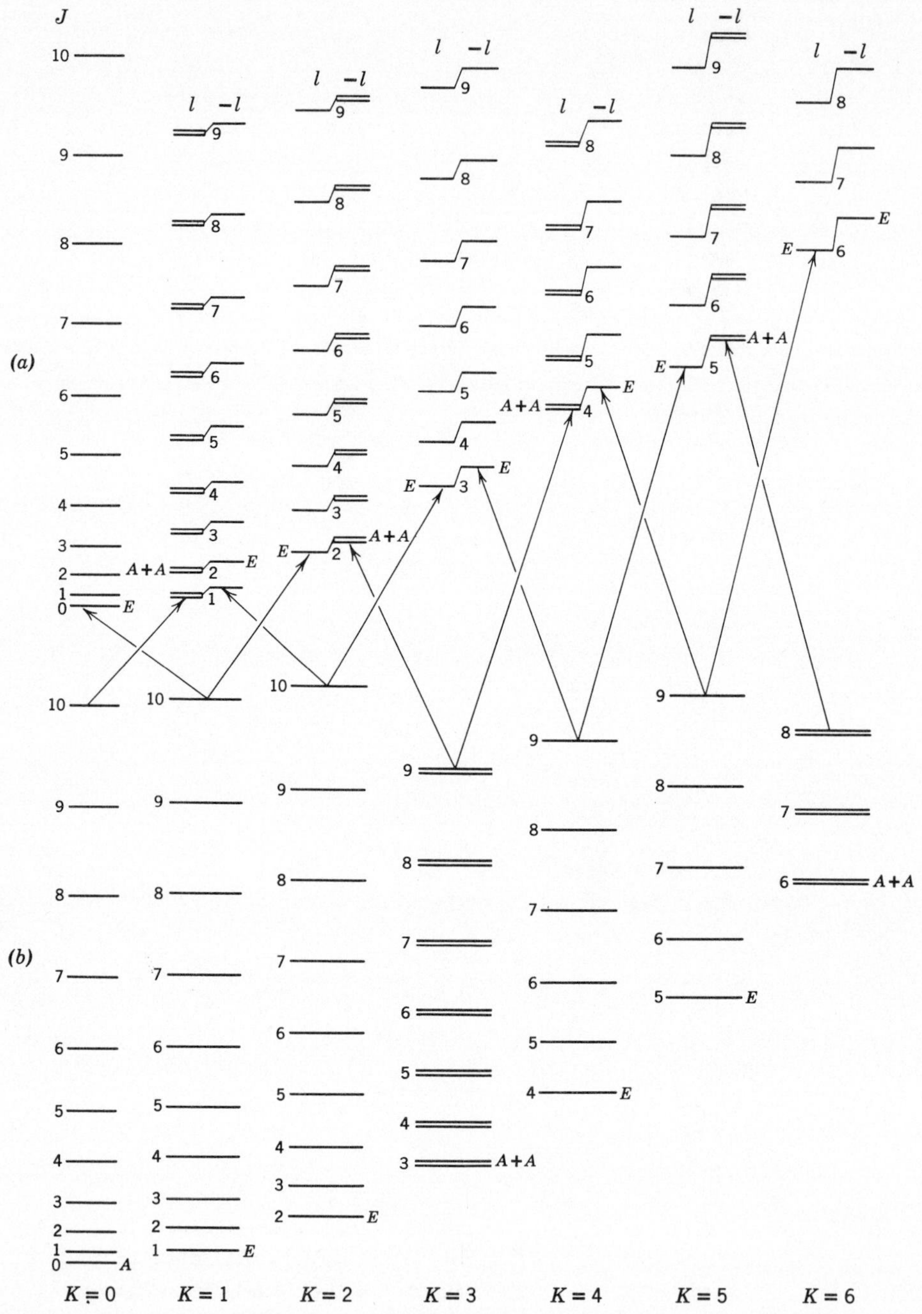

Fig. 7c2 (*a*) The rotational energy levels (positive K) of a degenerate vibrational state of a symmetric top. (*b*) The rotational levels of a nondegenerate totally symmetric vibrational state of a symmetric top. The symmetries of the rotational levels in both states are shown. Since all the levels in the same column have the same symmetry, only the symmetry of the lowest level is shown. The transitions of a perpendicular band are indicated by the arrows. A molecular point group of C_{3v} has been assumed (also applies to C_{3h} and C_3). For a dihedral group the two A levels are distinguished as A_1 and A_2 see [4, pp. 406–411]. (From Herzberg, *Infrared and Raman Spectra*, copyright 1945, D. Van Nostrand Company, Inc., Princeton, New Jersey.)

the ζ's is seldom available for molecules of interest, so some other method must be resorted to.

In Chapter 4, where the ζ-sum rule was discussed, it was shown that the sum of the ζ's for a given symmetry species can be calculated without regard to the potential function. This relation then should help to complete the analyses of perpendicular bands. The procedure requires the observation of all the bands of a given symmetry class for a given molecule with comparable resolution. The latter requirement is very difficult to fulfill since the wavelength differences in these bands require great sensitivity over a larger spectral range than is currently available. However, assuming the experimental problem is soluble, consider a molecule such as CH_3D which has C_{3v} symmetry. Elementary considerations indicate that this molecule has nine fundamental vibrations, of which three are totally symmetric and the other six are broken into three degenerate pairs of species E. Thus these are the three states which must be observed as part of a transition, preferably as the excited state of a fundamental band. The result of the analysis of these three bands leads to three derived constants

$$\begin{aligned}(A_4 - A_4\zeta_4 - B_4) &= \Delta\nu_4 \\ (A_5 - A_5\zeta_5 - B_5) &= \Delta\nu_5 \\ (A_6 - A_6\zeta_6 - B_6) &= \Delta\nu_6\end{aligned} \tag{9}$$

together with the ζ-sum rule

$$\zeta_4 + \zeta_5 + \zeta_6 = \frac{B}{2A}\,. \tag{10}$$

Since the B values can be determined in other ways such as from P- and R-branch transitions of the sub-branches of the perpendicular bands, this leaves six unknowns, A_4, A_5, A_6, ζ_4, ζ_5, ζ_6, to be determined from four pieces of information. Thus it is not possible to determine all of these quantities from the analysis of infrared spectra. Some usable information can be obtained by making the approximation that A and B are the same for all the excited states. The error involved in this assumption cannot be fairly estimated but should not exceed 5%. With this assumption (9) and (10) can be reduced to

$$\Delta\nu_4 + \Delta\nu_5 + \Delta\nu_6 = 3A - \tfrac{7}{2}B, \tag{11}$$

from which A is obtained if B is known. It is hard to assess the value of an A obtained in this manner, but it is certainly better than no A value at all and is probably as accurate as the value obtained by calculating ζ's on the harmonic approximation.

As an example of a perpendicular band which shows only the main features, consider ν_8 of methylacetylene located near 10 μ [25]. The data are shown in Fig. 7c3 and the measurements in Table 7c1. The intensity alternation for the Q-branches arising from the nuclear spin statistics is clearly evident. Since the intensity of the $K = 0$ Q-branch is normally the strongest and the K-branches fall off in intensity as K increases, the strongest Q-branch in the region must be the ${}^RQ_0$. Once this identification has been made the assignment of the other quantum numbers is straightforward, and it is clear that the ${}^PQ_3$ and ${}^RQ_3$ are strong as required by nuclear statistics. The analysis of this band can be carried out using (8) plus a sum relation which helps locate the band center

$$\begin{aligned}{}^RQ_k + {}^PQ_k = 2[\nu_0 + A'(1 - 2\zeta) - B'] + 2[(A' - B') \\ - (A'' - B'')]K^2 + 2(B' - B'')J(J + 1).\end{aligned} \tag{12}$$

For data of this type, in which the resolution is not great, graphical methods of solution would seem to be adequate. From this band the authors find $[A'(1 - \zeta) - B'] = 5.81\ \text{cm}^{-1}$ and $[(A' - B') - (A'' - B'')] = 0.021\ \text{cm}^{-1}$. The $CH_3—C{\equiv}CH$ ground state inertial constants could be calculated from microwave results; and using $A' = A''$

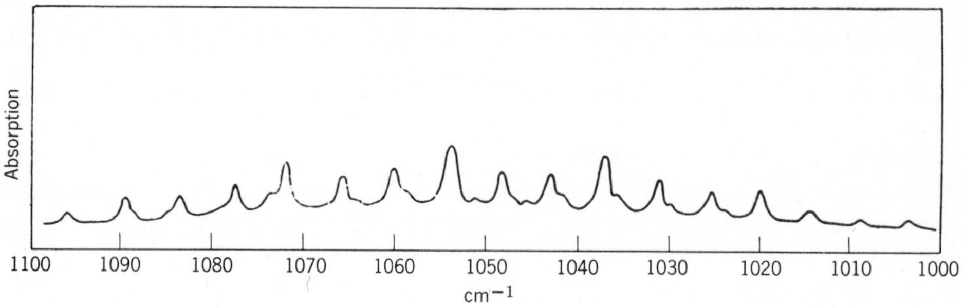

Fig. 7c3 Methylacetylene, band ν_8; path length 10 cm; pressure 520 mm; slit widths 0.6 cm^{-1} [25].

and $B' = B''$ it was found that $\zeta = 0.387$ and $\nu_0 = 1052.5\ \text{cm}^{-1}$. This band is typical of what one can expect to observe with medium resolution. The only recognizable features are the unresolved Q-branches and only a minimum amount of information can be obtained.

***Table 7c1—Perpendicular Band ν_8. Positions of Lines in cm^{-1}* [25]**

ν_{vac}	ASSIGNMENT	ν_{vac}	ASSIGNMENT
1096.1	${}^RQ_7$	1048.2	${}^PQ_1$
1089.7	${}^RQ_6$	1045.8	
1084.9		1043.0	${}^PQ_2$
1083.7	${}^RQ_5$	1041.8	
1077.7	${}^RQ_4$	1037.0	${}^PQ_3$
1073.9		1035.6	
1071.9	${}^RQ_3$	1031.2	${}^PQ_4$
1065.7	${}^RQ_2$	1029.6	
1059.6	${}^RQ_1$	1025.5	${}^PQ_5$
1053.8	${}^RQ_0$	1023.8	
1050.9		1020.0	${}^PQ_6$
		1014.4	${}^PQ_7$
		1008.8	${}^PQ_8$
		1003.2	${}^PQ_9$

As an example of a perpendicular band in which it is possible to obtain more information a methylfluoride band at 3010 cm^{-1} is considered [93]. The data are shown in Fig. 7c4. The sub-band assignments are shown above the strips of the diagram. The wave numbers and assignments of quantum numbers are given in Table 7c2. When this much information is available, there are several other combination and

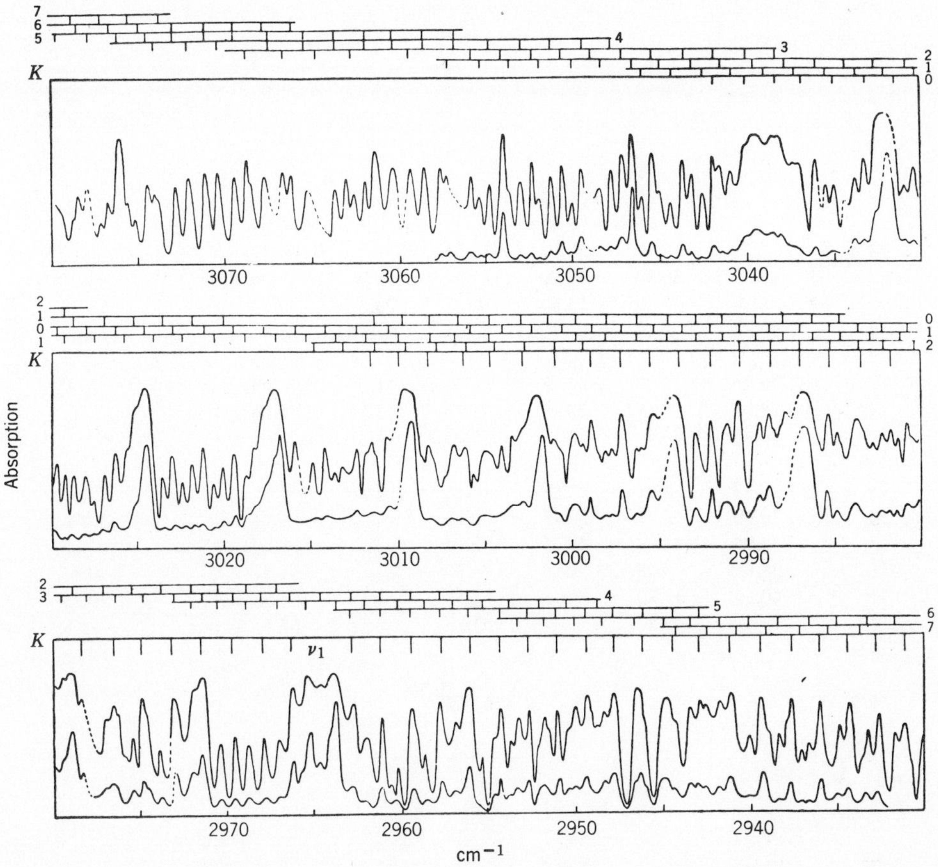

Fig. 7c4 The perpendicular band of methyl fluoride at 3030 cm^{-1}, and the parallel band at 2970 cm^{-1} [93].

difference relations which can be useful. It should be pointed out that when a degenerate excited state is present, it is not possible to separate completely the constants of the two states because the $\Delta K = -1$ and $\Delta K = +1$ transitions have different excited states. Thus, neglecting centrifugal distortion effects,

$$
\begin{aligned}
&(a)\ {}^RR_{(K-1)}(J-1) - {}^PP_{(K+1)}(J+1) = 4(A'' - A'\zeta - B'')K + 4B''(J+\tfrac{1}{2})\\
&(b)\ {}^RR_{(K-1)}(J) - {}^PP_{(K+1)}(J) = 4(A'' - A'\zeta - B'')K + 4B'(J+\tfrac{1}{2})\\
&(c)\ {}^RR_K(J-1) - {}^PP_K(J+1) = 4(A' - A'\zeta - B')K + 4B''(J+\tfrac{1}{2}) \qquad (13)\\
&(d)\ {}^RR_K(J) - {}^PP_K(J) = 4(A' - A'\zeta - B')K + 4B'(J+\tfrac{1}{2})\\
&(e)\ {}^RR_K(J-1) + {}^PP_K(J) = 2[\nu_0 + A'(1-2\zeta) - B']\\
&\qquad + 2[(A'-B') - (A''-B'')]K^2 + 2(B'-B'')J^2.
\end{aligned}
$$

Relations (8) and (12) are first applied to the maxima of the Q-branches from which one obtains preliminary values of $\nu_0 + [A'(1-2\zeta) - B']$, $[A'(1-\zeta) - B']$, and $[(A'-B') - (A''-B'')]$. These constants then enable the calculations of the centers

Table 7c2—Sub-Bands of the

J \ K	K = 0	K = 1		K = 2		K = 3		K = 4	
	$\Delta K = \pm 1$	$\Delta K = 1$	$\Delta K = -1$	$\Delta K = 1$	$\Delta K = -1$	$\Delta K = 1$	$\Delta K = -1$	$\Delta K = 1$	$\Delta K = -1$
$\Delta J = 0$	3009.28	3016.87	3001.78	3024.42	2994.07	3031.91	2986.59	3039.52	2978.84
$\Delta J = 1$									
0	3010.59								
1	12.42	3019.98	3004.89						
2	14.13	21.77	06.86	3029.27	2999.47				
3	15.82	23.46	08.57	30.99	3000.95	3038.60			
4	17.58	25.12	09.94	32.59		40.32		3047.80	
5	19.34	26.90	11.80	34.41	04.04	42.00		49.50	
6	21.10	28.69	13.64	36.17	06.24	43.75		51.33	
7	22.84	30.44		38.14	07.92	45.47		53.07	
8						47.23		54.90	
9	26.26	33.87	18.73	41.58	11.33	48.99	3003.49	56.60	
10	28.04	35.61	20.49	43.18	13.17	50.68		58.59	
11	29.85	37.42	22.27	45.05	14.81	52.45		60.46	
12						54.10		61.97	
13	33.37	40.97	25.69	48.48	18.19	55.91		63.63	
14	35.19	42.59	27.43	50.12		57.64		65.28	
15	37.04	44.39	29.27	52.01		59.44		67.07	
16	38.60	46.08		53.70		61.37		68.71	
17	40.32	47.80		55.52		63.06		70.39	
18		49.50		57.28		64.55		72.09	
19						66.62		74.09	
20						68.44		75.55	
21				62.62					
22									
23									
$\Delta J = -1$									
1			2999.91						
2	3005.57		98.17		2990.46				
3	04.04	3011.51	96.36		88.83		2981.31		
4		09.94		3017.58			79.47		2971.94
5	00.74	08.18	93.04	15.82	85.43		78.01		70.33
6	2998.93	06.45	91.46	14.13	83.72		76.26		68.66
7	97.21	04.89	89.68	12.42	82.28		74.60		67.09
8	95.46	03.16	88.06	10.59	80.42		72.87		
9									
10	92.05	2999.91		07.30	76.98		69.58		62.02
11	90.46	98.17	82.85	05.57	75.39		67.90		60.23
12	88.83	96.36	81.31	04.04	73.79		66.20		58.62
13			80.00		71.94		64.57		56.93
14	85.43	93.04	78.01		70.33		62.82		55.39
15	83.72	91.46	76.57		68.66		61.15		53.74
16	82.28	89.68	74.86		67.09		59.42		51.82
17	80.77		73.03				57.75		50.39
18									48.69
19	77.51						54.64		
20					60.58				
21									
22									
23									
24							46.33		
25							44.72		

CH_3F *Band at 3010 cm^{-1}* (units of cm^{-1})

$K = 5$		$K = 6$		$K = 7$		$K = 8$		$K = 9$	
$\Delta K = 1$	$\Delta K = -1$	$\Delta K = 1$	$\Delta K = -1$	$\Delta K = 1$	$\Delta K = -1$	$\Delta K = 1$	$\Delta K = -1$	$\Delta K = 1$	$\Delta K = -1$
3046.62	2971.35	3053.70	2963.75	3061.37	2956.08	3068.71	2947.76		2940.07
3056.60									
58.59		3066.08							
60.46		67.70		3075.13					
61.97		69.43		76.82		3084.59			
63.63		71.06		78.77		86.31		3093.70	
65.28		72.80		80.56		88.02		95.33	
67.01		74.39		82.29		89.70		97.05	
		76.82		84.03		91.43		98.76	
71.06		78.46		85.72		93.02			
72.80		80.16		87.41		94.76			
74.39		81.81		89.09		96.28			
76.03		83.51		90.80		98.05			
77.87		85.18							
79.60		86.97							
81.24		89.09							
		92.48							
		94.19							
		95.88							
	2962.82								
	61.15		2953.28						
	59.22		51.56						
	57.75		50.02		2942.53		2934.39		
	56.08		48.38				32.76		2925.27
	54.35		46.56		39.01		31.06		23.39
	52.66		45.06		37.12		29.40		21.92
	51.02		43.40		35.51		27.76		
	49.32		41.81				26.15		
	47.76		40.07				24.53		
	46.03		38.48		30.72		22.97		
	44.28		36.69		28.96				
	42.53		34.95		27.12				
	41.07		33.29		25.69				
			31.68		22.98				
			29.98		22.38				
			28.40						
			26.68						

of the sub-bands. In order to identify the P- and R-branch transitions of the sub-bands, some estimate of B is necessary. This is probably available from the previous analysis of a parallel-type band. The approximate positions of the P- and R-branch lines can then be calculated for each sub-band by the relation

$$\nu = \nu_{\text{sub}} + (B' + B'')m + (B' - B'')m^2. \tag{14}$$

Line positions estimated in this manner should enable one to make the assignments. This procedure for assigning quantum numbers was used in the analysis of the

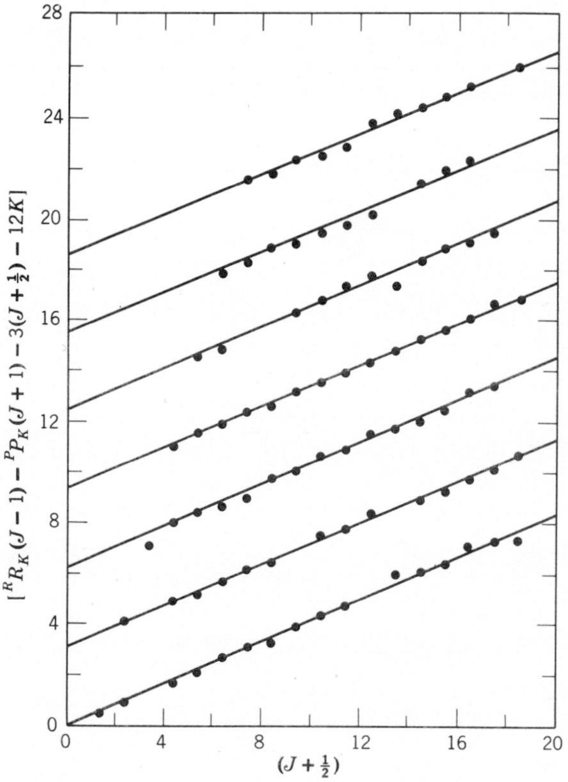

Fig. 7c5 Plots of ${}^R R_K(J-1) - {}^P P_K(J+1)$ against $(J+\frac{1}{2})$ for different values of K in the perpendicular band of CH_3F at 3030 cm^{-1}, lower state [93].

CH_3F band. Keeping K constant, values of the combination sum (13*e*) were plotted against J^2, the value of $B' - B''$ being determined by the slope of these plots. Then the intercepts of these plots were plotted against K^2 to obtain $[(A' - B') - (A'' - B'')]$ and $\nu_0 + A'(1 - 2\zeta) - B'$. The combination differences (13*c*) and (13*d*) were plotted against $(J + \frac{1}{2})$ for different values of K to determine B' and B''. These plots for the ground state are shown in Fig. 7c5. The intercepts of these plots give $4(A' - A'\zeta - B')K$, and plotting the intercepts against K, as shown in Fig. 7c6, yields a mean value of $[A'(1 - \zeta) - B']$. The P- and R-branch lines for the $K = 0$ sub-band can be treated by the usual combination relationships in order to obtain other estimates of B', B''. The mean values of the constants found in this manner are in cm^{-1}

$$B'' = 0.851_5 \qquad B' = 0.853_4$$
$$\nu_0 + A'(1 - 2\zeta) - B' = 3009.12$$
$$A'(1 - \zeta) - B' = 3.77_6$$
$$(A' - B') - (A'' - B'') = -0.01$$
$$B' - B'' = +0.0019.$$

When $K > 6$ those constants do not reproduce the observed spectrum well. There are deviations in both position and intensity which Pickworth and Thompson could not explain. On the whole the analysis is quite satisfactory, considering that centrifugal distortion effects were ignored.

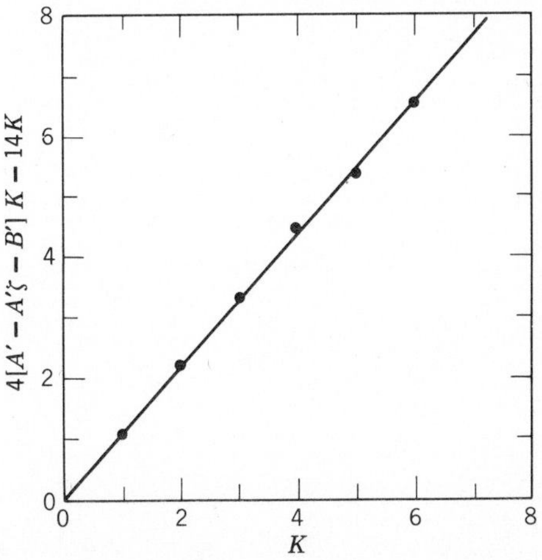

Fig. 7c6 Plot of the intercepts $4[A' - A'\zeta - B']K$ of Fig. 7c5 against K [93].

As a third example [16] of a perpendicular band the combination band of CH_3D at 2780 cm^{-1} is discussed. The observed absorption is shown in Fig. 7c7. In this band it was possible to resolve the individual transitions of the Q-branches. The K quantum numbers are readily assigned with the help of the intensity alternation due to the nuclear spin statistics. The strongest Q-branch is again assigned to $^RQ_0$. Because this molecule is comparatively light and because the individual Q-branch transitions are resolved, it is not possible to observe the Q-branches to as high quantum numbers as was possible for CH_3F where the Q-branch transition could not be resolved.

The manner of assigning the P- and R-branch transitions in the analysis of this band was by a different technique than was used previously. For each sub-band, the Q-branch transitions are given by (5) plus a correction for centrifugal distortion. The J quantum numbers are assigned in each Q-branch on the basis of intensity. One predicts that the transitions of maximum intensity is that for $J = 5$ or 6. Another help is the fact that as K increases, since $J > K$ there are less transitions to be assigned. This fact can readily be seen from the indicated assignments in Fig. 7c7. Using these assignments and (5) an effective band center for each K, as well as $B' - B''$ and $D_J' - D_J''$, were calculated* for each sub-band. Use was then made of the known ground state constants obtained from the analysis of the parallel band described in 7b. Values were calculated for $B' + B''$, and $D_J' + D_J''$, the results being used to calculate P- and R-branch transitions for each of the sub-bands according to the formula

$$\nu^{P,R} = \nu_0^{Q_K} + (B' + B'')m + (B' - B'')m^2 - 2(D_J' + D_J'')m^3 - (D_J' - D_J'')m^4 \qquad (15)$$

In (15) it is assumed that $|D_J'' - D_J'| \ll |B' - B''|$ and $D_{JK}' = D_{JK}'' = 0$. This method predicted the observed P- and R-branch transitions for nine sub-bands, $K > 0$, with the largest deviation between observed and calculated being 0.05 cm^{-1}. The

* Assuming $D_{JK}' = D_{JK}'' = 0$.

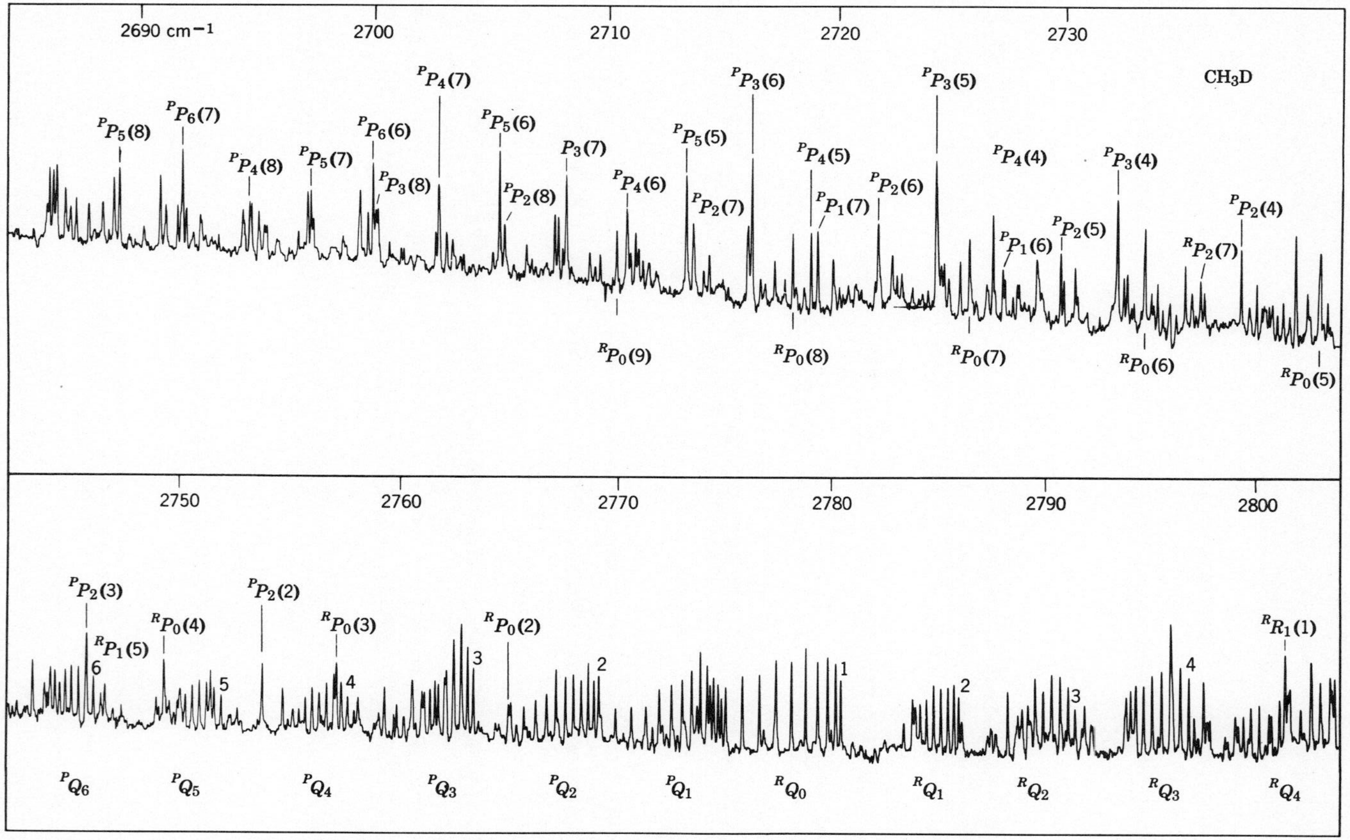

Fig. 7c7 The perpendicular band $\nu_3 + \nu_5$ of CH_3D. The transitions have been designated by the notation suggested by Herzberg. The pre-superscript P or R means $\Delta K = -1$ or $\Delta K = +1$, respectively. The basic symbol P, Q, or R has the usual meaning $\Delta J = -1, 0, +1$. The subscript gives the K value of the ground state sub-band, and the number in parentheses is the J value of the ground state [16].

assignments that could be made in each sub-band are given in Table 7c3. The P- and R-branch transitions of the $K = 0$ sub-band were found by trial and error using known ground state constants.

Excited state B-values for the $K = 0$ sub-band can be calculated in two ways. One method uses the P- and R-branch transitions and $\Delta F_2'$ values. The second method is to calculate $B' - B''$ from (5). For the $K = 0$ sub-band these two methods lead to considerably different values of B'. Such a circumstance is reminiscent of a perpendicular (Π–Σ) band of a linear molecule where two rotational levels in a degenerate vibrational state are split by the so-called l-type doubling. This perturbation was discussed in 4g and 6b and it is easily seen that the $K = 0$ sub-band of a symmetric rotor is entirely analogous to the Π–Σ transition of a linear molecule. Thus from the two B' values the interaction parameter q could be evaluated.

Sufficient lines in the P- and R-branches of sub-bands were assigned to enable a determination of the sub-band centers using the combination sum

$$^RR_K(J-1) + {^RP_K}(J) = 2\nu_0^{Q_K} + 2(B' - B'')J^2 - 2(D_J' - D_J'')J^4. \qquad (16)$$

The highest value of K for any of these sub-bands is three. Similarly, the band centers of these same sub-bands could be determined from (5). The sub-band centers found by these two methods are in excellent agreement. The results are compared in Table 7c4. In addition, it is possible to determine centers from three other sub-bands using only (16). These sub-band centers are then used together with (6) to determine the quantities

$$\nu_0 + [A'(1 - 2\zeta) - B'], \qquad [A'(1 - \zeta) - B'] \qquad \text{and} \qquad [(A' - B') - (A'' - B'')].$$

This treatment ignores the effect of centrifugal distortion, but this omission should not be serious since only small values of J and K are used. The main error arises because nowhere in the analysis is D_{JK} considered.

In view of the perturbation in the excited state, the usual combination relations (13) cannot be used in the analysis of this band. However, if this l-type doubling is the only perturbation present, the combination relation (12) for a constant K somewhat averages out this perturbation. Thus if one plots

$$\frac{1}{2}\left[{^RQ_K}(J) + {^PQ_K}(J)\right] \quad \text{versus} \quad J(J+1),$$

an approximate straight line should result. Actually (12) should have a centrifugal distortion correction term $(D_J' - D_J'')J^2(J+1)^2$, which should be subtracted from ${^RQ_K}(J) + {^PQ_K}(J)$ in order to obtain a better straight line. Such plots for three values of K give good straight lines and thus seems to rule out further perturbations.

The value of q obtained from the $K = 0$ sub-band does not appear to be entirely satisfactory for the other sub-bands. However, not enough transitions in the other sub-bands are observed to enable one to make meaningful higher order corrections. The constants derived from the analysis of this band are collected in Table 7c5. Unfortunately, the inadequacy of the data makes these constants subject to uncertainties which are larger than would be normal for measurements of this precision. The strong overlapping in the high frequency portion of the absorption sharply limits the number of ^RP and ^PR lines which can be assigned, and the general weakness of the ^RP and ^PR lines makes it impossible to assign many of these transitions. However,

Table 7c3—Sub-Bands of CH_3D

	$K = 0$	$K = 1$		$K = 2$	
	$\Delta K = \pm 1$	$\Delta K = 1$	$\Delta K = -1$	$\Delta K = 1$	$\Delta K = -1$
J					
$\Delta J = 0$					
1	2780.42 cm^{-1}		2774.85		
2	80.20	2785.96	74.70		2769.15
3	79.82	85.74	74.48	2791.43	68.91
4	79.35	85.44	74.19	91.11	68.64
5	78.78	85.11	73.85	90.75	68.33
6	78.13	84.78	73.45	90.34	67.98
7	77.40	84.43	73.00	89.95	67.60
8	76.61		72.52	89.57	67.18
9	75.82		71.95	89.23	66.73
10	75.04		71.32		66.23
$\Delta J = +1$					
0	2788.27				
1	95.99	2801.48			
2	2803.64	09.00	2797.76	2814.70	2792.19
3	11.28	16.47	2805.26	22.15	99.68
4	18.93	23.89	12.65	29.51	2807.12
5	26.61	31.35	20.00	36.85	14.50
6	34.32	38.71	27.29	44.20	21.83
$\Delta J = -1$					
1					
2	2764.99		2759.33		2753.76
3	57.15		51.44		45.85
4	49.35	2754.70	43.47		37.91
5	41.47	46.67	35.43	2752.36	29.89
6	33.63	38.61	27.34	44.27	21.83
7	25.85	30.54	19.20	36.12	13.76
8	18.12		11.05	27.97	05.64
9					
10					

Table 7c4—Sub-Band Centers of $\nu_3 + \nu_5$ *Band of* CH_3D [16]

K	ΔK	Q-BRANCH	R- AND P-BRANCH
0	1	2780.55 cm^{-1}	2780.55
1	1	2786.15	2786.17
1	−1	2774.90	2774.88
2	1	2791.90	2791.90
2	−1	2769.33	2769.31
3	1	2797.73	
3	−1	2763.76	2763.75
4	−1	2758.30	
5	−1	2752.77	

Perpendicular Band at 2780 cm^{-1}

$K = 3$		$K = 4$	$K = 5$	$K = 6$
$\Delta K = 1$	$\Delta K = -1$	$\Delta K = -1$	$\Delta K = -1$	$\Delta K = -1$
	2763.40			
2796.86	63.14	2757.66		
96.44	62.82	57.38	2751.91	2746.15
	62.47	57.04	51.59	45.85
95.54	62.09	56.71	51.27	45.50
95.11	61.64	56.37	50.92	45.19
94.70	61.14	56.05	50.61	44.90
	60.58		50.30	
2827.87	2794.16			
35.20	2801.62			
42.51	09.00			
	16.29			
	2740.35			
	32.39	2726.91		
2758.13	24.40	18.91	2713.46	
49.96	16.34	10.89	05.43	2700.03
41.78	08.27	02.85	2697.39	2691.97
33.63	00.14	2694.77	89.32	83.91
			81.27	
			73.17	

Table 7c5—Derived Constants from the $\nu_3 + \nu_5$ Band of CH_3D [16]

$\nu_0 + [A'(1 - 2\zeta) - B']$	$= 2780.55$ cm^{-1}
$A'(1 - \zeta) - B'$	$= 2.82$
$(A' - B') - (A'' - B'')$	$= 0.02$
B''	$= 3.88$
$B' - B''$	$= -0.041$
$D_J' - D_J''$	$= -0.00011$
q	$= 0.011$

it is felt that the analysis is correct in all essential features and that only the evaluation of the small higher order corrections needs improvement.

If A'' is calculated from the methane geometry, it is then possible to evaluate ζ and ν_0. Using $A = 5.24\ \text{cm}^{-1}$, one finds $\zeta = -0.26$. Simple theory predicts (Chapter 4) that the ζ of this band should be the same as for the fundamental ν_5. Herzberg [4] has estimated $\zeta_5 = -0.27$ in remarkable agreement, but this may be partly fortuitous. The preceding figures lead to a value of $\nu_0 = 2776.33\ \text{cm}^{-1}$.

Thus, in principle, the analysis of perpendicular bands can be accomplished in a straightforward manner using combination and difference relations. However, it is generally found that nature frustrates the experimenter in many ways, and some modifications ,of the textbook procedure must be introduced. This band of CH_3D serves to illustrate some of the difficulties which can arise. Further difficulties are seen in the discussion of hybrid bands.

7d HYBRID BANDS

In some cases it is possible that $M_x = M_y \neq M_z \neq 0$. When this is true for a transition, the resulting band has the features of both a parallel band and a perpendicular band. Such a band is called a hybrid band. This type of transition can only occur in accidentally symmetric rotors and in the higher overtones and combinations of a necessarily symmetric rotor. In this transition it may be that $M_x = M_y \gg M_z$, or conversely; then only the component with the large change in electric moment is observed. Under those circumstances one may miss the fact that the band under investigation is a hybrid. The only hybrid band which is discussed in detail is that due to a higher overtone of a degenerate fundamental in a necessarily symmetric rotor. Table 7d1 gives the symmetry species of the components of the overtones of degenerate fundamentals for a few common point groups. If the vibrational selection rules are such that transitions from the ground state are permitted to more than one of these species, then a hybrid band should be found.

For example, if the molecule belongs to the point group C_{3v}, the first overtone of a degenerate fundamental has two components, A_1 and E. The electric moment selection rules may permit transitions to both these components. Hence, if M_x, M_y, and M_z are sufficiently large, one should see a hybrid band in the region of absorption of this overtone.

Consider the molecule CD_3H which belongs to this point group. This molecule has two degenerate E fundamentals. The selection rules for this molecule allow transitions from the ground state to both A_1 and E levels. Thus one would expect to find a hybrid band in the overtone regions of the degenerate fundamentals.

In fact, a hybrid band of this molecule has been observed near $2600\ \text{cm}^{-1}$ [16]. The identifying features of this band may be seen in Fig. 7d1. The collected Q-branch of the parallel component is clearly evident near $2565\ \text{cm}^{-1}$, whereas the series of Q-branches characteristic of the perpendicular component is clearly discernible centered near $2590\ \text{cm}^{-1}$.

The analysis of a hybrid band follows the procedures outlined for parallel and perpendicular bands, each component being analyzed separately. We start the analysis by identifying the collected Q-branch of the parallel component near 2565 cm^{-1}. It is then possible to identify the P- and R-branch transitions of this band, which are labeled in Figs. 7d1 and 7d2. Unfortunately, since the absorption is very weak, the K-components of the P- and R-branch transitions could not be identified, only the general absorption peak representing the aggregate absorption of all the K-components. Thus precise molecular constants could not be determined for this component of the overtone. The band center of this component is 2564.6 cm^{-1}. The B_0 value is compatible with the previously determined values, 3.278 cm^{-1}, and the $B' - B''$ value is about 0.01 cm^{-1}.

Table 7d1—Symmetry Species of the Higher Vibrational Levels of Degenerate Vibrations for Some of the Point Groups *

POINT GROUP	SPECIES OF VIBRATIONS, FUND.	RESULTING STATES		
C_{3v}	$(E)^2$	$A_1 + E$		
	$(E)^3$	$A_1 + A_2 + E$		
	$(E)^4$	$A_1 + 2E$		
D_{3h}	$(E')^2$	$A_1' + E'$	$(E'')^2$	$A_1' + E'$
	$(E')^3$	$A_1' + A_2' + E'$	$(E'')^3$	$A_1'' + A_2'' + E''$
	$(E')^4$	$A_1' + 2E'$	$(E'')^4$	$A_1' + 2E'$
D_{4h}	$(E_g)^2$	$A_{1g} + B_{1g} + B_{2g}$	$(E_u)^2$	$A_{1g} + B_{1g} + B_{2g}$
	$(E_g)^3$	$2E_g$	$(E_u)^3$	$2E_u$
	$(E_g)^4$	$2A_{1g} + A_{2g} + B_{1g} + B_{2g}$	$(E_u)^4$	$2A_{1g} + A_{2g} + B_{1g} + B_{2g}$

* The method for determining these species is outlined in [9, p. 151ff].

The series of Q-branches of the sub-bands belonging to the E-component are clearly evident in Figs. 7d1 and 7d2, the strongest being near 2590 cm^{-1}. This strongest Q-branch belongs to the $K = 0$ sub-band, and the other K-numbering follows automatically, since for this molecule the sub-bands with K a multiple of three are strong due to nuclear spin statistics. Here, the ratio is 11:8. Unfortunately, the structure of these Q-branches cannot be resolved well because the band is very weak, and if the pressure in the absorption cell is increased, the pressure broadening eliminates the fine structure. Also the quantity $(B' - B'')$ is very small in this transition—approximately 0.007 cm^{-1}.

The assignments of the P- and R-branch transitions were made by calculating the approximate line positions using the relation

$$\nu^{P,R} = \nu_0 + [A'(1 - 2\zeta) - B'] \pm [A'(1 - \zeta) - B']K + [(A' - B') - (A'' - B'')]K^2 + (B' + B'')m + (B' - B'')m^2. \quad (1)$$

Here, as usual, the negative sign of the K dependent term applies for $\Delta K = -1$, and $m = -J$ for the P-branch and $m = J + 1$ for the R-branch. The coefficients of K and K^2 were estimated from the Q-branch positions, and the coefficients of m and m^2 were estimated from previous work on this molecule [23, 120]. The resulting assignments

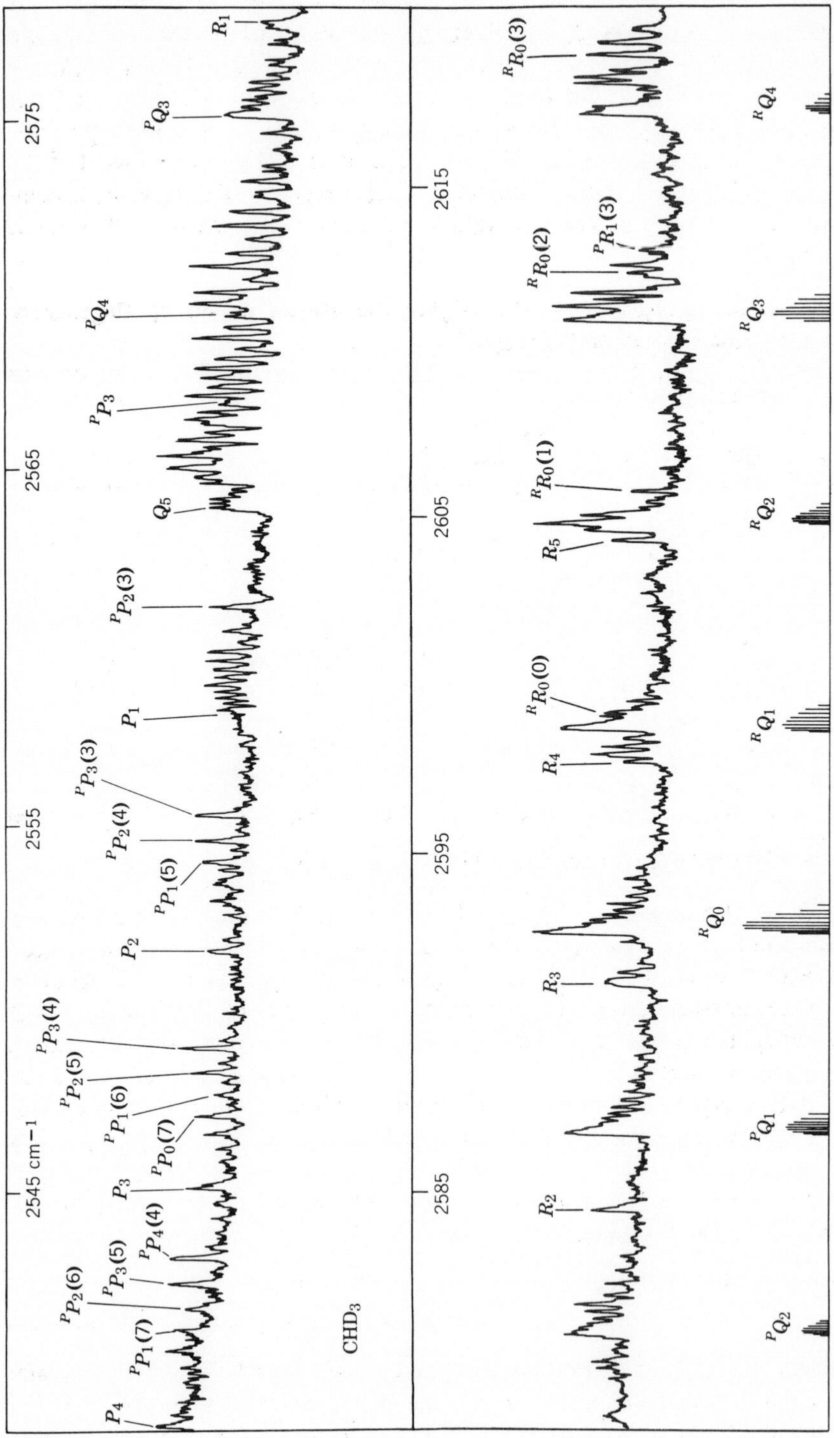

Fig. 7d1 A portion of the $2\nu_5$ band of CD_3H. The upper panel shows the *Q*-branch and some of the *P*-branch transitions of the *A*-component. The lower panel shows the series of *Q*-branches of the *E*-component, as well as some of the *R*-branch transitions of the *A*-component [16].

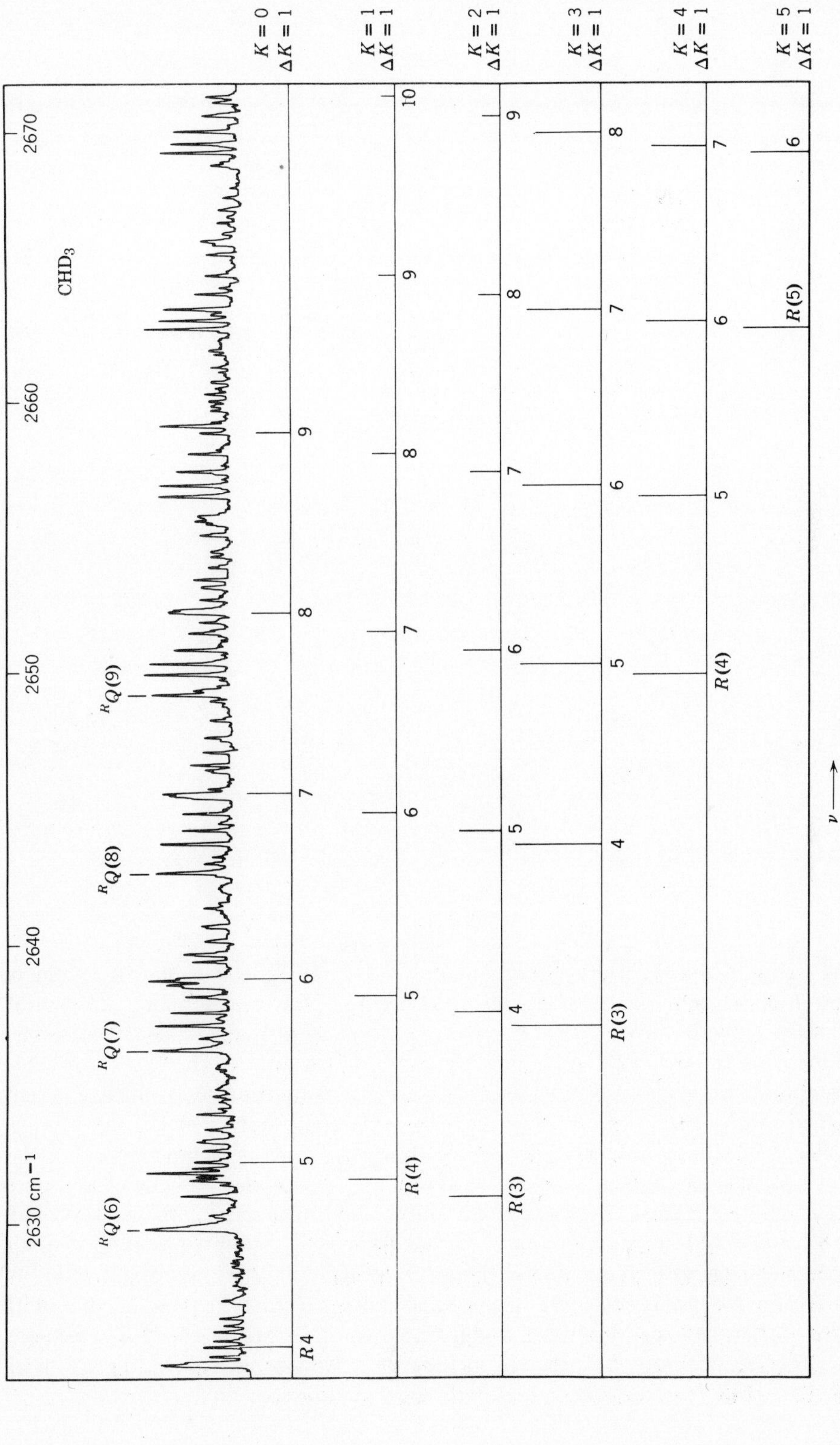

Fig. 7d2 The ${}^{R}R_{K}$ branches of the E-component of the hybrid band $2\nu_5$ of CD_3H [16].

Table 7d2—Sub-Band Assignments for the

		$K = 1$		$K = 2$	
J	$K = 0$	$\Delta K = +1$	$\Delta K = -1$	$\Delta K = 1$	$\Delta K = -1$
$\Delta J = +1$					
0	2599.19 cm^{-1}				
1	2605.76				
2	12.36				
3	18.98		2613.06	2624.40	
4	25.58		19.70	30.39	
5	32.20	2631.57	26.33	37.59	
6	38.84	38.16		44.19	
7	45.52	44.79		50.81	
8	52.17	51.43		57.41	
9		58.07		64.03	
10		64.72		70.66	
11		71.38		77.31	
12		78.07		83.98	
13					
14					
15					
$\Delta J = -1$					
1					
2					
3	2572.99		2567.08		2561.21
4		2572.45	60.58		54.71
5	59.98	65.96	54.09		48.24
6		59.50	47.62		41.80
7			41.21		35.39
8			34.83		
9					

are given in Table 7d2 and shown in Figs. 7d1 and 7d2. Assignments could be made in eleven sub-bands. Unfortunately, because of weakness of the absorption only lines in the branches $\Delta K = \Delta J$ could be assigned with any degree of assurance and these are the only ones used in subsequent calculations. In the $K = 0$ sub-band most of the P-branch lines were masked by other absorption and could not be uniquely assigned to absorption peaks. The measurements could not be extended to longer wavelengths because of the weakness of the absorption and the fact that the sensitivity of the PbS detector falls off rapidly beyond 3.5 μ. Since the Q-branch transitions were not resolved and since, in general, only the stronger branch of the sub-bands could be observed, it was not possible to use the usual combination relations for a perpendicular band in the analysis. Thus an alternate and less desirable method of deriving the molecular constants was necessary. This method involved the use of an expression for the line position deduced from the energy expression for the two energy levels in the transition. In order to include the effects of centrifugal distortion an expression of thirteen terms was essential. One would expect to fit almost any set of

E-Component of the Hybrid Band of CD_3H *

$K = 3$		$K = 4$		$K = 5$		$K = 6$
$\Delta K = 1$	$\Delta K = -1$	$\Delta K = 1$	$\Delta K = -1$	$\Delta K = 1$	$\Delta K = -1$	$\Delta K = 1$
2637.12		2649.89				
43.70		56.48		2662.74		
50.29		63.06		69.31		2675.64
56.89		69.64		75.86		82.18
63.48		76.21		82.42		88.71
70.08		82.80		88.99		95.27
76.69		89.39		95.55		2701.80
83.32		95.97		2702.12		08.36
89.93		2702.58		08.67		14.90
96.55		09.18				21.42
2703.19						27.99
09.83						
	2555.42					
	48.94		2543.21			
	42.48		36.74		2531.09	
	36.06		30.28		24.63	
			23.86		18.19	
					11.78	
					05.39	

* Lines not listed are obscured by another absorption in this band.

observations with this many disposable constants, and the use of this expression is justified only by the results obtained. The expression is

$$\nu = \nu_0 + [A'(1 - 2\zeta) - B'] + bK + cK^2 + dK^3 + eK^4 + fm + gm^2 + hm^3 + im^4 + jmK + kmK^2 + lm^2K + nm^2K^2, \quad (2)$$

in which the coefficients are readily related to the molecular constants from the expressions for the vibrational rotational terms (1*a*) and (4*c*). Eighty-one equations of the type (2) were fit by the method of least squares. The resulting values of the molecular constants, where they can be compared, are in remarkable agreement with those determined previously. The agreement between observed and calculated frequencies, about 0.01 cm^{-1}, and the uncertainties in the constants are very small. The derived constants are given in Table 7d3. The size of the higher order terms indicates that there is no *l*-type doubling interaction in the excited state of this band. Using the methane geometry it is possible to calculate a value for the ground state A of this molecule, $A'' = 2.628\ cm^{-1}$. With this value of A'' and the constants in Table 7d3 one finds a value $\zeta = -1.34$. Thompson [100] has estimated ζ from the fundamental to be 0.67. The simple theory in 4c predicts the ζ here should be the negative of twice the ζ of the fundamental. However, it is thought that the ζ value from the fundamental band is not sufficiently precise to justify comparisons at this time.

Table 7d3—Molecular Constants of CD_3H Derived from the E-Component of the Hybrid Band near 2600 cm^{-1} [16]

$A'(1 - \zeta) - B'$	=	$2.977 \pm .002\ cm^{-1}$
$(A' - B') - (A'' - B'')$	=	$+0.0379 \pm 0.0018$
B''	=	3.278 ± 0.001
B'	=	3.285 ± 0.001
$D_J'' - D_J'$	=	$-5.0 \pm 0.8 \times 10^{-5}$
D_{JK}''	=	-4×10^{-5}
D_{JK}'	=	-1×10^{-5}
$D_K' \sim D_K''$	$\sim$0	
$\nu_0 + [A'(1 - 2\zeta) - B']$	=	2592.637 ± 0.005

The band center ν_0 becomes 2586.09 if one uses the mentioned values of A'' and ζ. The separation of the parallel and perpendicular components of the overtone is given by $g_{55}[l_\perp^2 - l_\parallel^2]$. with $l_\perp = -2$ and $l_\parallel = 0$ in this case. Thus from the component separation $g_{55} = 5.37\ cm^{-1}$.

The analysis of this band is described in some detail since it involves still another way that the analysis of a perpendicular band can be attacked. Again one finds that the straightforward textbook method of analysis can be inadequate. The analysis of these bands requires experience for which it is very difficult to find a substitute.

7e MOLECULES WITH INVERSION SPLITTING

No discussion of symmetric rotor bands would be complete without a discussion of bands involving inversion splitting. These types of bands will probably never be important in vibrational-rotational spectra since inversion splitting has been observed only rarely in the infrared spectral region. Calculations quoted in Chapter 4 indicate that the splittings in most molecules are probably too small to ever be observed at the resolutions attainable. The rotational-vibrational band system of NH_3 has been studied in great detail and at high resolution by Benedict, Plyler, and Tidwell [21] and Mould, Price, and Wilkinson [83]. Many small interactions occur among the energy levels, some of which are still not well understood. However, it seems unlikely that their work will be improved upon for many years to come.

The energy expressions for inversion doubling are given in 4k9 to 4k12. The selection rules are the same as those quoted earlier for parallel and perpendicular bands, with additional restrictions on the parity of the inversion state. The selection rules for the bands analyzed by Benedict, Plyler, and Tidwell given in Table 7e1 are taken from reference [21].

Table 7e1—Selection Rules for NH_3 [21]

TYPE*	$\Delta \lvert l \rvert$	Δp†	ΔK	EXAMPLE‡
∥, *a-s*	0	$+ \leftarrow -$	0	$\nu_1 + \nu_2$, upper, $\nu_0 = 4{,}320$ cm^{-1}.
∥, *s-a*	0	$- \leftarrow +$	0	$\nu_1 + \nu_2$, lower, $\nu_0 = 4{,}294$ cm^{-1}.
⊥, *a-s*	$1 \leftarrow 0$	$- \leftarrow -$	± 1	$\nu_3 + \nu_2$, lower, $\nu_0 = 4{,}417$ cm^{-1}.
⊥, *s-a*	$1 \leftarrow 0$	$+ \leftarrow +$	± 1	$\nu_3 + \nu_2$, upper, $\nu_0 = 4{,}435$ cm^{-1}.
⊥	$2 \leftarrow 0$		± 1	$2\nu_3^2$ $\nu_0 = 6{,}850$ cm^{-1}.

*s and a refer to states whose vibrational wave functions are symmetric or antisymmetric with respect to inversion.
† p refers to the algebraic sign in 4k9.
‡ Upper and lower refer to the two components of the doublet.

The result of these selection rules and the inversion splitting is to give two bands for each change in vibrational quantum number. As is usually true for high resolution spectra, most of the observed bands are severely overlapped by a band that results from some other vibrational transition. However, Fig. 7e1 shows the ν_2 bands of NH_3 under a resolution of about 0.1 cm^{-1} [83]. Here the double character of the band is clearly shown. This is a parallel-type band, and the two components are marked with ground state identifications. In the state $\nu_2 = 1$ the inversion splitting is approximately 36 cm^{-1} compared to 0.8 cm^{-1} in the ground vibrational state.

Each component band from these transitions is analyzed by the method described earlier in this chapter using effective inertial constants for each component. Because of the coupling of inversion and rotation this method of analysis is effective only for low J, K transitions. In writing the expression for the expansion of the inertial constants in terms of vibrational quantum numbers, 3c5, the average of the inertial constants of the two components may be used, or an expansion of this type may be used for the series of components of the same type. In practice, it is extremely difficult

Table 7e2—Vibrational Levels of Ammonia (cm^{-1}*)* [21]

v_1	$v_3^{l_3}$	$v_4^{l_4}$	$v_2^{s,a} = 0^s$	0^a	1^s	1^a	2^s	2^a	3^s	3^a
						NH_3				
0	0^0	0^0	0.00	0.793	932.51	968.32	1597.6	1910	2383.46	2895.48
0	0^0	1^1	1626.1	1627.4	2539.6?	2585.0?				
0	0^0	2^0	(3216.4)	(3218.6)						
1	0^0	0^0	3336.18	3337.18	4294.51	4320.06				
0	1^1	0^0	3443.59	3443.94	4416.91	4435.40				
1	0^0	1^1	4955.94	4956.8						
0	1^1	1^1	5052.61	5053.18	6012.72	6036.40				
{1	1^1	0^0	6608.71							
{0	1^1	2^0	(6700?)							
{0	2^2	0^0	6849.96	6850.39						
						ND_3				
0	0^0	0^0	0.00	0.053	745.7	749.4	1359	1429	1830	2106.60
0	0^0	1^1	1191							
{0	0^0	2^0	(2359?)		3093.01	3099.46				
{1	0^0	0^0	2420.05	2420.64	3171.89	3175.87				
0	1^1	0^0	2563.96		3327.94	3329.56				
{0	1^1	2^0	4887.29	4887.67						
{1	1^1	0^0	4938.44							
0	2^2	0^0	5100.66							

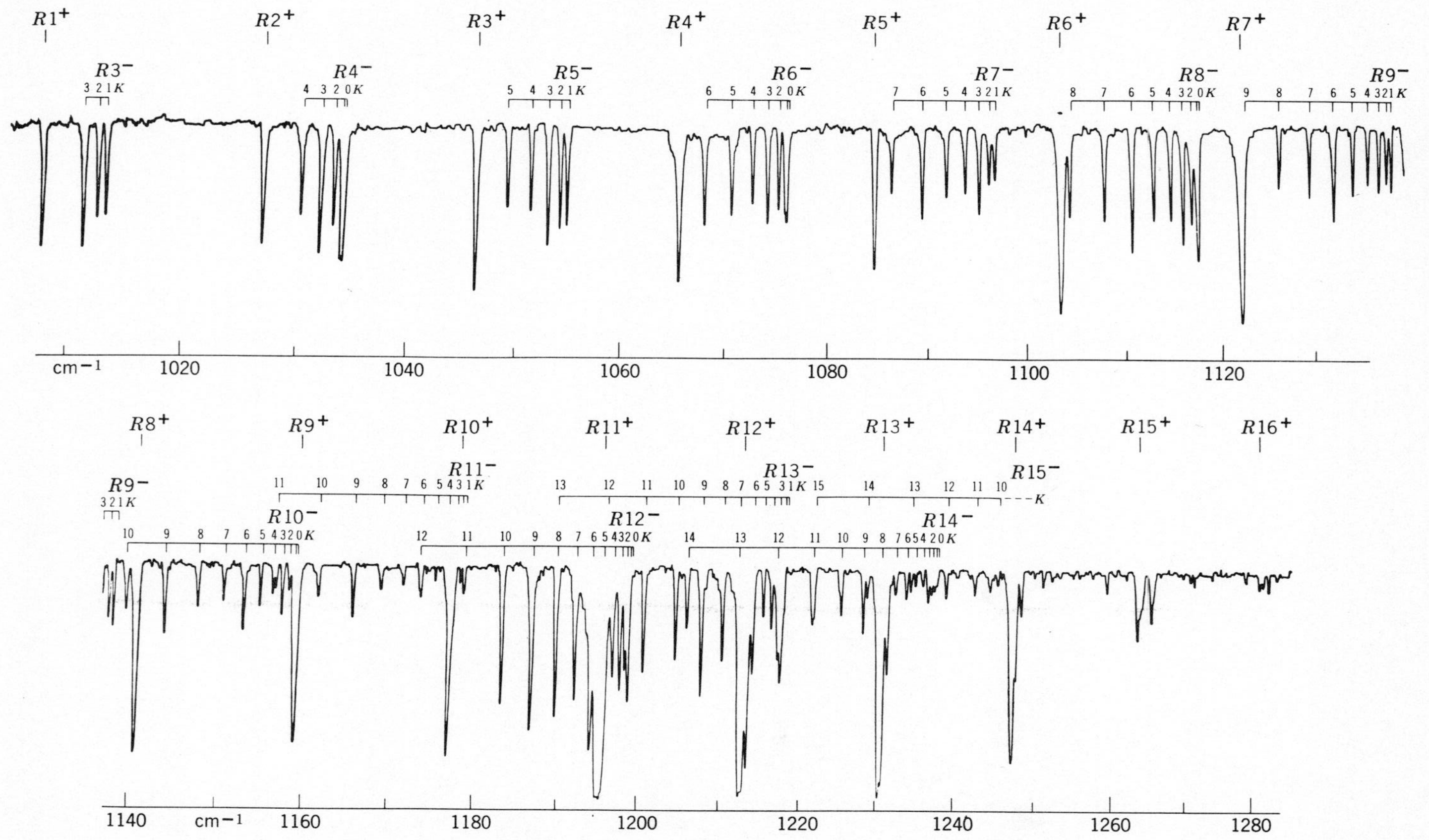

(a)

Fig. 7e1(*a*) The ν_2 vibration-rotation band of ammonia. (The upper assignments refer to the $0^s \rightarrow 1^a$ transition and the lower to the $0^a \rightarrow 1^s$ transition [83].)

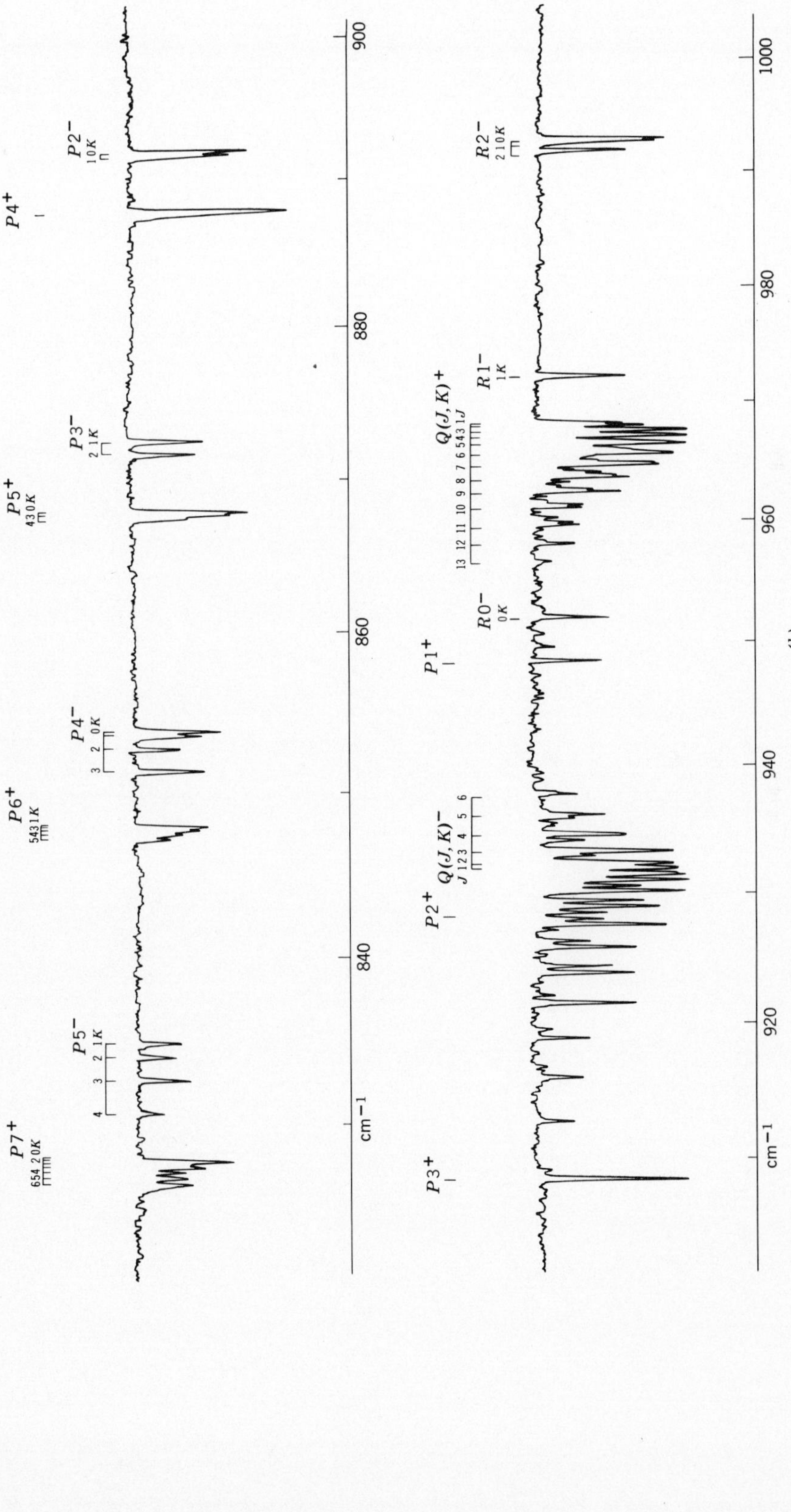

(*b*)

Fig. 7e1(*b*) See caption on opposite page.

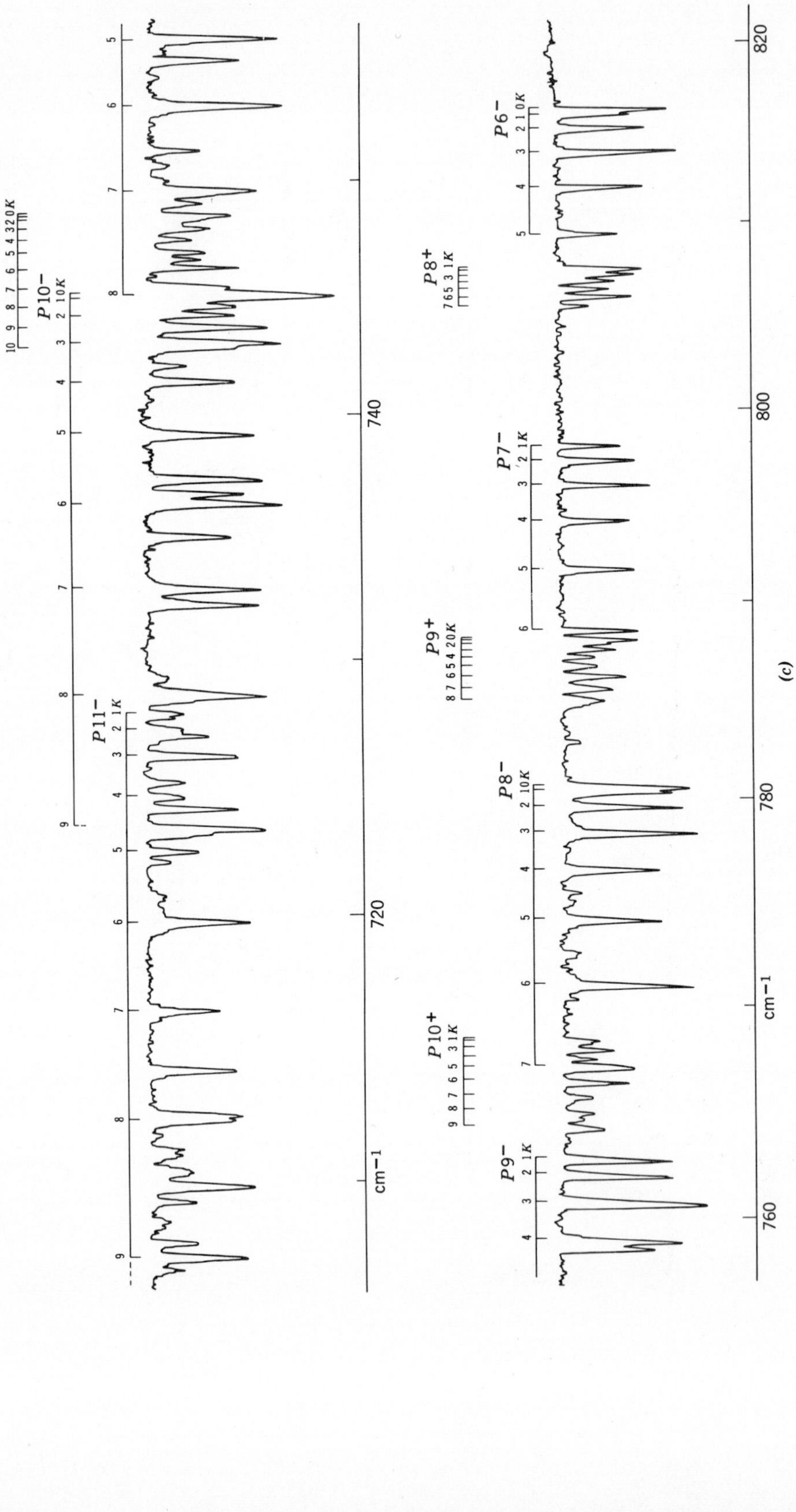

(c)

Fig. 7e1(c) See caption on p. 172.

to determine the constants in such an expansion because of the ever-present perturbations in the higher energy regions arising from the multiplicity of closely lying energy levels.

In NH_3, ν_2 is the vibration which is most like an inversion. Hence one would expect to find the largest inversion effects where ν_2 is excited—the higher ν_2 is excited the greater the effect. This result can be seen in Table 7e2, again taken directly from reference [21].

Because of the remoteness of observing inversion splitting in any other molecule, the details of the analysis of the NH_3 spectrum are not reproduced. The reader who is interested is referred to the excellent series of papers devoted to the analysis of the NH_3 rotational-vibrational spectrum by Benedict, Plyler, and Tidwell [21].

7f SYMMETRIC ROTOR MOLECULES WITH INTERNAL ROTATION

The energy levels and selection rules of a molecule with internal rotation were discussed in Chapter 4. So far no vibrational-rotational spectra of such a molecule have been studied under ultra-high resolution. For a low barrier but not free rotation the spectrum of such a molecule is extremely complicated, because not only is there a coupling between vibration and internal rotation, but there is also the further

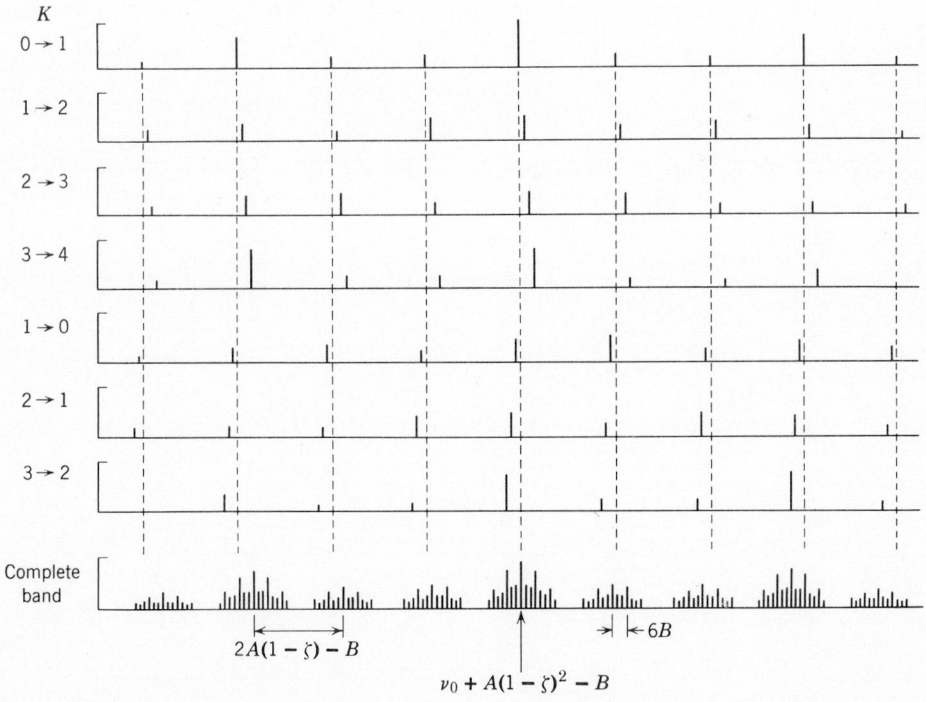

Fig. 7f1 Predicted Q-branch structure of a perpendicular-type fundamental in dimethylacetylene assuming free rotation [82].

complication of hot bands overlapping the main bands. These hot bands originate in the excited torsional states. Since these states are generally of the order of a few hundred cm^{-1} in energy, and since kT at room temperature is about 200 cm^{-1}, many of these torsional states have appreciable populations. The study of such bands would require measurements over a considerable temperature range, and thus none has been studied up to this time.

Free rotation in a symmetric top such as ethane or dimethylacetylene has been studied by several workers whose references were given in Chapter 4. The energy expression given by Mills and Thompson [82] for the case of free rotation was given in 4g7 and 4g8. The structure expected for a series of Q-branches in a perpendicular band is given in Fig. 7f1. They have applied this theory to dimethylacetylene and concluded that the internal rotation was essentially free.

8

THE STRUCTURE AND ANALYSIS OF ASYMMETRIC-ROTOR BANDS

8a INTRODUCTION

The rotational-vibrational bands of an asymmetric rotor are even more complex than those of a symmetric rotor. The introduction of asymmetry removes the K-type degeneracy which is present in the energy levels of a symmetric rotor to give $2J + 1$ levels for each J value of an asymmetric rotor. This increase in the number of energy levels is accompanied by a corresponding increase in the number of allowed transitions. In addition, the selection rules for an asymmetric rotor are considerably more relaxed than they are for symmetric rotors. Comparatively few asymmetric rotor bands have been fully analyzed, largely because the available spectral resolution has not been adequate until recently and because there has been no systematic method of attack. In the past the analytical method has been called "stochastic." When applied in this sense the word means a combination of shrewd guesses and trial calculations [11]. In this chapter methods are presented which should minimize the guess work and organize the calculations. If one understands the structure of an asymmetric rotor band, the process of assigning quantum numbers is greatly simplified, and this procedure is the most difficult part of any analysis. Once the quantum numbers have been assigned the methods of determining band centers and inertial constants are in principle quite analogous to those used for simpler types of rotors. The systematic classification of transitions into a sub-band structure is a natural outgrowth of the discussion in 5i and 5j.

8b THE STRUCTURE OF A-TYPE BANDS

If the change in electric moment during a transition lies along the axis of least moment of inertia, the resulting band is called an A-type band [4]. In the prolate symmetric limit, $I_B = I_C$, a change in electric moment along the axis of least moment of inertia is a change along the top axis, thus giving rise to a parallel-type band. For an A-type band, in this symmetric limit, the selection rules are given by 5d33 as

$$\begin{array}{lll} K_{-1} = 0 & \Delta J = \pm 1 & \Delta K_{-1} = 0 \\ K_{-1} \neq 0 & \Delta J = 0, \pm 1 & \Delta K_{-1} = 0. \end{array} \tag{1}$$

If there is to be a correlation between an asymmetric rotor band and the symmetric limit, we might expect that (1) would furnish the restrictions on the changes allowed in K_{-1}, at least for the strongest transitions. In the oblate symmetric limit ($I_A = I_B$) a change in electric moment along the axis of least moment of inertia is a change perpendicular to the top axis, thus giving rise to a perpendicular-type band. The selection rules for an A-type band in this symmetric limit are given by 5d35 as

$$\Delta J = 0, \pm 1 \qquad \Delta K_1 = \pm 1. \tag{2}$$

Again, if there is to be a correlation between the asymmetric rotor band and this symmetric limit, (2) should represent the allowed changes in K_1, at least for the strong transitions. Combining (1) and (2), we obtain

$$\begin{array}{llll} K_{-1} = 0 & \Delta J = \pm 1 & \Delta K_{-1} = 0 & \Delta K_1 = \pm 1 \\ K_{-1} \neq 0 & \Delta J = 0, \pm 1 & \Delta K_{-1} = 0 & \Delta K_1 = \pm 1 \end{array} \tag{3}$$

as the selection rules for the strong transitions in an asymmetric rotor A-type band. We would expect these transitions to be strong because they are "allowed" in both symmetric rotor limits. It is indeed true that the strongest transitions in an A-type band of a rigid asymmetric rotor are those allowed by (3). We can readily verify that the selection rules given in (3) are a special case of the general selection rules given in 5h, namely

$$ee \leftrightarrow eo \qquad oe \leftrightarrow oo. \tag{4}$$

Although (4) permits changes in K_{-1} and K_1 which are greater than $|1|$, transitions with changes greater than ± 1 are generally an order of magnitude weaker than those allowed by (3).

It should be remembered that for asymmetric rotors $K_{-1} + K_1 = J$ or $J + 1$. Therefore, for those transitions allowed by (3), the selection rule $\Delta K_1 = 1$ applies to R-branch transitions, and the selection rule $\Delta K_1 = -1$ applies to the P-branch transitions. Thus the selection rules for the strongest transitions of an A-type band (that is, those allowed in both symmetric limits) have been deduced from an examination of the behavior in the two symmetric limits.

The sub-branches which make up an asymmetric rotor band are discussed in 5h and 5j. These sub-branches are characterized by the changes in ΔJ, ΔK_{-1}, ΔK_1, and the parity of the ground state levels. They can then be further subdivided into wings, a wing being characterized by specifying a constant value of K_{-1} or K_1 (J varies).

The sub-branches which are of interest here are those allowed by (3). From the previous discussion it can be ascertained that these sub-branches are R_{01}, $P_{0\bar{1}}$, Q_{01}, and $Q_{0\bar{1}}$. For both the R_{01} and $P_{0\bar{1}}$ it is possible to subdivide the transitions, according to the parity of the ground state levels. Thus one obtains both an even and odd sub-branch for the *P*- and *R*-branches. The ground state levels of the Q_{01} sub-branch have even parity, whereas those of the $Q_{0\bar{1}}$ have odd parity.

For the purpose of understanding band structure it is now convenient to appropriately group the wings of the sub-branches into sub-bands. Each sub-band is composed of *R*-, *Q*-, and *P*-wings. The sub-band is completely characterized by a constant value of K_{-1} or K_1 and the parity of the ground state levels [11]. In the subsequent discussions those sub-bands whose ground state levels have even parity are called even sub-bands, whereas those whose ground state levels have odd parity are called odd sub-bands. The strong transitions of the even sub-bands of an *A*-type band are composed of the sub-branches ${}^eR_{01}$, ${}^eQ_{01}$, and ${}^eP_{0\bar{1}}$, and there is a sub-band for each value of K_{-1} in which the value of K_{-1} is held constant. Since K_{-1} does not change in the transitions of an *A*-type band, it is more convenient to specify the sub-bands by a constant value of K_{-1}. There are then even sub-bands for each value of $K_{-1} = 0, 1, 2, 3, \ldots$. The odd sub-band is composed of the sub-branches ${}^oR_{0\bar{1}}$, ${}^oQ_{0\bar{1}}$ and ${}^oP_{0\bar{1}}$. There is an odd sub-band for each value of $K_{-1} = 1, 2, 3, \ldots$. Thus for $K_{-1} = 0$ only the even sub-band exists, whereas for each $K_{-1} > 0$ there is both an even and an odd sub-band.

The way these sub-bands arise for the strong transitions is shown schematically in Fig. 8b1. Each column of the vibrational rotational levels is labeled by its constant value of K_{-1}. The value of K_1 for each level is shown to the right of the line representing that level in the figure. Since for the strongest transitions $\Delta K_{-1} = 0$, the *P*- and *R*-branch transitions for a given sub-band take place between the rotational levels of the two vibrational states in the same column. Since $\Delta K_{-1} = 0$, it is easily seen that there can be no *Q*-branch for the $K_{-1} = 0$ sub-band. However, since the asymmetry removes the *K* degeneracy, the *Q*-branches for the other sub-bands arise from transitions between the two components of the $K_{-1} > 0$ levels.

The sub-band structure is shown schematically in another way in Fig. 8b2. No simple closed expression for the transition positions can be expressed for an asymmetric rotor band as can be for simpler rotors. Thus, the energy levels for each state were calculated using 2j9 and values of the reduced energy $E(\kappa)$ for $\kappa = -0.9$ [Appendix IV]. The transition frequencies are then calculated by taking differences between the energy levels in the two vibrational states. Relative intensities of the transitions were calculated by combining the appropriate Boltzmann factor and nuclear spin factor with tabulated values of the line strengths [Appendix IX]. In Fig. 8b2 only the transitions allowed by (3) have been plotted and only the sub-bands with $K_{-1} < 3$ have been plotted separately. It can be seen that this sub-band structure bears an amazing similarity to that of a parallel band as shown in Fig. 7b2. In the lowest line of the figure the sub-bands have been superimposed to give the composite band. This composite band contains not only those sub-bands shown separately above it but also the transitions allowed by (3) belonging to sub-bands with $K_{-1} > 3$. A comparison with a composite parallel-type band shows that the composite *A*-type band is considerably more complex; however, it is comforting to know that even this complex spectrum can be broken down into a series of simple sub-bands. To analyze an *A*-type

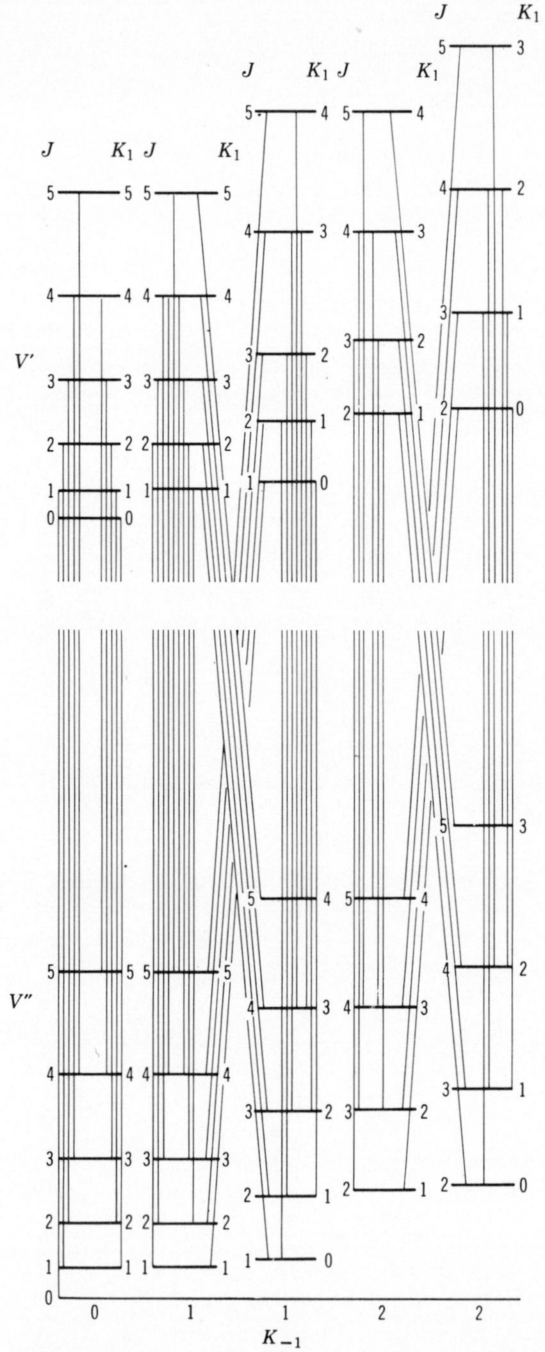

Fig. 8b1 The sub-band structure of an A-type band of an asymmetric rotor. Each column of the vibrational rotational levels is labeled by its constant value of K_{-1}. The value of K_1 for each level is shown alongside the line representing that level [11].

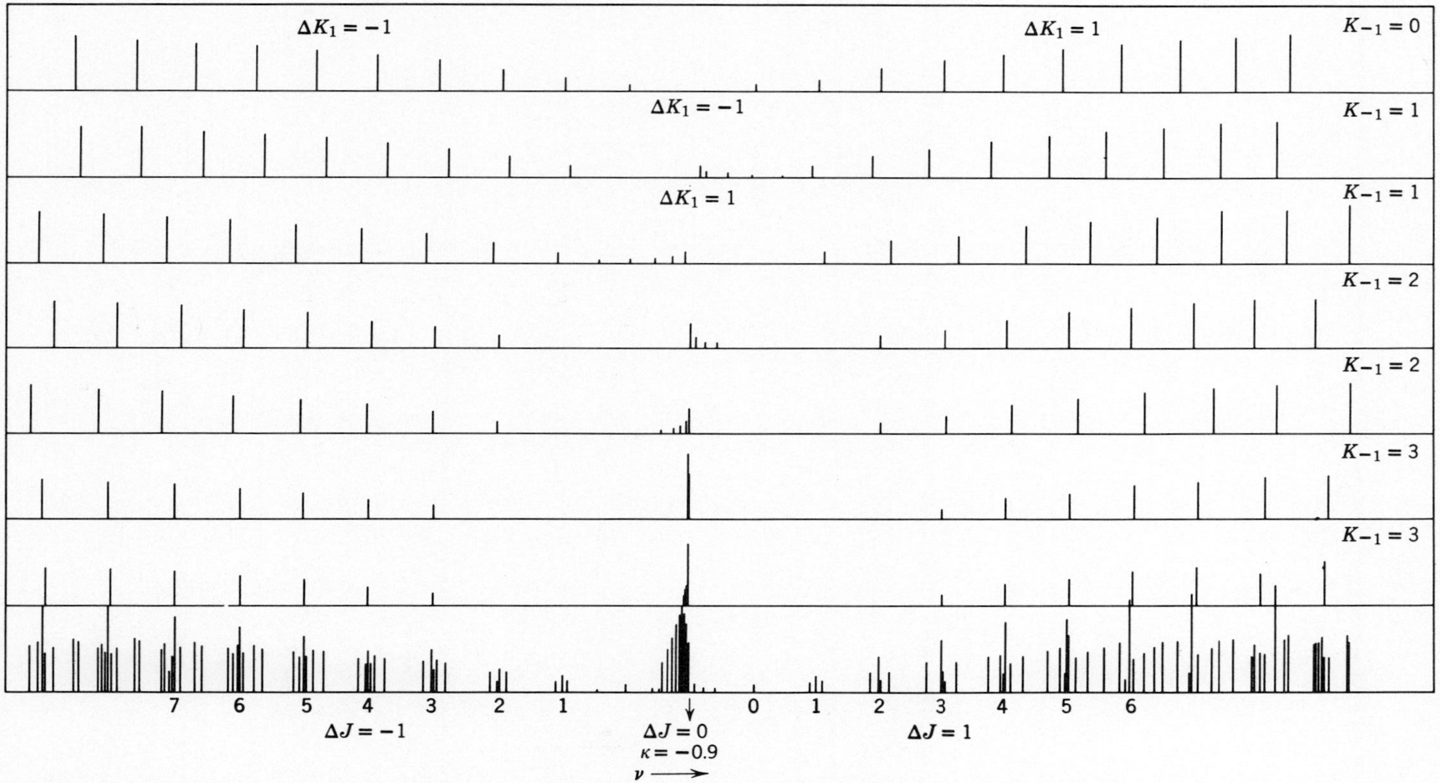

Fig. 8b2 The sub-bands of an A-type band of an asymmetric rotor with $\kappa = -0.9$. The constants used in calculating the energy levels were $A'' = 5.241$, $B'' = 3.946$, $C'' = 3.878$, $A' = 5.211$, $B' = 3.870$, and $C' = 3.855$. The intensities were calculated for $T = 298°$K. In the composite spectrum only the first transition of each sub-band Q-branch has been plotted. The parity of the ground state levels of the $K_{-1} = 0$ sub-band is e. For the sub-bands with $K_{-1} > 0$, the parity of the ground state levels in the upper sub-band is o, whereas it is e in the lower sub-band [11].

band it is essential to understand fully this sub-band structure, for it is a knowledge of this structure that provides the clues which enable one to assign quantum numbers to absorption peaks in an observed spectrum.

This spectrum represents a typical A-type band of a slightly asymmetric rotor. It should not be regarded as a pattern, however, since different choices of A', B', C' and A'', B'', C'' could give the band a quite different appearance even though the value of κ is not changed significantly. Equally important, a different value of κ can be expected to give the spectrum a quite different aspect. The value of $\kappa = -0.9$ is very near to the prolate symmetric limit where an A-type band corresponds to a parallel-type band; hence it is not surprising that a marked resemblance to a parallel-type band is found.

Figure 8b3 is a schematic diagram similar to Fig. 8b2 except that the asymmetry parameter κ has a value of -0.5 for this spectrum. There is still a marked resemblance to a parallel-type band, although the Q-branches of the sub-bands are now spread out somewhat. The P- and R-branches of the composite band still show a greater complexity than is present in the band of the limiting prolate rotor. Figure 8b4 is a schematic spectrum for an asymmetric rotor with $\kappa = 0.5$. Here the sub-band Q-branches are spread out considerably and only the first, and strongest, Q-branch transition for each sub-band remains near the band center. At this asymmetry we can see that the $K_{-1} = 0$ and the odd $K_{-1} = 1$ sub-bands are essentially degenerate for all J. This feature arises because the oblate symmetric limit is being approached, and for a given value of J the levels involved in these transitions have the same value of K_1. In the oblate limit these levels are exactly degenerate. As one draws closer to the prolate limit the even sub-band with a value $K_{-1} = 1$ becomes degenerate with the odd sub-band with a value $K_{-1} = 1$. Near the prolate limit the two sub-bands with the same value of K_{-1} are essentially degenerate. In Fig. 8b5 a schematic spectrum for $\kappa = 0.9$ is shown. At this asymmetry most of the resemblance to a parallel-type band has disappeared.

The composite spectrum has many features of a perpendicular band of a symmetric rotor. The sub-band Q-branches are now completely spread out. The first transition of each sub-band Q-branch is all that is left near ν_0. The Q-branch transitions with the same value of K_1 are now nearly superimposed, giving the composite spectrum a series of Q-branches characteristic of a perpendicular-type band. The Q-branches are labeled under the composite spectrum according to the notation suggested by Herzberg. The subscripts denote the values of K_1 of the ground state levels for the various Q-branches; the $\Delta K = +1$ Q-branches are at lower frequencies because $(C - B)$ is negative. From the series of Figs. 8b2 to 8b5, it is evident that the appearance of an A-type band can vary from great similarity to a parallel-type symmetric rotor band near the prolate limit to great similarity to a perpendicular symmetric rotor band near the oblate limit. In each of these figures the composite bands may be considered as typical, but the relative values of A', B', C' and A'', B'', C'' can cause nearly as drastic effects on the band appearance as the degree of asymmetry.

Figure 8b6 illustrates the part inertial constants can play in determining the appearance of bands. All the bands in this series have the same ground state energy levels. However, the excited state energy levels were calculated using the values of A', B', C' given at the left of each spectrum. The excited state constants used for these bands, although extreme, are not unrealistic. They represent values obtained from the analysis

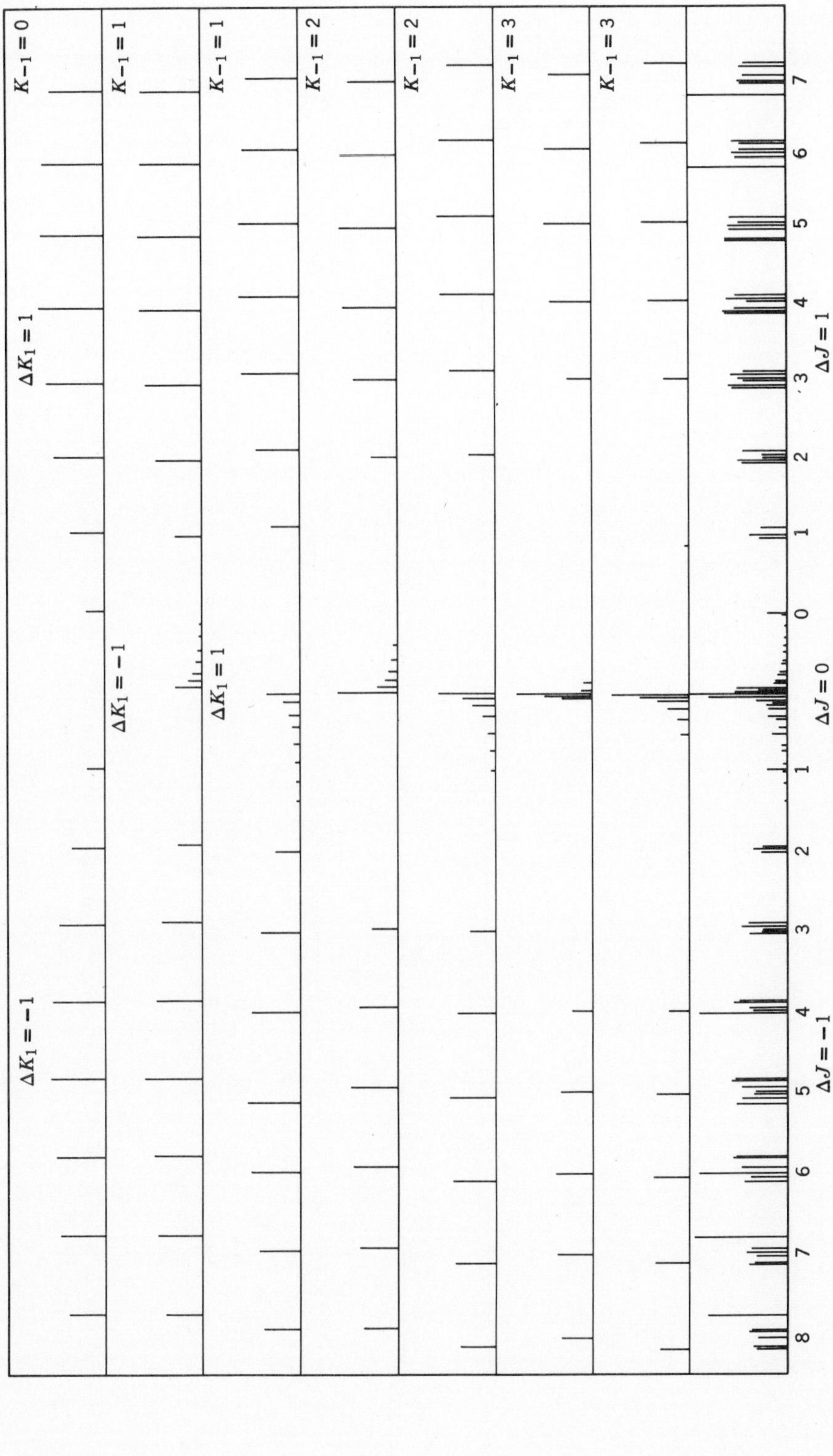

Fig. 8b3 The sub-band of an A-type band of an asymmetric rotor with $\kappa = -0.5$. The constants used in calculating the energy levels are $A'' = 5.241$, $B'' = 4.219$, $C'' = 3.878$, $A' = 5.211$, $B' = 4.194$, $C' = 3.855$ cm^{-1}. The intensities were calculated for a temperature of 298° K [11].

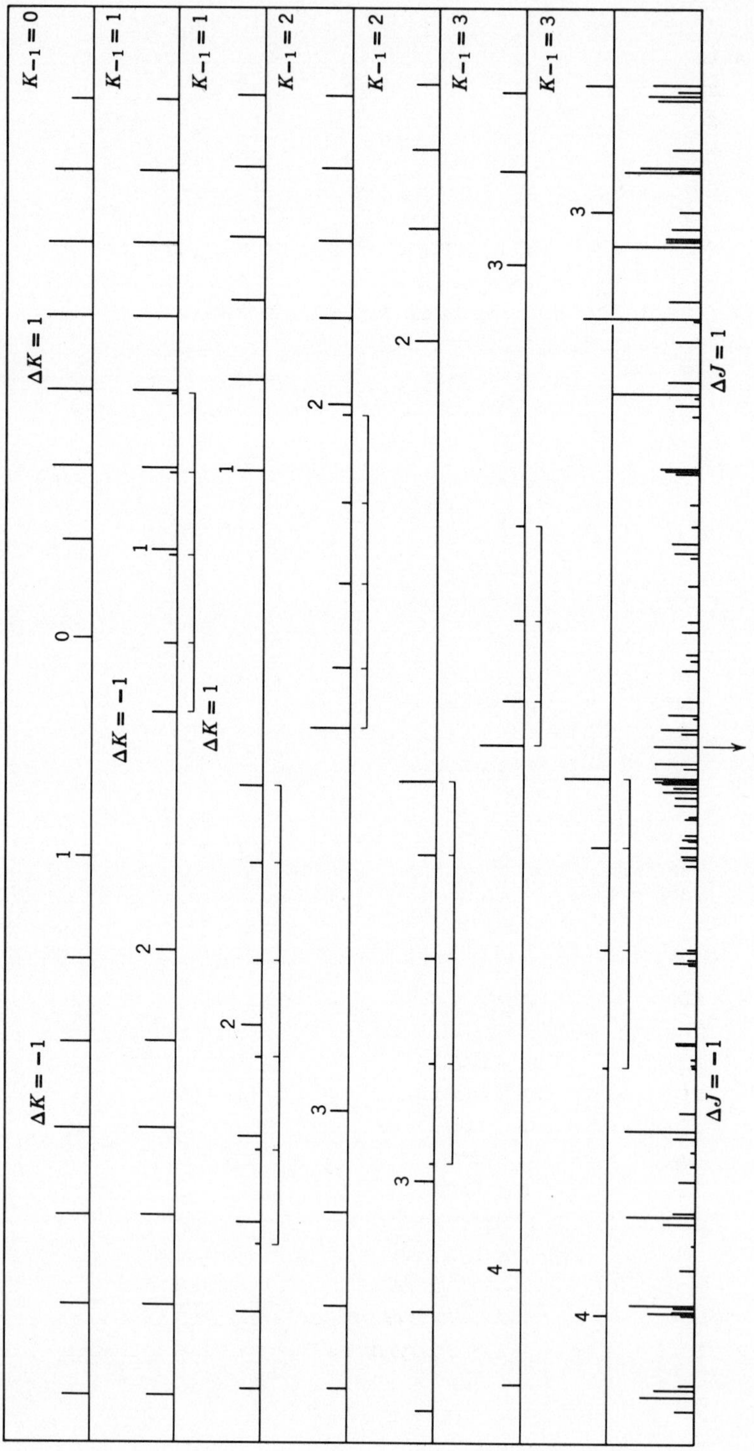

Fig. 8b4 The sub-bands and composite band of an A-type band of an asymmetric rotor with $\kappa = 0.5$. The values of A'', C'', A', and C' are the same as those used in Fig. 8b3. The intensities were calculated for $T = 298°$ K [11].

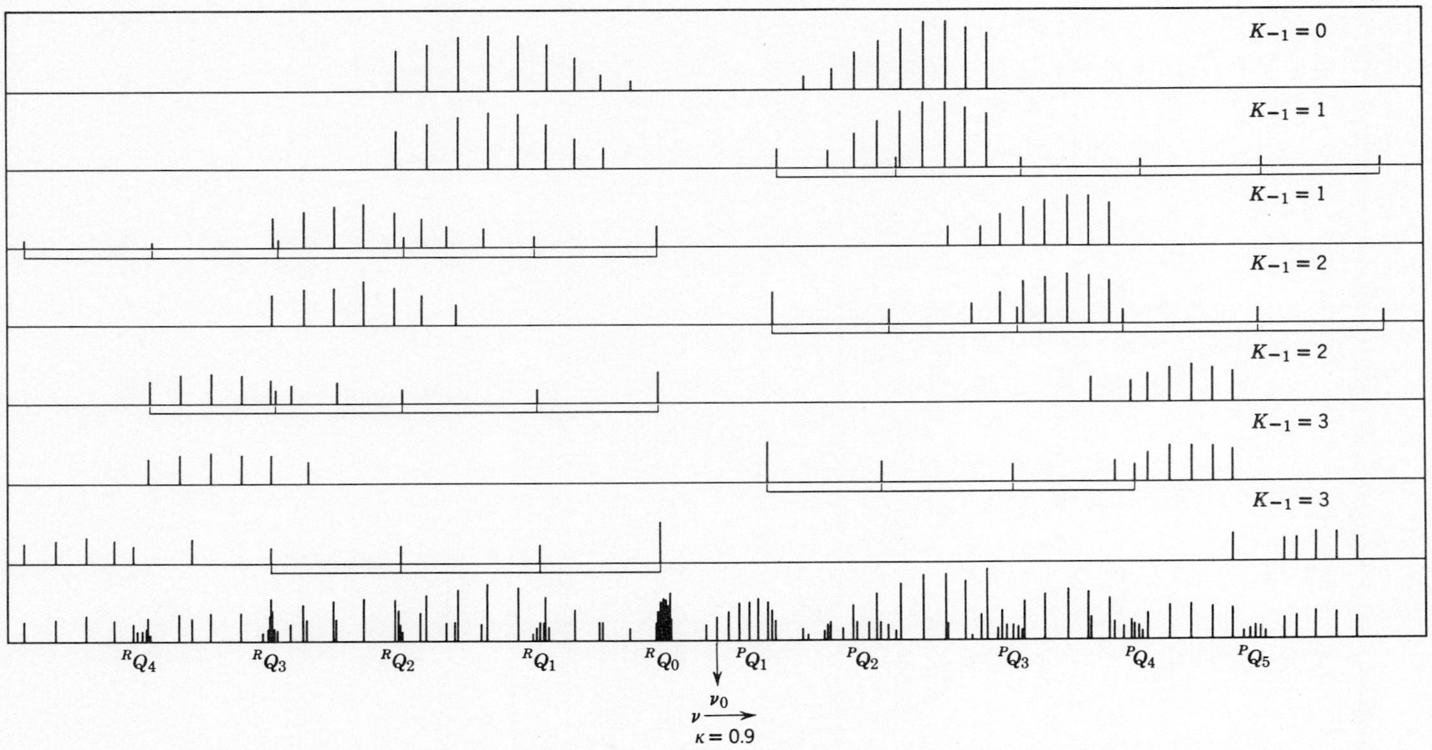

Fig. 8b5 The sub-bands are whole band of an A-type asymmetric rotor band $\kappa = 0.9$. The pictured sub-bands are still identified according to the K_{-1} value of the ground state level. At this asymmetry the band bears an amazing similarity to a perpendicular band of an oblate symmetric rotor. The vestiges of the series of Q-branches are clearly evident and labeled according to the notation of Herzberg. The inertial constants used for the calculation are $A'' = 4.861\ \text{cm}^{-1}$, $B'' = 4.660$, $C'' = 0.8340$, $A' = 4.841$, $B, = 4.640$, and $C' = 0.8140$. The intensities are calculated for 298° K [11].

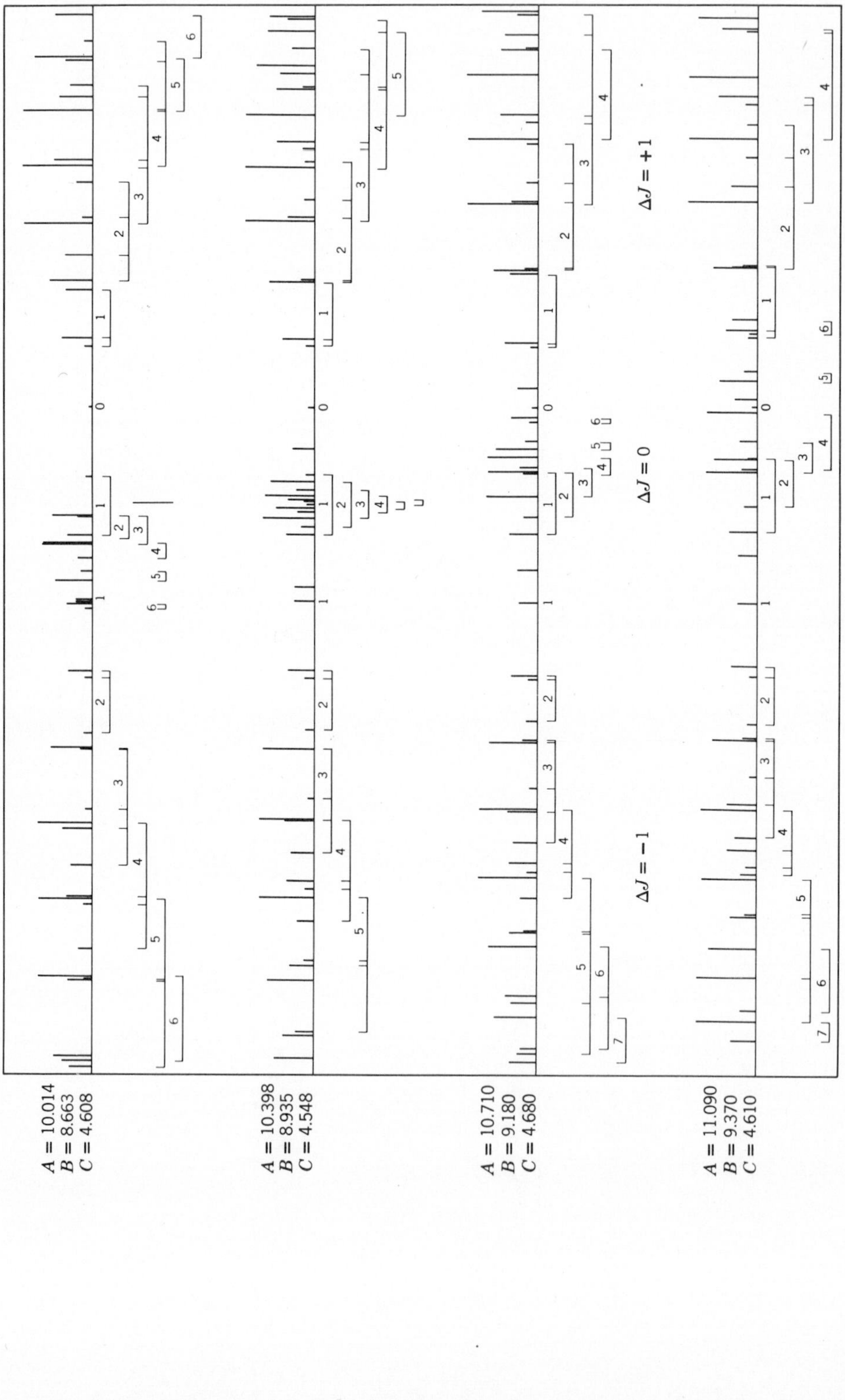
A = 10.014
B = 8.663
C = 4.608
A = 10.398
B = 8.935
C = 4.548
A = 10.710
B = 9.180
C = 4.680
A = 11.090
B = 9.370
C = 4.610
ΔJ = +1
ΔJ = 0
ΔJ = −1

Fig. 8b6

Fig. 8b6 Bands of H_2S plotted to show the effect of different excited state constants on the band appearance. The excited state inertial constants are given to the left of the figure. The ground state constants are $A'' = 10.373$, $B'' = 8.991$, and $C'' = 4.732$. The lines and numbers under the spectra indicate the ground state J value of the transitions. For the most part, only the first transition of a sub-band Q-branch has been plotted, although a few of the other transitions are plotted when the intensity warrants it. The intensities were calculated for 298° K and include the nuclear spin factor for H_2S [11].

of observed bands of H_2S [14], the ground state being that given by Cross [35]. The value of κ is about 0.5. Figure 8b6 should illustrate that such extreme variations in band appearance are to be expected in real spectra. The numbering below the bands, which gives the ground state J value of the transitions, is intended to help the reader follow the various transitions as the inertial constants are changed. In the Q-branch regions only the first and strongest member of the Q-branch of each sub-band has been plotted. The most drastic changes in these bands occur in the Q-branch region. As is seen in the following section, these changes arise because of the large changes in A'. The convergence of the strongest R- and P-branch lines also changes, largely as a result of the changes in C'.

8c THE DEPENDENCE OF THE ENERGY LEVELS ON THE INERTIAL CONSTANTS

As an alternative to 2j9 it is sometimes convenient to express the energy levels of a rigid asymmetric rotor as

$$E(A, B, C) = \alpha A + \beta B + \gamma C, \tag{1}$$

in which α, β, and γ are the derivatives of the energy defined in 3h5, 3h6, and 3h7. Approximate derivatives for various values of κ are given in Table 8c1. Here the increment of κ used was $\Delta\kappa = 0.1$, which is rather large. However, it is easily seen from the table that the change in these derivatives with κ usually is quite small. These derivatives have the further advantage that they do not depend explicitly on the values of A, B, and C but merely on their ratio as defined by κ or b.

By substituting the numerical values of α, β, and γ for a given level into (1), we can ascertain immediately the relative importance of the three inertial constants in determining the energy of that level. As an example, using the values of α, β, γ for the 6_{60} level with $\kappa = 0$, we find

$$E_{6_{60}}(A, B, C) = 35.652\, A + 4.264\, B + 2.084\, C. \tag{2}$$

Equation (2) shows that the value of A very nearly determines the energy of this level. Furthermore, we can see from Table 8c1 that this is true over the whole range of κ except very near the oblate limit where the level depends nearly equally on A and B. On the other hand, the level 6_{61} depends predominantly on A throughout the complete range of κ and, except very near the oblate limit, the dependence of this pair of levels on A is essentially the same. This pair of levels ($K_{-1} = J$; $K_1 = 0, 1$) in each J gives rise to the first Q-branch transitions in each of the sub-bands. Figures 8b2 to 8b5 show clearly that these transitions are the strongest transitions in the composite Q-branch. This series of transitions may be described symbolically by the expression

$$J_{J0} \rightleftharpoons J_{J1}, \tag{3}$$

it being understood that one of these rotational levels is in the ground vibrational state and the other in the excited vibrational state. When we neglect for the minute the small dependence of these levels on B and C, and assume the α values are the same in both vibrational states, the transition frequency becomes

$$\nu \sim \nu_0 + \alpha(A' - A''). \tag{4}$$

in which ν_0 is the frequency of vibration. Thus we find that the frequency of the strongest transitions in the Q-branch of an A-type band is determined essentially by the difference between the A constants of the two vibrational states. This effect shows up very clearly in Fig. 8b6. From the top spectrum to the bottom spectrum the quantity $(A' - A'')$ becomes increasingly positive and thus the Q-branch transitions shown are shifted to successively higher frequencies.

We may also consider the level 6_{06} at $\kappa = 0$. The expression for the energy of this level is

$$E_{6_{06}}(A, B, C) = 1.963\,A + 4.506\,B + 35.531\,C. \tag{5}$$

Here the dependence on C is an order of magnitude greater than it is on the other inertial parameters except near the prolate limit where the dependence on B and C is nearly equal. The dependence of the level 6_{16} is predominantly on C throughout the whole range of κ. These two levels in fact have almost identical dependence on C except very near the prolate limit. These properties are generally true for the two lowest levels in a given J. Now the R- and P-branch transitions in the $K_{-1} = 0$ and the odd $K_{-1} = 1$ sub-bands occur between levels of this type, that is,

(*a*) $$J_{0J} \leftrightharpoons (J + 1)_{0,J+1},$$

and (6)

(*b*) $$J_{1J} \leftrightharpoons (J + 1)_{1,J+1}.$$

Note in Table 8c1 that $\gamma \sim K^2$ for these two lowest levels in each J. Making use of this approximation and also the fact that the energy dependence on A and B is small, the transition frequencies can be expressed as

$$\begin{aligned} \nu \sim \nu_0 + (J + 1)^2C' - J^2C'' = \nu_0 + (2J + 1)C' + (C' - C'')J^2 \qquad \Delta J = 1 \\ \nu \sim \nu_0 - (2J - 1)C' + (C' - C'')J^2 \qquad \Delta J = -1, \end{aligned} \tag{7}$$

where J, as usual, is the ground state J.

Thus we see that the transition frequency of the P- and R-branch of these two sub-bands is essentially determined by the C''s, and the convergence or divergence of the P- and R-branches is, in this approximation, determined by the difference between the C values in the two vibrational states.

A study of Table 8c1 shows that the other levels do not generally depend predominantly on only one of the inertial parameters. Hence no generalizations can be made concerning them. However, these observations about the two highest and two lowest levels in a given J can be extremely useful in band analysis.

8d THE ΔF_2 VALUES OF AN A-TYPE BAND

The sub-band structure is very useful in the determination of ΔF_2 values. Each sub-band may be treated just as a band of a linear molecule. Ground state energy differences are given by

$$\Delta F_2'' = R(J - 1, K_{-1}, K_1 - 2) - P(J + 1, K_{-1}, K_1), \tag{1}$$

Table 8c1—Approximate Energy

	$\kappa \sim -0.9$			$\kappa \sim -0.5$		
$^JK_{-1}$	α	β	γ	α	β	γ
5_{50}	25.000	2.536	2.464	24.917	2.993	2.080
5_{51}	25.000	2.536	2.464	24.928	2.990	2.082
5_{41}	16.000	7.115	6.885	15.748	8.668	5.583
5_{42}	16.000	7.115	6.885	15.785	8.492	5.723
5_{32}	9.000	10.746	10.254	8.284	15.048	6.667
5_{33}	9.000	10.720	10.280	8.788	12.297	8.915
5_{23}	4.000	14.210	11.790	3.241	20.880	5.878
5_{24}	4.000	12.885	13.115	4.215	11.508	14.277
5_{14}	1.000	21.718	7.283	1.788	16.959	11.253
5_{15}	1.000	6.744	22.256	1.284	4.713	24.002
5_{05}	0.000	13.675	16.325	1.010	5.452	23.538
6_{60}	36.000	3.042	2.958	35.915	3.577	2.508
6_{61}	36.000	3.042	2.958	35.915	3.576	2.508
6_{51}	25.000	8.632	8.368	24.730	10.321	6.948
6_{52}	25.000	8.632	8.368	24.737	10.280	6.973
6_{42}	16.000	13.246	12.754	15.410	16.817	9.773
6_{43}	16.000	13.245	12.755	15.582	15.990	10.428
6_{33}	9.000	16.983	16.017	7.622	25.510	8.868
6_{34}	9.000	16.907	16.093	8.778	18.881	14.341
6_{24}	4.000	21.323	16.677	3.328	29.388	9.284
6_{25}	4.000	18.712	19.288	4.503	15.434	22.063
6_{15}	1.000	30.383	10.617	2.652	20.147	19.200
6_{16}	1.000	9.461	31.539	1.484	5.830	34.686
6_{06}	0.000	18.388	23.612	1.347	6.219	34.434

and analogously for the excited state we find

$$\Delta F_2' = R(J, K_{-1}, K_1) - P(J, K_{-1}, K_1). \qquad (2)$$

Equations (1) and (2) apply only to those transitions in the sub-bands that are allowed by 8b3; however, since these are by far the strongest transitions they are the most important ones. The dependence of the ΔF_2 values on the inertial parameters is readily found to be, with the help of 8c1.

$$\Delta F_2(J, K_{-1}) = [\alpha(J+1) - \alpha(J-1)]A + [\beta(J+1) - \beta(J-1)]B + [\gamma(J+1) - \gamma(J-1)]C. \quad (3)$$

The dependence on the inertial parameters of the first few ΔF_2 values in the sub-bands with low values of K_{-1} are shown in Table 8d1. From Table 8c1, one can see that over most of the κ range the values of $\alpha \sim K_{-1}^2$. Since in the ΔF_2 values under consideration $\Delta K_{-1} = 0$, the dependence of the ΔF_2 values on A would be expected to be small; this indeed turns out to be true. We can see that except very close to the oblate limit these ΔF_2 values have virtually no dependence on A. Thus from the experimental ΔF_2 values for these sub-bands we would not expect to obtain a very good value of A. However, these ΔF_2 values have a dependence on B and C so that good values of these parameters can be obtained.

To obtain any information about A from an A-type band, it is necessary to assign

Derivatives of a Rigid Asymmetric Rotor

$\kappa \sim 0.0$			$\kappa \sim 0.5$			$\kappa \sim 0.9$		
α	β	γ	α	β	γ	α	β	γ
24.695	3.598	1.707	23.538	5.452	1.010	16.325	13.675	0.000
24.715	3.549	1.736	24.002	4.713	1.284	22.256	6.744	1.000
14.799	11.128	4.073	11.253	16.959	1.788	7.283	21.718	1.000
15.283	9.827	4.890	14.277	11.508	4.215	13.115	12.885	4.000
6.699	19.340	3.961	5.878	20.880	3.241	11.790	14.210	4.000
8.640	12.720	8.640	8.915	12.797	8.788	10.280	10.720	9.000
3.605	20.053	6.342	6.667	15.048	8.284	10.254	10.746	9.000
4.717	10.173	15.110	5.723	8.492	15.785	6.885	7.115	16.000
3.634	12.062	14.360	5.583	8.668	15.748	6.885	7.115	16.000
1.645	3.731	24.624	2.082	2.990	24.928	2.464	2.536	25.000
1.596	3.819	24.585	2.080	2.993	24.917	2.464	2.536	25.000
35.652	4.264	2.084	34.434	6.219	1.347	23.612	18.388	0.000
35.657	4.251	2.091	34.686	5.830	1.484	31.539	9.461	1.000
23.810	12.710	5.480	19.200	20.147	2.652	10.617	30.383	1.000
24.002	12.225	5.773	22.063	15.434	4.563	19.288	18.712	4.000
13.145	22.724	6.131	9.284	29.388	3.328	16.677	21,323	4.000
14.861	17.941	9.198	14.341	18.881	8.778	16.093	16.907	9.000
6.176	29.648	6.176	8.868	25.510	7.622	16.017	16.983	9.000
9.021	18.295	14.684	10.428	15.990	15.582	12.755	13.245	16.000
5.240	24.506	12.254	9.773	16.817	15.410	12.754	13.246	16.000
5.482	12.807	23.711	6.973	10.290	24.737	8.368	8.632	25.000
5.014	13.641	23.344	6.948	10.321	24.730	8.368	8.632	25.000
1.977	4.480	35.543	2.508	3.576	33.915	2.958	3.042	36.000
1.963	4.506	35.531	2.508	3.577	33.915	2.958	3.042	36.000

transitions for which $\Delta K_{-1} \neq 0$. The strongest set of wings would be those for which $\Delta K_{-1} = |2|$, $\Delta K_1 = |1|$. The line strengths for these transitions are an order of magnitude smaller than those for the transitions allowed by 8b3. Thus in the observed spectrum these transitions will be much harder to find and assign correctly. However, if they can be assigned, then ΔF_2 values can be found for which $\Delta K_{-1} = |2|$. Some examples of these ΔF_2 values together with their dependence on the inertial parameters are given at the bottom of Table 8d1. It is readily seen that these ΔF_2 values have a sizable dependence on A; thus provided enough transitions with $\Delta K_{-1} = |2|$ can be assigned, we can obtain a meaningful value of A. However, these transitions are forbidden in the prolate limit and only have an appreciable intensity near the oblate limit. Even then the strongest of these transitions will be weak compared with the same transitions, except with $\Delta K_{-1} = 0$. Although this discussion has centered on the ΔF_2 values which can be determined from P- and R-branch transitions, in a practical sense these are really the only ones that can be determined from observed spectra. If it is possible uniquely to identify Q-branch transitions, it is then possible to determine new ΔF_2 values using these transitions. However, since $\Delta K_{-1} = 0$ is also true for the ΔF_2 values which can be obtained from strong Q-branch transitions, they are of no help in determining A. Usually, the Q-branches of all the sub-bands are so badly overlapped near the center of the band that it is hopeless to try to identify them uniquely. However, if the asymmetry parameter has a value near $\kappa = 1$, the possibility is improved as can be seen from Figs. 8b4 and 8b5. Also, if the change in A between

Table 8dI—Dependence of ΔF_2 Values on α, β,

ΔF_2	$\kappa \sim -0.9$			$\kappa \sim -0.5$		
	$\Delta\alpha$	$\Delta\beta$	$\Delta\gamma$	$\Delta\alpha$	$\Delta\beta$	$\Delta\gamma$
$K_{-1} = 0, e$						
2_{02}–0_{00}	0	2.962	3.038	0.046	2.605	3.349
3_{03}–1_{01}	0	4.803	5.192	0.202	3.192	6.606
4_{04}–2_{02}	0	6.465	7.535	0.597	2.113	11.290
5_{05}–3_{03}	0	7.867	10.133	0.808	1.260	15.932
6_{06}–4_{04}	0	8.961	13.043	0.704	1.501	19.795
$K_{-1} = 1, o$						
3_{13}–1_{11}	0	1.000	3.000	0.025	2.281	7.694
4_{14}–2_{12}	0	1.476	4.524	0.091	2.670	11.239
5_{15}–3_{13}	0	4.268	13.732	0.259	2.432	15.308
6_{16}–4_{14}	0	5.059	16.941	0.256	2.160	19.447
$K_{-1} = 1, e$						
3_{12}–1_{10}	0	4.476	1.524	0.037	7.203	2.760
4_{13}–2_{11}	0	10.396	3.604	0.166	9.172	4.662
5_{14}–3_{12}	0	13.242	4.759	0.751	8.756	8.493
6_{15}–4_{13}	0	15.987	6.013	1.486	6.975	13.538
$K_{-1} = 2, o$						
4_{23}–2_{21}	0	6.970	7.030	0.036	7.747	7.277
5_{24}–3_{22}	0	8.885	9.115	0.215	7.508	10.277
6_{25}–4_{23}	0	10.742	11.258	0.467	7.747	13.786
$K_{-1} = 2, e$						
4_{22}–2_{20}	0	7.504	6.495	−0.392	10.905	3.487
5_{23}–3_{21}	0	10.018	7.982	−0.557	15.072	3.484
6_{24}–4_{22}	0	12.780	9.220	−0.233	17.088	5.145
ΔF_2 Values with $\Delta K_{-1} = 2$						
3_{31}–1_{11}	8.000	1.524	0.476	7.975	1.719	0.306
4_{32}–2_{12}	8.000	4.598	1.402	7.910	5.330	0.760
5_{33}–3_{13}	8.000	7.244	1.756	7.763	10.016	0.321
6_{34}–4_{14}	8.000	12.505	1.495	7.767	15.211	−0.898

the two vibrational states is sufficiently great to spread the Q-branch out, it is sometimes possible to make unique assignments of Q-branch transitions. This can be seen best by the lowest spectrum in Fig. 8b6.

From this discussion one can see that there is a limit to the amount of information that can be obtained from the analysis of an A-type band.

8e A-TYPE BANDS OF NEARLY SYMMETRIC ROTORS

If a molecule is only slightly asymmetric although the sub-band picture is still valid, it is sometimes better to consider a group of lines with the same $\Delta J = \pm 1$, since this

and γ for Various Values of κ (A-Type Band)

$\kappa \sim 0.0$			$\kappa \sim 0.5$			$\kappa \sim 0.9$		
$\Delta\alpha$	$\Delta\beta$	$\Delta\gamma$	$\Delta\alpha$	$\Delta\beta$	$\Delta\gamma$	$\Delta\alpha$	$\Delta\beta$	$\Delta\gamma$
0.240	2.058	3.702	0.652	1.395	3.953	0.962	1.038	4.000
0.721	1.658	7.621	1.240	0.797	7.963	1.476	0.524	8.000
0.954	1.133	11.913	1.064	0.928	12.008	1.008	0.992	12.000
0.875	1.161	15.964	0.820	1.196	15.954	0.988	1.012	16.000
0.779	1.315	19.916	0.792	1.254	19.954	0.988	1.012	20.000
0.114	2.026	7.810	0.306	1.719	7.975	0.476	1.524	8.000
0.345	1.930	11.725	0.720	1.313	11.967	0.970	1.030	12.000
0.531	1.705	15.764	0.776	1.271	15.953	0.988	1.012	16.000
0.632	1.550	19.818	0.788	1.263	19.948	0.988	1.012	20.000
0.227	6.547	3.176	1.394	4.808	3.798	2.808	3.192	4.000
1.184	6.339	5.477	3.338	2.828	7.834	4.396	1.604	8.000
1.357	4.515	11.184	3.189	2.860	11.950	3.077	2.923	12.000
2.830	3.302	15.867	2.510	3.493	15.896	2.972	3.028	16.000
0.214	6.192	7.594	0.760	5.330	7.910	1.402	4.598	8.000
0.717	6.173	11.110	1.723	4.492	11.785	2.885	3.115	12.000
2.268	5.615	15.117	2.213	3.960	15.827	2.996	3.034	16.000
−0.677	11.890	3.787	0.790	9.695	3.515	4.419	5.581	4.000
0.326	12.711	4.963	3.907	6.845	4.723	7.730	2.270	8.000
2.156	10.674	9.170	5.634	4.515	11.849	5.297	4.703	12.000
7.886	1.974	0.140	7.694	2.281	0.025	7.524	2.476	0
7.655	6.070	0.275	7.277	6.697	0.036	7.030	6.470	0
7.526	10.694	−0.220	7.509	10.578	−0.187	8.804	9.196	0
7.676	15.365	−1.041	8.708	13.677	−0.385	10.785	11.785	0

group of lines forms a characteristic pattern. Consider a parallel band of a prolate symmetric rotor. If the effective moments of inertia are the same in both vibrational states, all the K-components of a given $\Delta J = \pm 1$ transition occur at precisely the same frequency and only one absorption line occurs. The first column of Fig. 8e1 illustrates the P_4 transition under these conditions. If B and C differ very slightly, the molecule becomes an asymmetric rotor, and the first result of the asymmetry is a splitting of the $K_{-1} = 1$ levels in each vibrational state similar to the case shown in the central columns of Fig. 8e1. As is shown at the bottom of the column, the result of this splitting is to produce a triplet structure for the group of transitions. Finally, if the value of $B - C$ becomes quite large, higher order spittings occur for values of $K_{-1} > 1$.

A possible set of energy levels and the resulting transitions for P_4 are shown in the

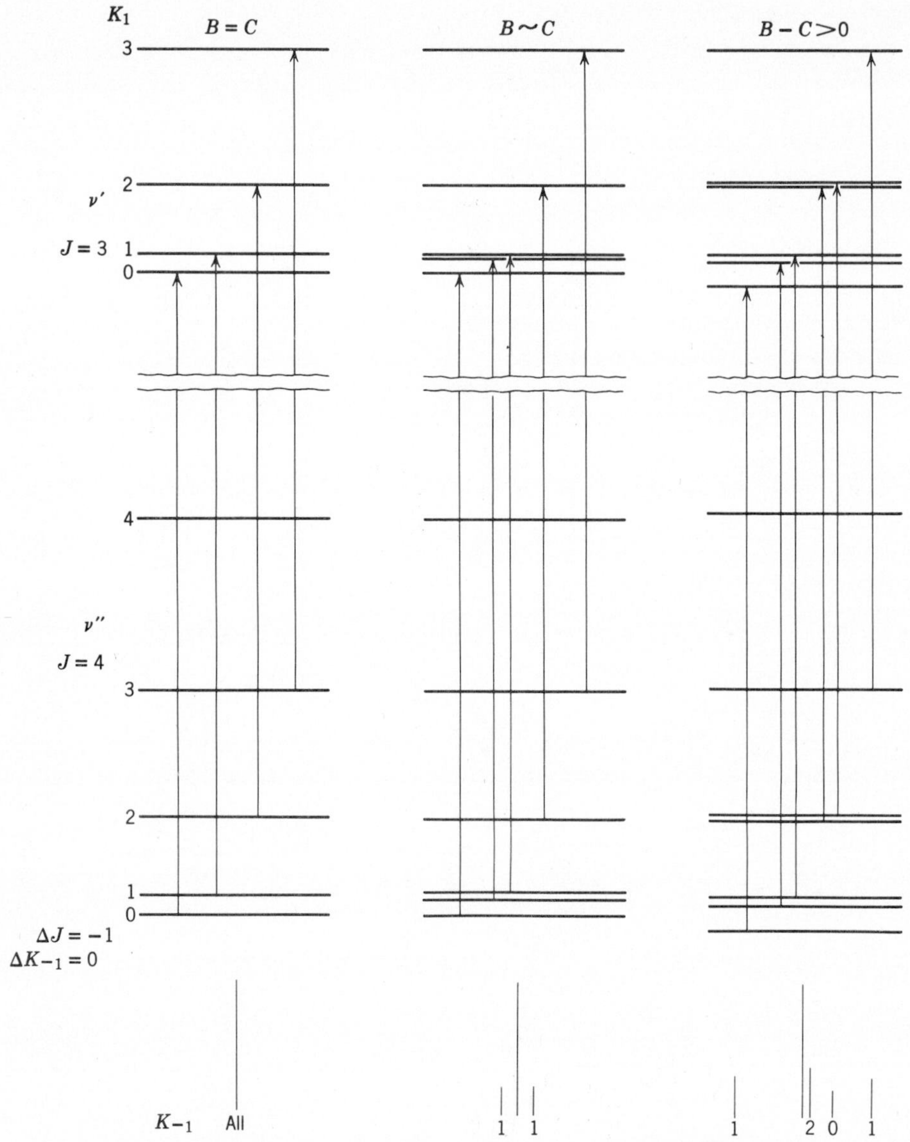

Fig. 8e1 The allowed transitions and resulting spectra for a nearly prolate symmetric rotor. The rotational constants have been assumed to be equal in both vibrational states. The left-hand column corresponds to the symmetric case. The center column corresponds to the nearly symmetric case, and the right-hand colunm has considerable asymmetry. For P-branch transitions $J = 4 \rightarrow J = 3$ only.

final column of the figure. This column shows that the $K_{-1} = 1$ transitions are split out even more than in the middle column and that the $K_{-1} = 0$ levels have been shifted. In addition, one of the $K_{-1} = 2$ transitions has split out. If one is confronted with well-defined patterns such as these, one can make assignments very readily, for the outer two transitions in the group are always the $K_{-1} = 1$ transitions. Such patterns also indicate a nearly symmetric rotor, especially if they persist to higher J values. The splitting between the $K_{-1} = 1$ transitions increases with J and eventually the patterns of adjacent J groups overlap. The same discussion can be applied equally well to a nearly oblate rotor. The transition patterns which result are the same. By referring to Fig. 8b2 we can see some of these characteristics for low J even at an asymmetry of $\kappa = -0.5$

8f EXAMPLES OF *A*-TYPE BANDS

As an example of an A-type band consider the absorption due to ethylene (C_2H_4) near 2990 cm^{-1}, which is shown in Fig. 8f1. The P, Q, R structure of the band suggests that this is either an A- or C-type band. The frequency of the vibrational transition is in the region where vibrations primarily due to C—H stretchings are observed but any further decision must be based on some molecular model.

Ethylene is generally considered to be a planar molecule with the configuration pictured in Fig. 8f2. If this structure is generally correct, the principal inertial axes are those indicated in the diagram. That is, the least inertial axis is along the C=C bond, whereas the intermediate axis of inertia passes through the center of mass perpendicular to the a-axis and in the plane of the molecule; the axis of largest moment passes through the center of mass and is perpendicular to the plane of the molecule. In addition it can be seen from Fig. 8f2 that the largest and intermediate moments of inertia are nearly equal. Usually there is some previous structural information available, which may have been obtained from a previous infrared study, a microwave study, or from an electron diffraction investigation. From this structural information one can calculate fairly good values of the moments of inertia. However, if no previous structural study has been made, it is still possible to estimate the moments of inertia by using bond distances estimated from covalent radii. These estimated moments of inertia can then be used to calculate the molecular rotational energy levels using 2j9 and tables of the reduced energy $E(\kappa)$ given in Appendix IV. These energy levels together with the appropriate selection rules can help in making the choice between an A- and a C-type band and can also be used as a rough guide in making quantum number assignments.

Another useful tool in the process of assigning quantum numbers is the relative intensities of the transitions. In addition, the nuclear spin statistics for C_2H_4, which arise from the identical hydrogen nuclei, can be very helpful. The relative intensities can be calculated using the tables of line strengths given in Appendix IX together with the Boltzmann factor determined by using the calculated values of the energy levels. In addition, we must apply the nuclear spin correction to the intensities of the appropriate transitions. In C_2H_4 the nuclear spins weights are seven for the totally symmetric rotational levels and three for all the others, assuming totally symmetric vibrational and electronic states. This means that in the $K_{-1} = 0$ sub-band transitions originating in levels with even J are stronger than those originating in levels of odd J by the ratio 7/3 because of the nuclear spin degeneracies. Thus the P- and R-branch transitions in this sub-band should be alternately strong and weak—a fact which should help in the assigning of quantum numbers. No transitions originate from totally symmetric levels in either of the $K_{-1} = 1$ sub-bands. However, each of the

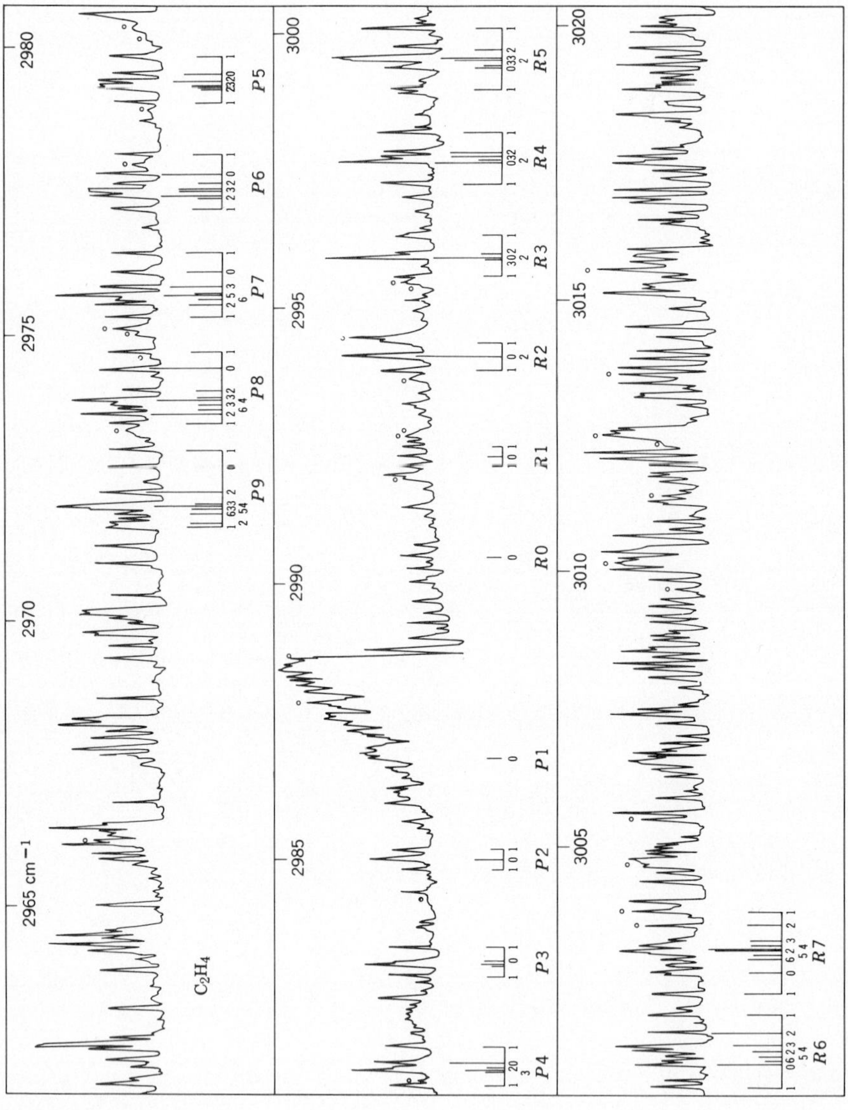

Fig. 8f1 The A-type band of ethylene near 2990 cm^{-1}. The lines marked with an "0" are due to the absorption of atmospheric H_2O. The calculated spectrum is shown beneath the observed spectrum. The value of K_{-1} for each transition is given below the calculated spectrum [12].

$K_{-1} = 2$ sub-bands again show the intensity alternation. For the even sub-band the stronger transitions are those originating in levels of even J, whereas in the odd sub-band the stronger transitions originate in levels of odd J. In general, one can say for this molecule that the sub-bands with odd K_{-1} values have no intensity alternation, whereas the sub-bands with even values of K_{-1} do have an intensity alternation.

Furthermore, the transitions originating in levels with even J are strong in the even sub-bands, whereas the transitions originating in levels of odd J are strong in the odd sub-bands. This intensity alternation furnishes valuable clues to aid in the assignment

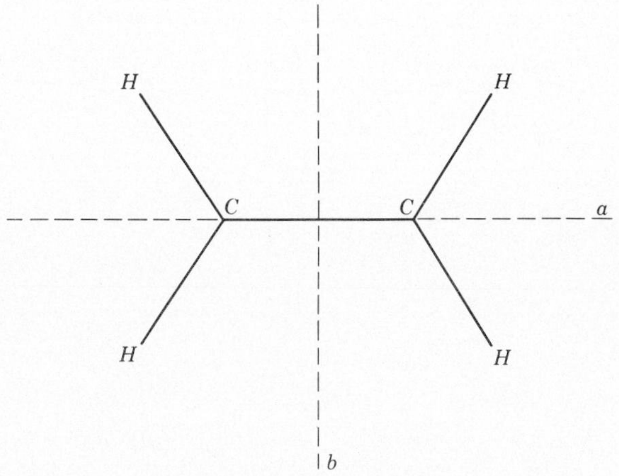

Fig. 8f2 The structure and principal inertial axes of C_2H_4.

of quantum numbers. With the estimated values of the inertial constants and the relative intensities of the transitions we are now ready to return to the observed spectrum in Fig. 8f1 and start the process of assigning quantum numbers. The decision on band type can be made by determining the frequency difference between two successive lines in a sub-band. These successive lines can be readily seen in the C_2H_4 spectrum. If differences are taken in both the P- and R-branches near the center of the band, the average $\Delta\nu$ should be approximately $B + C$ for an A-type band and $A + B$ for a C-type band for a nearly symmetric rotor. Estimates of the inertial constants of ethylene show B and C to be nearly equal and to have a value near 0.9 cm^{-1} thus $B + C$ will be about 1.8 cm^{-1}, which agrees well with the differences found. Thus we find this to be an A-type band, which determines the selection rules to be used. The value of $A + B$ is about 5.6 cm^{-1} which is clearly too large.

The groups of lines with the same value of the ground state J clearly form patterns similar to those given in Fig. 8b2 for a molecule with $\kappa \sim -0.9$. The outer two lines in these groups should have $K_{-1} = 1$ and should be approximately equally intense. Thus we can readily pick out the two $K_{-1} = 1$ sub-bands. After the first couple of transitions in the P- and R-branches of these sub-bands have been identified, the rest of the sub-band can easily be found. Using a pair of dividers set to the frequency difference of the first two transitions in the P- and R-branch, we can step off the outer transitions if an allowance is made for a slight convergence or divergence as J'' increases and retains in mind the relative intensity pattern of the sub-band. We should also be able to identify the $K_{-1} = 0$ sub-band. For $R(1)$ there can only be three transitions and the center one must be the $K_{-1} = 0$ transition. The same pattern must be true for $P(2)$. This gives one transition in both the P- and R-branch of this sub-band which, together with an approximate spacing from the $K_{-1} = 1$ sub-bands and the intensity alternation to be expected, should enable us to identify the $K_{-1} = 0$ sub-band. This procedure leads to the sub-bands listed in Table 8f1 for $K_{-1} = 0, 1$. From these sub-bands we can determine more precise values of B and C which enables us to calculate more precise energy levels. $\Delta F_2''$ values determined from these new energy levels enable us to identify the transitions in sub-bands with $K_{-1} > 1$.

By this method, transitions in the sub-bands up through $K_{-1} = 3$ were assigned in the ethylene band, and are also listed in Table 8f1. Assignments in sub-bands with $K_{-1} > 3$ were difficult to make because in general they were not resolved from each other. In any event sufficient transitions were assigned in the seven sub-bands with the lowest values of K_{-1} to enable one to determine very good values of B and C for both the ground and excited vibrational state. As indicated in 8d, the value of A is very poorly determined from the analysis of an A-type band. Methods for deducing the inertial constants from the assigned transitions are discussed in a later section since the methods apply equally well to all types of bands.

As a second example of an A-type asymmetric rotor band, consider the HDO band near 2725 cm^{-1} shown in Fig. 8f3 [22]. The molecule is considerably more asymmetric than C_2H_4 having $\kappa \sim -0.68$. The procedure in the analysis of this band is the same as in the previous case. In this molecule, there is the possibility of a component of the change in electric moment along both the principal inertial axes in the plane of the molecule. Fortunately, for the 2725 cm^{-1} vibration, the component along the b-axis is very small and the observed band is essentially pure A-type. There is no possibility of a C-type band in a planar triatomic molecule. In this molecule there can be no intensity alternation due to spin statistics. The first estimates of the moments of inertia can be made from the well-known structure of water. Using the energy levels calculated from these inertial parameters, we can obtain the Boltzmann factors to be used in the calculation of the relative intensities of the transitions. As shown in the figures the transitions in the P- and R-branches that originate in ground state levels with the same J are nicely grouped, especially for low J values. The asymmetry of this molecule is greater than in the C_2H_4 molecule and hence the splitting of the K_{-1} degeneracy is greater, especially for small values of K_{-1}. An added advantage for this molecule is that it is much lighter than C_2H_4 and hence the inertial constants A, B, and C are much larger. Thus the transitions are separated to a greater extent. Both these factors combine to enable a more complete resolution of the individual transitions, although the available instrument resolution was not as great as that available for the C_2H_4 work. The same principles apply to the groups of lines with common J. For $P(2)$ and $R(1)$ only three transitions are possible. The center one belongs to the $K_{-1} = 0$ sub-band, whereas the two outer ones belong to the two $K_{-1} = 1$ sub-bands.

For $P(3)$ and $R(2)$ there are only five allowed transitions. In principle the $K_{-1} = 0$ transition is the strongest and the two K_{-1} transitions are the next strongest and are also at the extremes of the group. The two $K_{-1} = 2$ transitions are weaker and about the same intensity. Thus these five transitions in the P- and R-branch are readily

Table 8f1—The Sub-Bands of the A-Type Band of C_2H_4 at 2988.66 cm^{-1} [12]

$J''_{K_{-1}K}$	cm^{-1}	$J''_{K_{-1}K}$	cm^{-1}	$J''_{K_{-1}K}$	cm^{-1}	$J''_{K_{-1}K}$	cm^{-1}	$J_{K_{-1}K}$	cm^{-1}	$J''_{K_{-1}K}$	cm^{-1}	$J''_{K_{-1}K}$	cm^{-1}
$K_{-1} = 0, e$		$K_{-1} = 1, o$		$K_{-1} = 1, e$		$K_{-1} = 2, e$		$K_{-1} = 2, o$		$K_{-1} = 3, e$		$K_{-1} = 3, o$	
$\Delta J = +1$		$\Delta J = +1$		$\Delta J = +1$		$\Delta J = +1$		$\Delta J = +1$		$\Delta J = +1$		$\Delta J = +1$	
0_{00}	2990.48												
1_{01}	92.29	1_{11}	2992.11	1_{10}	2992.46								
2_{02}	94.10	2_{12}	93.84			2_{20}	2994.10	2_{21}	2994.10				
3_{03}	95.90	3_{13}	95.57	3_{12}	96.28	3_{21}	95.97	3_{22}	95.90	3_{30}	2995.90	3_{31}	2995.90
4_{04}	97.63	4_{14}	97.26	4_{13}	98.18	4_{22}	97.84	4_{23}	97.70	4_{31}	97.84	4_{32}	97.84
5_{05}	99.36	5_{15}	98.97	5_{14}	3000.07	5_{23}	99.72	5_{24}	99.52	5_{32}	99.52	5_{33}	99.52
6_{06}	3001.05	6_{16}	3001.64	6_{15}	01.93	6_{24}	3001.62	6_{25}	3001.30	6_{33}	3001.37	6_{34}	3001.37
7_{07}	02.71	7_{17}	02.32	7_{16}	03.80	7_{25}		7_{26}	03.10	7_{34}	03.29	7_{35}	03.20
8_{08}	04.34	8_{18}	03.99	8_{17}	05.63	8_{26}	05.50	8_{27}	04.85	8_{35}	05.18	8_{36}	05.01
9_{09}	05.95	9_{19}	05.63	9_{18}	07.45	9_{27}	07.45	9_{28}	06.60	9_{36}	07.06	9_{37}	06.94
10_{010}	07.54	10_{110}	07.26	10_{19}	09.22								
11_{011}	09.12												
$\Delta J = -1$		$\Delta J = -1$		$\Delta J = -1$		$\Delta J = -1$		$\Delta J = -1$		$\Delta J = -1$		ΔJ -1	
1_{01}													
2_{02}	2985.01	2_{12}	2985.17	2_{11}	2984.82								
3_{03}	83.19	3_{13}	83.43	3_{12}	82.90	3_{21}	2983.14	3_{22}	2983.14				
4_{04}	81.39	4_{14}	81.67	4_{13}	80.97	4_{22}	81.26	4_{23}	81.31	4_{31}	2981.26	4_{32}	2981.26
5_{05}	79.59	5_{15}	79.92	5_{14}	79.10	5_{23}	79.38	5_{24}	79.47	5_{32}	79.41	5_{33}	79.41
6_{06}	77.82	6_{16}	78.18	6_{15}	77.20	6_{24}	77.46	6_{25}	77.64	6_{33}	77.54	6_{34}	77.54
7_{07}	76.08	7_{17}	76.42	7_{16}	75.30	7_{25}	75.55	7_{26}	75.81	7_{34}	75.68	7_{35}	75.68
8_{08}	74.34	8_{18}	74.68	8_{17}	73.41	8_{26}	73.56	8_{27}	73.99	8_{35}	73.82	8_{36}	73.82
9_{09}	72.62	9_{19}	72.95	9_{18}	71.55	9_{27}	71.00	9_{28}	72.16	9_{36}	71.94	9_{37}	71.94
10_{010}	70.90	10_{110}	71.19	10_{19}	69.70	10_{28}	69.70	10_{29}	70.33	10_{37}	70.04	10_{38}	70.04
11_{011}	69.25	11_{111}	69.48										
12_{012}	67.54	12_{112}	67.70										

assigned. With this start the assignment of the rest of the transitions in the sub-bands followed quite readily by inspection. Table 8f2 shows the assignments in the *P*-branches of the sub-bands which has been made by Benedict, Gailar, and Plyler [22].

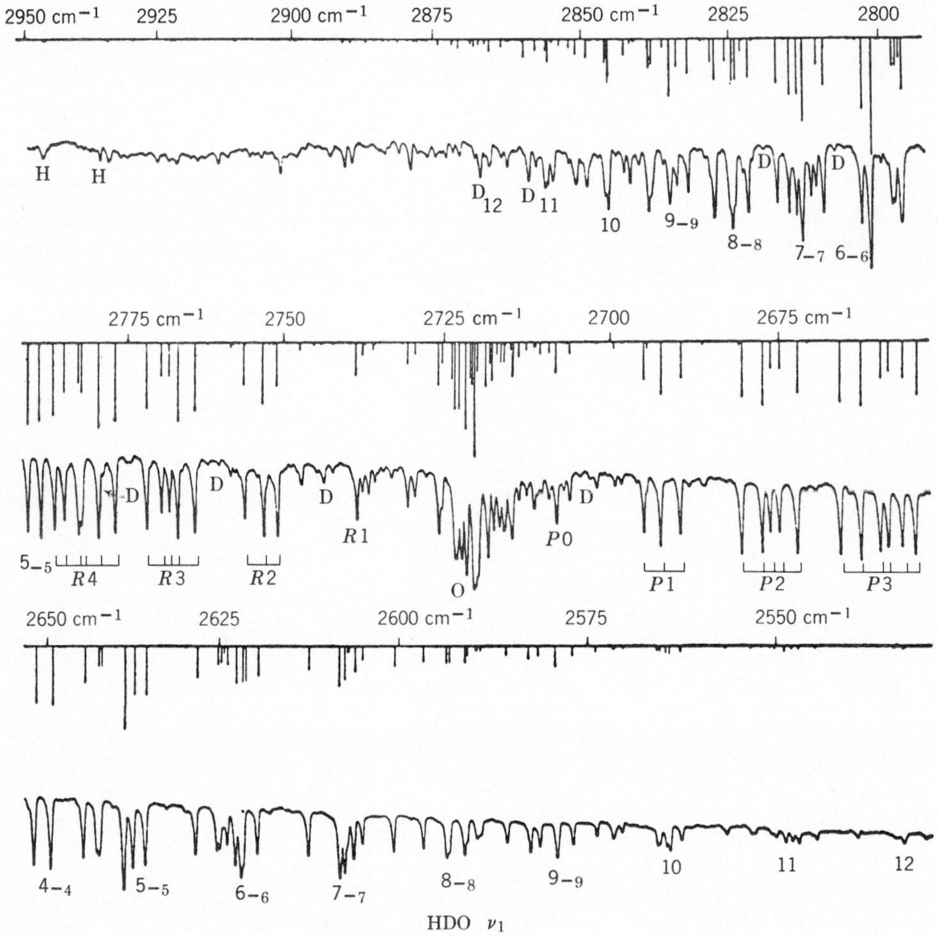

Fig. 8f3 Observed and calculated HDO spectra, 2950–2530 cm^{-1}. The observed spectrum is of an isotopic mixture with approximately 5% D_2O, 35% HDO, 60% H_2O. A few lines of the 020 band of H_2O, marked *H*, appear at high frequencies. The more prominent lines of D_2O are marked *D*. The HDO lines constitute principally the 100 band, a pure type-*A* band, with a strong *Q*-branch near 2720 cm^{-1}, and *P*- and *R*-branches whose splitting increases with increasing *J*. Some of the more prominent lines are marked by the *upper state J*. In the calculated spectrum $\kappa \sim -0.6$ and the heights of the lines are proportional to the intensity [22].

One can see from the table and the spectrum that a very complete assignment could be made both to high J and high K_{-1}.

Finally, as an example of a band in which the *Q*-branch transitions can be uniquely identified to a great extent, consider the absorption of H_2S near 6200 cm^{-1} [13]. This absorption represents a typical situation which is encountered more often than

Table 8f2—Series Structure in the P_{01} Branch of ν_1, HDO [Observed Frequency(cm^{-1}) and Calculated Relative Intensity Are Tabulated] [23]

$(J+\tau)$ / J'	0		1	2		3	4	5		6	7		8	9		10	11–12	13–14
0	2708.23 93																	
1	2692.68 161		2695.23 116	2689.76 111														
2	2677.70 196		2680.73 169	2672.55 157		2676.54 83	2675.22 83											
3	2663.21 197		2666.20 184	2655.53 163		2660.46 112	2657.39 110		2659.28 92									
4	2649.49 179		2651.92 172	2638.69 143		2644.55 112	2638.69 107	2642.58 59		2642.29 59		2642.29 42						
5	2635.73 147		2637.51 144	2621.92 109		2628.53 93	2619.84 87	2625.45 54		2624.07 53		2625.08 50			2624.73 14.8			
6	2621.92 111		2622.84 110	2606.40 77		2612.59 71	2600.94 60	2608.26 42		2605.28 41		2607.63 42			2606.40 17.3		2606.40 4.5	
7	2607.63 78		2608.26 77	2591.24 50		2596.82 47	2582.39 38	2591.24 28		2585.56 27	2589.49 15.7		2589.08 15.7		2589.49 14.2		2588.20 4.9	2587.73 1.3
8	2593.38 51		2593.71 51	2576.72 31.0		2581.11 30.6	2564.42 22.7	2573.66 18.5		2565.50 96.8	2571.48 10.0		2570.30 9.8		2571.48 9.5		2570.30 3.7	1.4
9		2578.79 62.2		2562.50 17.9		2565.50 17.8	2547.29 12.2	2556.74 9.8		2544.89 8.7	2553.37 6.0		2550.80 5.6	2553.37 3.0		2552.93 3.0	2.3	0.9
10		2563.96 35.4		2548.13 9.5		2550.17 9.5	2530.94 6.0	2539.56 5.3		2524.98 4.5	2534.71 3.2		2530.94 2.9	2535.02 1.5		2534.34 1.5	1.3	
11		2549.01 18.8		2533.46 4.7		2534.71 4.7	2515.43 2.8	2522.86 2.5		2505.15 1.9	2516.78 4.7		1.4	2516.78 0.7		0.7		
12		2533.81 9.4		2518.45 2.1		2519.27 2.1	1.3	2506.05 1.1		2486.41 0.8								
13		2518.45 4.2			2503.44 1.9													
14		2502.66 1.9																

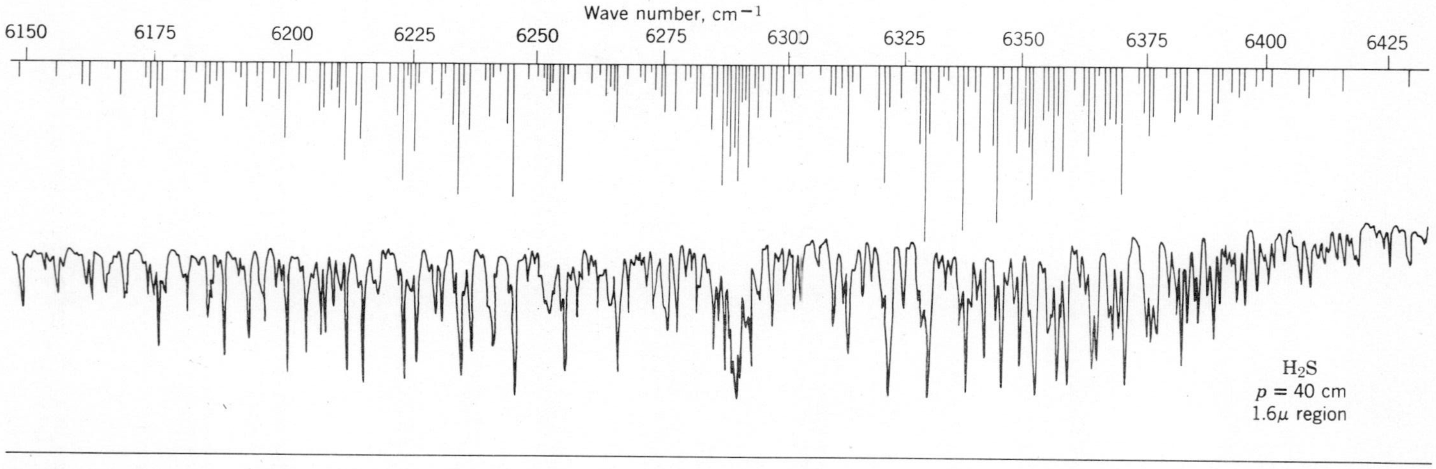

Fig. 8f4 Absorption bands (1,1,1) and 2,0,1) for H_2S in the region 6100 to 6500 cm^{-1}. The calculated positions of the lines are shown above the spectra [13].

not. Two bands are found in this region, the stronger band an *A*-type band and the weaker band arising from a *B*-type transition. Such overlapping of bands is a very common occurrence, and the complete analysis of such a region of absorption is necessarily a very long and tedious process. The over-all appearance of this band is that of an *A*-type band and indeed most of the strong absorption peaks arise from this type transition. The actual analysis of the *P*- and *R*-branch regions follows the procedure just outlined. However, in this band it is possible to identify unambiguously many of the strong *Q*-branch transitions. Figure 8f4 shows the general absorption

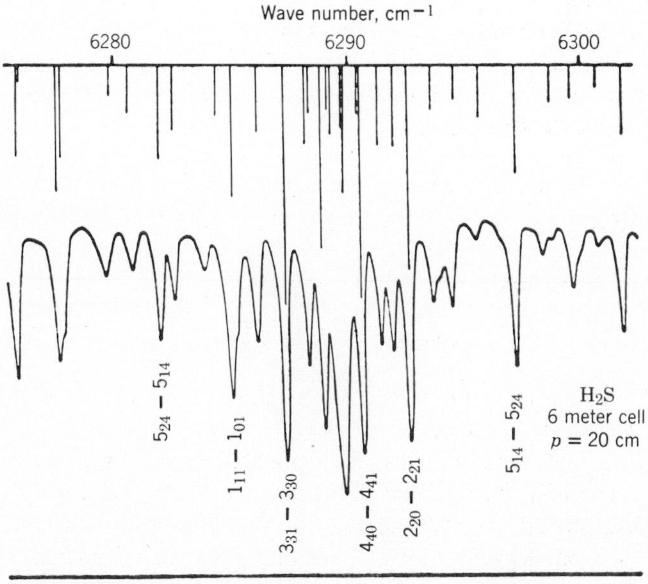

Fig. 8f5 An expanded run of the *Q*-branch region of the *A*-type band of H_2S near 6290 cm^{-1} [13].

in this region along with a computed spectrum. Figure 8f5 is an expanded run of the *Q*-branch region of the *A*-type band. The asymmetry of H_2S corresponds to a κ value of about 0.5; so the *Q*-branch region should be similar to that of the composite spectrum in Fig. 8b4. The intensity distribution is somewhat altered since H_2S bands show an intensity alternation due to nuclear spin statistics which was not included in Fig. 8b4. It can be seen from Fig. 8b4 that the strongest *Q*-branch transitions are the first transition in each sub-band *Q*-branch and these are the prominent ones in the observed *Q*-branch. The quantum number assignments are indicated in the figure. As was pointed out in 8e, when such unique assignments can be made, new independent ΔF_2 values can be obtained to help in the determination of the molecular constants. These new ΔF_2 values can also serve as a valuable check on the assignments made in the *R*- and *P*-branches. It should be emphasized that $\kappa \sim 0.5$ for the molecule; in the particular transition investigated here there is a considerable change in *A* between the two vibrational states. Both these conditions tend to help spread out the *Q*-branch so that it is more readily resolved experimentally.

8g BAND STRUCTURE OF C-TYPE BANDS

Since C-type bands are in many ways quite analogous to A-type bands, it is logical to discuss them next. In fact, there is a symmetry to be found between A- and C-type bands which makes much of the discussion of A-type bands applicable to C-type bands. As will be seen as the discussion proceeds we have only to interchange the roles of K_{-1} and K_1, A and C and make $\kappa = -\kappa$, in the A-type band discussion to achieve one of C-type bands.

A C-type band results when the electric moment change during the vibrational transition is along the principal axis of largest moment of inertia. Such a change corresponds to a parallel band in the oblate symmetric limit and a perpendicular band in the prolate symmetric limit. Thus in a manner entirely analogous to that used for the A-type band, we are led to the following selection rules for the strong transitions in the sub-bands of a C-type band:

$$\begin{aligned} &\Delta J = \pm 1 \qquad \Delta K_{-1} = \pm 1 \qquad \Delta K_1 = 0 \qquad K_1 = 0 \\ &\Delta J = 0, \pm 1 \qquad \Delta K_{-1} = \pm 1 \qquad \Delta K_1 = 0 \qquad K_1 \neq 0. \end{aligned} \tag{1}$$

The general selection rule given in 5h is

$$ee \leftrightarrow oe \qquad eo \leftrightarrow oo, \tag{2}$$

that is, the parity of K_1 does not change, whereas the parity of K_{-1} must change. It is readily verified that (1) is a special case of (2).

Since the value of K_1 does not change during a strong transition in a C-type band, it is convenient to classify the sub-bands according to a constant value of K_1 and the evenness or oddness of the ground state levels. Thus one finds that the $K_1 = 0$ sub-band is an even sub-band and for all values of $K_1 \neq 0$ there is both an even and an odd sub-band. Figures 8b1 to 8b5 show just how these bands are made up and the type of patterns to be expected provided the roles of K_{-1} and K_1 are interchanged and κ is read as $-\kappa$.

The strongest transitions in the composite Q-branch of a C-type band are those for which

$$J_{0J} \rightleftharpoons J_{1J}. \tag{3}$$

As shown in 8c these levels involved in (3) are those which depend almost entirely on C throughout the whole range of κ. The equation for a C-type band which is analogous to 8c4 is

$$\nu \sim \nu_0 + \gamma(C' - C''), \tag{4}$$

which shows that the frequencies of the strongest Q-branch transitions depend mainly on the difference between the effective values of C in the two vibrational states. The strong P- and R-branch transitions in the $K_1 = 0$ and odd $K_1 = 1$ sub-bands take place between the levels

$$\begin{aligned} &(a) \qquad J_{J0} \rightleftharpoons (J+1)_{J+1,0} \\ &(b) \qquad J_{J1} \rightleftharpoons (J+1)_{J+1,1}. \end{aligned} \tag{5}$$

and the approximate equation analogous to 8c7, for their frequency becomes

$$\begin{aligned} \nu &\sim \nu_0 + (2J + 1)A' + (A' - A'')J^2 \qquad \Delta J = 1 \\ \nu &\sim \nu_0 - (2J - 1)A' + (A' - A'')J^2 \qquad \Delta J = -1. \end{aligned} \tag{6}$$

By comparing 8c4 and 8c7 with (4) and (6), we find that the roles of A and C are interchanged in the two types of bands.

8h THE ΔF_2 VALUES OF A C-TYPE BAND

We can also obtain the ΔF_2 values of a C-type band in a manner entirely analogous to that used for A-type bands. The equations have the same form as 8d1 and 8d2 provided the conditions on K_1 and K_{-1} are interchanged. The dependence of the ΔF_2 values on the inertial parameters is illustrated in Table 8h1. We find that it is now the C-constant for which we obtain very little information from the $\Delta K_1 = 0$ transitions. We must seek transitions for which $\Delta K_1 > 0$ in order to obtain ΔF_2 values with a sizable dependence on C. These are the weaker transitions in a C-type band and hence subject to the same difficulties as the $\Delta K_{-1} > 0$ transitions in an A-type band. This discussion clearly shows the symmetrical nature of A- and C-type bands of an asymmetric rotor.

Because of the peculiar dependence of the ΔF_2 values on the inertial parameters, it is not possible to obtain a complete set of accurate inertial parameters from the analysis of either an A- or C-type band alone. However, by combining the $\Delta F_2''$ values obtained from both an A- and a C-type band, it is possible to determine completely the ground state constants with good accuracy. Once the ground state constants are well known, we can determine all the excited state constants in both type bands as is discussed later.

8i EXAMPLES OF C-TYPE BANDS

The number of C-type bands which have been observed under high resolution is very small. Figure 8i1 shows the absorption of CH_2D_2 near 4425 cm^{-1}. This absorption arises mainly because of a change in electric moment along the largest inertial axis. It is easily seen that this observed band bears a marked resemblance to an A-type band. The analysis of this band would proceed in a manner entirely analogous to that described for an A-type band. This molecule has two sets of identical nuclei which contribute to the nuclear spin statistics that result in the antisymmetric rotational levels having a weight of 21, whereas the symmetric rotational levels have a weight of 15, all for a totally symmetric vibrational state. The moments of inertia and energy levels for this molecule can be estimated from the well-known structure of methane. These energy levels together with the relative intensities can then be used in the manner described previously to make quantum number assignments. As a result of these considerations the assignments of the first few transitions in the low K_1 sub-bands are given in Table 8i1.

Other examples of C-type bands under high resolution have not been published as yet. Another example of a C-type band which has appeared in the literature is the

Table 8h1—Dependence of ΔF_2 Values on

ΔF_2	$\kappa \sim -0.9$ α	β	γ	$\kappa \sim -0.5$ α	β	γ
$K_1 = 0, e$						
2_{20}–0_{00}	4.000	1.038	0.962	3.953	1.395	0.652
3_{30}–1_{10}	8.000	0.524	1.476	7.963	0.797	1.240
4_{40}–2_{20}	12.000	0.992	1.008	12.008	0.928	1.064
5_{50}–3_{30}	16.000	1.012	0.988	15.954	1.196	0.840
6_{60}–4_{40}	20.000	1.012	0.988	19.954	1.254	0.792
$K_1 = 1, o$						
3_{31}–1_{11}	8.000	1.524	0.476	7.975	1.719	0.306
4_{41}–2_{21}	12.000	1.030	0.970	11.967	1.313	0.720
5_{51}–3_{31}	16.000	1.012	0.988	15.953	1.271	0.776
6_{61}–4_{41}	20.000	1.012	0.988	19.948	1.263	0.788
$K_1 = 1, e$						
3_{21}–1_{01}	4.000	3.192	2.808	3.798	4.808	1.394
4_{21}–2_{11}	8.000	1.604	4.396	7.834	2.828	3.330
5_{41}–3_{21}	12.000	2.923	3.077	11.950	2.860	3.189
6_{51}–4_{31}	16.000	3.028	2.972	15.896	3.493	2.510
$K_1 = 2, o$						
4_{32}–2_{12}	8.000	4.598	1.402	7.910	5.330	0.760
5_{42}–3_{22}	12.000	3.115	2.885	11.785	4.492	1.723
6_{52}–4_{32}	16.000	3.034	2.996	15.827	3.960	2.213
$K_1 = 2, e$						
4_{22}–2_{02}	4.000	5.584	4.419	3.515	9.695	0.790
5_{32}–3_{12}	8.000	2.270	7.730	4.723	6.845	3.907
6_{42}–4_{22}	12.000	4.703	5.297	11.849	4.517	5.634
ΔF_2 Values with $\Delta K = 2$						
3_{13}–1_{11}	0	2.476	7.524	0.025	2.281	7.694
4_{23}–2_{21}	0	6.970	7.030	0.036	6.687	7.277
5_{33}–3_{31}	0	9.196	8.804	−0.187	10.578	7.509
6_{43}–4_{41}	0	11.215	10.785	−0.385	13.677	8.708

C_2H_4 band near 950 cm^{-1}. This band is interesting since for it $\kappa \sim -0.9$. Since near this limit a *C*-type band corresponds to a perpendicular band, we would expect this band to have a marked resemblance to a perpendicular band. In Fig. 8i2 the spectrum is shown under rather low resolution, but the series of *Q*-branches characteristic of a perpendicular band are readily apparent.

8j THE STRUCTURE OF *B*-TYPE BANDS

A *B*-type band arises when the change in electric moment during a vibration is along the principal axis of intermediate moment of inertia. In both the prolate and the oblate symmetric limits the *B*-axis is perpendicular to the top axis; thus in both

α, β, γ for Various Values of κ (C-Type Band)

$\kappa \sim 0$			$\kappa \sim 0.5$			$\kappa \sim 0.9$		
α	β	γ	α	β	γ	α	β	γ
3.761	1.942	0.297	3.349	2.605	0.046	3.038	2.962	0
7.723	1.453	0.824	6.606	3.192	0.202	5.192	4.808	0
11.967	1.024	1.009	11.290	2.113	0.597	7.535	6.465	0
15.992	1.145	0.883	15.932	1.260	0.808	10.133	7.867	0
19.924	1.298	0.778	19.795	1.501	0.704	13.034	8.961	0
7.886	1.974	0.140	7.694	2.281	0.025	3.000	1.000	0
11.786	1.808	0.406	11.239	2.670	0.091	4.524	1.476	0
15.829	1.575	0.596	15.308	2.432	0.259	13.732	4.268	0
19.871	1.443	0.685	19.447	2.160	0.256	16.941	5.059	0
3.729	6.342	0.379	2.760	7.203	0.037	1.524	4.476	0
6.816	5.661	1.523	4.662	9.172	0.166	3.604	10.396	0
11.520	3.786	2.694	8.493	8.756	0.751	4.759	13.242	0
15.994	3.049	2.957	13.538	6.975	1.486	6.013	15.987	0
7.655	6.070	0.275	7.277	7.747	0.036	7.030	6.970	0
11.283	5.827	0.890	10.277	7.508	0.215	9.115	8.885	0
15.347	5.155	1.498	13.786	7.747	0.467	11.258	10.742	0
2.844	11.774	−0.618	3.487	10.905	−0.392	6.495	7.504	0
5.422	11.793	0.785	3.484	15.072	−0.557	7.982	10.018	0
10.061	8.892	3.047	5.145	17.088	−0.233	9.220	12.780	0
0.114	2.026	7.860	0.306	1.719	7.975	0.476	1.524	8.000
0.214	6.192	7.594	0.760	5.330	7.910	1.402	4.598	8.000
−0.246	10.746	7.500	0.321	10.016	7.763	1.756	7.244	8.000
−0.915	15.133	7.792	−0.898	15.211	7.767	1.495	12.505	8.000

the symmetric rotor limits a *B*-type band corresponds to a perpendicular band. In a manner entirely analogous to the previous two cases we can deduce the selection rules for the strong transitions of a *B*-type band from those in the two symmetric limits. Thus with the help of 5d34, we obtain

$$\Delta J = 0, \pm 1 \qquad \Delta K_{-1} = \pm 1 \qquad \Delta K_1 = \pm 1. \tag{1}$$

It is readily verified that (1) is a special case of the general selection rule

$$ee \leftrightarrow oo \qquad oe \leftrightarrow eo. \tag{2}$$

Equation (2) merely restricts the changes in K_{-1} and K_1 to 1, 3, 5, ... However, just as in the previous type bands transitions with large changes in K_{-1} or K_1 are very much weaker than those in which K_{-1} and K_1 change by only one. Another important

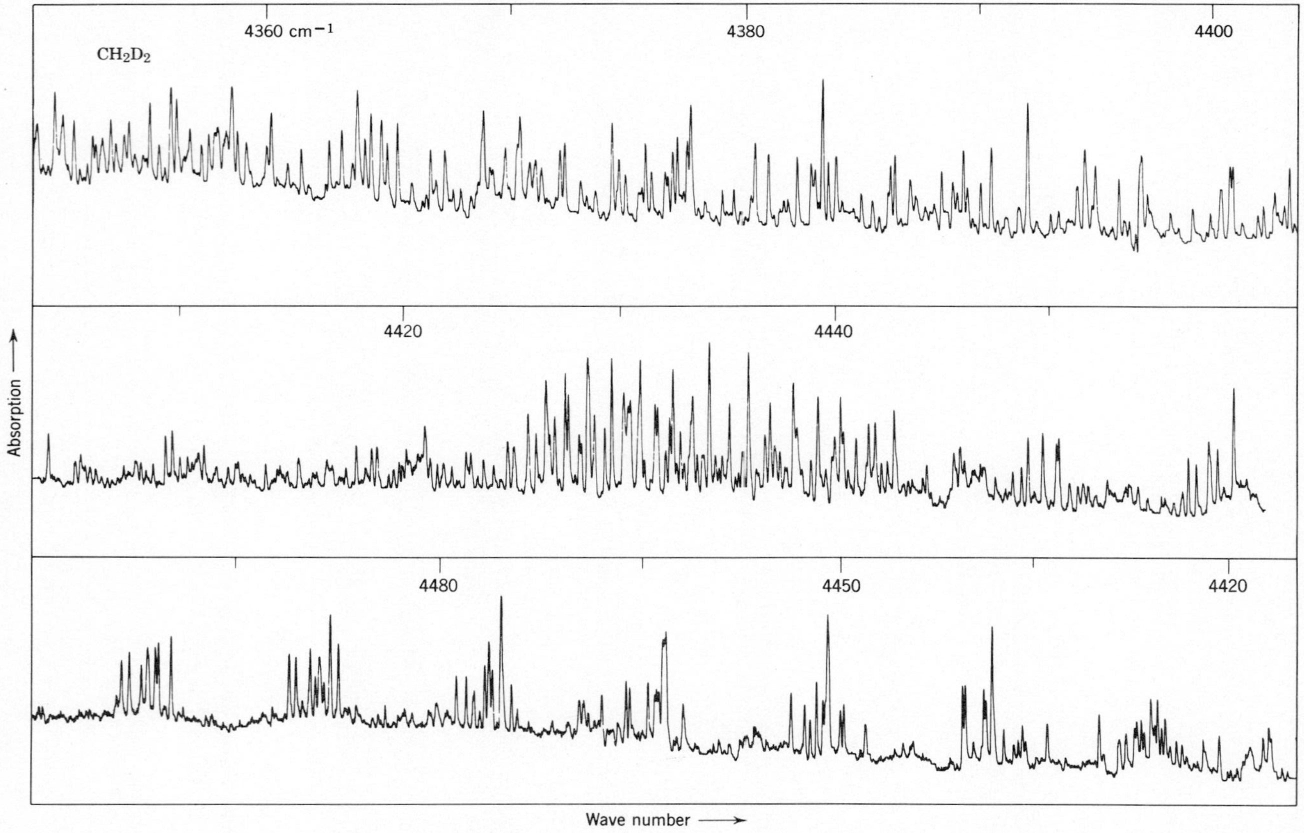

Fig. 8i1 The *C*-type band of CH_2D_2 near 4400 cm^{-1}. For the low values of *J*, the *P*- and *R*-branch transitions are completely resolved (W. B. Olson, H. C. Allen and E. K. Plyler, *J. Research*, *N.B.S.* (in press.)

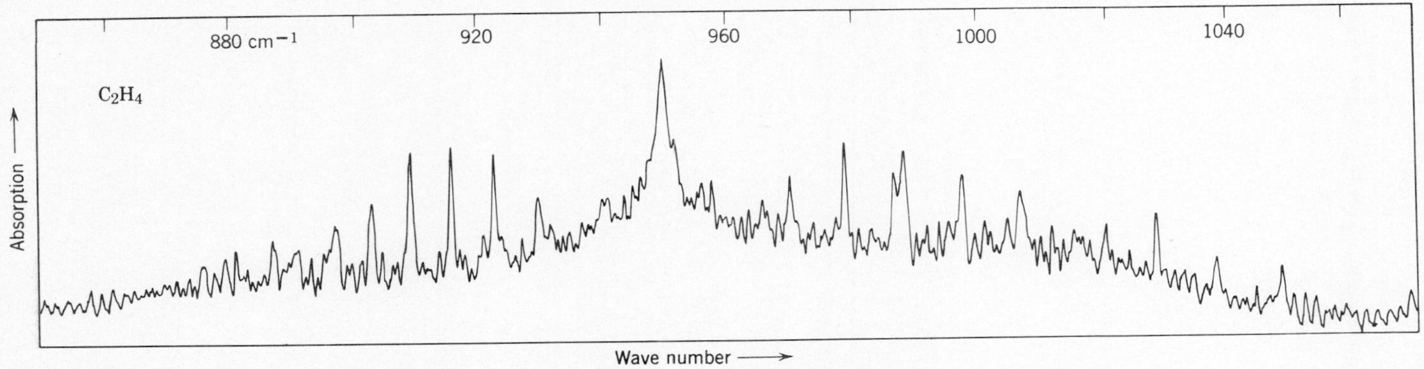

Fig. 8i2 The *C*-type band of C_2H_4 at 950 cm^{-1}. The series of unresolved *Q*-branches characteristic of a perpendicular band of a nearly symmetric rotor are clearly evident (H. C. Allen and E. K. Plyler, in press.)

Table 8il—Sub-Band Assignments in C-Type Band of CH_2D_2 near 4425 cm^{-1}

TRANSITION	OBSERVATION, cm^{-1}	TRANSITION	OBSERVATION, cm^{-1}
$K_1 = 0, e$			
1_{10}–0_{00}	4433.45	0_{00}–1_{10}	4417.79
2_{20}–1_{10}	41.72	1_{10}–2_{20}	09.54
3_{30}–2_{20}	50.22	2_{20}–3_{30}	00.84
4_{40}–3_{30}	58.64	3_{30}–4_{40}	4392.04
5_{50}–4_{40}	66.86	4_{40}–5_{50}	83.20
6_{60}–5_{50}	75.04	5_{50}–6_{60}	74.27
$K_1 = 1, o$			
2_{21}–1_{11}	4442.02	1_{11}–2_{21}	
3_{31}–2_{21}	50.22	2_{21}–3_{31}	4400.71
4_{41}–3_{31}	58.64	3_{31}–4_{41}	4392.04
5_{51}–4_{41}	66.86	4_{41}–5_{51}	83.19
6_{61}–5_{51}	75.04	5_{51}–6_{61}	74.27
$K_1 = 1, e$			
2_{11}–1_{01}	4440.80	1_{01}–2_{11}	
3_{21}–2_{11}	48.89	2_{11}–3_{21}	4403.27
4_{31}–3_{21}	57.45	3_{21}–4_{31}	4395.06
5_{41}–4_{31}	66.13	4_{31}–5_{41}	86.32
6_{51}–5_{41}	74.65	5_{41}–6_{51}	77.59
$K_1 = 2, o$			
3_{22}–2_{12}	4449.58	2_{12}–3_{22}	
4_{32}–3_{22}	57.87	3_{22}–4_{32}	4394.56
5_{42}–4_{32}	66.26	4_{32}–5_{42}	86.12
6_{52}–5_{42}	74.65	5_{42}–6_{52}	77.54
$K_1 = 2, e$			
3_{12}–2_{02}	4448.57	2_{02}–3_{12}	
4_{22}–3_{12}	56.48	3_{12}–4_{22}	4396.94
5_{32}–4_{22}	64.86	4_{22}–5_{32}	89.28
6_{42}–5_{32}	73.67	5_{32}–6_{42}	80.90

Table 8jl—Sub-Band Structure of B-Type Bands

R-WING	*Q*-WING	*P*-WING	
	${}^{be}Q_{1\bar{1}}$	${}^{be}P_{1\bar{1}}$	
${}^{be}R_{11}$			Even sub-bands
	${}^{be}Q_{\bar{1}1}$	${}^{be}P_{\bar{1}1}$	
${}^{bo}R_{1\bar{1}}$	${}^{bo}Q_{1\bar{1}}$		
		${}^{bo}P_{\bar{1}\bar{1}}$	Odd sub-bands
${}^{bo}R_{\bar{1}1}$	${}^{bo}Q_{\bar{1}1}$		

property of perpendicular bands which plays an important part in the structure of B-type bands concerns the relative intensities of the branches of a sub-band. A wing in a sub-band is strong when $\Delta J = \Delta K$; for example, for a sub-band of an oblate rotor with $\Delta K_1 = 1$ the R-branch is stronger than the P-branch, whereas the converse is true when $\Delta K_1 = -1$.

A sub-band is defined in the same manner as previously, that is, by a constant value of K_{-1} or K_1 and the evenness or oddness of the parity of the ground state levels. We find that there is an even band for K_{-1} (or K_1) $= 0, 1, 2, \ldots$ and an odd sub-band for K_{-1} (or K_1) $= 1, 2, 3, \ldots$. The proper grouping of wings into sub-bands is given in Table 8j1. For the even sub-bands there is only one R-branch wing, where the change in both K_{-1} and K_1 is $+1$. This wing satisfies the criterion $\Delta J = \Delta K$ in both symmetric limits; hence we would expect the R-branches of the even sub-bands to be relatively strong regardless of the value of κ. On the other hand, if K_{-1} or $K_1 \geqslant 2$ there are two P-branch wings for each even sub-band. In one of these wings $\Delta J = \Delta K$ condition is satisfied only in the oblate limit, we would expect this wing to be very weak near the prolate limit and increase in intensity as the oblate limit is approached. For the other P-branch wing, $\Delta K_{-1} = -1$ and $\Delta K_1 = 1$; thus we conclude that this wing is strong near the prolate limit and decreases in intensity to extreme weakness near the oblate limit. The even sub-band with K_{-1} or $K_1 = 1$ has only one P-branch wing because K_{-1} and K_1 must both be less than J.

In the odd sub-bands there is only one P-branch wing for which both K's change by -1. Since this wing satisfies the condition $\Delta J = \Delta K$ at both symmetric limits, it is relatively intense throughout the whole range of κ. There are, however, two R-branch wings for the odd sub-bands, one with $\Delta K_{-1} = 1$, $\Delta K_1 = -1$, and the other with $\Delta K_{-1} = -1$, $\Delta K_1 = 1$. By applying the intensity condition as before, we find that the former is strong near the prolate limit and weak near the oblate limit, whereas the converse is true in the latter case. All these predictions are amply borne out in actual spectra.

To discuss the properties of the sub-bands further we must decide whether to classify the sub-bands on the basis of K_{-1} or K_1. This choice, however, can have no effect on the general conclusions. Therefore for convenience the sub-bands discussed here have been chosen so that the value of K_{-1} is constant. On this basis the sub-bands with $K_{-1} < 3$ have been plotted in Fig. 8j1 for an asymmetry of $\kappa = -0.9$. It can be seen that all the wings in the even sub-band $K_{-1} = 0$ are reasonably intense. The fact that the ${}^eP_{1\bar{1}}$ wing seems to be strong in this sub-band apparently violates the intensity condition. It must be remembered that this condition applies only to the relative intensity, and it becomes apparent that this wing does indeed increase in intensity as the oblate symmetric limit is approached. The ${}^eP_{1\bar{1}}$ wing is, however, very weak in the even sub-bands with $K_{-1} > 0$. They are, in fact, so weak that we would have difficulty identifying them in an observed spectrum. This fact can be seen from the composite spectrum given on the lowest line of Fig. 8j1. The transitions in the even sub-bands that belong to the wings ${}^eP_{\bar{1}1}$ are relatively quite intense, so one has little difficulty identifying them in the composite spectrum. The even sub-bands have only one R-branch wing, ${}^eR_{11}$, which as predicted earlier is quite intense. In the P-branches of the odd sub-bands the one wing ${}^oP_{\bar{1}\bar{1}}$ is strong as predicted. Of the two R-branch wings the ${}^oR_{1\bar{1}}$ is much stronger than the ${}^oR_{\bar{1}1}$ as one would expect. Again it is very difficult to identify the transitions of the ${}^oR_{\bar{1}1}$ wings in the composite spectrum. The Q-branches

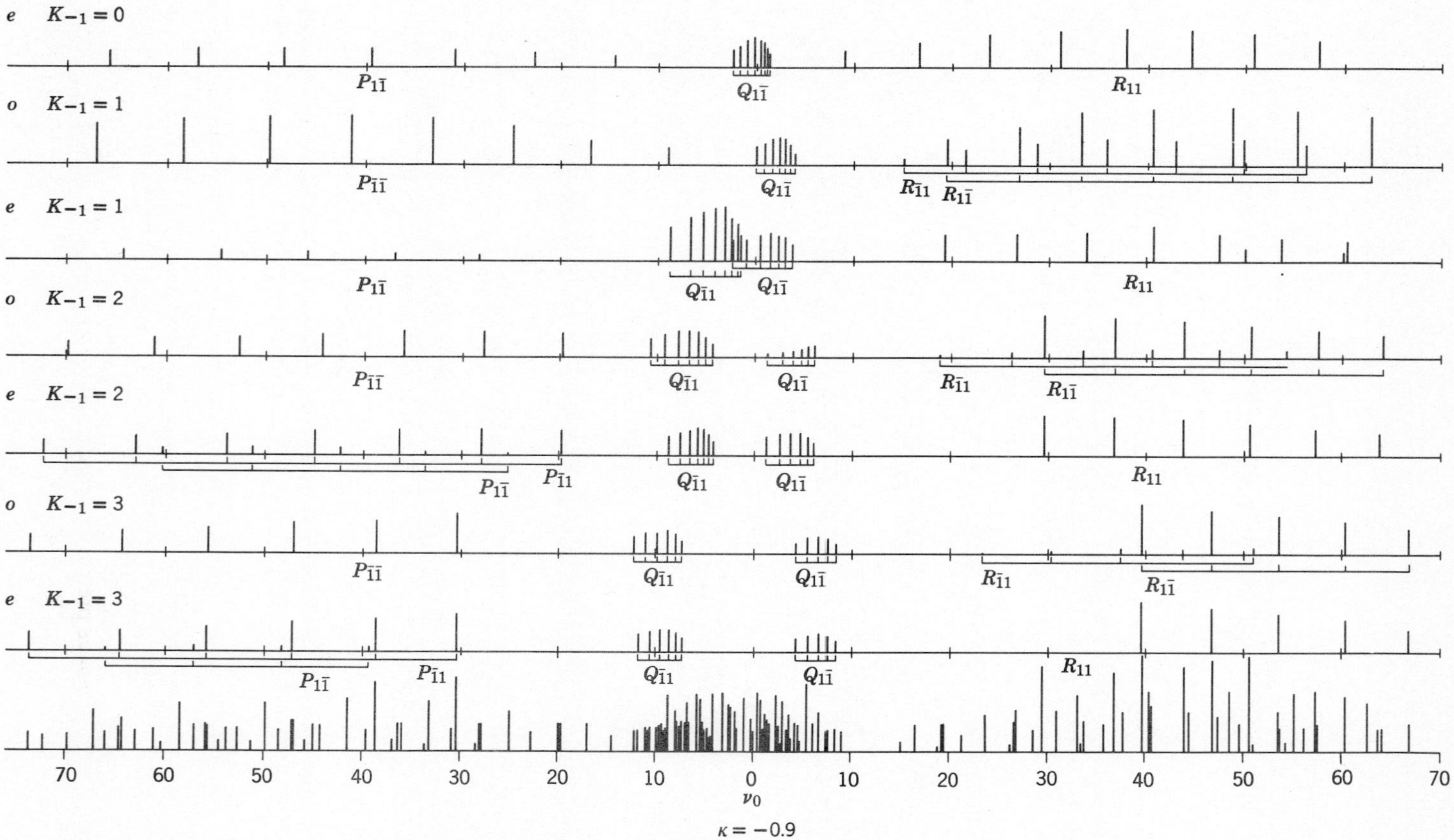

Fig. 8j1 A calculated B-type band for an asymmetric rotor with $\kappa = -0.9$. The top seven lines show the individual sub-band with $K_{-1} \leq 3$, whereas the lowest line shows the composite spectrum obtained by superimposing the sub-bands. The inertial constants used to calculate the spectrum are $A'' = 5.241$, $B'' = 3.946$, $C'' = 3.878$, $A' = 5.200$, $B' = 3.870$, and $C' = 3.800$. The relative intensities were calculated for a temperature of 300° K [11].

are quite intense in each sub-band and for the most part quite well spread out. However, in the composite spectrum the Q-branches are so badly overlapped that it is extremely difficult to identify the individual transitions with any degree of confidence. This situation makes the Q-branch transitions of little use in analysis.

Figure 8j2 is a plot similar to 8j1 except that the value of κ is -0.5. At this degree of asymmetry we find that the wings ${}^eP_{1\bar{1}}$ and ${}^oR_{\bar{1}1}$ are increasing in intensity, especially for $K_{-1} = 0, 1, 2$. For the higher values of K_{-1} the increase is not nearly so marked. Again the Q-branch transitions of the various sub-bands are badly overlapped in the regions on each side of the band center. The particular patterns assumed by the Q-branches in this figure are very sensitive to the relative values of the inertial constants in the two vibrational states. In general, the Q-branch transitions spread out from the center as shown, even for excited state constants considerably different from those used in the figures.

Figure 8j3 shows the situation when $\kappa = 0.5$. At this degree of asymmetry the wings ${}^oR_{\bar{1}1}$ and ${}^eP_{1\bar{1}}$ are seen to be slightly stronger than the wings ${}^oR_{1\bar{1}}$ and ${}^eP_{\bar{1}1}$. This condition is just what one would expect from an application of the intensity criterion. In fact, the relative intensity of these two sets of wings are just the reverse of that for an asymmetry $\kappa = -0.5$. The Q-branch transitions are now spread out considerably, which would seem to make them easily identifiable in an observed spectrum. Figure 8j4 presents the situation for a value of $\kappa = 0.9$. This asymmetry is as close to the oblate limit as $\kappa = -0.9$ is to the prolate limit. The relative intensities of the two R-wings in the odd sub-bands and the two P-wings in the even sub-bands are quite the reverse in Figs. 8j1 and 8j4. It is thus seen that the relative intensities of the wings in the sub-bands behave exactly as predicted from considering their behavior in the two symmetric-rotor limits.

8k DEPENDENCE OF TRANSITIONS ON INERTIAL CONSTANTS

The dependence of the energy levels on the inertial constants was discussed in 8c. This dependence can be applied to determine the principal inertial dependence of a few of the transitions in a B-type band just as was done for the other two band types. The P- and R-branch transitions of the $K_{-1} = 0$ and the odd $K_{-1} = 1$ sub-bands involve the lowest two levels in adjacent J sets as follows:

$$\begin{aligned} J_{0,J} &\rightleftharpoons (J+1)_{1,J+1} \\ J_{1,J} &\rightleftharpoons (J+1)_{0,J+1}, \end{aligned} \qquad (1)$$

where one of these rotational levels is in the ground vibrational state and the other is in the excited vibrational state. These energy levels are the ones which were shown to depend mainly on C, except in the prolate limit, and the transitions between them are approximately represented by 8c7. Thus these transition frequencies are essentially determined by the values of C' and C'', whereas the convergence of the R- or P-branch is determined by the difference $(C' - C'')$ just as was true for A-type bands. If the sub-bands had been classified according to the value of K_1, then it would be necessary

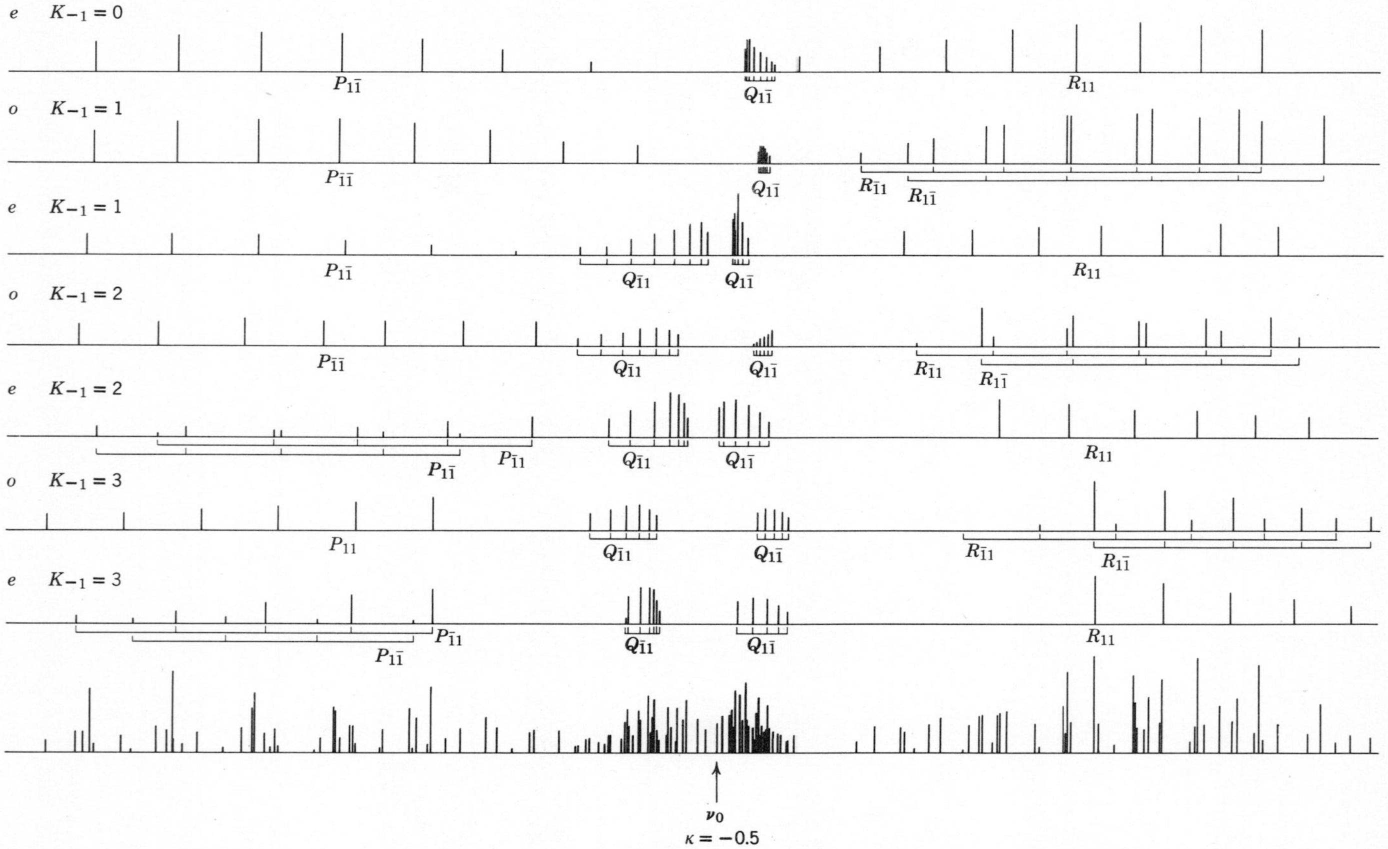

Fig. 8j2 A calculated *B*-type band for an asymmetric rotor with $\kappa = -0.5$. The arrangement is the same as Fig. 8j1. The inertial constants used to calculate the spectrum are $A'' = 4.241$, $B'' = 4.218$, $C'' = 3.878$, $A' = 5.200$, $B' = 4.150$, and $C' = 3.800$ [11].

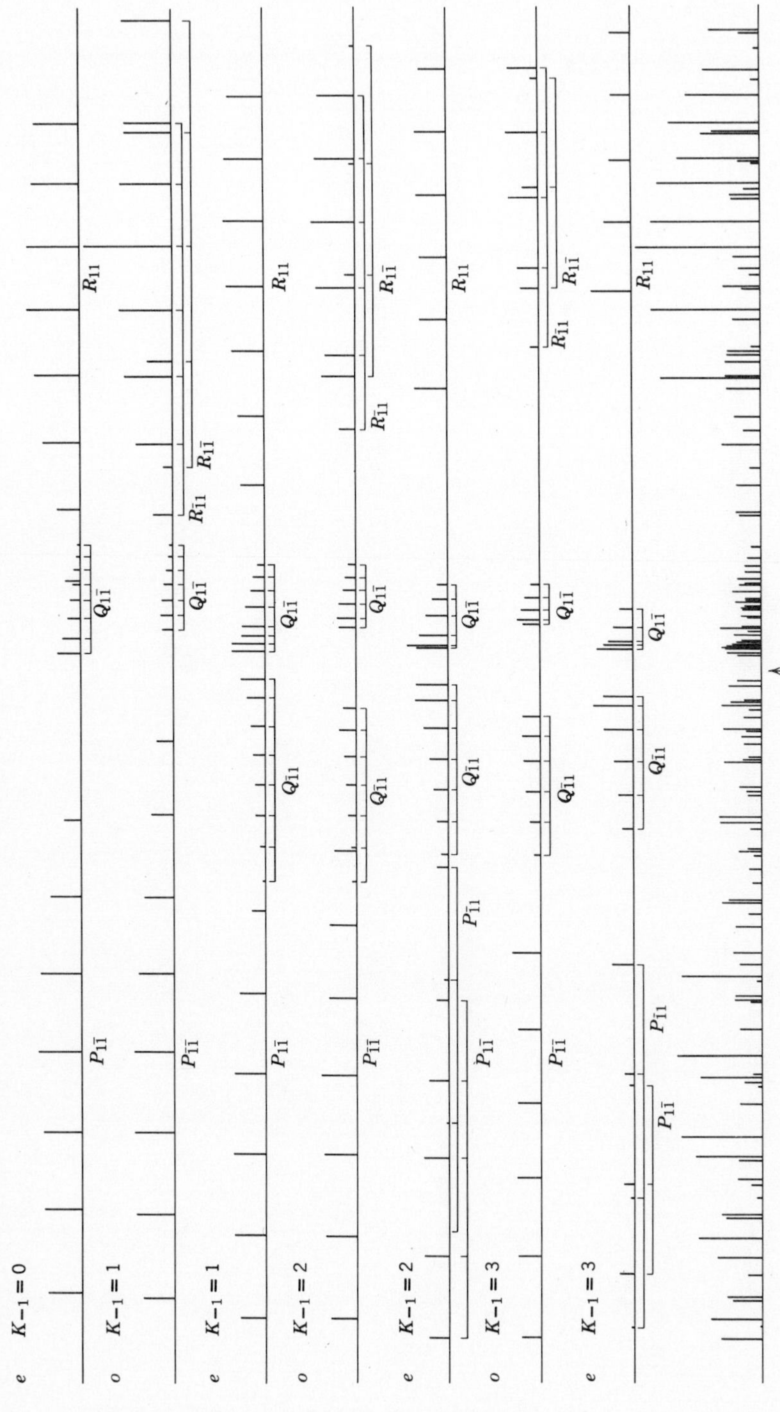

Fig. 8j3 A calculated *B*-type band for an asymmetric rotor with $\kappa = 0.5$. The arrangement is the same as in Fig. 8j1. The inertial constants used to calculate the spectrum are $A'' = 5.241$, $B'' = 4.900$, $C'' = 3.878$, $A' = 5.200$, $B' = 4.850$, and $C' = 3.800$ [11].

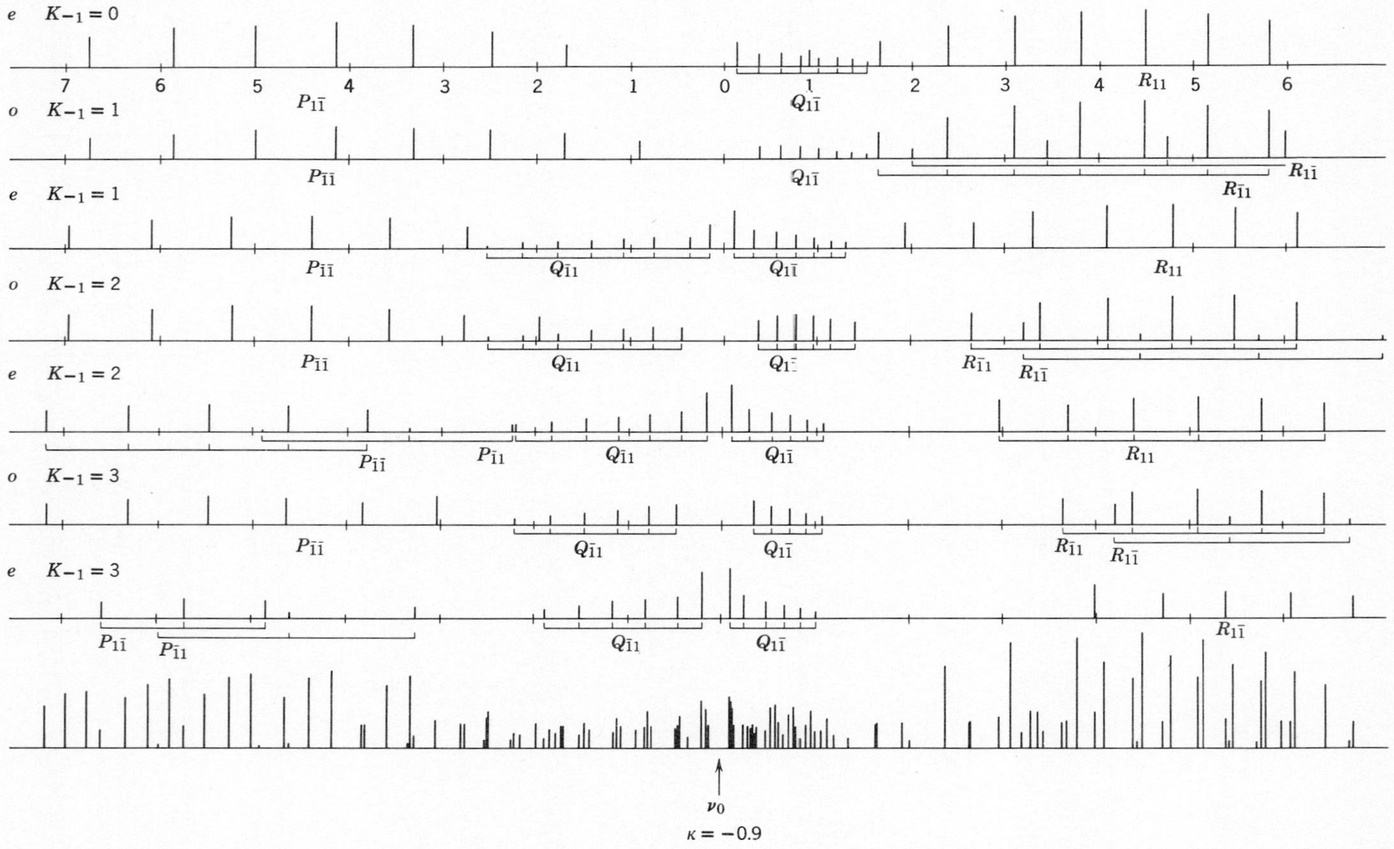

Fig. 8j4 A calculated *B*-type band for an asymmetric rotor with $\kappa = 0.9$. The arrangement is the same as in Fig. 8j1. The inertial constants used to calculate the spectrum are $A'' = 5.241$, $B'' = 5.173$, $C'' = 3.878$, $A' = 5.200$, $B' = 5.130$, and $C' = 3.800$ [11].

to use the arguments appropriate to the transitions of the $K_1 = 0$ and odd $K_1 = 1$ sub-bands of a C-type band. The selection rules would give

$$J_{J0} \rightleftharpoons (J+1)_{J+1,1}$$
$$J_{J1} \rightleftharpoons (J+1)_{J+1,0}, \qquad (2)$$

and the transition frequencies would be given approximately by 8g6. Here we find the dependence is on A', and $A' - A''$ rather than on the C's.

8*l* THE ΔF_2 VALUES OF A *B*-TYPE BAND

The sub-band structure is again useful in determining ΔF_2 values from the transitions of a B-type band, but the process is not quite as straightforward as it was for A- and

Table 8l1—The Combination of Wings To Be Used To Determine ΔF_2 Values from The R- and P-Branches of the Sub-Bands of a B-Type Band [12]

EXCITED STATE	GROUND STATE
${}^{be}R_{11}(K_{-1})$–${}^{be}P_{1\bar{1}}(K_{-1})$	${}^{bo}R_{\bar{1}1}(K_{-1})$–${}^{bo}P_{\bar{1}\bar{1}}(K_{-1})$
${}^{be}R_{11}(K_{-1})$–${}^{be}P_{\bar{1}1}(K_{-1})$	${}^{bo}R_{11}(K_{-1})$–${}^{bo}P_{\bar{1}\bar{1}}(K_1+2)$
${}^{be}P_{1\bar{1}}(K_{-1})$–${}^{be}P_{\bar{1}1}(K_{-1})$	${}^{bo}R_{\bar{1}1}(K_{-1}+2)$–${}^{bo}R_{1\bar{1}}(K_{-1})$
${}^{bo}R_{\bar{1}1}(K_{-1})$–${}^{bo}P_{\bar{1}\bar{1}}(K_{-1})$	${}^{be}R_{11}(K_{-1})$–${}^{be}P_{\bar{1}1}(K_{-1}+2)$
${}^{bo}R_{1\bar{1}}(K_{-1})$–${}^{bo}P_{\bar{1}\bar{1}}(K_{-1})$	${}^{be}R_{11}(K_{-1})$–${}^{be}P_{1\bar{1}}(K_{-1})$
${}^{bo}R_{1\bar{1}}(K_{-1})$–${}^{bo}R_{\bar{1}1}(K_{-1})$	${}^{be}P_{\bar{1}1}(K_{-1}+2)$–${}^{be}P_{1\bar{1}}(K_{-1})$

C-type bands. Since the sub-bands have been classified according to their ground state values of K_{-1}, the $\Delta F_2'$ values may be readily obtained from each sub-band. However, only a part of the $\Delta F_2''$ values may be determined from the transitions in one sub-band. This is a direct result of the selection rules which require the parity of both K_{-1} and K_1 change during a transition. The remaining $\Delta F_2''$ values are readily obtained by combining the transitions of the K_{-1} sub-band with those of the $K_{-1} + 2$ sub-band, provided both these sub-bands have the same parity. The wings which should be used to obtain the ΔF_2 values for both vibrational states are summarized and correlated in Table 8*l*1.

Since it is difficult to assign complete Q-branch wings, only the ΔF_2 values which can be obtained from P- and R-branch transitions are discussed. According to Table 8*l*1 there are six possible types of ΔF_2 values in this category, three which can be obtained from the even sub-bands and three from the odd sub-bands. Four of these possible combinations involve both the P- and R-branch transitions. These ΔF_2 values are summarized in Table 8*l*2. The wings which would be used to obtain the $\Delta F_2'$ values are indicated in this table. To find the wings which would yield the corresponding $\Delta F_2''$ values we have only to make use of the correlation given in Table 8*l*1.

An examination of these ΔF_2 values reveals that they fall into two groups. One group is identical with that obtained from an A-type band, and the other group is identical with that obtained from a C-type band. The dependence of these two groups of ΔF_2 values on the inertial constants is contained in Tables 8d1 and 8h1. The A-type ΔF_2 values have practically no dependence on A, whereas the C-type ΔF_2 values have practically no dependence on C. The fact that both A- and C-type ΔF_2 values may be obtained from the transitions of one B-type band indicates clearly that more information can be obtained from the analysis of a B-type band than one can obtain from the analysis of either an A- or C-type band alone. In fact, since both type ΔF_2 values are available, all the inertial parameters can be obtained with good precision.

A third group of $\Delta F_2'$ and $\Delta F_2''$ values can be obtained from the two R-wings of the odd sub-bands and the two P-wings of the even sub-bands. These ΔF_2 values involve precisely the same transitions as the other two groups of ΔF_2 values which have been

Table 8l2—Summary of $\Delta F_2'$ Values Which Can Be Obtained from the R- and P-Branches of a B-Type Band [11]

	A-TYPE $\Delta F_2'$ VALUES	C-TYPE $\Delta F_2'$ VALUES
$K_{-1} = 0, e$	${}^eR_{11}$–${}^eP_{1\bar{1}}$	
	3_{13}–1_{11}	
	4_{14}–2_{12}	
	5_{15}–3_{13}	
	6_{16}–4_{14}	
	7_{17}–5_{15}	
	8_{18}–6_{16}	
$K_{-1} = 1, o$	${}^oR_{\bar{1}1}$–${}^oP_{\bar{1}\bar{1}}$	${}^oR_{1\bar{1}}$–${}^oP_{\bar{1}\bar{1}}$
	2_{02}–0_{00}	2_{20}–0_{00}
	3_{03}–1_{01}	3_{21}–1_{01}
	4_{04}–2_{02}	4_{22}–2_{02}
	5_{05}–3_{03}	5_{23}–3_{03}
	6_{06}–4_{04}	6_{24}–4_{04}
	7_{07}–5_{05}	7_{25}–5_{05}
	8_{08}–6_{06}	8_{26}–6_{06}
$K_{-1} = 1, e$	${}^eR_{11}$–${}^eP_{1\bar{1}}$	
	4_{23}–2_{21}	
	5_{24}–3_{22}	
	6_{25}–4_{23}	
	7_{26}–5_{24}	
	8_{27}–6_{25}	
$K_{-1} = 2, o$	${}^oR_{\bar{1}1}$–${}^oP_{\bar{1}\bar{1}}$	${}^oR_{1\bar{1}}$–${}^oP_{\bar{1}\bar{1}}$
	3_{12}–1_{10}	3_{30}–1_{10}
	4_{13}–2_{11}	4_{31}–2_{11}
	5_{14}–3_{12}	5_{32}–3_{12}
	6_{15}–4_{13}	6_{33}–4_{13}
	7_{16}–5_{14}	7_{34}–5_{14}
	8_{17}–6_{15}	8_{35}–6_{15}

Table 8l2—(Continued)

$K_{-1} = 2, e$	${}^eR_{11}$–${}^eP_{1\bar{1}}$	${}^eR_{11}$–${}^eP_{\bar{1}1}$
		3_{31}–1_{11}
		4_{32}–2_{12}
	5_{33}–3_{31}	5_{33}–3_{13}
	6_{34}–4_{32}	6_{34}–4_{14}
	7_{35}–5_{33}	7_{35}–5_{15}
	8_{36}–6_{34}	8_{36}–6_{16}
$K_{-1} = 3, o$	${}^oR_{\bar{1}1}$–${}^oP_{\bar{1}\bar{1}}$	${}^oR_{1\bar{1}}$–${}^oP_{\bar{1}\bar{1}}$
	4_{22}–2_{20}	4_{40}–2_{20}
	5_{23}–3_{21}	5_{41}–3_{21}
	6_{24}–4_{22}	6_{42}–4_{22}
	7_{25}–5_{23}	7_{43}–5_{23}
	8_{26}–6_{24}	8_{44}–6_{24}
$K_{-1} = 3, e$	${}^eR_{11}$–${}^eP_{1\bar{1}}$	${}^eR_{11}$–${}^eP_{\bar{1}1}$
		4_{41}–2_{21}
		5_{42}–3_{22}
	6_{43}–4_{41}	6_{43}–4_{23}
	7_{44}–5_{42}	7_{44}–5_{24}
	8_{45}–6_{43}	8_{45}–6_{25}

The wings designated in this table lead to excited state ΔF_2 values. The ΔF_2 values for the ground state are found using the wings given in Table 811.

discussed. Thus this group of ΔF_2 values is not independent of the first two groups and can, in fact, be deduced by properly combining the *A*- and *C*-type ΔF_2 values. In any least squares treatment for the determination of the inertial constants, the inclusion of all three sets of ΔF_2 values gives the *P*- and *R*-branch transitions a double weight. Furthermore, the third set of ΔF_2 values does not even furnish an independent check on the *P*- and *R*-branch assignments.

Although in principle one can determine accurately six inertial parameters of a molecule (*A*, *B*, *C* in both vibrational states) from the analysis of a *B*-type band, there are in practice certain limitations. From Fig. 8j1 it is apparent that the ${}^oR_{\bar{1}1}$ and ${}^eP_{1\bar{1}}$ wings are extremely weak near the prolate symmetric limit, so that the chance of identifying them in the observed spectrum is very small. Unless these transitions can be assigned, the ΔF_2 values characteristic of an *A*-type band cannot be determined. Thus the available information is reduced to that obtained from a *C*-type band and the *C* constant is poorly determined. On the other hand, it can be seen from Fig. 8j4 that the wings ${}^oR_{1\bar{1}}$ and ${}^eP_{\bar{1}1}$ are extremely weak near the oblate limit. For this reason the *C*-type ΔF_2 values are difficult, if not impossible, to obtain, thus allowing only the *A*-type ΔF_2 values. Then the *A* constant is very poorly determined.

In Figs. 8j2 and 8j3 we find all the *P*- and *R*-branch wings are strong enough to be identified, especially for the lower values of K_{-1}. Thus in the region $-0.5 > \kappa > 0.5$ we should be able to obtain accurate values for the three inertial parameters in both vibrational states.

If the Q-branch region is sufficiently resolved so that the transitions can be uniquely assigned, it is possible to obtain additional ΔF_2 values. Since these ΔF_2 values involve new transition assignments, they are independent of the ΔF_2 values obtained from the P- and R-branches alone. However, the proper combination of these new ΔF_2 values eliminates the effect of the Q-branch transition and can thus furnish checks on the ΔF_2 values obtained from the P- and R-branch transitions.

8m EXAMPLES OF *B*-TYPE BAND

Examples of B-type bands observed under high resolution are difficult to find. The bending mode of H_2O and H_2S have been observed, but the latter under rather low resolution and with only one-half of the band seen. Figure 8m1 is a reproduction of the bending fundamental of H_2O which occurs near 6 μ. It has the typical appearance of a B-type band which usually has a gap at the center, especially for light molecules. Then the strong lines on each side of the gap are usually the $1_{01} \leftarrow 1_{10}$ and $1_{10} \leftarrow 1_{01}$ transitions. Assignments in the P- and R-branches of the sub-bands are made similarly to those described for A- and C-type bands. The results of such assignments are given for the first few sub-bands in Table 8m1. It can be seen that for this B-type band assignments can be made so that both types of ΔF_2 values can be obtained. Furthermore, since the molecule is light and the asymmetry is large, unique assignments can be made for many Q-branch transitions in the sub-bands listed. These assignments enable the determination of additional ΔF_2 values which help in getting more precise values of the molecular constants. They also furnish useful checks on assignments. A set of assignments such as this represents a very complete analysis, more complete than is usually able to be made.

8n THE DETERMINATION OF BAND CONSTANTS

The preceding sections of this chapter have been concerned with the process of assigning quantum numbers to the observed absorption peaks. This is the first and most difficult part of the analytical procedure. It is now assumed that the quantum numbers have been assigned and that we want to determine the best set of molecular parameters for each band. In principle, there are seven parameters to be determined for each band if a rigid molecule is assumed. These are the three inertial constants in each vibrational state plus the band center. The band center gives the frequency of the vibrational transition. However, as was shown in the previous discussion the information which can be obtained from an A- or C-type band is limited and sometimes this is also true of B-type bands.

The method of determining the inertial constants varies according to the way assignments are made and the number and type of assignments which can be made.

If sufficient assignments have been made in a B-type band or in both an A- and a C-type band so that most of the $\Delta F_2''$ values of each type can be determined, then the ground state inertial parameters may be determined by using 8d3. One such equation is set up for each $\Delta F_2''$ value, and the resulting set of equations is solved by the method of least squares for the best estimates of A'', B'', and C''. Only transitions involving

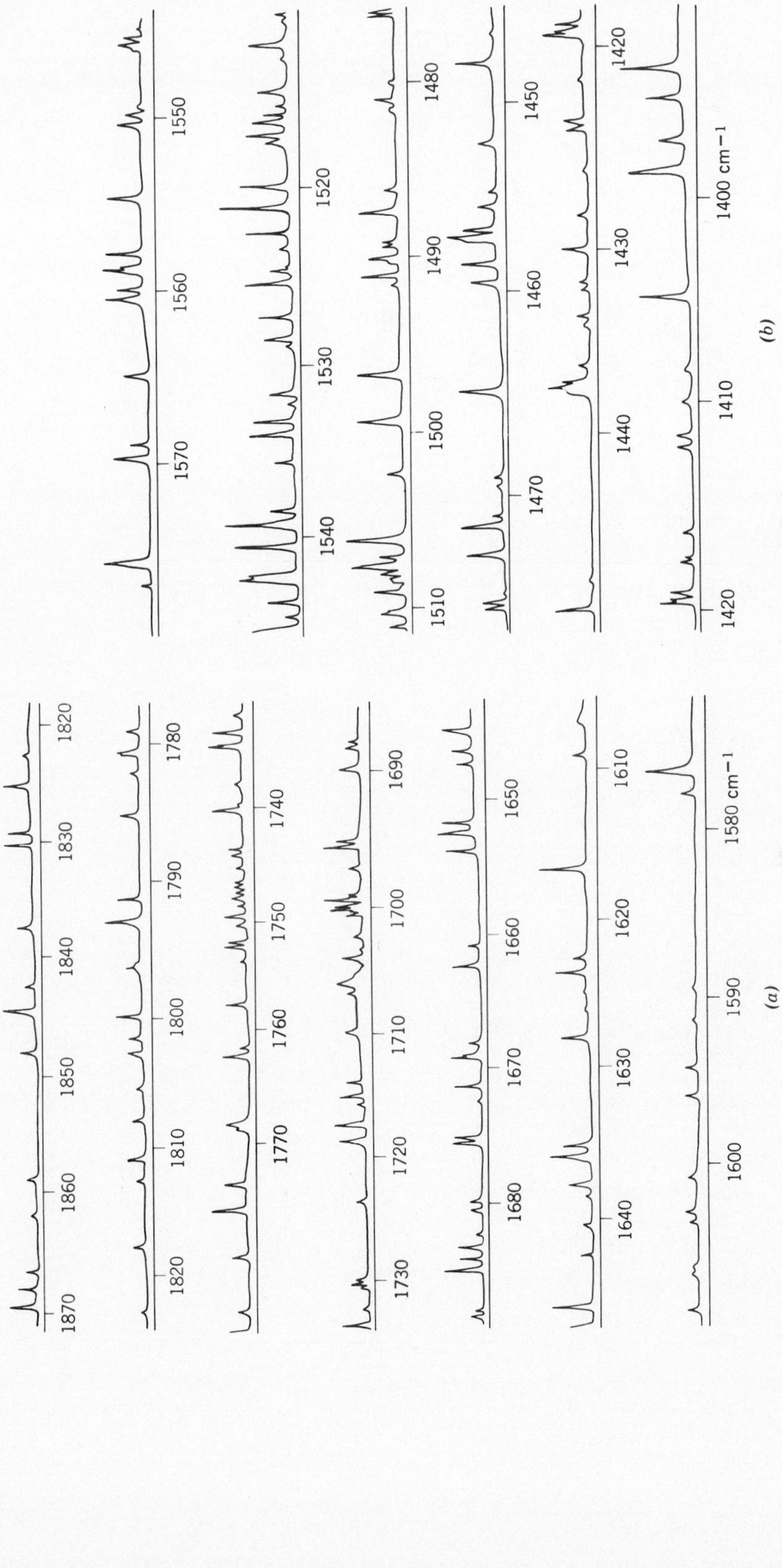

Fig. 8m1 *B*-type bending fundamental of H_2O near 6μ [F. W. Dalby and H. H. Nielsen, *J. Chem. Phys.* **25**, 934 (1956)].

Table 8ml—Sub-Bands of the 6.3 μ Band of H_2O (B-Type) (cm^{-1})

$K_{-1} = 0, e$		$K_{-1} = 1, o$		$K_{-1} = 1, e$		$K_{-1} = 2, o$		$K_{-1} = 2, e$	
$1_{10}-1_{01}$	1616.76	$2_{21}-2_{12}$	1662.87	$1_{01}-1_{10}$	1576.27	$3_{31}-3_{22}$	1701.19	$3_{30}-3_{21}$	1695.56
$2_{11}-2_{02}$	23.61	$3_{22}-3_{13}$	71.60	$2_{02}-2_{11}$	69.88	$4_{32}-4_{23}$	04.52	$4_{31}-4_{22}$	90.23
$3_{12}-3_{03}$	35.75	$4_{23}-4_{14}$	83.27	$3_{03}-3_{12}$	58.59	$5_{33}-5_{24}$	10.30	$5_{32}-5_{23}$	84.06
$4_{13}-4_{04}$	53.38	$5_{24}-5_{15}$	97.59	$4_{04}-4_{13}$	42.02	$6_{34}-6_{25}$	18.91	$6_{33}-6_{24}$	79.91
$5_{14}-5_{05}$	77.63	$6_{25}-6_{16}$	1714.12	$5_{05}-5_{14}$	21.33				
$6_{15}-6_{06}$	99.69			$6_{06}-6_{15}$	1498.92	$2_{12}-2_{21}$	1542.19	$2_{11}-2_{20}$	1557.61
						$3_{13}-3_{22}$	33.27	$3_{12}-3_{21}$	60.30
$1_{11}-0_{00}$	1635.06	$2_{02}-1_{11}$	1627.90	$2_{20}-2_{11}$	1648.39	$4_{14}-4_{23}$	21.33	$4_{13}-4_{22}$	59.75
$2_{12}-1_{01}$	53.38	$3_{03}-2_{12}$	52.50	$3_{21}-3_{12}$	46.05	$5_{15}-5_{24}$	06.70	$5_{14}-5_{23}$	54.41
$3_{13}-2_{02}$	69.49	$4_{04}-3_{13}$	75.30	$4_{22}-4_{13}$	47.49	$6_{16}-6_{25}$	1489.93	$6_{15}-6_{24}$	43.55
$4_{14}-3_{03}$	84.92	$5_{05}-4_{14}$	96.00	$5_{23}-5_{14}$	54.59				
$5_{15}-4_{04}$	1700.79	$6_{06}-5_{15}$	1715.24	$6_{24}-6_{15}$	68.38	$3_{12}-2_{21}$	1637.63	$3_{31}-2_{20}$	1771.37
$6_{16}-5_{05}$	17.48					$4_{13}-3_{22}$	69.29	$4_{32}-3_{21}$	92.72
						$5_{14}-4_{23}$	1700.59	$5_{33}-4_{22}$	1810.71
$1_{11}-2_{02}$	1564.96	$2_{20}-1_{11}$	1706.40	$2_{21}-1_{10}$	1700.01	$6_{15}-5_{24}$	30.13	$6_{34}-5_{23}$	25.30
$2_{12}-3_{03}$	40.36	$3_{21}-2_{12}$	39.90	$3_{22}-2_{11}$	18.70				
$3_{13}-4_{04}$	17.47	$4_{22}-3_{13}$	80.70	$4_{23}-3_{12}$	34.67	$3_{30}-2_{21}$	1772.78	$3_{31}-4_{22}$	1591.76
$4_{14}-5_{05}$	1496.31	$5_{23}-4_{14}$	1829.20	$5_{24}-4_{13}$	48.70	$4_{31}-3_{22}$	99.67	$4_{32}-5_{23}$	58.39
$5_{15}-6_{06}$	76.21	$6_{24}-5_{15}$	84.60	$6_{25}-5_{14}$	61.95	$5_{32}-4_{23}$	1830.21	$5_{33}-6_{24}$	23.68
						$6_{33}-5_{24}$	66.46		
		$0_{00}-1_{11}$	1557.61	$2_{21}-3_{12}$	1569.02				
		$1_{01}-2_{12}$	39.12	$3_{22}-4_{13}$	38.34	$1_{10}-2_{21}$	1505.68	$1_{11}-2_{20}$	1498.92
		$2_{02}-3_{13}$	22.74	$4_{23}-5_{14}$	08.63	$2_{11}-3_{22}$	1487.37	$2_{12}-3_{21}$	65.01
		$3_{03}-4_{14}$	07.13	$5_{24}-6_{15}$	1481.30	$3_{12}-4_{23}$	72.14	$3_{13}-4_{22}$	23.77
		$4_{04}-5_{15}$	1490.92			$4_{13}-5_{24}$	59.33	$4_{14}-5_{33}$	1575.18
		$5_{05}-6_{16}$	73.59			$5_{14}-6_{25}$	48.04	$5_{15}-6_{34}$	20.15

These sub-bands are taken from the assignments of F. W. Dalby and H. H. Nielsen, *J. Chem. Phys.* **25**, 934 (1956).

low J values should be used in this process to minimize the effects of centrifugal distortion. It is well to point out here that even if the excited state of the transitions is perturbed, this can have no effect on the ground state energy differences.

If the assignments have been made by calculating a trial spectrum and the constants have been adjusted by means of the energy derivatives to give a close fit, the best values of the inertial constants can be obtained by finding the corrections to A, B, and C which minimize the differences between the observed and calculated energy differences. An equation of the type

$$\Delta E = E_{\text{obs}} - E_{\text{calc}} = [\alpha(J+1) - \alpha(J-1)]\,\Delta A + [\beta(J+1) - \beta(J-1)]\,\Delta B + [\gamma(J+1) - \gamma(J-1)]\Delta C \quad (1)$$

is set up for each $\Delta F_2''$ value and the ΔE's are minimized by solving the set of equations by the method of least squares. One should always use the $\Delta F_2''$ values to determine the ground state inertial constants when it is possible, and these $\Delta F_2''$ values should be the averages of the $\Delta F_2''$ values obtained from as many bands as possible.

Once the ground state inertial constants have been obtained, the excited state constants are readily obtained. We can form an equation

$$\nu_{\text{obs}} - E_g = \nu_0 + \alpha' A' + \beta' B' + \gamma' C' \quad (2)$$

for each assigned transition in a given band. E_g is the energy of the ground state level involved in the transition and is calculated from the values of the ground state inertial constants determined previously. The resulting set of equations can be solved by the method of least squares for the best estimates of ν_0, A', B', and C'.

Alternatively, if the analysis has been carried out by adjusting a calculated spectrum to the observed spectrum by means of the energy derivatives, we form equations analogous to (1) for each transition

$$\Delta E = E_{\text{obs}} - E_{\text{calc}} = \Delta\nu_0 + \alpha'\Delta A' + \beta'\Delta B' + \gamma'\Delta C' \quad (3)$$

and calculate the corrections to the constants by solving this set of equations by the method of least squares.

These procedures are all valid for the calculation of the rigid constants if sufficient assignments of the correct type have been made. However, the effects of centrifugal distortion must be included in all precise work. For low J levels the effect is small. Thus we must in general determine the centrifugal distortion constants from transitions involving high values of J.

If sufficient structural information and force constant information is available, the constants $A_1, A_2, \ldots A_3$ can be calculated allowing the correction to be calculated and applied level by level by the methods described in Chapter 3.

If the corrections can be calculated and applied to the levels, the final values of the band constants are determined by using (1) and (3) as described before. If the constants cannot be calculated, they must be determined from the observed data.

We first determine preliminary values of the constants from low J transitions as previously described. These constants are then used to calculate the high J transitions which will in general not agree with the observed values The difference is ascribed to centrifugal distortion, which is used to determine preliminary values of the centrifugal distortion constants. The preliminary values of the centrifugal distortion constants can then be used to calculate the corrections to the energy levels. The final values of all the constants are obtained by repeating this process until the best fit between the observed and calculated spectra is obtained.

We can also include the centrifugal distortion constants as parameters in the least square calculations from the beginning. However, unless an electronic computer is available for the least squares treatment the resulting problem is impractical to solve.

The problem of determining the ground state molecular parameters is quite straightforward when a sufficient number of assignments can be made which enable the determination of the appropriate $\Delta F_2''$ values. Unfortunately, since one seldom finds such a favorable situation, it is necessary to resort to less desirable methods of obtaining molecular constants. It may be impossible to make assignments in a portion of the band because of overlapping absorption either from another band of the same molecule or from an atmospheric constituent. This situation makes it impossible to determine a sufficient number of $\Delta F_2''$ values, especially if the overlapping absorption makes either the *R*- or *P*-branch unaccessible as often happens. Nonlinear triatomic molecules cannot have *C*-type bands—another fact which leads to trouble in determining molecular constants. It may be that one-half of the band is too weak to observe readily, which is what happens in two of the H_2S fundamentals, *B*-type bands.

In these situations there is no alternative but to match an observed spectrum with a calculated spectrum. Energy levels for each vibrational state are calculated from estimated constants, and the agreement is improved by adjusting the calculated spectrum by means of the energy derivatives. The process requires judgment, but this is acquired with a little experience. It is made easier if we work with sub-bands that are particularly sensitive to only one constant. Suppose that only the *R*-branch region and a portion of the *P*-branch region of an *A*-type band is available for analysis and that assignments have been made using a trial spectrum. This has allowed the assignment of several sub-bands on the basis of frequency and intensity. The transitions of the two $K_{-1} = 1$ sub-bands together with $\Delta = E_{\text{obs}} - E_{\text{calc}}$ and the necessary derivatives are shown in Table 8n1. It is readily seen that the main dependence of the transitions in the first sub-band is on C' and C''. Suppose we decide to bring the observed and calculated into agreement by adjusting the value of C''. The amount of the adjustment can be determined by dividing the Δ for a transition by the value of γ'' for the same transition. It is usually best to make this calculation on the highest J transition since this minimizes the effect of the experimental uncertainty in the measured frequency. Thus using the 7_{17}–8_{18} transitions we find that $\Delta C'' = 0.031$. We should increase the value of C'' by 0.031 cm^{-1} since all the calculated transitions are too high. The result of this calculation is shown under the appropriate column of Table 8n1.

This correction obviously did not help the situation as much as hoped for. From an examination of the results we find the *R*-branch is not corrected enough, whereas the *P*-branch is overcorrected for all but the transition used to calculate the correction. Thus another correction should be applied. Inspection of the transitions shows that transitions with the same excited state level have about the same error; thus a correction to C' rather than C'' seems to be reasonable. Using the 8_{18}–7_{17} transition we find C' should be reduced by 0.040 cm^{-1}. The results of this correction are shown in

Table 8nI

$J'K_{-1}K_1 - J''K_{-1}K_1$	Δ	$-0.031\gamma''$	$-0.040\gamma'$	$0.035\beta''$	α'	β'	γ'	α''	β''	γ''
$K_{-1} = 1, o$										
2_{12}–1_{11}	0.16	−0.03	−0.16	0.00	1.007	1.000	3.993	1.000	0.000	1.000
3_{13}–2_{12}	0.32	−0.12	−0.34	0.04	1.008	2.476	8.506	1.007	1.000	3.993
4_{14}–3_{14}	0.61	−0.26	−0.58	0.09	1.032	4.402	14.566	1.018	2.476	8.506
5_{15}–4_{14}	0.88	−0.45	−0.89	0.15	1.049	6.744	22.207	1.032	4.402	14.566
6_{16}–5_{15}	1.22	−0.69	−1.25	0.24	1.069	9.461	31.470	1.049	6.744	22.207
7_{17}–6_{16}	1.68	−0.98	−1.70	0.33	1.088	12.512	42.397	1.069	9.461	31.470
8_{18}–7_{17}	2.22	−1.31	−2.20	0.44	1.116	15.852	55.032	1.088	12.512	42.397
1_{11}–2_{12}	0.03	−0.12	−0.04	0.04	1.000	0.000	1.000	1.007	1.000	3.993
2_{12}–3_{13}	0.15	−0.26	−0.16	0.09	1.007	1.000	3.993	1.018	2.476	8.506
3_{13}–4_{14}	0.33	−0.45	−0.34	0.15	1.018	2.476	8.506	1.032	4.402	14.566
4_{14}–5_{15}	0.57	−0.69	−0.58	0.24	1.032	4.402	14.566	1.049	6.744	22.207
5_{15}–6_{16}	0.88	−0.98	−0.89	0.33	1.049	6.744	22.207	1.069	9.461	31.470
6_{16}–7_{17}	1.24	−1.31	−1.25	0.44	1.069	9.461	31.470	1.088	12.512	42.397
7_{17}–8_{18}	1.69	−1.70	−1.70	0.56	1.088	12.512	42.397	1.116	15.852	55.032
$K_{-1} = 1, e$	Δ	$\Delta -0.04\gamma'$	$-0.035\beta''$	$0.035\beta''$						
2_{11}–1_{10}	0.00	−0.04	−0.04	0.04	1.029	4.000	0.071	1.000	1.000	0.000
3_{12}–2_{11}	−0.03	−0.13	−0.14	0.14	1.062	8.476	2.463	1.029	4.000	0.971
4_{13}–3_{12}	−0.14	−0.32	−0.30	0.30	1.105	14.396	4.499	1.062	8.476	2.463
5_{14}–4_{13}	−0.22	−0.50	−0.50	0.50	1.158	21.718	7.124	1.105	14.396	4.499
6_{15}–5_{14}	−0.33	−0.75	−0.76	0.76	1.222	30.384	10.395	1.158	21.718	7.124
7_{16}–6_{15}	−0.46	−1.04	−1.06	1.06	1.294	40.320	14.387	1.222	30.384	10.395
8_{17}–7_{16}	−0.66	−1.43	−1.41	1.41	1.375	51.432	19.193	1.294	40.320	14.387
1_{10}–2_{11}	−0.13	−0.13	−0.14	0.14	1.000	1.000	0	1.029	0.971	0.971
2_{11}–3_{12}	−0.25	−0.29	−0.30	0.30	1.029	4.000	0.971	1.062	8.476	2.463
3_{12}–4_{13}	−0.38	−0.50	−0.50	0.50	1.062	8.476	2.463	1.105	14.396	4.499
4_{13}–5_{14}	−0.59	−0.77	−0.76	0.76	1.105	14.396	4.499	1.158	21.718	7.124
5_{14}–6_{15}	−0.80	−1.08	−1.06	1.06	1.158	21.718	7.124	1.222	30.384	10.395
6_{15}–7_{16}	−1.00	−1.42	−1.41	1.41	1.222	30.384	10.395	1.294	40.320	14.387
7_{16}–8_{17}	−1.20	−1.78	−1.80	1.80	1.294	40.320	14.387	1.375	51.432	19.193

Table 8n1, where it is readily seen that the observed and calculated values are now in good agreement.

We could have saved time and bother by studying the Δ values first to discover that C' and not C'' should have been corrected. Now that this sub-band is in good agreement and the correction has been applied to all sub-bands, it may be found that the agreement for the other sub-bands is worse. Such is true for the even $K_{-1} = 1$ sub-band shown in Table 8n1. The original Δ and the original Δ plus the correction are shown in the second and third columns of the table. Fortunately, this sub-band depends predominately on B. We should examine these Δ's carefully to see whether to correct B' or B''. The table shows that transitions with the same ground state levels are in error by about the same amount; thus the ground state B'' should be corrected. Proceeding as before we find a trial correction of B'' of $-0.035\ \text{cm}^{-1}$. The result of this correction on the even sub-band is indicated in the fourth column of the table. If we examine the first sub-band again we find that the agreement there is now not quite so good and a further adjustment of C' or C'' must be made. The Δ values for other sub-bands may also give clues as to the best way to adjust the constants. The procedure should be continued until the agreement between the observed and calculated transitions is as good as possible. Final corrections to the constants should be made by solving the set of equations

$$\Delta = E_{\text{obs}} - E_{\text{calc}} = \Delta\nu_0 + \alpha'\,\Delta A' + \beta'\,\Delta B' + \gamma'\,\Delta C' - \alpha''\,\Delta A'' - \beta''\,\Delta B'' - \gamma''\,\Delta C'' \quad (4)$$

by the method of least squares. The centrifugal distortion correction should be treated as described previously.

When both P- and R-branch transitions are available, it is possible to ascertain whether the ground or excited state constant should be corrected. However, with only one branch available this is not possible. When working with only one branch of the sub-bands in Table 8c1, it is found that corrections with the opposite sign can be applied to the inertial constants of the other vibrational state to achieve fairly good agreement for the one branch chosen. Thus for this method to give unambiguous results, it is necessary to assign several sets of transitions, each set having the same ground state level or excited state level. Only in this way is it possible to tell in which vibrational state to correct the inertial constant. However, it is not necessary to assign every transition in each sub-band, as is necessary in order to get the best results from the ΔF_2 method.

Whenever possible, one should always use ΔF_2 values to determine the molecular constants; the second method should be resorted to only when it is not feasible to use ΔF_2 values.

Once a reliable set of $\Delta F_2''$ values is available, the analysis of further bands is greatly simplified, since the $\Delta F_2''$ values can be used to help identify the transitions in the new band. Using relative intensities as a guide, energy differences in the observed spectrum can be determined and compared to known $\Delta F_2''$ values, thus making the assignment process easier. The $\Delta F_2''$ values can be calculated precisely if the microwave spectrum has been analyzed. It cannot be overemphasized too much that an analysis is made considerably easier by taking advantage of all available information, including that deduced from other experimental techniques.

8o THE BAND ENVELOPE METHOD

The procedures for band analysis thus far described have assumed that sufficient resolution was available to allow the identification of individual transitions. In general, this requires resolution of 0.1 cm^{-1} or better except for very light molecules. Such a degree of resolution is at present available in a limited number of laboratories. On the other hand, commercial instruments are now available which can achieve a resolution of 0.5 to 1.0 cm^{-1} throughout a large portion of the near infrared. Even

Table 8o1—Comparison of Inertial Constants Determined from Band Envelope Analysis and from Analysis of High Resolution Spectra. All Bands Are Those of D_2O

	A	B	C
	PUNCH CARD MACHINES		
000			
BE[a]	15.38	7.25	4.84
HR[c]	15.385	7.272	4.846
010			
BE[a]	16.50	7.33	4.79
HR[c]	16.546	7.354	4.796
	WITHOUT MACHINES		
001			
BE[b]	14.88	7.22	4.79
HR[c]	14.792	7.227	4.791
101			
BE	14.65	7.12	4.72
HR[c]	14.543	7.133	4.714

[a] G. W. King, *J. Chem. Phys.* **15,** 85 (1947).
[b] K. K. Innes, P. C. Cross, and P. Giguere, *J. Chem. Phys.* **19,** 1086 (1951).
[c] W. S. Benedict, N. Gailar, and E. K. Plyler, *J. Chem. Phys.* **24,** 1139 (1956).

at this resolution it is possible to obtain a considerable amount of information from a band analysis, provided the molecule is not too heavy. This is particularly true if the ground state molecular constants are known from some other source such as microwave spectroscopy, ultraviolet spectroscopy, or even a previous infrared investigation. In these cases one must use what has been called the "band envelope" method. One cannot obtain the molecular constants to such a high precision by the method, and a considerably greater degree of judgment and experience is necessary. However, much useful information can be obtained from such an analysis.

The procedure is not greatly different from the procedures previously described. It is necessary to estimate the molecular constants in some manner and from these

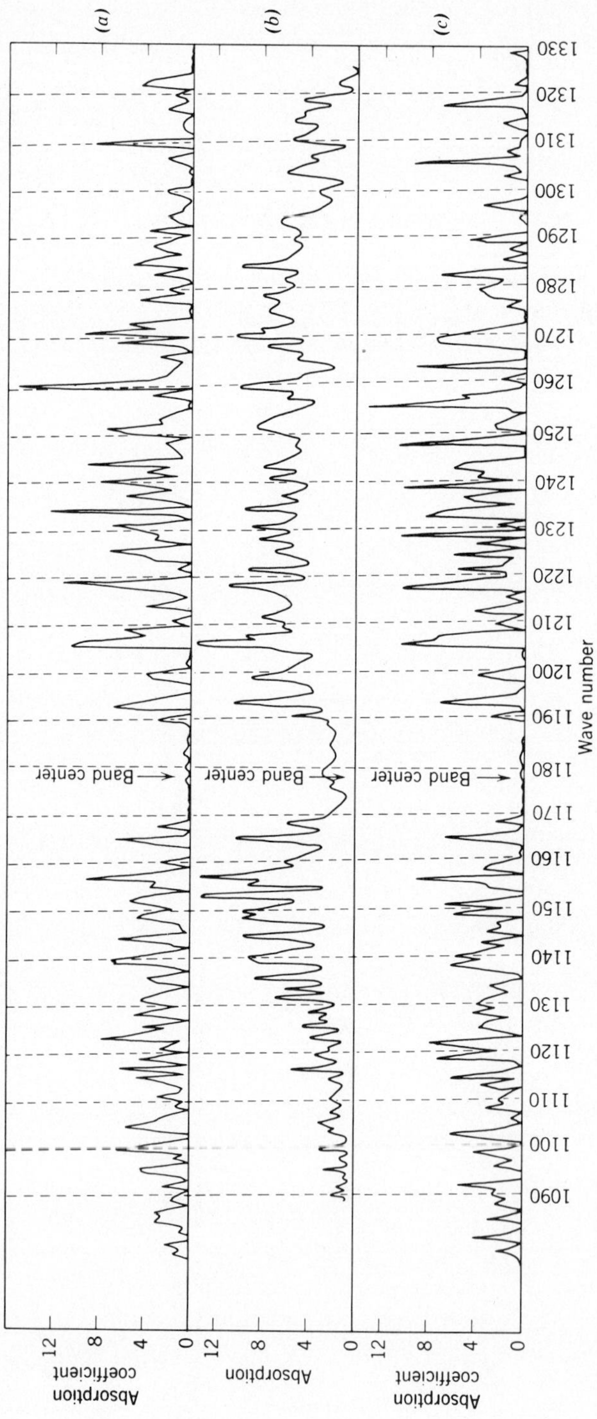

Fig. 8o1 An analysis of the 8.5μ band of D_2O by punched-card techniques. The experimental curve is shown in the middle. The upper curve is based on estimated molecular constants for the lower and upper states. By using slightly different upper state constants, the lower curve is obtained [65].

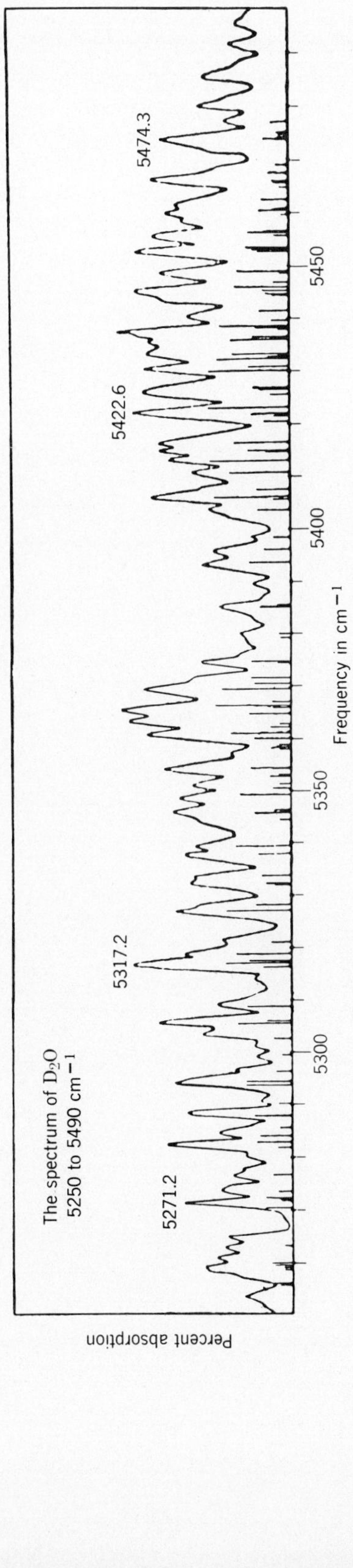

Fig. 8·2 Calculated and observed combination band $\nu_1 + \nu_3$ of D_2O at 5373.2 cm^{-1} [59].

calculate a trial spectrum as described before. We then proceed to adjust the trial spectrum by means of the energy derivatives until the calculated spectrum reproduces the observed spectrum as well as possible. Since with reduced resolution a given absorption peak may represent a blend of several transitions, we must take into account the integrating effect of the slit on the transitions. The absorption peak observed does not correspond uniquely to any of the calculated transitions and its appearance depends critically on the effective instrumental slit width and to some extent on the slit function which is assumed in "integrating" the trial spectrum. The effective experimental slit width can be determined experimentally or by calculation, but the slit function must in general be assumed. In the past a Gaussian shape has frequently been assumed for the slit function. Once the slit function has been assumed and the effective slit width has been determined, we must then integrate the calculated transitions in order to calculate the spectrum which the instrument produces. The constants can be altered and the resulting spectrum integrated again, this procedure being repeated until the best reproduction of the observed spectrum is obtained. Some guide is furnished in adjusting the constants, for the dependence of the transitions on the inertial parameters is readily obtained from a table of derivatives.

This method of analysis has been adapted to IBM punch card machines with a fair degree of success [52, 65, 66], and has also been applied without the aid of computing machines [59]. Figure 8o1 shows the sort of precision that can be obtained using machine techniques. The inertial constants determined by the band envelope method are compared to those obtained from the analysis of data of higher resolution in Table 8o1.

As an example of the band envelope method applied without the aid of a computing machine, the D_2O band in Fig. 8o2 is presented. The results of the band envelope method and the more precise method are also compared in Table 8o1.

Appendix I

PROOF OF INTEGRAL VALUES OF K

The fact that K must take on integral values may be seen most readily by considering the wave functions $\Psi_{J,K,M}(\theta, \Phi, \chi)$ of 2*l*1

$$\Psi_{J,K,M,}(\theta, \Phi, \chi) = \theta_{JKM}(\theta)\, e^{iK\chi} e^{iM\phi}$$

The possible values of K are dictated by the boundary condition $\Psi_{J,K,M}(\theta, \Phi, \chi) = \Psi_{J,K,M}(\theta, \chi + 2\pi n, \phi)$, where n is any integer. Thus $e^{iK\chi}$ must equal $e^{iK(\chi+2\pi n)}$. This condition is fulfilled for all n only if K is integral.

Similarly, M may assume only integral values.

Appendix II

VALUES OF $f(J, n)$ [67]

n \ J	1	2	3	4	5	6	7	8	9	10	11	12	13	14	15
0	1^2	3^2	6^2	10^2	15^2	21^2	28^2	36^2	45^2	55^2	66^2	78^2	91^2	105^2	120^2
1		**12**	**60**	**180**	**420**	**840**	**1512**	**2520**	**3960**	**5940**	**8580**	**12012**	**16380**	**21840**	**28560**
2			15	63	168	360	675	1155	1848	2808	4095	5775	7920	10608	13923
3				28	108	270	550	990	1638	2548	3780	5400	7480	10098	13338
4					45	165	396	780	1365	2205	3360	4896	6885	9405	12540
5						66	234	546	1050	1800	2856	4284	6156	8550	11550
6							91	315	720	1360	2295	3591	5320	7560	10395
7								120	408	918	1710	2850	4410	6468	9108
8									153	513	1140	2100	3465	5313	7728
9										190	630	1386	2530	4140	6300
10											231	759	1656	3000	4875
11												276	900	1950	3510
12													325	1053	2268
13														378	1218
14															435
15															
16															
17															
18															
19															
20															
21															
22															
23															
24															
25															
26															
27															
28															
29															

n \ J	16	17	18	19	20	21	22	23	24	25	26	27	28	29	30	J / n
0	136^2	153^2	171^2	190^2	210^2	231^2	253^2	276^2	300^2	325^2	351^2	378^2	406^2	435^2	465^2	0
1	**36720**	**46512**	**58140**	**71820**	**87780**	**106260**	**127512**	**151800**	**179400**	**210600**	**245700**	**285012**	**328860**	**377580**	**431520**	1
2	17955	22800	28560	35343	43263	52440	63000	75075	88803	104328	121800	141375	163215	187488	214368	2
3	17290	22050	27720	34408	42228	51300	61750	73710	87318	102718	120060	139500	161200	185328	212058	3
4	16380	21021	26565	33120	40800	49725	60021	71820	85260	100485	117645	136896	158400	182325	208845	4
5	15246	19734	25116	31500	39000	47736	57834	69426	82650	97650	114576	133584	154836	178500	204750	5
6	13915	18216	23400	29575	36855	45360	55216	66555	79515	94240	110880	129591	150535	173880	199800	6
7	12420	16500	21450	27378	34398	42630	52200	63240	75888	90288	106590	124950	145530	168498	194028	7
8	10800	14625	19305	24948	31668	39585	48825	59520	71808	85833	101745	119700	139860	162393	187473	8
9	9100	12636	17010	22330	28710	36270	45136	55440	67320	80920	96390	113886	133570	155610	180180	9
10	7371	10584	14616	19575	25575	32736	41184	51051	62475	75600	90576	107559	126711	148200	172200	10
11	5670	8526	12180	16740	22320	29040	37026	46410	57330	69930	84360	100776	119340	140220	163590	11
12	4060	6525	9765	13888	19008	25245	32725	41580	51948	63973	77805	93600	111520	131733	154413	12
13	2610	4650	7440	11088	15708	21420	28350	36630	46398	57798	70980	86100	103320	122808	144738	13
14	1395	2976	5280	8415	12495	17640	23976	31635	40755	51480	63960	78351	94815	113520	134640	14
15	496	1584	3366	5950	9450	13986	19684	26676	35100	45100	56826	70434	86086	103950	124200	15
16		561	1785	3780	6660	10545	15561	21840	29520	38745	49665	62436	77220	94185	113505	16
17			630	1998	4218	7410	11700	17220	24108	32508	42570	54450	68310	84318	102648	17
18				703	2223	4680	8200	12915	18963	26488	35640	46575	59455	74448	91728	18
19					780	2460	5165	9030	14190	20790	28980	38916	50760	64680	80850	19
20						861	2709	5676	9900	15525	22701	31584	42336	55125	70125	20
21							946	2970	6210	10810	16920	24696	34300	45900	59670	21
22								1035	3243	6768	11760	18375	26775	37128	49608	22
23									1128	3528	7350	12750	19890	28938	40068	23
24										1225	3825	7956	13780	21465	31185	24
25											1326	4134	8586	14850	23100	25
26												1431	4455	9240	15960	26
27													1540	4788	9918	27
28														1653	5133	28
29															1770	29

Values of

$$f(J, n)=\tfrac{1}{4}\,[J(J+1)-n(n+1)]\,[J(J+1)-n(n-1)]$$
$$=\tfrac{1}{4}(J-n)(J-n+1)(J+n)(J+n+1).$$

The factor 2 which always occurs with $f(J, 1)$ in the factored secular equations has been included in calculating the entries in the table. These values are therefore set apart by the use of bold face type.

Appendix III

SYMMETRY OF THE WANG FUNCTIONS

The functions $S(J, K, M, \gamma)$ which occur in any submatrix must all belong to the same irreducible representation of the Four group, $V(x, y, z)$. The classification may be carried out by determining the effect of each of the operations E, C_2^x, C_2^y, and C_2^z on the functions $S(J, K, M, \gamma)$ defined in 2/3.

The identity leaves all coordinates unchanged; hence

$$E\,S(J, K, M, \gamma) = S(J, K, M, \gamma). \tag{1}$$

The operations C_2^z and C_2^y have the following effect on the coordinates:

$$\begin{aligned} C_2^z{:}\quad \theta' &= \theta, \qquad \chi' = \chi - \pi,\ \phi' = \phi \\ C_2^y{:}\quad \theta'' &= \pi - \theta, \qquad \chi'' = -\chi,\ \phi'' = \phi - \pi, \end{aligned} \tag{2}$$

whereas the effect of C_2^x is the same as C_2^z followed by C_2^y.

The effect of C_2^z is easily found as

$$\begin{aligned} C_2^z S(J, K, M, \gamma) &= C_2^z 2^{-1/2}[(-1)^{\beta} e^{iK\chi} e^{iM\phi}\Theta_{J,K,M}(\theta) \\ &\quad + (-1)^{\gamma+\beta'} e^{-iK\chi} e^{iM\phi}\Theta_{J,K,M}(\theta)] = e^{i\pi K}[S(J, K, M, \gamma)] = (-1)^K S(J, K, M, \gamma), \end{aligned} \tag{3}$$

where β is the larger of K and M.

The effect of C_2^y is found in the same manner, although the process is somewhat more tedious and it is necessary to make use of the relation

$$\theta_{J,K,M}(\pi - \theta) = (-1)^n \Theta_{J,-K,M}(\theta) \tag{4}$$

with

$$n = J - \tfrac{1}{2}|K - M| - \tfrac{1}{2}|K + M|$$

Thus

$$\begin{aligned} &C_2^y S(J, K, M, \gamma) \\ &\quad = (2^{-1/2})[\Psi^{*\times}_{JKM}(\theta'', \phi'', \chi'') + (-1)^{\gamma}\Psi^{*\times}_{J,-K,M}(\theta'', \phi'', \chi'')] \\ &\quad = 2^{-1/2}[(-1)^{\beta+n}\theta_{J,-K,M} e^{-iK\chi} e^{iM(\phi-\pi)} + (-1)^{\beta'+\gamma+n}\theta_{J,K,M} e^{iK\chi} e^{iM(\phi-\pi)}] \\ &\quad = 2^{-1/2}(-1)^{n+M+\gamma+\beta+\beta'}[\Psi^{*\times}_{J,K,M}(\theta, \phi, \chi) + (-1)^{\gamma}\Psi^{*\times}_{J,-K,M}(\theta, \phi, \chi)] \\ &\quad = (-1)^{n+M+\gamma} S(J, K, M, \gamma). \end{aligned} \tag{5}$$

The reduction of (5) to the required form is most easily seen by considering the special cases

$$|K| > |M|, K > 0; \qquad |K| > |M|, K < 0; \qquad |M| > |K|, M > 0;$$
$$|M| > |K|, M < 0.$$

In each case it is found that

$$C_2^{\,y} S(J, K, M, \gamma) = (-1)^{J+\gamma} S(J, K, M, \gamma). \tag{6}$$

The product of (3) and (6) gives

$$C_2^{\,x} S(J, K, M, \gamma) = (-1)^{J+K+\gamma} S(J, K, M, \gamma).$$

Appendix IV

TABLE OF RIGID ROTOR ENERGY LEVEL PATTERNS $E(\kappa)$

The asymmetry parameter $\kappa=(2b-a-c)/(a-c)$, where a, b, c, equal $\hbar^2/2I_a$, $\hbar^2/2I_b$, $\hbar^2/2I_c$, respectively, and where the condition $I_a \leqslant I_b \leqslant I_c$ is applied in assigning the moments of inertia.

Energy level patterns for $0 \leqslant \kappa \leqslant 1$ may be readily obtained from the table by the use of the relation

$$E_\tau^J(\kappa) = -E_{-\tau}^J(-\kappa).$$

Rotational energy levels are given by

$$E_\tau^J(a, b, c) = [(a+c)/2]J(J+1) + [(a-c)/2]E_\tau^J(\kappa).$$

Symmetries are included in terms of the J_{K_{-1}, K_1} notation.

J_{K_{-1},K_1} \ δ	0	0.05	0.1	0.15	0.2	0.25	0.3	0.35	0.4	0.45	0.5	δ
κ	−1	−0.9	−0.8	−0.7	−0.6	−0.5	−0.4	−0.3	−0.2	−0.1	0	κ
$0_{0,0}$	0	0	0	0	0	0	0	0	0	0	0	
$1_{1,0}$	0	0.1	0.2	0.3	0.4	0.5	0.6	0.7	0.8	0.9	1	
$1_{1,1}$	0	0	0	0	0	0	0	0	0	0	0	
$1_{0,1}$	−2	−1.9	−1.8	−1.7	−1.6	−1.5	−1.4	−1.3	−1.2	−1.1	−1	
$2_{2,0}$	2	2.10384	2.21576	2.33631	2.46606	2.60555	2.75528	2.91568	3.08712	3.26987	3.46410	
$2_{2,1}$	2	2.1	2.2	2.3	2.4	2.5	2.6	2.7	2.8	2.9	3	
$2_{1,1}$	−4	−3.6	−3.2	−2.8	−2.4	−2.0	−1.6	−1.2	−0.8	−0.4	0	
$2_{1,2}$	−4	−3.9	−3.8	−3.7	−3.6	−3.5	−3.4	−3.3	−3.2	−3.1	−3	
$2_{0,2}$	−6	−5.70384	−5.41576	−5.13631	−4.86606	−4.60555	−4.35528	−4.11568	−3.88712	−3.66987	−3.46410	
$3_{3,0}$	6	6.15245	6.31027	6.47424	6.64530	6.82496	7.01332	7.21314	7.42586	7.65364	7.89898	
$3_{3,1}$	6	6.15236	6.30949	6.47147	6.63837	6.81025	6.98717	7.16917	7.35629	7.54855	7.74597	
$3_{2,1}$	−4	−3.58082	−3 12186	−2.62186	−2.08082	−1.5	−0.88175	−0.22917	0.45421	1.16476	1.89898	
$3_{2,2}$	−4	−3.6	−3.2	−2.8	−2.4	−2.0	−1.6	−1.2	−0.8	−0.4	0	
$3_{1,2}$	−10	−9.15245	−8.31027	−7.47424	−6.64530	−5.82496	−5.01332	−4.21314	−3.42586	−2.65364	−1.89898	
$3_{1,3}$	−10	−9.75236	−9.50949	−9.27147	−9.03837	−8.81025	−8.58717	−8.36917	−8.15629	−7.94855	−7.74597	
$3_{0,3}$	−12	−11.41918	−10.87814	−10.37814	−9.91918	−9.5	−9.11825	−8.77083	−8.45421	−8.16476	−7.89898	
$4_{4,0}$	12	12.20299	12.41230	12.62852	12.85233	13.08461	13.32641	13.57912	13.84441	14.12449	14.42221	
$4_{4,1}$	12	12.20299	12.41227	12.62834	12.85173	13.08301	13.32279	13.57174	13.83056	14.10000	14.38083	
$4_{3,1}$	−2	−1.43958	−0.85577	−0.24426	0.40000	1.08276	1.81033	2.58928	3.42587	4.32541	5.29150	
$4_{3,2}$	−2	−1.44022	−0.86120	−0.26350	0.35224	0.98528	1.63481	2.30000	2.97998	3.67388	4.38083	
$4_{2,2}$	−12	−11.14570	−10.18216	−9.11536	−7.95859	−6.72860	−5.44216	−4.11432	−2.75776	−1.38320	0.00000	
$4_{2,3}$	−12	−11.20299	−10.41227	−9.62834	−8.85173	−8.08301	−7.32279	−6.57174	−5.83056	−5.10000	−4.38083	
$4_{1,3}$	−18	−16.56042	−15.14423	−13.75574	−12.40000	−11.08276	−9.81033	−8.58928	−7.42587	−6.32541	−5.29150	
$4_{1,4}$	−18	−17.55978	−17.13880	−16.73650	−16.35224	−15 98528	−15.63481	−15.30000	−14.97998	−14.67388	−14.38083	
$4_{0,4}$	−20	−19.05729	−18.23015	−17.51316	−16.89374	−16.35601	−15.88425	−15.46480	−15.08665	−14.74129	−14.42221	

J_{K_{-1},K_1}	δ 0	0.05	0.1	0.15	0.2	0.25	0.3	0.35	0.4	0.45	0.5 δ
	κ −1	−0.9	−0.8	−0.7	−0.6	−0.5	−0.4	−0.3	−0.2	−0.1	0 κ
$5_{5,0}$	20	20.25361	20.51481	20.78427	21.06273	21.35107	21.65036	21.96188	22.28725	22.62851	22.98829
$5_{5,1}$	20	20.25361	20.51481	20.78426	21.06267	21.35092	21.64990	21.96077	22.28477	22.62338	22.97825
$5_{4,1}$	2	2.71155	3.44756	4.21049	5.00354	5.83082	6.69763	7.61079	8.57900	9.61304	10.72586
$5_{4,2}$	2	2.71153	3.44727	4.20892	4.99818	5.81665	6.66583	7.54704	8.46136	9.40961	10.39230
$5_{3,2}$	−12	−10.92537	−9.79317	−8.58944	−7.29895	−5.90702	−4.40221	−2.77918	−1.04053	0.80359	2.73757
$5_{3,3}$	−12	−10.92796	−9.81475	−8.66510	−7.48387	−6.27602	−5.04632	−3.79947	−2.53993	−1.27204	0.00000
$5_{2,3}$	−22	−20.57904	−18.92805	−17.09089	−15.12406	−13.07747	−10.98945	−8.88912	−6.80000	−4.74287	−2.73757
$5_{2,4}$	−22	−20.71153	−19.44727	−18.20892	−16.99818	−15.81665	−14.66583	−13.54704	−12.46136	−11.40961	−10.39230
$5_{1,4}$	−28	−25.82824	−23.72164	−21.69483	−19.76378	−17.94405	−16.24815	−14.68270	−13.24672	−11.93210	−10.72586
$5_{1,5}$	−28	−27.32565	−26.70006	−26.11916	−25.57880	−25.07490	−24.60358	−24.16130	−23.74484	−23.35134	−22.97825
$5_{0,5}$	−30	−28.63251	−27.51951	−26.61960	−25.87948	−25.25335	−24.70818	−24.22167	−23.77900	−23.37017	−22.98829
$6_{6,0}$	30	30.30423	30.61738	30.94020	31.27357	31.61849	31.97615	32.34798	32.73569	33.14140	33.56782
$6_{6,1}$	30	30.30423	30.61738	30.94020	31.27357	31.61847	31.97610	32.34782	32.73527	33.14039	33.56554
$6_{5,1}$	8	8.86322	9.75433	10.67572	11.63026	12.62142	13.65350	14.73201	15.86409	17.05926	18.33030
$6_{5,2}$	8	8.86322	9.75431	10.67561	11.62975	12.61966	13.64861	14.72012	15.83798	17.00616	18.22865
$6_{4,2}$	−10	−8.67533	−7.29809	−5.86179	−4.35735	−2.77237	−1.09071	0.70756	2.64470	4.74200	7.01437
$6_{4,3}$	−10	−8.67542	−7.29951	−5.86955	−4.38365	−2.84099	−1.24202	0.41154	2.11663	3.86897	5.66311
$6_{3,3}$	−24	−22.30159	−20.48431	−18.51309	−16.35698	−14.00000	−11.44897	−8.73201	−5.88946	−2.96481	0.00000
$6_{3,4}$	−24	−22.30932	−20.54826	−18.73335	−16.88024	−15.00321	−13.11511	−11.22733	−9.35010	−7.49261	−5.66311
$6_{2,4}$	−34	−31.86770	−29.33128	−26.53537	−23.61166	−20.64988	−17.71113	−14.84213	−12.08230	−9.46494	−7.01437
$6_{2,5}$	−34	−32.12881	−30.31787	−28.57065	−26.88992	−25.27748	−23.73408	−22.25936	−20.85190	−19.50936	−18.22865
$6_{1,5}$	−40	−36.96163	−34.07002	−31.36263	−28.87328	−26.62142	−24.60453	−22.80000	−21.17463	−19.69445	−18.33030
$6_{1,6}$	−40	−39.05391	−38.20605	−37.44226	−36.74951	−36.11645	−35.53350	−34.99279	−34.48788	−34.01355	−33.56554
$6_{0,6}$	−42	−40.16120	−38.78801	−37.74304	−36.90456	−36.19624	−35.57431	−35.01341	−34.49809	−34.01846	−33.56782
$7_{7,0}$	42	42.35486	42.71997	43.09619	43.48453	43.88613	44.30236	44.73481	45.18543	45.65657	46.15116
$7_{7,1}$	42	42.35486	42.71997	43.09619	43.48453	43.88613	44.30235	44.73479	45.18537	45.65638	46.15066
$7_{6,1}$	16	17.01500	18.06163	19.14256	20.26091	21.42032	22.62515	23.88066	25.19347	26.57205	28.02767
$7_{6,2}$	16	17.01500	18.06162	19.14255	20.26087	21.42013	22.62446	23.87866	25.18825	26.55961	28.00000
$7_{5,2}$	−6	−4.42312	−2.78954	−1.09415	0.66943	2.50949	4.43711	6.46744	8.62113	10.92575	13.41608
$7_{5,3}$	−6	−4.42313	−2.78962	−1.09480	0.66638	2.49918	4.40873	6.39965	8.47556	10.63855	12.88860
$7_{4,3}$	−24	−21.95606	−19.81731	−17.56836	−15.18606	−12.63901	−9.88997	−6.90390	3.66093	−0.16798	3.53956
$7_{4,4}$	−24	−21.95636	−19.82250	−17.59639	−15.27966	−12.87783	−10.40000	−7.85829	−5.26703	−2.64204	0.00000
$7_{3,4}$	−38	−35.56370	−32.90450	−29.95229	−26.67345	−23.09517	−19.29241	−15.35524	−11.36816	−7.40649	−3.53957
$7_{3,5}$	−38	−35.58293	−33.06125	−30.47630	−27.86342	−25.25166	−22.66461	−20.12130	−17.63693	−15.22320	−12.88860
$7_{2,5}$	−48	−44.99890	−41.38121	−37.47756	−33.51139	−29.61371	−25.87353	−22.35690	−19.10561	−16.13111	−13.41608
$7_{2,6}$	−48	−45.45864	−43.03912	−40.74616	−38.58120	−36.54230	−34.62446	−32.82037	−31.12122	−29.51757	−28.00000
$7_{1,6}$	−54	−49.96804	−46.22593	−42.84975	−39.88051	−37.30045	−35.04706	−33.04701	−31.23840	−29.57583	−28.02767
$7_{1,7}$	−54	−52.74881	−51.66910	−50.72509	−49.88749	−49.13365	−48.44647	−47.81314	−47.22400	−46.67173	−46.15066
$7_{0,7}$	−56	−53.66004	−52.06311	−50.89664	−49.96344	−49.16761	−48.46165	−47.81986	−47.22692	−46.67296	−46.15116

J_{K_{-1},K_1}	δ 0 κ −1	0.05 −0.9	0.1 −0.8	0.15 −0.7	0.2 −0.6	0.25 0.5	0.3 −0.4	0.35 −0.3	0.4 −0.2	0.45 −0.1	0.5 δ 0 κ
$8_{8,0}$	56	56.40550	56.82257	57.25221	57.69554	58.15386	58.62871	59.12189	59.63558	60.17242	60.73566
$8_{8,1}$	56	56.40550	56.82257	57.25221	57.69554	58.15386	58.62871	59.12189	59.63557	60.17238	60.73555
$8_{7,1}$	26	27.16683	28.36913	29.60991	30.89263	32.22133	33.60075	35.03654	36.53557	38.10638	39.75991
$8_{7,2}$	26	27.16683	28.36913	29.60991	30.89263	32.22131	33.60066	35.03622	36.53460	38.10367	39.75292
$8_{6,2}$	0	1.82951	3.72124	5.68052	7.71361	9.82803	12.03308	14.34073	16.76690	19.33369	22.07216
$8_{6,3}$	0	1.82951	3.72123	5.68047	7.71330	9.82667	12.02844	14.32726	16.73241	19.25327	21.89858
$8_{5,3}$	−22	−19.60431	−17.11206	−14.51381	−11.79659	−8.94181	−5.92261	−2.70132	0.77076	4.54704	8.66888
$8_{5,4}$	−22	−19.60432	−17.11240	−14.51661	−11.80954	−8.98498	−6.03922	−2.97225	0.21113	3.50079	6.88121
$8_{4,4}$	−40	−37.12871	−34.10050	−30.88066	−27.41611	−23.64282	−19.50903	−15.00742	−10.18985	−5.15070	0.00000
$8_{4,5}$	−40	−37.12961	−34.11595	−30.96275	−27.68384	24.30158	−20.84414	−17.34256	−13.82839	−10.33195	−6.88121
$8_{3,5}$	−54	−50.70574	−47.02209	−42.84192	−38.18998	−33.21092	−28.07787	−22.94146	−17.92750	−13.14274	−8.66888
$8_{3,6}$	−54	−50.74775	−47.35653	−43.91096	−40.47535	−37.09670	−33.80861	−30.63410	−27.58741	−24.67532	−21.89857
$8_{2,6}$	−64	−59.96219	−55.08643	−49.97550	−44.94942	−40.18548	−35.79447	−31.82321	−28.25352	−25.02598	−22.07216
$8_{2,7}$	−64	−60.70539	−57.62785	−54.76993	−52.12499	−49.67895	−47.41301	−45.30659	−43.33959	−41.49370	−39.75292
$8_{1,7}$	−70	−64.85679	−60.23498	−56.25418	−52.90606	−50.06860	−47.60027	−45.39376	−43.37883	−41.51068	−39.75991
$8_{1,8}$	−70	−68.41477	−67.10020	−65.98234	−65.00774	−64.13963	−63.35283	−62.62987	−61.95832	−61.32914	−60.73555
$8_{0,8}$	−72	−69.14410	−67.35688	−66.07651	−65.04361	−64.15359	−63.35829	−62.63199	−61.95912	−61.32943	−60.73566
$9_{9,0}$	72	72.45613	72.92518	73.40825	73.90658	74.42165	74.95515	75.50910	76.08592	76.68854	77.32058
$9_{9,1}$	72	72.45613	72.92518	73.40825	73.90658	74.42165	74.95515	75.50910	76.08591	76.68853	77.32056
$9_{8,1}$	38	39.31868	40.67675	42.07753	43.52486	45.02321	46.57781	48.19485	49.88181	51.64782	53.50433
$9_{8,2}$	38	39.31868	40.67675	42.07753	43.52486	45.02321	46.57780	48.19481	49.88164	51.64726	53.50266
$9_{7,2}$	8	10.08234	12.23284	14.45731	16.76247	19.15619	21.64787	24.24891	26.97372	29.84111	32.87689
$9_{7,3}$	8	10.08234	12.23284	14.45731	16.76244	19.15603	21.64717	24.24648	26.96631	29.82081	32.82576
$9_{6,3}$	−18	−15.25136	−12.40015	−9.43737	−6.35217	−3.13080	0.24515	3.80150	7.57555	11.62086	16.00923
$9_{6,4}$	−18	−15.25136	−12.40016	−9.43761	−6.35370	−3.13740	0.22283	3.73803	7.41715	11.26493	15.27959
$9_{5,4}$	−40	−36.67904	−33.20766	−29.56890	−25.73632	−21.66848	−17.30456	−12.56784	−7.38796	−1.74182	4.31610
$9_{5,5}$	−40	−36.67907	−33.20883	−29.57856	−25.78034	−21.81221	−17.68086	−13.40321	−9.00616	−4.52484	0.00000
$9_{4,5}$	−58	−54.19095	−50.13492	−45.75802	−40.95556	−35.63960	−29.81240	−23.58911	−17.14554	−10.66402	−4.31610
$9_{4,6}$	−58	−54.19330	−50.17453	−45.96402	−41.60316	−37.14601	−32.65039	−28.17155	−23.75826	−19.45078	−15.27959
$9_{3,6}$	−72	−67.72006	−62.80114	−57.13828	−50.92760	−44.48261	−38.07036	−31.89729	−26.11873	−20.82173	−16.00923
$9_{3,7}$	−72	−67.80326	−63.44066	−59.06273	−54.76962	−50.62458	−46.66453	−42.90671	−39.35371	−35.99794	−32.82576
$9_{2,7}$	−82	−76.75116	−70.47091	−64.10763	−58.08158	−52.60633	−47.75172	−43.46093	−39.61925	−36.11803	−32.87689
$9_{2,8}$	−82	−77.87402	−74.10206	−70.67590	−67.56800	−64.73980	−62.15024	−59.76129	−57.54053	−55.46141	−53.50266
$9_{1,8}$	−88	−81.63937	−76.14922	−71.65838	−68.00513	−64.92675	−62.22810	−59.79288	−57.55295	−55.46610	−53.50433
$9_{1,9}$	−88	−86.05614	−84.50853	−83.22427	−82.11906	−81.14089	−80.25693	−79.44566	−78.69235	−77.98656	−77.32056
$9_{0,9}$	−90	−86.62521	−84.67077	−83.27451	−82.13555	−81.14648	−80.25884	−79.44631	−78.69257	−77.98663	−77.32058
$10_{10,0}$	90	90.50677	91.02780	91.56430	92.11765	92.68947	93.28164	93.89638	94.53636	95.20481	95.90573
$10_{10,1}$	90	90.50677	91.02780	91.56430	92.11765	92.68947	93.28164	93.89638	94.53636	95.20481	95.90573
$10_{9,1}$	52	53.47056	54.98444	56.54530	58.15739	59.82560	61.55565	63.35435	65.22987	67.19213	69.25347
$10_{9,2}$	52	53.47056	54.98444	56.54530	58.15739	59.82560	61.55564	63.35434	65.22984	67.19202	69.25308

J_{K_{-1},K_1} \ δ	0	0.05	0.1	0.15	0.2	0.25	0.3	0.35	0.4	0.45	0.5	δ
κ	−1	−0.9	−0.8	−0.7	−0.6	−0.5	−0.4	−0.3	−0.2	−0.1	0	κ
$10_{8,2}$	18	20.33528	22.74489	25.23517	27.81341	30.48810	33.26924	36.16881	39.20139	42.38529	45.74432	
$10_{8,3}$	18	20.33527	22.74489	25.23517	27.81341	30.48808	33.26914	36.16840	39.19991	42.38053	45.73037	
$10_{7,3}$	−12	−8.89789	−5.68593	−2.35463	1.10694	4.71202	8.47676	12.42191	16.57564	20.97829	25.68964	
$10_{7,4}$	−12	−8.89789	−5.68593	−2.35468	1.10677	4.71111	8.47295	12.40876	16.53640	20.87367	25.43593	
$10_{6,4}$	−38	−34.22662	−30.29839	−26.20121	−21.91700	−17.42058	−12.67461	−7.62217	−2.18038	3.75474	10.27364	
$10_{6,5}$	−38	−34.22663	−30.29846	−26.20216	−21.92298	−17.44611	−12.75906	−7.85527	−2.73819	2.57523	8.05291	
$10_{5,5}$	−60	−55.64584	−51.06976	−46.24088	−41.10481	−35.57362	−29.53245	−22.88752	−15.64523	−7.94325	0.00000	
$10_{5,6}$	−60	−55.64594	−51.07323	−46.26924	−41.23155	−35.97438	−30.53144	−24.95372	−19.30370	−13.64837	−8.05291	
$10_{4,6}$	−78	−73.14002	−67.90347	−62.14553	−55.70142	−48.53948	−40.84180	−32.89913	−24.99790	−17.38959	−10.27364	
$10_{4,7}$	−78	−73.14548	−67.99406	−62.60214	−57.06078	−51.47118	−45.92680	−40.50444	−35.26085	−30.23225	−25.43593	
$10_{3,7}$	−92	−86.59725	−80.20810	−72.83972	−64.98059	−57.11839	−49.60411	−42.67271	−36.41928	−30.79761	−25.68964	
$10_{3,8}$	−92	−86.74965	−81.32404	−75.96389	−70.80642	−65.92196	−61.33674	−57.04805	−53.03606	−49.27319	−45.73037	
$10_{2,8}$	−102	−95.36420	−87.56942	−79.96900	−73.07902	−67.07495	−61.87341	−57.28235	−53.13293	−49.31110	−45.74432	
$10_{2,9}$	−102	−96.96993	−92.48017	−88.49517	−84.94730	−81.76026	−78.86492	−76.20507	−73.73723	−71.42832	−69.25308	
$10_{1,9}$	−108	−100.32958	−94.02065	−89.11005	−85.17893	−81.84560	−78.89585	−76.21603	−73.74100	−71.42956	−69.25347	
$10_{1,10}$	−108	−105.67708	−103.90124	−102.45750	−101.22619	−100.14037	−99.16041	−98.26134	−97.42648	−96.64413	−95.90573	
$10_{0,10}$	−110	−106,11121	−104.00141	−102.48373	−101.23362	−100.14256	−99.16106	−98.26154	−97.42654	−96.64415	−95.90573	
$11_{11,0}$	110	110.55741	111.13042	111.72035	112.32873	112.95732	113.60817	114.28371	114.98687	115.72118	116.49101	
$11_{11,1}$	110	110.55741	111.13042	111.72035	112.32873	112.95732	113.60817	114.28371	114.98687	115.72118	116.49101	
$11_{10,1}$	68	69.62244	71.29217	73.01320	74.79013	76.62830	78.53398	80.51457	82.57895	84.73792	87.00484	
$11_{10,2}$	68	69.62244	71.29217	73.01320	74.79013	76.62830	78.53398	80.51457	82.57895	84.73790	87.00476	
$11_{9,2}$	30	32.58828	35.25722	38.01369	40.86557	43.82203	46.89387	50.09389	53.43757	56.94391	60.63695	
$11_{9,3}$	30	32.58828	35.25722	38.01369	40.86557	43.82203	46.89386	50.09383	53.43728	56.94285	60.63335	
$11_{8,3}$	−4	−0.54413	3.02952	6.73101	10.57202	14.56622	18.72994	23.08308	27.65082	32.46668	37.57816	
$11_{8,4}$	−4	−0.54413	3.02952	6.73101	10.57200	14.56612	18.72935	23.08058	27.64194	32.43881	37.49896	
$11_{7,4}$	−34	−29.77306	−25.38384	−20.81819	−16.05925	−11.08615	−5.87160	−0.37697	5.45566	11.71501	18.52846	
$11_{7,5}$	−34	−29.77306	−25.38384	−20.81827	−16.05996	−11.09013	−5.88800	−0.43240	5.29497	11.30427	17.59094	
$11_{6,5}$	−60	−55.09517	−49.96884	−44.59948	−38.95660	−32.99222	−26.62837	−19.74690	−12.20554	−3.91237	5.07342	
$11_{6,6}$	−60	−55.09517	−49.96909	−44.60266	−38.97639	−33.07488	−26.89374	−20.44706	−13.77235	−6.93039	0.00000	
$11_{5,6}$	−82	−76.50325	−70.69043	−64.50488	−57.83704	−50.52275	−42.41746	−33.54235	−24.13221	−14.52682	−5.07342	
$11_{5,7}$	−82	−76.50333	−70.69956	−64.57838	−58.15601	−51.48395	−44.64497	−37.73949	−30.87020	−24.12952	−17.59094	
$11_{4,7}$	−100	−93.97251	−87.38351	−79.97747	−71.57973	−62.37910	−52.82106	−43.35259	−34.34210	−26.04216	−18.52846	
$11_{4,8}$	−100	−93.98416	−87.57242	−80.88745	−74.09626	−67.35804	−60.79826	−54.50178	−48.51452	−42.85004	−37.49896	
$11_{3,8}$	−114	−107.32617	−99.22088	−89.98920	−80.49073	−71.37141	−63.02979	−55.60829	−49.01277	−43.05668	−37.57816	
$11_{3,9}$	−114	−107.58790	−101.02059	−94.65135	−88.64681	−83.06597	−77.90522	−73.12851	−68.68815	−64.53694	−60.63335	
$11_{2,9}$	−124	−115.80441	−106.42498	−97.66776	−90.09101	−83.68305	−78.15043	−73.22096	−68.72135	−64.54822	−60.63695	
$11_{2,10}$	−124	−117.99898	−112.78018	−108.25410	−104.28948	−100.76150	−97.57133	−94.64631	−91.93402	−89.39628	−87.00476	
$11_{1,10}$	−130	−120.94321	−113.89249	−108.62177	−104.40728	−100.79904	−97.58319	−94.64999	−91.93512	−89.39660	−87.00484	
$11_{1,11}$	−130	−127.28140	−125.28365	−123.68604	−122.33152	−121.13930	−120.06384	−119.07714	−118.16077	−117.30184	−116.49101	
$11_{0,11}$	−132	−127.60622	−125.34436	−123.69950	−122.33481	−121.14015	−120.06406	−119.07720	−118.16078	−117.30185	−116.49101	
$12_{12,0}$	132	132.60804	133.23304	133.87642	134.53983	135.22519	135.93472	136.67109	137.43744	138.23763	139.07639	
$12_{12,1}$	132	132.60804	133.23304	133.87642	134.53983	135.22519	135.93472	136.67109	137.43744	138.23763	139.07639	
$12_{11,1}$	86	87.77433	89.59994	91.48116	93.42300	95.43124	97.51266	99.67528	101.92873	104.28467	106.75754	
$12_{11,2}$	86	87.77433	89.59994	91.48116	93.42300	95.43124	97.51266	99.67528	101.92873	104.28466	106.75752	
$12_{10,2}$	44	46.84133	49.76974	52.79264	55.91853	59.15729	62.52052	66.02198	69.67821	73.50943	77.54083	
$12_{10,3}$	44	46.84133	49.76974	52.79264	55.91853	59.15729	62.52052	66.02197	69.67816	73.50920	77.53995	
$12_{9,3}$	6	9.80982	13.74573	17.81847	22.04046	26.42621	30.99284	35.76089	40.75549	46.00848	51.56206	
$12_{9,4}$	6	9.80982	13.74573	17.81847	22.04046	26.42620	30.99276	35.76044	40.75362	46.00157	51.53914	
$12_{8,4}$	−28	−23.31885	−18.46658	−13.42845	−8.18721	−2.72237	2.99115	8.98537	15.30406	22.01179	29.20920	
$12_{8,5}$	−28	−23.31885	−18.46658	−13.42846	−8.18729	−2.72292	2.98828	8.97354	15.26295	21.88669	28.86916	
$12_{7,5}$	−58	−52.54227	−46.85722	−40.92453	−34.71915	−28.20792	−21.34247	−14.04652	−6.19852	2.37386	11.83982	
$12_{7,6}$	−58	−52.54227	−46.85724	−40.92484	−34.72181	−28.22242	−21.40126	−14.23932	−6.73320	1.09462	9.18884	
$12_{6,6}$	−84	−77.85578	−71.40626	−64.61849	−57.43904	−49.77366	−41.46701	−32.32446	−22.23803	−11.34297	0.00000	
$12_{6,7}$	−84	−77.85579	−71.40700	−64.62784	−57.49624	−50.00573	−42.17943	−34.07848	−25.79945	−17.46142	−9.18884	
$12_{5,7}$	−106	−99.24886	−92.05990	−84.32677	−75.84376	−66.37094	−55.87896	−44.69269	−33.31136	−22.22538	−11.83982	
$12_{5,8}$	−106	−99.24950	−92.08176	−84.49852	−76.55839	−68.37935	−60.11399	−51.91698	−43.92037	−36.21961	−28.86916	
$12_{4,8}$	−124	−116.68433	−108.54615	−99.19126	−88.59068	−77.32074	−66.08323	−55.43986	−45.74204	−37.05142	−29.20920	
$12_{4,9}$	−124	−116.70744	−108.91030	−100.83935	−92.76304	−84.89902	−77.38842	−70.30021	−63.64716	−57.40662	−51.53914	
$12_{3,9}$	−138	−129.89446	−119.83602	−108.66278	−97.63069	−87.50774	−78.60398	−70.80785	−63.84208	−57.47608	−51.56206	
$12_{3,10}$	−138	−130.31993	−122.54720	−115.16409	−108.34710	−102.11751	−96.42279	−91.18633	−86.33359	−81.80156	−77.53995	
$12_{2,10}$	−148	−138.07886	−127.08806	−117.31187	−109.20383	−102.42723	−96.52870	−91.22101	−86.34443	−81.80478	−77.54083	
$12_{2,11}$	−148	−140.96729	−135.01890	−129.97341	−125.61179	−121.75481	−118.27567	−115.08791	−112.13194	−109.36548	−106.75752	
$12_{1,11}$	−154	−143.49856	−135.79253	−130.18555	−125.66986	−121.77085	−118.28009	−115.08911	−112.13226	−109.36555	−106.75754	
$12_{1,12}$	−154	−150.87245	−148.65947	−146.91218	−145.43616	−144.13816	−142.96738	−141.89309	−140.89519	−139.95968	−139.07639	
$12_{0,12}$	−156	−151.11155	−148.69573	−146.91899	−145.43760	−144.13848	−142.96745	−141.89311	−140.89521	−139.95968	−139.07639	

J = 13

0.13	-182.00000	-176.62675	-174.05264	-172.14059	-170.54123	-169.13722	-167.87108	-166.70918	-165.62972	-164.61760	-163.66183
1.13	-180.00000	-176.45315	-174.03124	-172.13718	-170.54061	-169.13710	-167.87105	-166.70917	-165.62972	-164.61760	-163.66183
1.12	-180.00000	-168.01370	-159.73225	-153.78725	-148.95244	-144.75236	-140.98191	-137.53101	-134.33116	-131.33574	-128.51113
2.12	-174.00000	-165.88121	-159.21118	-153.66822	-148.92452	-144.74566	-140.98030	-137.53062	-134.33107	-131.33572	-128.51113
2.11	-174.00000	-162.19770	-149.61538	-138.98786	-130.43517	-123.26733	-116.96351	-111.25160	-105.98416	-101.07095	-96.45076
3.11	-164.00000	-154.94857	-145.92298	-137.54037	-129.95461	-123.11921	-116.91967	-111.23910	-105.98074	-101.07007	-96.45054
3.10	-164.00000	-154.28945	-142.07023	-128.95839	-116.59016	-105.73889	-96.40933	-88.21348	-80.80521	-73.97192	-67.58808
4.10	-150.00000	-141.31362	-132.01189	-122.48585	-113.12367	-104.18710	-95.80364	-87.99803	-80.73395	-73.94996	-67.58182
4.9	-150.00000	-141.27042	-131.35706	-119.75043	-106.81982	-93.60593	-81.01402	-69.60670	-59.49031	-50.43434	-42.15984
5.9	-132.00000	-123.88257	-115.21409	-106.02715	-96.45870	-86.72523	-77.05949	-67.65782	-58.65379	-50.11378	-42.04875
5.8	-132.00000	-123.88113	-115.16588	-105.66009	-95.01995	-83.03646	-70.04085	-56.73316	-43.81238	-31.81930	-20.98868
6.8	-110.00000	-102.50714	-94.60672	-86.26611	-77.46776	-68.23437	-58.64784	-48.84666	-39.00167	-29.28432	-19.83904
6.7	-110.00000	-102.50711	-94.60472	-86.24134	-77.32005	-67.65892	-56.99471	-45.15424	-32.31884	-19.00188	-5.81546
7.7	-84.00000	-77.20457	-70.10207	-62.66487	-54.86144	-46.65997	-38.03764	-28.99565	-19.57521	-9.86592	0.00000
7.6	-84.00000	-77.20457	-70.10202	-62.66386	-54.85280	-46.61387	-37.85622	-28.42643	-18.10039	-6.67519	5.81546
8.6	-54.00000	-47.98814	-41.74037	-35.23609	-28.45111	-21.35702	-13.92120	-6.10845	2.11386	10.76557	19.83904
8.5	-54.00000	-47.98814	-41.74037	-35.23606	-28.45078	-21.35475	-13.90967	-6.06211	2.26982	11.21873	20.98868
9.5	-20.00000	-14.86424	-9.54767	-4.03480	1.69238	7.65498	13.87801	20.39119	27.22974	34.43429	42.04875
9.4	-20.00000	-14.86424	-9.54767	-4.03480	1.69238	7.65506	13.87848	20.39352	27.23940	34.46903	42.15984
10.4	18.00000	22.16389	26.46243	30.90709	35.51108	40.28980	45.26139	50.44741	55.87384	61.57235	67.58182
10.3	18.00000	22.16389	26.46243	30.90709	35.51108	40.28981	45.26140	50.44748	55.87422	61.57397	67.58808
11.3	60.00000	63.09440	66.28239	69.57189	72.97205	76.49345	80.14854	83.95205	87.92168	92.07896	96.45054
11.2	60.00000	63.09440	66.28239	69.57189	72.97205	76.49345	80.14854	83.95206	87.92169	92.07900	96.45076
12.2	106.00000	107.92623	109.90773	111.94919	114.05598	116.23434	118.49159	120.83635	123.27899	125.83207	128.51113
12.1	106.00000	107.92623	109.90773	111.94919	114.05598	116.23434	118.49159	120.83635	123.27899	125.83207	128.51113
13.1	156.00000	156.65868	157.33566	158.03248	158.75093	159.49307	160.26130	161.05849	161.88805	162.75413	163.66183
13.0	156.00000	156.65868	157.33566	158.03248	158.75093	159.49307	160.26130	161.05849	161.88805	162.75413	163.66183

J = 14

0,14	-210.00000	-204.15050	-201.41309	-199.36339	-197.64535	-196.13621	-194.77487	-193.52536	-192.36434	-191.27559	-190.24733
1,14	-208.00000	-204.02590	-201.40059	-199.36170	-197.64509	-196.13616	-194.77486	-193.52536	-192.36434	-191.27559	-190.24733
1,13	-208.00000	-194.50688	-185.71135	-179.41414	-174.24643	-169.73913	-165.68650	-161.97460	-158.53122	-155.30678	-152.26537
2,13	-202.00000	-192.74710	-185.36961	-179.34883	-174.23327	-169.73638	-165.68592	-161.97448	-158.53120	-155.30677	-152.26537
2,12	-202.00000	-188.17345	-174.06657	-162.74395	-153.76358	-146.16584	-139.42846	-133.29790	-127.63277	-122.34271	-117.36450
3,12	-192.00000	-181.47756	-171.16835	-161.81532	-153.50551	-146.09757	-139.41091	-133.29353	-127.63173	-122.34248	-117.36445
3,11	-192.00000	-180.49959	-165.95700	-150.98956	-137.54566	-126.16040	-116.40566	-107.74142	-99.83587	-92.50425	-85.63475
4,11	-178.00000	-167.80136	-156.88522	-145.86205	-135.24402	-125.30570	-116.12344	-107.65507	-99.81112	-92.49762	-85.63312
4,10	-178.00000	-167.72457	-155.78011	-141.66364	-126.41543	-111.51943	-97.97450	-86.08454	-75.57164	-66.04701	-57.24198
5,10	-160.00000	-150.40055	-140.09177	-129.16870	-117.89359	-106.60726	-95.61436	-85.12103	-75.22412	-65.93355	-57.20830
5,9	-160.00000	-150.39754	-139.99252	-128.44501	-115.27713	-100.57176	-85.20100	-70.17667	-56.27538	-43.81848	-32.64884
6,9	-138.00000	-129.04778	-119.56238	-109.50640	-98.88277	-87.77757	-76.37034	-64.89810	-53.59894	-42.66510	-32.21755
6,8	-138.00000	-129.04771	-119.55745	-109.44641	-98.53686	-86.50620	-73.04178	-58.27542	-42.87343	-27.66402	-13.37474
7,8	-112.00000	-103.75890	-95.11394	-86.02802	-76.46070	-66.37956	-55.78257	-44.72455	-33.33072	-21.78478	-10.29596
7,7	-112.00000	-103.75889	-95.11378	-86.02503	-76.43570	-66.24959	-55.29160	-43.28190	-29.94939	-15.34913	0.00000
8,7	-82.00000	-74.55123	-66.78853	-58.68404	-50.20471	-41.31213	-31.96466	-22.12495	-11.77548	-0.94033	10.29596
8,6	-82.00000	-74.55123	-66.78853	-58.68394	-50.20355	-41.30411	-31.92483	-21.96979	-11.27668	0.41049	13.37474
9,6	-48.00000	-41.43326	-34.62038	-27.54004	-20.16735	-12.47289	-4.42175	4.02713	12.92019	22.30472	32.21755
9,5	-48.00000	-41.43326	-34.62038	-27.54004	-20.16732	-12.47257	-4.41969	4.03724	12.96107	22.44659	32.64884
10,5	-10.00000	-4.40936	1.37233	7.36140	13.57677	20.04055	26.77891	33.82312	41.21089	48.98786	57.20830
10,4	-10.00000	-4.40936	1.37233	7.36140	13.57677	20.04056	26.77898	33.82355	41.21302	48.99686	57.24198
11,4	32.00000	36.51804	41.17949	45.99653	50.98321	56.15587	61.53374	67.13965	73.00111	79.15171	85.63312
11,3	32.00000	36.51804	41.17949	45.99653	50.98321	56.15587	61.53374	67.13966	73.00118	79.15208	85.63475
12,3	78.00000	81.34751	84.79514	88.35137	92.02597	95.83027	99.77756	103.88356	108.16715	112.65130	117.36445
12,2	78.00000	81.34751	84.79514	88.35137	92.02597	95.83027	99.77756	103.88356	108.16715	112.65131	117.36450
13,2	128.00000	130.07813	132.21554	134.41725	136.68903	139.03757	141.47070	143.99769	146.62961	149.37996	152.26537
13,1	128.00000	130.07813	132.21554	134.41725	136.68903	139.03757	141.47070	143.99769	146.62961	149.37996	152.26537
14,1	182.00000	182.70932	183.43829	184.18856	184.96204	185.76096	186.58790	187.44592	188.33869	189.27067	190.24733
14,0	182.00000	182.70932	183.43829	184.18856	184.96204	185.76096	186.58790	187.44592	188.33869	189.27067	190.24733

J = 15

0,15	-240.00000	-233.68124	-230.77579	-228.58688	-226.74977	-225.13537	-223.67877	-222.34164	-221.09903	-219.93363	-218.83287
1,15	-238.00000	-233.59268	-230.76856	-228.58605	-226.74966	-225.13535	-223.67877	-222.34164	-221.09903	-219.93363	-218.83287
1,14	-238.00000	-222.99412	-213.72356	-207.05701	-201.54699	-196.72896	-192.39283	-188.41935	-184.73213	-181.27846	-178.02009
2,14	-232.00000	-221.57126	-213.50426	-207.02182	-201.54089	-196.72786	-192.39262	-188.41931	-184.73213	-181.27846	-178.02009
2,13	-232.00000	-216.02069	-200.49761	-188.59073	-179.15900	-171.09791	-163.90979	-157.35302	-151.28681	-145.61818	-140.28092
3,13	-222.00000	-209.91152	-198.30413	-188.01903	-179.02503	-171.06736	-163.90295	-157.35153	-151.28650	-145.61812	-140.28091
3,12	-222.00000	-208.51605	-191.54181	-174.87861	-160.61754	-148.76465	-138.52461	-129.33518	-120.90097	-113.05582	-105.69330
4,12	-208.00000	-196.16976	-183.54222	-171.00778	-159.18736	-148.32160	-138.39956	-129.30205	-120.89270	-113.05390	-105.69289
4,11	-208.00000	-196.03925	-181.78441	-164.98672	-147.56366	-131.35029	-117.18399	-104.86896	-93.86345	-83.77697	-74.38634
5,11	-190.00000	-178.80137	-166.71161	-153.93608	-140.91485	-128.12272	-115.90518	-104.43213	-93.72914	-83.73925	-74.37670
5,10	-190.00000	-178.79540	-166.51935	-152.61360	-136.59088	-119.16293	-101.75220	-85.54790	-71.11111	-58.28328	-46.64203
6,10	-168.00000	-157.47615	-146.26788	-134.33942	-121.74372	-108.67811	-95.45466	-82.40766	-69.80781	-57.81835	-46.49581
6,9	-168.00000	-157.47599	-146.25656	-134.20500	-121.00323	-106.16917	-89.59590	-72.01996	-54.55244	-38.17327	-23.40019
7,9	-142.00000	-132.20412	-121.88806	-111.00320	-99.50195	-87.36576	-74.64785	-61.50263	-48.17207	-34.93303	-22.03515
7,8	-142.00000	-132.20412	-121.88764	-110.99515	-99.43621	-87.03591	-73.47300	-58.36317	-41.68982	-24.09345	-6.54491
8,8	-112.00000	-103.00724	-93.60747	-83.76373	-73.43199	-62.56272	-51.10969	-39.05104	-26.42102	-13.33697	0.00000
8,7	-112.00000	-103.00724	-93.60746	-83.76342	-73.42830	-62.53758	-50.98842	-38.59854	-25.06185	-10.02641	6.54491
9,7	-78.00000	-69.89659	-61.46959	-52.69065	-43.52631	-33.93676	-23.87499	-13.28798	-2.12326	9.65444	22.03515
9,6	-78.00000	-69.89659	-61.46959	-52.69065	-43.52617	-33.93549	-23.86713	-13.25040	-1.97641	10.13920	23.40019
10,6	-40.00000	-32.87789	-25.49832	-17.83916	-9.87461	-1.57433	7.09779	16.18509	25.73945	35.82140	46.49581
10,5	-40.00000	-32.87789	-25.49832	-17.83916	-9.87461	-1.57429	7.09814	16.18714	25.74936	35.86211	46.64203
11,5	2.00000	8.04569	14.29309	20.75937	27.46444	34.43150	41.68793	49.26644	57.20657	65.55676	74.37670
11,4	2.00000	8.04569	14.29309	20.75937	27.46444	34.43150	41.68794	49.26651	57.20701	65.55897	74.38634
12,4	48.00000	52.87226	57.89680	63.08656	68.45642	74.02372	79.80878	85.83575	92.13369	98.73805	105.69289
12,3	48.00000	52.87226	57.89680	63.08656	68.45642	74.02372	79.80878	85.83576	92.13370	98.73813	105.69330
13,3	98.00000	101.60063	105.30795	109.13101	113.08019	117.16758	121.40731	125.81612	130.41408	135.22561	140.28091
13,2	98.00000	101.60063	105.30795	109.13101	113.08019	117.16758	121.40731	125.81612	130.41408	135.22561	140.28092
14,2	152.00000	154.23003	156.52337	158.88535	161.32215	163.84090	166.44996	169.15922	171.98052	174.92823	178.02009
14,1	152.00000	154.23003	156.52337	158.88535	161.32215	163.84090	166.44996	169.15922	171.98052	174.92823	178.02009
15,1	210.00000	210.75996	211.54091	212.34463	213.17316	214.02886	214.91450	215.83336	216.78935	217.78724	218.83287
15,0	210.00000	210.75996	211.54091	212.34463	213.17316	214.02886	214.91450	215.83336	216.78935	217.78724	218.83287

J = 16

0.16	-272.00000	-265.21746	-262.13992	-259.81081	-257.85438	-256.13465	-254.58277	-253.15798	-251.83377	-250.59172	-249.41844
1.16	-270.00000	-265.15502	-262.13577	-259.81041	-257.85433	-256.13464	-254.58277	-253.15798	-251.83377	-250.59172	-249.41844
1.15	-270.00000	-253.48798	-243.76110	-236.70981	-230.85145	-225.72073	-221.10036	-216.86496	-212.93369	-209.25064	-205.77520
2.15	-264.00000	-252.35974	-243.62287	-236.69114	-230.84867	-225.72029	-221.10028	-216.86495	-212.93369	-209.25064	-205.77520
2.14	-264.00000	-245.75583	-228.95288	-216.51600	-206.59692	-198.04923	-190.40063	-183.41367	-176.94456	-170.89634	-165.19936
3.14	-254.00000	-240.25584	-227.35058	-216.17556	-206.52923	-198.03588	-190.39803	-183.41318	-176.94447	-170.89632	-165.19935
3.13	-254.00000	-238.33419	-218.87882	-200.74223	-185.84894	-173.50180	-162.71451	-152.96439	-143.98518	-135.61904	-127.75969
4.13	-240.00000	-226.41858	-211.99844	-197.96487	-185.00923	-173.28256	-162.66122	-152.95212	-143.98251	-135.61850	-127.75959
4.12	-240.00000	-226.20539	-209.35296	-189.81108	-170.47127	-153.32251	-138.67832	-125.86122	-114.26923	-103.56609	-93.56318
5.12	-222.00000	-209.08291	-195.07276	-180.35126	-165.58555	-151.36955	-138.03746	-125.67544	-114.22006	-103.55415	-93.56054
5.11	-222.00000	-209.07161	-194.72024	-178.10520	-159.02800	-139.08094	-120.11616	-103.21641	-88.37712	-75.05584	-62.80531
6.11	-200.00000	-187.79058	-174.71704	-160.75932	-146.06745	-131.00486	-116.03264	-101.55070	-87.80580	-74.88395	-62.75925
6.10	-200.00000	-187.79022	-174.69262	-160.47838	-144.61348	-126.57261	-106.86392	-86.90111	-68.06026	-51.12302	-36.02146
7.10	-174.00000	-162.53903	-150.41928	-137.57888	-123.97010	-109.61767	-94.68168	-79.46096	-64.31987	-49.58929	-35.49486
7.9	-174.00000	-162.53902	-150.41821	-137.55889	-123.81131	-108.85862	-92.19540	-73.58456	-53.68052	-33.71556	-14.88358
8.9	-144.00000	-133.35525	-122.19330	-110.46586	-98.11564	-85.08350	-71.33285	-56.89399	-41.90744	-26.62856	-11.37915
8.8	-144.00000	-133.35525	-122.19327	-110.46493	-98.10496	-85.01240	-71.00279	-55.73612	-38.76096	-19.96012	0.00000
9.8	-110.00000	-100.25348	-90.09229	-79.47947	-68.37091	-56.71393	-44.44793	-31.51196	-17.86612	-3.52954	11.37915
9.7	-110.00000	-100.25348	-90.09229	-79.47944	-68.37042	-56.70958	-44.42149	-31.38959	-17.41002	-2.13089	14.88358
10.7	-72.00000	-63.24109	-54.14709	-44.68890	-34.83242	-24.53715	-13.75460	-2.42678	9.51349	22.13621	35.49486
10.6	-72.00000	-63.24109	-54.14709	-44.68890	-34.83240	-24.53696	-13.75317	-2.41843	9.55273	22.29133	36.02146
11.6	-30.00000	-22.32216	-14.37483	-6.13492	2.42437	11.33448	20.63272	30.36388	40.58238	51.35449	62.75925
11.5	-30.00000	-22.32216	-14.37483	-6.13492	2.42437	11.33449	20.63277	30.36427	40.58464	51.36543	62.80531
12.5	16.00000	22.50088	29.21439	36.15862	43.35448	50.82632	58.60284	66.71823	75.21379	84.14020	93.56054
12.4	16.00000	22.50088	29.21439	36.15862	43.35448	50.82632	58.60284	66.71824	75.21388	84.14072	93.56318
13.4	66.00000	71.22652	76.61430	82.17702	87.93045	93.89289	100.08582	106.53473	113.27025	120.32975	127.75959
13.3	66.00000	71.22652	76.61430	82.17702	87.93045	93.89289	100.08582	106.53473	113.27026	120.32976	127.75969
14.3	120.00000	123.85376	127.82082	131.91077	136.13465	140.50527	145.03763	149.74949	154.66212	159.80141	165.19935
14.2	120.00000	123.85376	127.82082	131.91077	136.13465	140.50527	145.03763	149.74949	154.66212	159.80141	165.19936
15.2	178.00000	180.38194	182.83120	185.35347	187.95531	190.64430	193.42933	196.32092	199.33164	202.47678	205.77520
15.1	178.00000	180.38194	182.83120	185.35347	187.95531	190.64430	193.42933	196.32092	199.33164	202.47678	205.77520
15.1	240.00000	240.81060	241.64354	242.50071	243.38428	244.29676	245.24112	246.22082	247.24004	248.30383	249.41844
16.0	240.00000	240.81060	241.64354	242.50071	243.38428	244.29676	245.24112	246.22082	247.24004	248.30383	249.41844

J = 17

0.17	-306.00000	-298.75784	-295.50497	-293.03502	-290.95913	-289.13403	-287.48684	-285.97438	-284.56895	-283.24984	-282.00404
1.17	-304.00000	-298.71415	-295.50259	-293.03482	-290.95911	-289.13402	-287.48684	-285.97438	-284.56855	-283.24984	-282.00404
1.16	-304.00000	-285.99690	-275.81694	-268.36880	-262.15838	-256.71383	-251.80878	-247.31124	-243.13576	-239.22322	-235.53062
2.16	-298.00000	-285.11825	-275.73110	-268.35903	-262.15712	-256.71366	-251.80876	-247.31123	-243.13576	-239.22322	-235.53062
2.15	-298.00000	-277.39708	-259.45955	-246.50025	-236.06095	-227.01198	-218.89751	-211.47817	-204.60504	-198.17654	-192.11934
3.15	-288.00000	-272.51656	-258.32657	-246.30288	-236.02749	-227.00626	-218.89653	-211.47801	-204.60501	-198.17653	-192.11934
3.14	-288.00000	-269.95434	-248.02893	-228.67057	-213.22544	-200.32369	-188.94471	-178.61399	-169.08131	-160.19018	-151.83174
4.14	-274.00000	-258.54832	-242.27240	-226.77441	-212.75435	-200.21906	-188.92270	-178.60957	-169.08047	-160.19003	-151.83172
4.13	-274.00000	-258.21239	-238.48893	-216.25123	-195.34000	-177.53547	-162.39074	-148.97426	-136.73572	-125.38772	-114.75959
5.13	-256.00000	-241.24310	-225.17741	-208.44478	-191.97450	-176.43603	-162.08806	-148.89899	-136.71847	-125.38409	-114.75889
5.12	-256.00000	-241.22259	-224.56322	-204.88957	-182.73378	-160.63911	-140.64393	-123.28131	-107.94603	-93.99458	-81.05337
6.12	-234.00000	-219.98926	-204.90394	-188.76565	-171.88723	-154.84767	-138.24272	-122.47753	-107.71408	-93.93496	-81.03966
6.11	-234.00000	-219.98852	-204.85409	-188.21514	-169.26715	-147.78308	-125.21255	-103.49508	-83.92581	-66.62910	-51.03956
7.11	-208.00000	-194.76231	-180.70205	-165.74363	-149.85582	-133.15595	-115.97192	-98.77037	-82.00358	-65.98609	-50.85544
7.10	-208.00000	-194.76229	-180.69953	-165.69727	-149.50087	-131.56915	-111.32588	-89.17055	-66.57443	-45.11978	-25.77050
8.10	-178.00000	-165.59427	-152.54183	-138.78035	-124.23764	-108.85265	-92.62980	-75.70963	-58.39459	-41.08908	-24.18739
8.9	-178.00000	-165.59427	-152.54175	-138.77781	-124.20918	-108.66919	-91.82162	-73.11597	-52.22553	-29.80393	-7.26370
9.9	-144.00000	-132.50318	-120.48523	-107.89874	-94.68632	-80.77989	-66.10676	-50.61330	-34.31386	-17.34787	0.00000
9.8	-144.00000	-132.50318	-120.48523	-107.89864	-94.68477	-80.76641	-66.02705	-50.25970	-33.08172	-13.96250	7.26370
10.8	-106.00000	-95.49836	-84.57139	-73.18172	-61.28508	-48.82845	-35.74808	-21.96946	-7.41449	7.97453	24.18739
10.7	-106.00000	-95.49836	-84.57139	-73.18171	-61.28502	-48.82776	-35.74283	-21.93963	-7.27924	8.48158	25.77050
11.7	-64.00000	-54.58501	-44.82214	-34.68139	-24.12773	-13.11979	-1.60816	10.46695	23.17792	36.61024	50.85544
11.6	-64.00000	-54.58501	-44.82214	-34.68139	-24.12773	-13.11977	-1.60791	10.46868	23.18767	36.65597	51.03956
12.6	-18.00000	-9.76619	-1.25032	7.57171	16.72782	26.25067	36.17889	46.55878	57.44660	68.91163	81.03966
12.5	-18.00000	-9.76619	-1.25032	7.57171	16.72782	26.25067	36.17890	46.55885	57.44709	68.9142	81.05337
13.5	32.00000	38.95617	46.13610	53.55882	61.24628	69.22402	77.52212	86.17633	95.22979	104.73535	114.75889
13.4	32.00000	38.95617	46.13610	53.55882	61.24628	69.22402	77.52212	86.17633	95.22981	104.73547	114.75959
14.4	86.00000	91.58081	97.33194	103.26783	109.40510	115.76307	122.36439	129.23590	136.40986	143.92555	151.83172
14.3	86.00000	91.58081	97.33194	103.26783	109.40510	115.76307	122.36439	129.23590	136.40986	143.92556	151.83174
15.3	144.00000	148.10690	152.33373	156.69063	161.18929	165.84326	170.66840	175.68349	180.91103	186.37838	192.11934
15.2	144.00000	148.10690	152.33373	156.69063	161.18929	165.84326	170.66840	175.68349	180.91103	186.37838	192.11934
16.2	206.00000	208.53385	211.13904	213.82161	216.58850	219.44776	222.40879	225.48274	228.68294	232.02557	235.53062
16.1	206.00000	208.53385	211.13904	213.82161	216.58850	219.44776	222.40879	225.48274	228.68294	232.02557	235.53062
17.1	272.00000	272.86124	273.74617	274.65679	275.59540	276.56468	277.56774	278.60829	279.69074	280.82045	282.00404
17.0	272.00000	272.86124	273.74617	274.65679	275.59540	276.56468	277.56774	278.60829	279.69074	280.82045	282.00404

J = 18

0.18	-342.00000	-334.30136	-330.87060	-328.25942	-326.06399	-324.13348	-322.39096	-320.79083	-319.30337	-317.90799	-316.58966
1.18	-340.00000	-334.27096	-330.86925	-328.25932	-326.06398	-324.13348	-322.39096	-320.79083	-319.30337	-317.90799	-316.58966
1.17	-340.00000	-320.52527	-309.88557	-302.03174	-295.46697	-289.70792	-284.51791	-279.75804	-275.33824	-271.19611	-267.28628
2.17	-334.00000	-319.85201	-309.83294	-302.02668	-295.46641	-289.70785	-284.51790	-279.75804	-275.33824	-271.19611	-267.28628
2.16	-334.00000	-310.96415	-292.02756	-278.52552	-267.54104	-257.98197	-249.39855	-241.54552	-234.26764	-227.45835	-221.04055
3.16	-324.00000	-306.70025	-291.24892	-278.41357	-267.52482	-257.97956	-249.39820	-241.54547	-234.26763	-227.45835	-221.04055
3.15	-324.00000	-303.38182	-279.05755	-258.71185	-242.71058	-229.19695	-217.19864	-206.27644	-196.18568	-186.76716	-177.90811
4.15	-310.00000	-292.56034	-274.38487	-257.47417	-242.45561	-229.14846	-217.18979	-206.27489	-196.18542	-186.76712	-177.90810
4.14	-310.00000	-292.04831	-269.21672	-244.43630	-222.32522	-203.97780	-188.24253	-174.15505	-161.23673	-149.22953	-137.96933
5.14	-292.00000	-275.27993	-257.03132	-238.25418	-220.15014	-203.39359	-188.10590	-174.12571	-161.23089	-149.22846	-137.96915
5.13	-292.00000	-275.24409	-256.01201	-232.98482	-207.90316	-184.14800	-163.50635	-145.65973	-129.69002	-115.02072	-101.34780
6.13	-270.00000	-254.07032	-236.82333	-218.36516	-199.25234	-180.30806	-162.21212	-145.29855	-129.60111	-115.00105	-101.34390
6.12	-270.00000	-254.06885	-236.72649	-217.35123	-194.91465	-170.00832	-145.08550	-122.29535	-102.30785	-84.52821	-68.26817
7.12	-244.00000	-228.87257	-212.73055	-195.48699	-177.15936	-158.02731	-138.63994	-119.61583	-101.42242	-84.28127	-68.20826
7.11	-244.00000	-228.87253	-212.72491	-195.38594	-176.42209	-155.01939	-130.93087	-105.60206	-81.12708	-58.98475	-39.33666
8.11	-214.00000	-199.72326	-184.64856	-168.69642	-151.78031	-133.85702	-115.02667	-95.60979	-76.09980	-57.00800	-38.71134
8.10	-214.00000	-199.72326	-184.64834	-168.68997	-151.70997	-133.42215	-113.24625	-90.54990	-65.72197	-40.38742	-16.37013
9.10	-180.00000	-166.64485	-152.64496	-137.94006	-122.45645	-106.10913	-88.82235	-70.58577	-51.53606	-32.00053	-12.44200
9.9	-180.00000	-166.64485	-152.64495	-137.93977	-122.45197	-106.07092	-88.60455	-69.67407	-48.65617	-25.17373	0.00000
10.9	-142.00000	-129.64903	-116.76845	-103.31101	-89.22003	-74.42695	-58.84994	-42.39982	-25.00582	-6.67329	12.44200
10.8	-142.00000	-129.64903	-116.76845	-103.31100	-89.21983	-74.42463	-58.83265	-42.30477	-24.59512	-5.24674	16.37013
11.8	-100.00000	-88.74230	-77.04657	-64.87469	-52.18181	-38.91455	-25.00851	-10.38562	5.04660	21.38709	38.71134
11.7	-100.00000	-88.74230	-77.04657	-64.87469	-52.18181	-38.91445	-25.00754	-10.37890	5.08330	21.55228	39.33666
12.7	-54.00000	-43.92850	-33.49543	-22.66975	-11.41532	0.31034	12.55781	25.38810	38.87583	53.11262	68.20826
12.6	-54.00000	-43.92850	-33.49543	-22.66975	-11.41532	0.31035	12.55784	25.38844	38.87811	53.12527	68.26817
13.6	-4.00000	4.78997	13.87496	23.28013	33.03459	43.17231	53.73339	64.76578	76.32758	88.49041	101.34390
13.5	-4.00000	4.78997	13.87496	23.28013	33.03459	43.17231	53.73339	64.76579	76.32768	88.49109	101.34780
14.5	50.00000	57.41153	65.05812	72.95974	81.13941	89.62391	98.44472	107.63922	117.25248	127.33958	137.96915
14.4	50.00000	57.41153	65.05812	72.95974	81.13941	89.62391	98.44472	107.63922	117.25248	127.33960	137.96933
15.4	108.00000	113.93514	120.04970	126.35890	132.88024	139.63405	146.64415	153.93878	161.55183	169.52456	177.90810
15.3	108.00000	113.93514	120.04970	126.35890	132.88024	139.63405	146.64415	153.93878	161.55183	169.52456	177.90811
16.3	170.00000	174.36006	178.84668	183.47058	188.24408	193.18148	198.29952	203.61800	209.16064	214.95628	221.04055
16.2	170.00000	174.36006	178.84668	183.47058	188.24408	193.18148	198.29952	203.61800	209.16064	214.95628	221.04055
17.2	236.00000	238.68576	241.44690	244.28977	247.22173	250.25126	253.38832	256.64467	260.03438	263.57455	267.28628
17.1	236.00000	238.68576	241.44690	244.28977	247.22173	250.25126	253.38832	256.64467	260.03438	263.57455	267.28628
18.1	306.00000	306.91188	307.84880	308.81287	309.80652	310.83259	311.89437	312.99577	314.14145	315.33708	316.58966
18.0	306.00000	306.91188	307.84880	308.81287	309.80652	310.83259	311.89437	312.99577	314.14145	315.33708	316.58966

J = 19

0.19	-380.00000	-371.84718	-368.23662	-365.48395	-363.16893	-361.13300	-359.29514	-357.60731	-356.03823	-354.56616	-353.17530
1.19	-378.00000	-371.82616	-368.23586	-365.48391	-363.16893	-361.13301	-359.29514	-357.60731	-356.03823	-354.56616	-353.17530
1.18	-378.00000	-357.07426	-345.96299	-337.69726	-330.77674	-324.70277	-319.22759	-314.20528	-309.54106	-305.16926	-301.04214
2.18	-372.00000	-356.56570	-345.93106	-337.69467	-330.77650	-324.70274	-319.22759	-314.20528	-309.54106	-305.16926	-301.04214
2.17	-372.00000	-346.47779	-326.65428	-312.57822	-301.03132	-290.95689	-281.90270	-273.61510	-265.93192	-258.74145	-251.96275
3.17	-362.00000	-342.81385	-326.13192	-312.51586	-301.02358	-290.95589	-281.90257	-273.61508	-265.93192	-258.74145	-251.96275
3.16	-362.00000	-338.62638	-312.03041	-290.87435	-274.26927	-260.10140	-247.46755	-235.94779	-225.29617	-215.34865	-205.98786
4.16	-348.00000	-328.45691	-308.35790	-290.09667	-274.13518	-260.07946	-247.46407	-235.94726	-225.29610	-215.34864	-205.98786
4.15	-348.00000	-327.70046	-301.57840	-274.50101	-251.50283	-232.59083	-216.17753	-201.37533	-187.75947	-175.08530	-163.18888
5.15	-330.00000	-311.19162	-290.64415	-269.82169	-250.17499	-232.29401	-216.11803	-201.36425	-187.75755	-175.08499	-163.18883
5.14	-330.00000	-311.13107	-289.03051	-262.45869	-234.75770	-209.84384	-188.69454	-170.23674	-153.53080	-138.09626	-123.67112
6.14	-308.00000	-290.03176	-270.47122	-249.57321	-228.22505	-207.48818	-188.04360	-170.08292	-153.49822	-138.09003	-123.67006
6.13	-308.00000	-290.02895	-270.29142	-247.81677	-221.59661	-193.54617	-166.92401	-143.54231	-123.09721	-104.65007	-87.60605
7.13	-282.00000	-264.86832	-246.49866	-226.80074	-205.89435	-184.30452	-162.82643	-142.16852	-122.72130	-104.56119	-87.58761
7.12	-282.00000	-264.86823	-246.48669	-226.59270	-204.47023	-179.13962	-151.29856	-123.48088	-97.97794	-75.49296	-55.36381
8.12	-252.00000	-235.74109	-218.50867	-200.20275	-180.72818	-160.09876	-138.58801	-116.75480	-95.26141	-74.63459	-55.13935
8.11	-252.00000	-235.74109	-218.50813	-200.18738	-180.56586	-159.14941	-135.07576	-108.12426	-79.88167	-52.67139	-28.10524
9.11	-218.00000	-202.67762	-186.56774	-169.59439	-151.66410	-132.67691	-112.57837	-91.46541	-69.67816	-47.76291	-26.30189
9.10	-218.00000	-202.67762	-186.56772	-169.59360	-151.65202	-132.57707	-112.03576	-89.37325	-63.90418	-36.13428	-7.97329
10.10	-180.00000	-165.69239	-150.73533	-135.06970	-118.62362	-101.30946	-83.02612	-63.68151	-43.25632	-21.90212	0.00000
10.9	-180.00000	-165.69239	-150.73533	-135.06967	-118.62298	-101.30232	-82.97443	-63.40888	-42.15393	-18.48021	7.97329
11.9	-138.00000	-124.79345	-111.04571	-96.70911	-81.72708	-66.03149	-49.53953	-32.15171	-13.75848	5.72834	26.30189
11.8	-138.00000	-124.79345	-111.04571	-96.70911	-81.72705	-66.03112	-49.53608	-32.12842	-13.63599	6.24972	28.10524
12.8	-92.00000	-79.98556	-67.51894	-54.56104	-41.06610	-26.97996	-12.23758	3.24013	19.55008	36.80828	55.13935
12.7	-92.00000	-79.98556	-67.51894	-54.56104	-41.06610	-26.97994	-12.23741	3.24155	19.55938	36.85806	55.36381
13.7	-42.00000	-31.27167	-20.16742	-8.65505	3.30279	15.74990	28.73824	42.33028	56.60230	71.64882	87.58761
13.6	-42.00000	-31.27167	-20.16742	-8.65505	3.30279	15.74990	28.73825	42.33034	56.60281	71.65214	87.60605
14.6	12.00000	21.34626	31.00082	40.98992	51.34388	62.09808	73.29420	84.98193	97.22141	110.08657	123.67006
14.5	12.00000	21.34626	31.00082	40.98992	51.34388	62.09808	73.29420	84.98194	97.22143	110.08673	123.67112
15.5	70.00000	77.86695	85.98039	94.36123	103.03360	112.02552	121.36990	131.10584	141.28035	151.95087	163.18883
15.4	70.00000	77.86695	85.98039	94.36123	103.03360	112.02552	121.36990	131.10584	141.28035	151.95088	163.18888
16.4	132.00000	138.28948	144.76755	151.45018	158.35577	165.50566	172.92486	180.64302	188.69568	197.12610	205.98786
16.3	132.00000	138.28948	144.76755	151.45018	158.35577	165.50566	172.92486	180.64302	188.69568	197.12610	205.98786
17.3	198.00000	202.61321	207.35965	212.25058	217.29899	222.51990	227.93093	233.55291	239.41081	245.53494	251.96275
17.2	198.00000	202.61321	207.35965	212.25058	217.29899	222.51990	227.93093	233.55291	239.41081	24553494	251.96275
18.2	268.00000	270.83768	273.75475	276.75795	279.85498	283.05481	286.36791	289.80668	293.38593	297.12368	301.04214
18.1	268.00000	270.83768	273.75475	276.75795	279.85498	283.05481	286.36791	289.80668	293.38593	297.12368	301.04214
19.1	342.00000	342.96252	343.95142	344.96895	346.01765	347.10051	348.22101	349.38325	350.59217	351.85373	353.17530
19.0	342.00000	342.96252	343.95142	344.96895	346.01765	347.10051	348.22101	349.38325	350.59217	351.85373	353.17530

J = 20

0.20	-420.00000	-411.39470	-407.60291	-404.70859	-402.27395	-400.13256	-398.19936	-396.42383	-394.77310	-393.22435	-391.76096
1.20	-418.00000	-411.38022	-407.60248	-404.70857	-402.27394	-400.13256	-398.19936	-396.42383	-394.77310	-393.22435	-391.76096
1.19	-418.00000	-395.64279	-384.04642	-375.36457	-368.08742	-361.69823	-355.93774	-350.65287	-345.74415	-341.14264	-336.79817
2.19	-412.00000	-395.26342	-384.02723	-375.36325	-368.08731	-361.69822	-355.93774	-350.65287	-345.74415	-341.14264	-336.79817
2.18	-412.00000	-383.95883	-363.33066	-348.64903	-336.52836	-325.93540	-316.40926	-307.68647	-299.59757	-292.02561	-284.88577
3.18	-402.00000	-380.86451	-362.98725	-348.61483	-336.52471	-325.93499	-316.40922	-307.68647	-299.59757	-292.02561	-284.88577
3.17	-402.00000	-375.70128	-347.00672	-325.14182	-307.87562	-293.02548	-279.74669	-267.62576	-256.41144	-245.93371	-236.07032
4.17	-388.00000	-366.24127	-344.21384	-324.66823	-307.80676	-293.01576	-279.74534	-267.62558	-256.41142	-245.93371	-236.07032
4.16	-388.00000	-365.15635	-335.62870	-306.56981	-282.87475	-263.31650	-246.16230	-230.62044	-216.29727	-202.95146	-190.41600
5.16	-370.00000	-348.97665	-326.02941	-303.19147	-282.10151	-263.17092	-246.13715	-230.61637	-216.29665	-202.95137	-190.41599
5.15	-370.00000	-348.87749	-323.59035	-293.41652	-263.52185	-237.82992	-216.12058	-196.93005	-179.42753	-163.20299	-148.01456
6.15	-348.00000	-327.87152	-305.84547	-282.41418	-258.87567	-236.48043	-215.81006	-196.86725	-179.41602	-163.20108	-148.01428
6.14	-348.00000	-327.86633	-305.52551	-279.55289	-249.44576	-218.73635	-191.05066	-167.21561	-146.13816	-126.89305	-109.00585
7.14	-322.00000	-302.74799	-282.00021	-259.68059	-236.09038	-212.08127	-188.67326	-166.56483	-145.98795	-126.86264	-109.00043
7.13	-322.00000	-302.74781	-281.97593	-259.27459	-233.53613	-204.00932	-172.85847	-143.39550	-117.40887	-94.48538	-73.64315
8.13	-292.00000	-273.64658	-254.11700	-233.28781	-211.07075	-187.60202	-163.41722	-139.33227	-116.09778	-94.14442	-73.56806
8.12	-292.00000	-273.64658	-254.11575	-233.25323	-210.71953	-185.69594	-157.23111	-126.24403	-95.49675	-67.42015	-42.60189
9.12	-258.00000	-240.60056	-222.24963	-202.85197	-182.29136	-160.46272	-137.38190	-113.34270	-88.95059	-64.93378	-41.87492
9.11	-258.00000	-240.60056	-222.24958	-202.84994	-182.26097	-160.22074	-136.14860	-109.10322	-78.98145	-47.68496	-17.83730
10.11	-220.00000	-203.62771	-186.46889	-168.45020	-149.48103	-129.45120	-108.24436	-85.79727	-62.22070	-37.90682	-13.48723
10.10	-220.00000	-203.62771	-186.46888	-168.45011	-149.47917	-129.43094	-108.10268	-85.09016	-59.62100	-30.98730	0.00000
11.10	-178.00000	-162.73786	-146.81701	-130.17857	-112.75185	-94.45064	-75.16975	-54.78783	-33.19497	-10.37199	13.48723

11.9	-178.00000	-162.73786	-146.81701	-130.17857	-112.75176	-94.44942	-75.15859	-54.71491	-32.82962	-8.93422	17.83730
12.9	-132.00000	-117.93685	-103.31871	-88.09714	-72.21507	-55.60432	-38.18223	-19.84753	-0.47779	20.06172	41.87492
12.8	-132.00000	-117.93685	-103.31871	-88.09714	-72.21507	-55.60426	-38.18159	-19.84222	-0.44412	20.23413	42.60189
13.8	-82.00000	-69.22831	-55.98923	-42.24252	-27.94125	-13.03014	2.55684	18.89973	36.09702	54.27075	73.56806
13.7	-82.00000	-69.22831	-55.98923	-42.24252	-27.94125	-13.03014	2.55687	18.90002	36.09925	54.28487	73.64315
14.7	-28.00000	-16.61461	-4.83842	7.36196	20.02521	33.19658	46.92960	61.28844	76.35121	92.21461	109.00043
14.6	-28.00000	-16.61461	-4.83842	7.36196	20.02521	33.19658	46.92960	61.28846	76.35131	92.21544	109.00585
15.6	30.00000	39.90267	50.12714	60.70076	71.65514	83.02708	94.85990	107.20518	120.12520	133.69637	148.01428
15.5	30.00000	39.90267	50.12714	60.70076	71.65514	83.02708	94.85990	107.20518	120.12520	133.69641	148.01456
16.5	92.00000	100.32242	108.90285	117.76318	126.92862	136.42848	146.29714	156.57541	167.31233	178.56775	190.41599
16.4	92.00000	100.32242	108.90285	117.76318	126.92862	136.42848	146.29714	156.57541	167.31233	178.56775	190.41600
17.4	158.00000	164.64385	171.48547	178.54164	185.83161	193.37778	201.20634	209.34835	217.84104	226.72968	236.07032
17.3	158.00000	164.64385	171.48547	178.54164	185.83161	193.37778	201.20634	209.34835	217.84104	226.72968	236.07032
18.3	228.00000	232.86638	237.87265	243.03064	248.35399	253.85847	259.56257	265.48816	271.66144	278.11421	284.88577
18.2	228.00000	232.86638	237.87265	243.03064	248.35399	253.85847	259.56257	265.48816	271.66144	278.11421	284.88577
19.2	302.00000	304.98959	308.06261	311.22613	314.48825	317.85839	321.34756	324.96876	328.73759	332.67295	336.79817
19.1	302.00000	304.98959	308.06261	311.22613	314.48825	317.85839	321.34756	324.96876	328.73759	332.67295	336.79817
20.1	380.00000	381.01316	382.05405	383.12503	384.22878	385.36843	386.54765	387.77074	389.04290	390.37039	391.76096
20.0	380.00000	381.01316	382.05405	383.12503	384.22878	385.36843	386.54765	387.77074	389.04290	390.37039	391.76096

J = 21

0.21	-462.00000	-452.94345	-448.96939	-445.93332	-443.37902	-441.13218	-439.10361	-437.24037	-435.50800	-433.88256	-432.34663
1.21	-460.00000	-452.93353	-448.96915	-445.93331	-443.37902	-441.13218	-439.10361	-437.24037	-435.50800	-433.88256	-432.34663
1.20	-460.00000	-436.22861	-424.13397	-415.03315	-407.39881	-400.69420	-394.64828	-389.10076	-383.94748	-379.11619	-374.55435
2.20	-454.00000	-435.94866	-424.12254	-415.03249	-407.39877	-400.69420	-394.64828	-389.10076	-383.94748	-379.11619	-374.55435
2.19	-454.00000	-423.42690	-402.04594	-386.73188	-374.03013	-362.91667	-352.91778	-343.75931	-335.26435	-327.31064	-319.80945
3.19	-444.00000	-420.85945	-401.82397	-386.71337	-374.02844	-362.91650	-352.91776	-343.75931	-335.26435	-327.31064	-319.80945
3.18	-444.00000	-414.62256	-384.03151	-361.48988	-343.51267	-327.96265	-314.03332	-301.30890	-289.53054	-278.52169	-268.15499
4.18	-430.00000	-405.91759	-381.97440	-361.20874	-343.47800	-327.95842	-314.03281	-301.30884	-289.53054	-278.52169	-268.15499
4.17	-430.00000	-404.40514	-371.43124	-340.73597	-316.40374	-296.11316	-278.17826	-261.88261	-246.84630	-232.82572	-219.64914
5.17	-412.00000	-388.63396	-363.20419	-338.40674	-315.97016	-296.04379	-278.16789	-261.88115	-246.84611	-232.82569	-219.64914
5.16	-412.00000	-388.47612	-359.67922	-325.98762	-294.38490	-268.09161	-245.69884	-225.69205	-207.35957	-190.33151	-174.37305
6.16	-390.00000	-367.58745	-342.94652	-316.92095	-291.27651	-267.36014	-245.55669	-225.66727	-207.35563	-190.33094	-174.37298
6.15	-390.00000	-367.57822	-342.39962	-312.53513	-278.65950	-245.90980	-217.58143	-193.18847	-171.32311	-151.20595	-132.44526
7.15	-364.00000	-342.50990	-319.22915	-294.12799	-267.79353	-241.46290	-216.30643	-192.89800	-171.26586	-151.19597	-132.44373
7.14	-364.00000	-342.50955	-319.18188	-293.37571	-263.54689	-229.85632	-196.09488	-165.74038	-139.35770	-115.76448	-94.05777
8.14	-334.00000	-313.43847	-291.46806	-267.94047	-242.80610	-216.41713	-189.64770	-163.52898	-138.77734	-115.63767	-94.03400
8.13	-334.00000	-313.43846	-291.46533	-267.86659	-242.09204	-212.91173	-179.86433	-145.50804	-113.30428	-84.89287	-59.62370
9.13	-300.00000	-280.41270	-259.68638	-237.70238	-214.32049	-189.45550	-163.27392	-136.36217	-109.59656	-83.77809	-59.35674
9.12	-300.00000	-280.41270	-259.68627	-237.69743	-214.24864	-188.90982	-160.72757	-128.80801	-94.47006	-60.83829	-30.40883
10.12	-262.00000	-243.45419	-223.96578	-203.44433	-181.77627	-158.82672	-134.48002	-108.76521	-82.02861	-54.96283	-28.38338
10.11	-262.00000	-243.45419	-223.96578	-203.44409	-181.77122	-158.77327	-134.12234	-107.11025	-76.70354	-43.08437	-8.67481
11.11	-220.00000	-202.57488	-184.35776	-165.27657	-145.24352	-124.15020	-101.86567	-78.25431	-53.24995	-27.00257	0.00000
11.10	-220.00000	-202.57488	-184.35776	-165.27656	-145.24325	-124.14646	-101.83251	-78.04637	-52.27418	-23.57651	8.67481
12.10	-174.00000	-157.78185	-140.89250	-123.27273	-104.85208	-85.54548	-65.24864	-43.83379	-21.15422	2.91762	28.38338
12.9	-174.00000	-157.78185	-140.89250	-123.27273	-104.85207	-85.54528	-65.24640	-43.81585	-21.04474	3.44659	30.40883
13.9	-124.00000	-109.07950	-93.58857	-77.47778	-60.68912	-43.15377	-24.78885	-5.49289	14.85962	36.42114	59.35674
13.8	-124.00000	-109.07950	-93.58857	-77.47778	-60.68912	-43.15377	-24.78873	-5.49175	14.86829	36.47410	59.62370
14.8	-70.00000	-56.47069	-42.45795	-27.92029	-12.80949	2.93113	19.36897	36.58535	54.68005	73.77755	94.03400
14.7	-70.00000	-56.47069	-42.45795	-27.92029	-12.80949	2.93113	19.36898	36.58541	54.68056	73.78136	94.05777
15.7	-12.00000	0.04265	12.49136	25.38079	38.75099	52.64877	67.12939	82.25892	98.11753	114.80423	132.44373
15.6	-12.00000	0.04265	12.49136	25.38079	38.75099	52.64877	67.12939	82.25892	98.11755	114.80443	132.44526
16.6	50.00000	60.45917	71.25382	82.41245	93.96796	105.95863	118.42947	131.43401	145.03681	159.31691	174.37298
16.5	50.00000	60.45917	71.25382	82.41245	93.96796	105.95863	118.42947	131.43401	145.03681	159.31692	174.37305
17.5	116.00000	124.77793	133.82547	143.16549	152.82431	162.83253	173.22603	184.04735	195.34760	207.18909	219.64914
17.4	116.00000	124.77793	133.82547	143.16549	152.82431	162.83253	173.22603	184.04735	195.34760	207.18909	219.64914
18.4	186.00000	192.99822	200.20346	207.63323	215.30772	223.25032	231.48846	240.05459	248.98764	258.33493	268.15499
18.3	186.00000	192.99822	200.20346	207.63323	215.30772	223.25032	231.48846	240.05459	248.98764	258.33493	268.15499
19.3	260.00000	265.11955	270.38566	275.81075	281.40907	287.19717	293.19441	299.42369	305.91245	312.69399	319.80945
19.2	260.00000	265.11955	270.38566	275.81075	281.40907	287.19717	293.19441	299.42369	305.91245	312.69399	319.80945
20.2	338.00000	341.14150	344.37048	347.69432	351.12154	354.66200	358.32724	362.13090	366.08932	370.22231	374.55435
20.1	338.00000	341.14150	344.37048	347.69432	351.12154	354.66200	358.32724	362.13090	366.08932	370.22231	374.55435
21.1	420.00000	421.06380	422.15668	423.28111	424.43991	425.63636	426.87429	428.15824	429.49364	430.88706	432.34663
21.0	420.00000	421.06380	422.15668	423.28111	424.43991	425.63636	426.87429	428.15824	429.49364	430.88706	432.34663

J = 22

0.22	-506.00000	-496.49310	-492.33601	-489.15812	-486.48414	-484.13183	-482.00789	-480.05694	-478.24292	-476.54078	-474.93231
1.22	-504.00000	-496.48632	-492.33588	-489.15811	-486.48414	-484.13183	-482.00789	-480.05694	-478.24292	-476.54078	-474.93231
1.21	-504.00000	-478.82898	-466.22439	-456.70271	-448.71079	-441.69060	-435.35913	-429.54890	-424.15101	-419.08990	-414.31064
2.21	-498.00000	-478.62434	-466.21763	-456.70237	-448.71077	-441.69060	-435.35913	-429.54890	-424.15101	-419.08990	-414.31064
2.20	-498.00000	-464.89902	-442.79030	-426.82291	-413.53541	-401.90014	-391.42789	-381.83338	-372.93208	-364.59640	-356.73370
3.20	-488.00000	-462.80579	-442.64888	-426.81301	-413.53463	-401.90007	-391.42789	-381.83338	-372.93208	-364.59640	-356.73370
3.19	-488.00000	-455.40846	-423.13099	-399.89573	-381.16988	-364.90911	-350.32576	-336.99620	-324.65278	-313.11205	-302.24148
4.19	-474.00000	-447.49092	-421.65977	-399.73233	-381.15272	-364.90730	-350.32557	-336.99618	-324.65278	-313.11205	-302.24148
4.18	-474.00000	-445.43912	-409.05572	-377.04751	-352.04541	-330.95473	-312.21511	-295.15736	-279.40413	-264.70647	-250.88714
5.18	-456.00000	-430.16302	-402.18849	-375.50727	-351.80955	-330.92246	-312.21093	-295.15685	-279.40407	-264.70646	-250.88714
5.17	-456.00000	-429.91838	-397.30663	-360.31507	-327.46048	-300.56215	-277.37064	-256.49705	-237.31606	-219.47638	-202.74327
6.17	-434.00000	-409.17743	-381.77801	-353.13331	-325.49598	-300.18248	-277.30770	-256.48755	-237.31475	-219.47621	-202.74326
6.16	-434.00000	-409.16147	-380.87912	-346.79057	-309.47102	-275.31472	-246.46742	-221.34618	-198.59240	-177.56349	-157.91274
7.16	-408.00000	-384.15230	-358.17990	-330.15183	-301.06430	-272.55534	-245.82463	-221.22236	-198.57141	-177.56033	-157.91232
7.15	-408.00000	-384.15162	-358.09140	-328.82816	-294.51261	-257.00344	-221.45095	-190.60907	-163.64981	-139.20395	-116.55126
8.15	-378.00000	-355.11544	-330.55609	-304.15094	-275.94463	-246.62049	-217.42704	-189.50546	-163.40890	-139.15915	-116.54406
8.14	-378.00000	-355.11542	-330.55036	-304.00048	-274.58102	-240.73190	-203.34116	-166.56698	-133.72195	-104.94546	-78.94033
9.14	-344.00000	-322.11301	-298.87351	-274.13459	-247.73560	-219.66015	-190.33511	-160.70716	-131.85068	-104.48893	-78.84883
9.13	-344.00000	-322.11301	-298.87324	-274.12317	-247.57524	-218.51519	-185.59150	-148.77144	-111.18161	-76.44330	-45.82303
10.13	-306.00000	-285.17100	-263.22250	-240.04333	-215.49226	-189.41186	-161.72515	-132.65386	-102.88044	-73.37680	-44.99183
10.12	-306.00000	-285.17100	-263.22249	-240.04270	-215.47938	-189.28005	-160.89228	-129.18676	-93.43833	-55.61202	-19.28739
11.12	-264.00000	-244.30382	-223.66510	-201.99617	-179.18845	-155.10661	-129.59361	-102.52519	-73.96379	-44.35235	-14.51698
11.11	-264.00000	-244.30382	-223.66510	-201.99614	-179.18769	-155.09598	-129.50249	-101.98334	-71.64421	-37.39782	0.00000
12.11	-218.00000	-199.51997	-180.23793	-160.08212	-138.96624	-116.78466	-93.40669	-68.67532	-42.43374	-14.62609	14.51698
12.10	-218.00000	-199.51997	-180.23793	-160.08212	-138.96620	-116.78402	-93.39955	-68.61992	-42.11193	-13.19096	19.28739
13.10	-168.00000	-150.82474	-132.96344	-114.35610	-94.93186	-74.60587	-53.27485	-30.81113	-7.05765	18.16336	44.99183
13.9	-168.00000	-150.82474	-132.96344	-114.35610	-94.93186	-74.60584	-53.27443	-30.80700	-7.02724	18.34052	45.82303
14.9	-114.00000	-98.22157	-81.85606	-64.85282	-47.15267	-28.68569	-9.36818	10.90178	32.25038	54.83546	78.84883
14.8	-114.00000	-98.22157	-81.85606	-64.85282	-47.15267	-28.68569	-9.36816	10.90201	32.25250	54.85080	78.94033
15.8	-56.00000	-41.71277	-26.92544	-11.59520	4.32762	20.90125	38.19471	56.29100	75.29152	95.32253	116.54406
15.7	-56.00000	-41.71277	-26.92544	-11.59520	4.32762	20.90125	38.19471	56.29101	75.29163	95.32352	116.55126
16.7	6.00000	18.70005	31.82176	45.40105	59.47943	74.10532	89.33583	105.23906	121.89751	139.41272	157.91232
16.6	6.00000	18.70005	31.82176	45.40105	59.47943	74.10532	89.33583	105.23906	121.89751	139.41277	157.91274
17.6	72.00000	83.01574	94.38079	106.12482	118.28204	130.89223	144.00214	157.66733	171.95469	186.94602	202.74326
17.5	72.00000	83.01574	94.38079	106.12482	118.28204	130.89223	144.00214	157.66733	171.95469	186.94603	202.74327
18.5	142.00000	151.23347	160.74822	170.56810	180.72056	191.23749	202.15628	213.52123	225.38556	237.81407	250.88714
18.4	142.00000	151.23347	160.74822	170.56810	180.72056	191.23749	202.15628	213.52123	225.38556	237.81407	250.88714
19.4	216.00000	223.35262	230.92149	238.72494	246.78403	255.12321	263.77109	272.76158	282.13527	291.94155	302.24148
19.3	216.00000	223.35262	230.92149	238.72494	246.78403	255.12321	263.77109	272.76158	282.13527	291.94155	302.24148
20.3	294.00000	299.37272	304.89869	310.59089	316.46422	322.53598	328.82642	335.35946	342.16378	349.27420	356.73370
20.2	294.00000	299.37272	304.89869	310.59089	316.46422	322.53598	328.82642	335.35946	342.16378	349.27420	356.73370
21.2	376.00000	379.29342	382.67835	386.16252	389.75485	393.46564	397.30696	401.29309	405.44112	409.77178	414.31064
21.1	376.00000	379.29342	382.67835	386.16252	389.75485	393.46564	397.30696	401.29309	405.44112	409.77178	414.31064
22.1	462.00000	463.11444	464.25931	465.43720	466.65104	467.90429	469.20094	470.54575	471.94438	473.40374	474.93231
22.0	462.00000	463.11444	464.25931	465.43720	466.65104	467.90429	469.20094	470.54575	471.94438	473.40374	474.93231

J = 23

0.23	-552.00000	-542.04341	-537.70274	-534.38297	-531.58930	-529.13151	-526.91219	-524.87353	-522.97785	-521.19902	-519.51800
1.23	-550.00000	-542.03879	-537.70267	-534.38297	-531.58930	-529.13151	-526.91219	-524.87353	-522.97785	-521.19902	-519.51800
1.22	-550.00000	-523.44119	-510.31686	-500.37302	-492.02325	-484.68735	-478.07027	-471.99726	-466.35470	-461.06374	-456.06704
2.22	-544.00000	-523.29284	-510.31288	-500.37285	-492.02324	-484.68735	-478.07027	-471.99726	-466.35470	-461.06374	-456.06704
2.21	-544.00000	-508.38837	-485.55581	-468.91969	-455.04342	-442.88542	-431.93935	-421.90847	-412.60060	-403.88278	-395.65843
3.21	-534.00000	-506.71039	-485.46682	-468.91445	-455.04306	-442.88539	-431.93935	-421.90847	-412.60060	-403.88278	-395.65843
3.20	-534.00000	-498.07903	-464.31320	-440.34171	-420.84085	-403.86257	-388.62286	-374.68693	-361.77762	-349.70441	-338.32950
4.20	-520.00000	-490.96712	-463.28798	-440.24841	-420.83249	-403.86181	-388.62279	-374.68693	-361.77762	-349.70441	-338.32950
4.19	-520.00000	-488.25508	-448.57493	-415.51024	-389.76337	-367.82569	-348.26700	-330.44191	-313.96904	-298.59252	-284.12914
5.19	-502.00000	-473.56395	-443.00449	-414.52747	-389.63822	-367.81099	-348.26534	-330.44173	-313.96902	-298.59252	-284.12914
5.18	-502.00000	-473.19419	-436.50432	-396.54473	-362.77619	-335.17415	-311.10092	-289.33119	-269.29077	-250.63408	-233.12291
6.18	-480.00000	-452.63934	-422.34720	-391.09559	-361.59364	-334.98377	-311.07382	-289.32764	-269.29034	-250.63403	-233.12291
6.17	-480.00000	-452.61248	-420.92618	-382.39823	-342.12452	-307.05840	-277.60491	-251.61503	-227.91493	-205.95244	-185.40146
7.17	-454.00000	-427.67333	-398.84779	-367.76973	-335.97329	-305.45467	-277.29593	-251.56118	-227.90748	-205.95147	-105.40135
7.16	-454.00000	-427.67206	-398.68799	-365.56039	-326.54437	-285.81515	-249.20522	-217.89618	-190.14562	-164.73747	-141.09626
8.16	-424.00000	-398.67608	-371.37505	-341.91208	-310.51208	-278.30935	-246.89862	-217.38071	-190.05045	-164.72230	-141.09416
8.15	-424.00000	-398.67604	-371.36348	-341.61935	-308.07201	-269.24793	-228.14690	-189.95587	-156.76598	-127.35576	-100.41944
9.15	-390.00000	-365.70040	-339.80620	-312.13718	-282.52506	-251.10398	-218.68329	-186.57385	-155.90410	-127.18089	-100.38974
9.14	-390.00000	-365.70040	-339.80561	-312.11204	-282.18635	-248.87567	-210.73074	-169.55241	-129.93472	-94.84451	-63.82640
10.14	-352.00000	-328.77729	-304.23527	-278.23779	-250.61119	-221.18734	-189.99928	-157.58808	-125.02035	-93.42833	-63.51500
10.13	-352.00000	-328.77729	-304.23525	-278.23624	-250.58013	-220.88260	-188.21560	-151.14392	-110.32907	-69.62880	-32.68479
11.13	-310.00000	-287.92398	-264.73600	-240.32994	-214.57191	-187.29488	-158.32376	-127.60301	-95.44951	-62.69561	-30.43567
11.12	-310.00000	-287.92398	-264.73600	-240.32987	-214.56982	-187.26663	-158.09130	-126.31451	-90.60245	-50.65320	-9.36915
12.12	-264.00000	-243.15063	-221.35251	-198.51932	-174.54585	-149.30143	-122.62386	-94.33027	-64.29558	-32.65156	0.00000
12.11	-264.00000	-243.15063	-221.35251	-198.51932	-174.54574	-149.29949	-122.60276	-94.17306	-63.43954	-29.24860	9.36915

13.11	-214.00000	-194.46354	-174.11173	-152.87248	-130.65993	-107.37011	-82.87490	-57.01499	-29.60093	-0.45636	30.43567
13.10	-214.00000	-194.46354	-174.11173	-152.87248	-130.65992	-107.37000	-82.87347	-57.00131	-29.50411	0.07453	32.68479
14.10	-160.00000	-141.86680	-123.03094	-103.43136	-82.99631	-61.64047	-39.26072	-15.73024	9.10953	35.45322	63.51500
14.9	-160.00000	-141.86680	-123.03094	-103.43136	-82.99631	-61.64046	-39.26065	-15.72934	9.11729	35.50852	63.82640
15.9	-102.00000	-85.36319	-68.12170	-50.22353	-31.60811	-12.20415	8.07338	29.32742	51.68487	75.30513	100.38974
15.8	-102.00000	-85.36319	-68.12170	-50.22353	-31.60811	-12.20415	8.07339	29.32747	51.68536	75.30936	100.41944
16.8	-40.00000	-24.95461	-9.39195	6.73217	23.46896	40.87832	59.03112	78.01231	97.92531	118.89840	141.09416
16.7	-40.00000	-24.95461	-9.39195	6.73217	23.46896	40.87832	59.03112	78.01231	97.92533	118.89865	141.09626
17.7	26.00000	39.35757	53.15265	67.42246	82.21000	97.56538	113.54759	130.22693	147.68837	166.03625	185.40135
17.6	26.00000	39.35757	53.15265	67.42246	82.21000	97.56538	113.54759	130.22693	147.68837	166.03626	185.40146
18.6	96.00000	107.57236	119.50800	131.83775	144.59715	157.82751	171.57736	185.90431	200.87766	216.58209	233.12291
18.5	96.00000	107.57236	119.50800	131.83775	144.59715	157.82751	171.57736	185.90431	200.87766	216.58209	233.12291
19.5	170.00000	179.68903	189.67107	199.97096	210.61727	221.64320	233.08765	244.99672	257.42574	270.44206	284.12914
19.4	170.00000	179.68903	189.67107	199.97096	210.61727	221.64320	233.08765	244.99672	257.42574	270.44206	284.12914
20.4	248.00000	255.70702	263.63957	271.81675	280.26053	288.99639	298.05417	307.46919	317.28376	327.54934	338.32950
20.3	248.00000	255.70702	263.63957	271.81675	280.26053	288.99639	298.05417	307.46919	317.28376	327.54934	338.32950
21.3	330.00000	335.62590	341.41174	347.37106	353.51943	359.87489	366.45856	373.29543	380.41539	387.85478	395.65843
21.2	330.00000	335.62590	341.41174	347.37106	353.51943	359.87489	366.45856	373.29543	380.41539	387.85478	395.65843
22.2	416.00000	419.44534	422.98622	426.63073	430.38816	434.26929	438.28671	442.45533	446.79298	451.32132	456.06704
22.1	416.00000	419.44534	422.98622	426.63073	430.38816	434.26929	438.28671	442.45533	446.79298	451.32132	456.06704
23.1	506.00000	507.16508	508.36194	509.59328	510.86218	512.17222	513.52759	514.93325	516.39513	517.92042	519.51800
23.0	506.00000	507.16508	508.36194	509.59328	510.86218	512.17222	513.52759	514.93325	516.39513	517.92042	519.51800

J = 24

0.24	-600.00000	-589.59419	-585.06956	-581.60788	-578.69450	-576.13122	-573.81652	-571.69013	-569.71280	-567.85727	-566.10370
1.24	-598.00000	-589.59106	-585.06952	-581.60788	-578.69450	-576.13122	-573.81652	-571.69013	-569.71280	-567.85727	-566.10370
1.23	-598.00000	-570.06282	-556.41082	-546.04395	-537.33612	-529.68441	-522.78164	-516.44581	-510.55854	-505.03770	-499.82352
2.23	-592.00000	-569.95605	-556.40849	-546.04387	-537.33612	-529.63441	-522.78164	-516.44581	-510.55854	-505.03770	-499.82352
2.22	-592.00000	-553.90368	-530.33653	-513.02068	-498.55361	-485.87222	-474.45195	-463.98443	-454.26981	-445.16969	-436.58356
3.22	-582.00000	-552.57972	-530.28114	-513.01793	-498.55345	-485.87221	-474.45195	-463.98443	-454.26981	-445.16969	-436.58356
3.21	-582.00000	-542.65585	-507.57312	-482.81527	-462.52169	-444.82154	-428.92380	-414.38053	-400.90465	-388.29845	-376.41881
4.21	-568.00000	-536.35273	-506.87444	-482.76281	-462.51766	-444.82122	-428.92377	-414.38053	-400.90465	-388.29845	-376.41881
4.20	-568.00000	-532.85505	-490.05971	-456.10407	-429.53213	-406.71696	-386.33034	-367.73434	-350.53978	-334.48297	-319.37448
5.20	-550.00000	-518.83769	-485.67554	-455.49530	-429.46711	-406.71038	-386.32969	-367.73428	-350.53977	-334.48297	-319.37448
5.19	-550.00000	-518.29209	-477.32235	-434.80799	-400.30154	-371.87804	-346.86969	-324.18658	-303.27975	-283.80213	-265.51023
6.19	-528.00000	-497.97114	-464.66528	-430.85392	-399.61737	-371.78523	-346.85829	-324.18528	-303.27962	-283.80212	-265.51023
6.18	-528.00000	-497.92709	-462.50502	-419.47645	-376.84074	-341.12333	-310.90431	-283.95322	-259.27416	-236.36507	-214.90692
7.18	-502.00000	-473.07105	-441.22964	-407.00833	-372.59551	-340.23963	-310.76129	-283.93297	-259.27159	-236.36478	-214.90689
7.17	-502.00000	-473.06876	-440.95085	-403.51535	-359.83428	-316.64185	-279.41701	-247.45936	-218.76226	-192.33154	-167.67826
8.17	-472.00000	-444.11893	-413.91880	-381.22103	-346.55083	-311.59261	-278.18567	-247.23046	-218.72614	-192.32658	-167.67767
8.16	-472.00000	-444.11885	-413.89624	-380.67624	-342.47234	-298.70653	-254.79406	-215.92195	-182.25878	-151.97120	-123.99551
9.16	-438.00000	-411.17374	-382.47938	-351.69872	-318.68480	-283.84071	-248.46225	-214.14335	-181.88797	-151.90761	-123.98629
9.15	-438.00000	-411.17373	-382.47809	-351.64579	-318.00713	-279.83926	-236.38576	-191.83165	-151.28562	-115.92425	-84.16790
10.15	-400.00000	-374.27214	-347.00014	-318.01766	-287.11528	-254.14397	-219.35869	-183.74061	-148.69593	-115.32783	-84.05887
10.14	-400.00000	-374.27214	-347.00008	-318.01398	-287.04424	-253.48272	-215.86922	-173.11295	-128.18610	-86.06639	-49.00481
11.14	-358.00000	-333.43462	-307.56729	-280.26997	-251.37802	-220.68966	-188.03717	-153.53507	-117.89479	-82.34546	-48.06714
11.13	-358.00000	-333.43462	-307.56729	-280.26978	-251.37264	-220.61902	-187.48552	-150.75823	-109.06905	-64.17113	-20.72222
12.13	-312.00000	-288.67320	-264.23360	-238.57794	-211.57832	-183.07358	-152.86626	-120.76613	-86.76658	-51.34115	-15.53295
12.12	-312.00000	-288.67320	-264.23360	-238.57793	-211.57801	-183.06806	-152.80819	-120.35498	-84.71729	-44.40211	0.00000
13.12	-262.00000	-239.99536	-217.03121	-193.02162	-167.86312	-141.42880	-113.56025	-84.06163	-52.72213	-19.43605	15.53295
13.11	-262.00000	-239.99536	-217.03121	-193.02162	-167.86310	-141.42847	-113.55572	-84.01988	-52.44097	-18.01500	20.72222
14.11	-208.00000	-187.40594	-165.98071	-143.65143	-120.33193	-95.91879	-70.28567	-43.27536	-14.69144	15.69454	48.06714
14.10	-208.00000	-187.40594	-165.98071	-143.65143	-120.33193	-95.91877	-70.28540	-43.27219	-14.66430	15.87429	49.00481
15.10	-150.00000	-130.90822	-111.09578	-90.50034	-69.04898	-46.65534	-23.21573	1.39655	27.33936	54.81199	84.05887
15.9	-150.00000	-130.90822	-111.09578	-90.50034	-69.04898	-46.65534	-23.21572	1.39674	27.34134	54.82831	84.16790
16.9	-88.00000	-70.50446	-52.38588	-33.59082	-14.05718	6.28796	27.53126	49.77721	73.15398	97.82205	123.98629
16.8	-88.00000	-70.50446	-52.38588	-33.59082	-14.05718	6.28796	27.53126	49.77722	73.15409	97.82318	123.99551
17.8	-22.00000	-6.19626	10.14232	27.06137	44.61369	62.86098	81.87606	101.74608	122.57679	144.49894	167.67767
17.7	-22.00000	-6.19626	10.14232	27.06137	44.61369	62.86098	81.87606	101.74608	122.57680	144.49900	167.67826
18.7	48.00000	62.01519	76.48395	91.44482	106.94233	123.02829	139.76368	157.22108	175.48803	194.67189	214.90689
18.6	48.00000	62.01519	76.48395	91.44482	106.94233	123.02829	139.76368	157.22108	175.48803	194.67190	214.90692
19.6	122.00000	134.12904	146.63541	159.55114	172.91311	186.76419	201.15468	216.14430	231.80482	248.22386	265.51023
19.5	122.00000	134.12904	146.63541	159.55114	172.91311	186.76419	201.15468	216.14430	231.80482	248.22386	265.51023
20.5	200.00000	210.14462	220.59402	231.37403	242.51436	254.04953	266.01996	278.47356	291.46778	305.07255	319.37448
20.4	200.00000	210.14462	220.59402	231.37403	242.51436	254.04953	266.01996	278.47356	291.46778	305.07255	319.37448
21.4	282.00000	290.06143	298.35768	306.90864	315.73718	324.86982	334.33762	344.17733	354.43298	365.15810	376.41881
21.3	282.00000	290.06143	298.35768	306.90864	315.73718	324.86982	334.33762	344.17733	354.43298	365.15810	376.41881
22.3	368.00000	373.87907	379.92479	386.15126	392.57469	399.21388	406.09083	413.23156	420.66723	428.43567	436.58356
22.2	368.00000	373.87907	379.92479	386.15126	392.57469	399.21388	406.09083	413.23156	420.66723	428.43567	436.58356
23.2	458.00000	461.59725	465.29409	469.09894	473.02149	477.07296	481.26649	485.61760	490.14490	494.87092	499.82352
23.1	458.00000	461.59725	465.29409	469.09894	473.02149	477.07296	481.26649	485.61760	490.14490	494.87092	499.82352
24.1	552.00000	553.21572	554.46458	555.74936	557.07331	558.44015	559.85424	561.32076	562.84589	564.43712	566.10370
24.0	552.00000	553.21572	554.46458	555.74936	557.07331	558.44015	559.85424	561.32076	562.84589	564.43712	566.10370

J = 25

0.25	-650.00000	-639.14534	-634.43645	-630.83283	-627.79974	-625.13096	-622.72087	-620.50675	-618.44775	-616.51552	-614.68940
1.25	-648.00000	-639.14322	-634.43642	-630.83283	-627.79974	-625.13096	-622.72087	-620.50675	-618.44775	-616.51552	-614.68940
1.24	-648.00000	-618.69182	-604.50589	-593.71539	-584.64935	-576.68174	-569.49321	-562.89451	-556.76251	-551.01175	-545.58008
2.24	-642.00000	-618.61547	-604.50454	-593.71535	-584.64935	-576.68174	-569.49321	-562.89451	-556.76251	-551.01175	-545.58008
2.23	-642.00000	-601.44936	-577.12818	-559.12486	-544.06561	-530.86031	-518.96552	-508.06115	-497.93959	-488.45705	-479.50904
3.23	-632.00000	-600.41975	-577.09402	-559.12343	-544.06554	-530.86030	-518.96552	-508.06115	-497.93959	-488.45705	-479.50904
3.22	-632.00000	-589.16169	-552.89903	-527.30799	-506.20993	-487.78499	-471.22794	-456.07655	-442.03354	-428.89392	-416.50921
4.22	-618.00000	-583.65485	-552.43180	-527.27888	-506.20801	-487.78486	-471.22793	-456.07655	-442.03354	-428.89392	-416.50921
4.21	-618.00000	-579.24636	-533.57179	-498.80010	-471.33513	-447.62305	-426.40279	-407.03327	-389.11538	-372.37711	-356.62262
5.12	-600.00000	-565.98601	-530.22520	-498.43210	-471.30195	-447.62015	-426.40254	-407.03325	-389.11538	-372.37711	-356.62262
5.20	-600.00000	-565.19950	-519.82392	-475.20084	-439.98590	-410.64184	-384.66548	-361.05837	-339.28027	-318.97870	-299.90392
6.20	-578.00000	-545.17097	-508.74727	-472.45337	-439.60251	-410.59766	-384.66079	-361.05791	-339.28023	-318.97870	-299.90392
6.19	-578.00000	-545.10040	-505.58982	-458.16817	-413.77330	-377.43529	-346.30597	-318.33770	-292.66063	-268.79638	-246.42592
7.19	-552.00000	-520.34344	-485.32436	-447.90269	-411.00420	-376.96839	-346.24178	-318.32985	-292.65976	-268.79630	-246.42591
7.18	-552.00000	-520.33942	-484.85387	-442.67421	-394.62283	-349.74348	-312.00360	-279.19300	-249.45508	-221.96836	-196.28857
8.18	-522.00000	-491.44243	-458.18120	-422.08102	-384.11875	-346.57968	-311.38182	-279.09536	-249.44183	-221.96678	-196.28841
8.17	-522.00000	-491.44227	-458.13871	-421.11097	-377.76027	-329.45403	-283.70965	-244.42954	-210.02861	-178.70855	-149.63598
9.17	-488.00000	-458.53182	-426.88765	-392.80845	-356.22207	-317.95011	-279.82458	-243.56048	-209.87717	-178.68639	-149.63323
9.16	-488.00000	-458.53181	-426.88495	-392.70155	-354.93871	-311.34546	-262.99450	-216.25396	-175.35915	-139.43949	-106.70005
10.16	-450.00000	-421.65460	-391.51288	-359.37225	-324.98818	-288.28920	-249.89970	-211.31027	-174.11961	-139.20481	-106.66386
10.15	-450.00000	-421.65460	-391.51275	-359.36396	-324.83361	-286.94434	-243.70898	-195.56693	-147.89194	-105.36184	-67.97779
11.15	-408.00000	-380.83494	-352.15560	-321.80778	-289.58984	-255.26735	-218.73552	-180.42419	-141.54096	-103.59517	-67.62020
11.14	-408.00000	-380.83494	-352.15560	-321.80730	-289.57666	-255.10076	-217.51945	-175.05794	-127.44269	-79.04965	-34.93596
12.14	-362.00000	-336.08703	-308.87844	-280.25118	-250.05009	-218.07721	-184.10123	-147.97245	-109.94276	-70.96680	-32.46180
12.13	-362.00000	-336.08703	-308.87844	-280.25116	-250.04924	-218.06243	-183.95184	-146.98208	-105.58190	-58.83903	-10.05705
13.13	-312.00000	-287.41965	-261.71958	-234.79794	-206.53053	-176.76277	-145.29958	-111.90802	-76.39355	-38.85104	0.00000
13.12	-312.00000	-287.41965	-261.71958	-234.79794	-206.53049	-176.76177	-145.28626	-111.79005	-75.64815	-35.49391	10.05705
14.12	-258.00000	-234.83852	-210.70337	-185.50832	-159.15051	-131.50517	-102.41794	-71.69471	-39.09795	-4.39268	32.46180
14.11	-258.00000	-234.83852	-210.70337	-185.50832	-159.15050	-131.50512	-102.41703	-71.68437	-39.01309	-3.86457	34.93596
15.11	-200.00000	-178.34746	-155.84597	-132.42160	-107.98734	-82.43946	-55.65227	-27.47048	2.30142	33.90797	67.62020
15.10	-200.00000	-178.34746	-155.84597	-132.42160	-107.98733	-82.43945	-55.65222	-27.46979	2.30861	33.96488	67.97779
16.10	-138.00000	-117.94914	-97.15851	-75.56439	-53.09239	-29.65479	-5.14673	20.55918	47.62072	76.23747	106.66386
16.9	-138.00000	-117.94914	-97.15851	-75.56439	-53.09239	-29.65479	-5.14673	20.55922	47.62119	76.24206	106.70005
17.9	-72.00000	-53.64545	-34.64888	-14.95536	5.49888	26.78852	49.00213	72.24613	96.65053	122.37736	149.63323
17.8	-72.00000	-53.64545	-34.64888	-14.95536	5.49888	26.78852	49.00213	72.24613	96.65056	122.37765	149.63598
18.8	-2.00000	14.56224	31.67723	49.39206	67.76120	86.84817	106.72792	127.48992	149.24253	172.11932	196.28841
18.7	-2.00000	14.56224	31.67723	49.39206	67.76120	86.84817	106.72792	127.48992	149.24253	172.11934	196.28857
19.7	72.00000	86.67289	101.81558	117.46797	133.67610	150.49357	167.98336	186.22038	205.29491	225.[illegible]1745	246.42591
19.6	72.00000	86.67289	101.81558	117.46797	133.67610	150.49357	167.98336	186.22038	205.29491	225.31746	246.42592
20.6	150.00000	162.68575	175.76299	189.26493	203.22979	217.70203	232.73376	248.38682	264.73549	281.87037	299.90392
20.5	150.00000	162.68575	175.76299	189.26493	203.22979	217.70203	232.73376	248.38682	264.73549	281.87037	299.90392
21.5	232.00000	242.60022	253.51704	264.77727	276.41178	288.45640	300.95308	313.95154	327.51139	341.70516	356.62262
21.4	232.00000	242.60022	253.51704	264.77727	276.41178	288.45640	300.95308	313.95154	327.51139	341.70516	356.62262
22.4	318.00000	326.41584	335.07582	344.00061	353.21397	362.74347	372.62139	382.88592	393.58283	404.76770	416.50921
22.3	318.00000	326.41584	335.07582	344.00061	353.21397	362.74347	372.62139	382.88592	393.58283	404.76770	416.50921
23.3	408.00000	414.13226	420.43786	426.93148	433.62999	440.55294	447.72320	455.16785	462.91927	471.01682	479.50904
23.2	408.00000	414.13226	420.43786	426.93148	433.62999	440.55294	447.72320	455.16785	462.91927	471.01682	479.50904
24.2	502.00000	505.74917	509.60197	513.56716	517.65483	521.87665	526.24629	530.77991	535.49685	540.42059	545.58008
24.1	502.00000	505.74917	509.60197	513.56716	517.65483	521.87665	526.24629	530.77991	535.49685	540.42059	545.58008
25.1	600.00000	601.26636	602.56721	603.90545	605.28445	606.70808	608.18090	609.70828	611.29665	612.95381	614.68940
25.0	600.00000	601.26636	602.56721	603.90545	605.28445	606.70808	608.18090	609.70828	611.29665	612.95381	614.68940

J = 26

0.26	-702.00000	-690.69674	-685.80339	-682.05782	-678.90500	-676.13072	-673.62523	-671.32338	-669.18272	-667.17379	-665.27512
1.26	-700.00000	-690.69531	-685.80338	-682.05782	-678.90500	-676.13072	-673.62523	-671.32338	-669.18272	-667.17379	-665.27512
1.25	-700.00000	-669.32652	-654.60184	-643.38727	-633.96289	-625.67930	-618.20497	-611.34336	-604.96658	-598.98589	-593.33671
2.25	-694.00000	-669.27225	-654.60106	-643.38725	-633.96289	-625.67930	-618.20497	-611.34336	-604.96658	-598.98589	-593.33671
2.24	-694.00000	-651.02623	-625.92773	-607.23156	-591.57914	-577.84950	-565.47992	-554.13852	-543.60988	-533.74481	-524.43483
3.24	-684.00000	-650.23589	-625.90684	-607.23082	-591.57911	-577.84950	-565.47992	-554.13852	-543.60988	-533.74481	-524.43483
3.23	-684.00000	-637.61982	-600.27746	-573.81431	-551.90397	-532.75218	-515.53482	-499.77464	-485.16401	-471.49062	-458.60056
4.23	-670.00000	-632.88094	-599.96999	-573.79835	-551.90307	-532.75213	-515.53481	-499.77464	-485.16401	-471.49062	-458.60056
4.22	-670.00000	-627.44109	-579.15647	-543.57114	-515.16191	-490.54055	-468.48272	-448.33766	-429.69509	-412.27439	-395.87314
5.22	-652.00000	-615.01164	-576.67627	-543.35317	-515.14524	-490.53929	-468.48262	-448.33765	-429.69509	-412.27439	-395.87314
5.21	-652.00000	-613.90341	-564.08109	-517.77082	-481.78234	-451.44602	-424.48145	-399.94334	-377.29031	-356.16238	-336.30295
6.21	-630.00000	-594.23722	-554.61167	-515.93537	-481.57305	-451.42542	-424.47955	-399.94318	-377.29030	-356.16238	-336.30295
6.20	-630.00000	-594.12666	-550.17297	-498.62924	-452.98372	-415.91814	-383.77405	-354.75543	-328.06840	-303.24286	-279.95605
7.20	-604.00000	-569.48842	-531.13361	-490.49466	-451.26544	-415.67961	-383.74599	-354.75246	-328.06811	-303.24283	-279.95605
7.19	-604.00000	-569.48153	-530.36516	-483.07271	-431.17063	-385.23136	-346.85044	-313.03304	-282.20005	-253.63753	-226.92138
8.19	-574.00000	-540.64496	-504.15644	-464.50294	-423.28571	-383.37000	-346.54977	-312.99275	-282.19533	-253.63705	-226.92133
8.18	-574.00000	-540.64466	-504.07890	-462.85214	-414.01761	-361.87836	-315.11638	-275.32680	-239.96512	-207.52478	-177.32327
9.18	-540.00000	-507.77339	-473.02532	-435.45725	-395.15867	-353.53283	-312.91341	-274.92527	-239.90581	-207.51733	-177.32248
9.17	-540.00000	-507.77338	-473.01989	-435.24977	-392.86451	-343.49704	-291.08968	-243.23504	-201.99748	-165.21028	-131.34818
10.17	-502.00000	-470.92367	-437.76904	-402.29037	-364.21716	-323.65327	-281.75340	-240.49376	-201.44569	-165.12268	-131.33668
10.16	-502.00000	-470.92366	-437.76876	-402.27244	-363.89673	-321.10039	-271.80700	-219.18249	-170.13486	-127.43583	-89.33267

11.16	-460.00000	-430.12414	-398.49738	-364.93431	-329.18966	-291.00979	-250.45172	-208.42716	-166.64501	-126.67218	-89.20505
11.15	-460.00000	-430.12414	-398.49737	-364.93316	-329.15889	-290.63868	-247.97445	-199.18689	-146.50760	-96.30005	-52.15115
12.15	-414.00000	-385.39143	-355.28412	-323.53185	-289.94652	-254.28718	-216.30243	-175.97696	-133.99692	-91.84678	-51.10498
12.14	-414.00000	-385.39143	-355.28412	-323.53180	-289.94429	-254.24979	-215.94248	-173.79175	-125.84851	-73.36372	-22.14330
13.14	-364.00000	-336.73584	-308.17439	-278.19551	-246.65050	-213.35116	-178.05955	-140.51689	-100.62948	-58.87654	-16.53656
13.13	-364.00000	-336.73584	-308.17439	-278.19551	-246.65037	-213.34831	-178.02282	-140.20743	-98.83419	-51.99685	0.00000
14.13	-310.00000	-284.16402	-257.19684	-228.99706	-199.44245	-168.38292	-135.63007	-100.94621	-64.06000	-24.80227	16.53656
15.13	-306.00000	-278.90678	-250.66740	-221.18019	-190.32373	-157.95051	-123.87747	-87.87249	-49.64477	-8.89062	34.46433
15.12	-306.00000	-278.90678	-250.66740	-221.18019	-190.32373	-157.95048	-123.87689	-87.86473	-49.57092	-8.36918	37.16473
16.12	-244.00000	-218.52145	-192.03359	-164.44835	-135.66192	-105.55028	-73.96277	-40.71249	-5.56239	31.78817	71.67736
16.11	-244.00000	-218.52145	-192.03359	-164.44835	-135.66192	-105.55028	-73.96274	-40.71195	-5.55597	31.84603	72.08274
17.11	-178.00000	-154.22850	-129.56827	-103.94259	-77.26178	-49.41989	-20.28995	10.28277	42.49082	76.57939	112.86391
17.10	-178.00000	-154.22850	-129.56827	-103.94259	-77.26178	-49.41989	-20.28995	10.28280	42.49126	76.58428	112.90713
18.10	-108.00000	-86.02981	-63.27922	-39.68133	-15.15840	10.38079	37.04462	64.96366	94.29796	125.24796	158.07094
18.9	-108.00000	-86.02981	-63.27922	-39.68133	-15.15840	10.38079	37.04462	64.96366	94.29799	125.24828	158.07446
19.9	-34.00000	-13.92676	6.82782	28.32189	50.62273	73.80897	97.97348	123.22736	149.70554	177.57490	207.04646
19.8	-34.00000	-13.92676	6.82782	28.32189	50.62273	73.80897	97.97348	123.22736	149.70554	177.57492	207.04669
20.8	44.00000	62.07961	80.74855	100.05695	120.06269	140.83318	162.44779	185.00101	208.60696	233.40568	259.57249
20.7	44.00000	62.07961	80.74855	100.05695	120.06269	140.83318	162.44779	185.00101	208.60696	233.40569	259.57250
21.7	126.00000	141.98850	158.47965	175.51612	193.14709	211.42975	230.43122	250.23124	270.92573	292.63188	315.49546
21.6	126.00000	141.98850	158.47965	175.51612	193.14709	211.42975	230.43122	250.23124	270.92573	292.63188	315.49546
22.6	212.00000	225.79928	240.01856	254.69343	269.86489	285.58053	301.89616	318.87792	336.60522	355.17479	374.70650
22.5	212.00000	225.79928	240.01856	254.69343	269.86489	285.58053	301.89616	318.87792	336.60522	355.17479	374.70650
23.5	302.00000	313.51148	325.36327	337.58420	350.20743	363.27142	376.82125	390.91027	405.60244	420.97552	437.12571
23.4	302.00000	313.51148	325.36327	337.58420	350.20743	363.27142	376.82125	390.91027	405.60244	420.97552	437.12571
24.4	396.00000	405.12470	414.51219	424.18471	434.16786	444.49128	455.18972	466.30424	477.88406	489.98895	502.69273
24.3	396.00000	405.12470	414.51219	424.18471	434.16786	444.49128	455.18972	466.30424	477.88406	489.98895	502.69273
25.3	494.00000	500.63862	507.46402	514.49199	521.74070	529.23122	536.98820	545.04078	553.42386	562.17980	571.36089
25.2	494.00000	500.63862	507.46402	514.49199	521.74070	529.23122	536.98820	545.04078	553.42386	562.17980	571.36089
26.2	596.00000	600.05301	604.21773	608.50362	612.92153	617.48406	622.20594	627.10460	632.20088	637.52007	643.09340
26.1	596.00000	600.05301	604.21773	608.50362	612.92153	617.48406	622.20594	627.10460	632.20088	637.52007	643.09340
27.1	702.00000	703.36764	704.77247	706.21762	707.70672	709.24395	710.83422	712.48331	714.19818	715.98722	717.86083
27.0	702.00000	703.36764	704.77247	706.21762	707.70672	709.24395	710.83422	712.48331	714.19818	715.98722	717.86083

J = 27

0.27	-756.00000	-744.24835	-739.17039	-735.28284	-732.01028	-729.13049	-726.52961	-724.14002	-721.91770	-719.83206	-717.86083
1.27	-754.00000	-744.24739	-739.17038	-735.28284	-732.01028	-729.13049	-726.52961	-724.14002	-721.91770	-719.83206	-717.86083
1.26	-754.00000	-721.96561	-706.69848	-695.05953	-685.27669	-676.67706	-668.91688	-661.79232	-655.17075	-648.96011	-643.09340
2.26	-748.00000	-721.92723	-706.69803	-695.05952	-685.27669	-676.67706	-668.91688	-661.79232	-655.17075	-648.96011	-643.09340
2.25	-748.00000	-702.63261	-676.73309	-657.34032	-641.09397	-626.83965	-613.99506	-602.21645	-591.28062	-581.03290	-571.36089
3.25	-738.00000	-702.03291	-676.72042	-657.33994	-641.09396	-626.83965	-613.99506	-602.21645	-591.28062	-581.03290	-571.36089
3.24	-738.00000	-688.05307	-649.69608	-622.33058	-599.60273	-579.72253	-561.84403	-545.47451	-530.29586	-516.08839	-502.69273
4.24	-724.00000	-684.03872	-649.49652	-622.32192	-599.60231	-579.72251	-561.84403	-545.47451	-530.29586	-516.08839	-502.69273
4.23	-724.00000	-677.45516	-626.83869	-590.39566	-561.00590	-535.46718	-512.56889	-491.64666	-472.27831	-454.17434	-437.12571
5.23	-706.00000	-665.91822	-625.04993	-590.26877	-560.99764	-535.46664	-512.56886	-491.64666	-472.27831	-454.17434	-437.12571
5.22	-706.00000	-664.39142	-610.17141	-562.52125	-525.65558	-494.27880	-466.31324	-440.83916	-417.30833	-395.35205	-374.70650
6.22	-684.00000	-645.16865	-602.27987	-561.33583	-525.54380	-494.26936	-466.31249	-440.83911	-417.30833	-395.35205	-374.70650
6.21	-684.00000	-644.99902	-596.27016	-541.01565	-494.45716	-456.51462	-423.28767	-393.19855	-365.49341	-339.70187	-315.49546
7.21	-658.00000	-620.50386	-578.66236	-534.83052	-493.43427	-456.39609	-423.27568	-393.19745	-365.49332	-339.70186	-315.49546
7.20	-658.00000	-620.49230	-577.44773	-524.80076	-469.72643	-423.08549	-383.86641	-348.94317	-316.98349	-287.33234	-259.57250
8.20	-628.00000	-591.72479	-551.83932	-508.50649	-464.12827	-422.04541	-383.72598	-348.92700	-316.98185	-287.33219	-259.57249
8.19	-628.00000	-591.72425	-551.70202	-505.82678	-451.42250	-396.34864	-349.01900	-308.47426	-272.00659	-238.39684	-207.04669
9.19	-594.00000	-558.89714	-520.88647	-479.63901	-435.53290	-390.70108	-347.84788	-308.29667	-271.98416	-238.39441	-207.04646
9.18	-594.00000	-558.89712	-520.87589	-479.25162	-431.67655	-376.55864	-321.19763	-272.86211	-231.00057	-193.13461	-158.07446
10.18	-556.00000	-522.07830	-485.76389	-446.76058	-404.79617	-360.29301	-315.07221	-271.46004	-230.76808	-193.10328	-158.07094
10.17	-556.00000	-522.07830	-485.76329	-446.72325	-404.16295	-355.79688	-300.49166	-244.67718	-195.17324	-152.02947	-112.90713
11.17	-514.00000	-481.30136	-446.58889	-409.63988	-370.15963	-327.90957	-283.25754	-237.73865	-193.43869	-151.72147	-112.86391
11.16	-514.00000	-481.30136	-446.58887	-409.63721	-370.09106	-327.12869	-278.63215	-223.48568	-167.19123	-116.45741	-72.08274
12.16	-468.00000	-436.58571	-403.44760	-368.41226	-331.25186	-291.67822	-249.45780	-204.85961	-159.16397	-114.28697	-71.67736
12.15	-468.00000	-436.58571	-403.44760	-368.41212	-331.24635	-291.58863	-248.64589	-200.48798	-145.79035	-89.10628	-37.16473
13.15	-418.00000	-387.94332	-356.39310	-323.20809	-288.21047	-251.17140	-211.80632	-169.86736	-125.50900	-79.78105	-34.46433
13.14	-418.00000	-387.94332	-356.39310	-323.20808	-288.21013	-251.16373	-211.71118	-169.11408	-121.62585	-67.63954	-10.73911
14.14	-364.00000	-335.38194	-305.45897	-274.11241	-241.19755	-206.53400	-169.89208	-130.98635	-89.54385	-45.60265	0.00000
14.13	-364.00000	-335.38194	-305.45897	-274.11241	-241.19753	-206.53349	-169.88372	-130.89837	-88.89891	-42.31006	10.73911
14.12	-310.00000	-284.16402	-257.19684	-228.99706	-199.44244	-168.38275	-135.62722	-100.91496	-63.81604	-23.40464	22.14330
15.12	-252.00000	-227.68046	-202.37047	-175.98303	-148.41510	-119.54275	-89.21419	-57.23941	-23.37795	12.65721	51.10498
15.11	-252.00000	-227.68046	-202.37047	-175.98303	-148.41510	-119.54274	-89.21401	-57.23700	-23.35396	12.83771	52.15115
16.11	-190.00000	-167.28826	-143.70830	-119.18485	-93.62974	-66.93832	-38.98462	-9.61414	21.36670	54.20455	89.20505
16.10	-190.00000	-167.28826	-143.70830	-119.18485	-93.62974	-66.93832	-38.98461	-9.61399	21.36851	54.22162	89.33267
17.10	-124.00000	-102.98964	-81.21956	-58.62447	-35.12840	-10.64197	14.94128	41.75009	69.94298	99.71960	131.33668
17.9	-124.00000	-102.98964	-81.21956	-58.62447	-35.12840	-10.64197	14.94128	41.75009	69.94309	99.72084	131.34818
18.9	-54.00000	-34.78620	-14.91093	5.68234	27.05909	49.29593	72.48350	96.73045	122.16909	148.96359	177.32248
18.8	-54.00000	-34.78620	-14.91093	5.68234	27.05909	49.29593	72.48350	96.73045	122.16910	148.96366	177.32327
19.8	20.00000	37.32087	55.21266	73.72399	92.91099	112.83911	133.58548	155.24203	177.91992	201.75585	226.92133
19.7	20.00000	37.32087	55.21266	73.72399	92.91099	112.83911	133.58548	155.24203	177.91992	201.75585	226.92138
20.7	98.00000	113.33067	129.14749	145.49177	162.41109	179.96082	198.20603	217.22400	237.10780	257.97123	279.95605

20.6	98.00000	113.33067	129.14749	145.49177	162.41109	179.96082	198.20603	217.22400	237.10780	257.97123	279.95605
21.6	180.00000	193.24250	206.89072	220.97904	235.54708	250.64086	266.31433	282.63146	299.66910	317.52088	336.30295
21.5	180.00000	193.24250	206.89072	220.97904	235.54708	250.64086	266.31433	282.63146	299.66910	317.52088	336.30295
22.5	266.00000	277.05584	288.44013	300.18067	312.30949	324.86371	337.88687	351.43049	365.55634	380.33957	395.87314
22.4	266.00000	277.05584	288.44013	300.18067	312.30949	324.86371	337.88687	351.43049	365.55634	380.33957	395.87314
23.4	356.00000	364.77027	373.79399	383.09263	392.69086	402.61730	412.90543	423.59491	434.73321	446.37801	458.60056
23.3	356.00000	364.77027	373.79399	383.09263	392.69086	402.61730	412.90543	423.59491	434.73321	446.37801	458.60056
24.3	450.00000	456.38544	462.95094	469.71173	476.68533	483.89205	491.35566	499.10426	507.17149	515.59821	524.43483
24.2	450.00000	456.38544	462.95094	469.71173	476.68533	483.89205	491.35566	499.10426	507.17149	515.59821	524.43483
25.2	548.00000	551.90109	555.90985	560.03539	564.28817	568.68035	573.22611	577.94224	582.84885	587.97031	593.33671
25.1	548.00000	551.90109	555.90985	560.03539	564.28817	568.68035	573.22611	577.94224	582.84885	587.97031	593.33671
26.1	650.00000	651.31700	652.66984	654.06154	655.49558	656.97601	658.50756	660.09579	661.74741	663.47052	665.27512
26.0	650.00000	651.31700	652.66984	654.06154	655.49558	656.97601	658.50756	660.09579	661.74741	663.47052	665.27512

J = 28

0.28	-812.00000	-799.80011	-794.53743	-790.50789	-787.11559	-784.13029	-781.43400	-778.95668	-776.65269	-774.49034	-772.44656
1.28	-810.00000	-799.79947	-794.53743	-790.50789	-787.11559	-784.13029	-781.43400	-778.95668	-776.65269	-774.49034	-772.44656
1.27	-810.00000	-776.60808	-760.79569	-748.73212	-738.59072	-729.67500	-721.62893	-714.24140	-707.37501	-700.93439	-694.85014
2.27	-804.00000	-776.58106	-760.79544	-748.73211	-738.59072	-729.67500	-721.62893	-714.24140	-707.37501	-700.93439	-694.85014
2.26	-804.00000	-756.26535	-729.54281	-709.45079	-692.60994	-677.83063	-664.51083	-652.29487	-640.95173	-630.32130	-620.28718
3.26	-794.00000	-755.81494	-729.53518	-709.45059	-692.60994	-677.83063	-664.51083	-652.29487	-640.95173	-630.32130	-620.28718
3.25	-794.00000	-740.48253	-701.14477	-672.85441	-649.30540	-628.69560	-610.15527	-593.17593	-577.42891	-562.68707	-548.78561
4.25	-780.00000	-737.13593	-701.01679	-672.84975	-649.30521	-628.69559	-610.15527	-593.17593	-577.42891	-562.68707	-548.78561
4.24	-780.00000	-729.30755	-676.62444	-639.25808	-608.86291	-582.40133	-558.66038	-536.95963	-516.86454	-498.07660	-480.38004
5.24	-762.00000	-718.71037	-675.36507	-639.18531	-608.85888	-582.40110	-558.66036	-536.95963	-516.86454	-498.07660	-480.38004
5.23	-762.00000	-716.65318	-658.17407	-609.42836	-571.58157	-539.13289	-510.15790	-483.74409	-459.33312	-436.54682	-415.11390
6.23	-740.00000	-697.96446	-651.77528	-608.68407	-571.52298	-539.12863	-510.15760	-483.74407	-459.33312	-436.54682	-415.11390
6.22	-740.00000	-697.70942	-643.92024	-585.46508	-538.14180	-499.18699	-464.83445	-433.66192	-404.93269	-378.17138	-353.04271
7.22	-714.00000	-673.38760	-627.91932	-580.95817	-537.55302	-499.12944	-464.82943	-433.66151	-404.93267	-378.17138	-353.04271
7.21	-714.00000	-673.36862	-626.06254	-567.98899	-510.48360	-463.22469	-422.99202	-386.90276	-353.79692	-323.04802	-294.23879
8.21	-684.00000	-644.68016	-601.22579	-554.12045	-506.72357	-462.66681	-422.92830	-386.89643	-353.79637	-323.04798	-294.23879
8.20	-684.00000	-644.67918	-600.98962	-549.97981	-490.21695	-433.13624	-385.30789	-343.77910	-306.11933	-271.31123	-238.79902
9.20	-650.00000	-611.90169	-570.46499	-525.35223	-477.39934	-429.56756	-384.71585	-343.70330	-306.11110	-271.31046	-238.79896
9.19	-650.00000	-611.90165	-570.44497	-524.65624	-471.32261	-410.89889	-353.71164	-305.00891	-262.22694	-223.15865	-186.85810
10.19	-612.00000	-575.11741	-535.49242	-492.77159	-446.72936	-398.29144	-350.01187	-304.33507	-262.13315	-223.14785	-186.85706
10.18	-612.00000	-575.11741	-535.49117	-492.69668	-445.53779	-390.97732	-330.26631	-272.62257	-222.89765	-178.93481	-138.61737
11.18	-570.00000	-534.36572	-496.42619	-455.91417	-412.48300	-365.97644	-317.26435	-268.56489	-222.09960	-178.81704	-138.60336
11.17	-570.00000	-534.36572	-496.42613	-455.90827	-412.33676	-364.42806	-309.40297	-248.58591	-190.30035	-139.49291	-94.44034
12.17	-524.00000	-489.66911	-453.36568	-414.88423	-373.94934	-330.22728	-283.57985	-234.75716	-185.70660	-138.53152	-94.29316
12.16	-524.00000	-489.66911	-453.36568	-414.88387	-373.93625	-330.02342	-281.87113	-226.90845	-166.13710	-107.15361	-55.26532
13.16	-474.00000	-441.04146	-406.37305	-369.82907	-331.19697	-290.19927	-246.50902	-199.96991	-151.18836	-101.88668	-54.10883
13.15	-474.00000	-441.04146	-406.37305	-369.82905	-331.19607	-290.17970	-246.27693	-198.27517	-143.75153	-83.19029	-23.55191
14.15	-420.00000	-388.49174	-355.48749	-320.84889	-284.40493	-245.93883	-205.17160	-161.77464	-115.55232	-66.96127	-17.52895
14.14	-420.00000	-388.49174	-355.48749	-320.84889	-284.40487	-245.93737	-205.14852	-161.54331	-113.99085	-60.17868	0.00000
15.14	-362.00000	-332.02595	-300.73482	-268.00840	-233.70418	-197.64694	-159.61590	-119.32857	-76.44716	-30.72515	17.52895
15.13	-362.00000	-332.02595	-300.73482	-268.00840	-233.70418	-197.64685	-159.61412	-119.30533	-76.23672	-29.35847	23.55191
16.13	-300.00000	-271.64827	-242.13268	-211.35083	-179.18122	-145.47751	-110.05998	-72.70263	-33.11619	9.05295	54.10883
16.12	-300.00000	-271.64827	-242.13268	-211.35083	-179.18122	-145.47751	-110.05987	-72.70081	-33.09518	9.23266	55.26532
17.12	-234.00000	-207.36167	-179.69353	-150.90616	-120.89460	-89.53405	-56.67391	-22.12885	14.33460	53.01658	94.29316
17.11	-234.00000	-207.36167	-179.69353	-150.90616	-120.89460	-89.53405	-56.67390	-22.12873	14.33624	53.03419	94.44034
18.11	-164.00000	-139.16828	-113.42634	-86.69583	-58.88543	-29.88751	0.42639	32.21206	65.66213	101.02049	138.60336
18.10	-164.00000	-139.16828	-113.42634	-86.69583	-58.88543	-29.88751	0.42639	32.21207	65.66223	101.02184	138.61737
19.10	-90.00000	-67.06971	-43.33775	-18.73553	6.81654	33.41171	61.16049	90.19567	120.67943	152.81384	186.85706
19.9	-90.00000	-67.06971	-43.33775	-18.73553	6.81654	33.41171	61.16049	90.19567	120.67943	152.81392	186.85810
20.9	-12.00000	8.93284	30.56723	52.96299	76.18924	100.32672	125.47063	151.73469	179.25675	208.20678	238.79896
20.8	-12.00000	8.93284	30.56723	52.96299	76.18924	100.32672	125.47063	151.73469	179.25675	208.20678	238.79902
21.8	70.00000	88.83844	108.28481	128.39078	149.21600	170.82989	193.31409	216.76576	241.30208	267.06663	294.23879
21.7	70.00000	88.83844	108.28481	128.39078	149.21600	170.82989	193.31409	216.76576	241.30208	267.06663	294.23879
22.7	156.00000	172.64637	189.81200	207.54095	225.88398	244.90009	264.65855	285.24156	306.74794	329.29835	353.04271
22.6	156.00000	172.64637	189.81200	207.54095	225.88398	244.90009	264.65855	285.24156	306.74794	329.29835	353.04271
23.6	246.00000	260.35609	275.14651	290.40807	306.18314	322.52093	339.47908	357.12594	375.54349	394.83159	415.11391
23.5	246.00000	260.35609	275.14651	290.40807	306.18314	322.52093	339.47908	357.12594	375.54349	394.83159	415.11391
24.5	340.00000	351.96713	364.28647	376.98785	390.10558	403.67947	417.75614	432.39078	447.64953	463.61281	480.38004
24.4	340.00000	351.96713	364.28647	376.98785	390.10558	403.67947	417.75614	432.39078	447.64953	463.61281	480.38004
25.4	438.00000	447.47913	457.23040	467.27684	477.64494	488.36541	499.47421	511.01386	523.03532	535.60043	548.78561
25.3	438.00000	447.47913	457.23040	467.27684	477.64494	488.36541	499.47421	511.01386	523.03532	535.60043	548.78561
26.3	540.00000	546.89181	553.97711	561.27227	568.79610	576.57044	584.62081	592.97740	601.67636	610.76157	620.28718
26.2	540.00000	546.89181	553.97711	561.27227	568.79610	576.57044	584.62081	592.97740	601.67636	610.76157	620.28718
27.2	646.00000	650.20493	654.52561	658.97185	663.55489	668.28778	673.18580	678.26698	683.55294	689.06988	694.85014
27.1	646.00000	650.20493	654.52561	658.97185	663.55489	668.28778	673.18580	678.26698	683.55294	689.06988	694.85014
28.1	756.00000	757.41828	758.87510	760.37371	761.91786	763.51189	765.16088	766.87084	768.64895	770.50394	772.44656
28.0	756.00000	757.41828	758.87510	760.37371	761.91786	763.51189	765.16088	766.87084	768.64895	770.50394	772.44656

J = 29

0.29	-870.00000	-857.35199	-851.90451	-847.73296	-844.22091	-841.13009	-838.33841	-835.77334	-833.38768	-831.14862	-829.03228
1.29	-868.00000	-857.35156	-851.90451	-847.73296	-844.22091	-841.13009	-838.33841	-835.77334	-833.38768	-831.14862	-829.03228
1.28	-868.00000	-833.25314	-816.89338	-804.40499	-793.90497	-784.67309	-776.34110	-768.69057	-761.57935	-754.90873	-748.60692
2.28	-862.00000	-833.23420	-816.89324	-804.40499	-793.90497	-784.67309	-776.34110	-768.69057	-761.57935	-754.90873	-748.60692
2.27	-862.00000	-811.92071	-784.35590	-763.56271	-746.12690	-730.82234	-717.02716	-704.37373	-692.62319	-681.60997	-671.21367
3.27	-852.00000	-811.58545	-784.35132	-763.56262	-746.12690	-730.82234	-717.02716	-704.37373	-692.62319	-681.60997	-671.21367
3.26	-852.00000	-794.92619	-754.61577	-725.38416	-701.01141	-679.67101	-660.46826	-642.87869	-626.56300	-611.28657	-596.87912
4.26	-838.00000	-792.18025	-754.53456	-725.38168	-701.01132	-679.67101	-660.46826	-642.87869	-626.56300	-611.28657	-596.87912
4.25	-838.00000	-783.01954	-728.50628	-690.14763	-658.73019	-631.34183	-606.75641	-584.27601	-563.45339	-543.98087	-525.63590
5.25	-820.00000	-773.39358	-727.63790	-690.10644	-658.72823	-631.34173	-606.75641	-584.27601	-563.45339	-543.98087	-525.63590
5.24	-820.00000	-770.68186	-708.16425	-658.45906	-619.54464	-586.00363	-556.01326	-528.65676	-503.36370	-479.74599	-457.52464
6.24	-798.00000	-752.62448	-703.12238	-658.00256	-619.51444	-586.00174	-556.01315	-528.65675	-503.36370	-479.74599	-457.52464
6.23	-798.00000	-752.24838	-693.18094	-632.07582	-583.98133	-543.91165	-508.40674	-476.14192	-446.38397	-418.64981	-392.59662
7.23	-772.00000	-728.13752	-678.91710	-628.92430	-583.65161	-543.88427	-508.40467	-476.14178	-446.38396	-418.64981	-392.59662
7.22	-772.00000	-728.10699	-676.17440	-612.79314	-553.54381	-505.56454	-464.19130	-426.89956	-392.63462	-360.78105	-330.91781
8.22	-742.00000	-699.50917	-652.31348	-601.38192	-551.14344	-505.27496	-464.16310	-426.89713	-392.63443	-360.78104	-330.91781
8.21	-742.00000	-699.50747	-651.91855	-595.29796	-530.67417	-472.35663	-423.86199	-381.18651	-342.28422	-306.25924	-272.57506
9.21	-708.00000	-666.78561	-621.75475	-572.60180	-520.82617	-470.23481	-423.57430	-381.15512	-342.28127	-306.25900	-272.57504
9.20	-708.00000	-666.78554	-621.71790	-571.39942	-511.84997	-446.92814	-388.79133	-339.50841	-295.59383	-255.25330	-217.68636
10.20	-670.00000	-630.03986	-586.94935	-540.31302	-490.03458	-437.75238	-386.71365	-339.19971	-295.55735	-255.24969	-217.68606
10.19	-670.00000	-630.03986	-586.94684	-540.16792	-487.90516	-426.75525	-361.70115	-303.27061	-253.08894	-208.02924	-166.42021
11.19	-628.00000	-589.31629	-548.00510	-503.74632	-456.14601	-405.24338	-352.61514	-301.09560	-252.74252	-207.98611	-166.41584
11.18	-628.00000	-589.31629	-548.00497	-503.73371	-455.84731	-402.36405	-340.44942	-275.24038	-216.24318	-165.13811	-119.04704
12.18	-582.00000	-544.64088	-505.03499	-462.93903	-418.02132	-369.91725	-318.71563	-265.85376	-213.87333	-164.74132	-118.99630
12.17	-582.00000	-544.64088	-505.03499	-462.93818	-417.99155	-369.47660	-315.38352	-253.22832	-187.84049	-128.14265	-76.14537
13.17	-532.00000	-496.02963	-458.11144	-418.05143	-375.59549	-330.40950	-282.14583	-230.88206	-177.89285	-125.51088	-75.69073
13.16	-532.00000	-496.02963	-458.11143	-418.05139	-375.59321	-330.36193	-281.61248	-227.38893	-165.34787	-99.80293	-39.37308
14.16	-478.00000	-443.49286	-407.28003	-369.20070	-329.05295	-286.57619	-241.43474	-193.28213	-142.14795	-89.14224	-36.44538
14.15	-478.00000	-443.49286	-407.28003	-369.20069	-329.05281	-286.57224	-241.37460	-192.71411	-138.72151	-77.05193	-11.41584
15.15	-420.00000	-387.03749	-352.57069	-316.46275	-278.54686	-238.61500	-196.40085	-151.56425	-103.74627	-52.90772	0.00000
15.14	-420.00000	-387.03749	-352.57069	-316.46275	-278.54686	-238.61474	-196.39563	-151.49899	-103.19129	-49.69489	11.41584
16.14	-358.00000	-326.66833	-294.00381	-259.88807	-224.17954	-186.70601	-147.25326	-105.54794	-61.24090	-13.94964	36.44538
16.13	-358.00000	-326.66833	-294.00381	-259.88807	-224.17954	-186.70600	-147.25290	-105.54216	-61.17704	-13.43804	39.37308
17.13	-292.00000	-262.38876	-231.59376	-199.51151	-166.01985	-130.97258	-94.19164	-55.45536	-14.48069	29.09608	75.69073
17.12	-292.00000	-262.38876	-231.59376	-199.51151	-166.01985	-130.97258	-94.19162	-55.45495	-14.47501	29.15433	76.14537
18.12	-222.00000	-194.20126	-165.35088	-135.35789	-104.11586	-71.49873	-37.35514	-1.50005	36.29777	76.33488	118.99630
18.11	-222.00000	-194.20126	-165.35088	-135.35789	-104.11586	-71.49873	-37.35514	-1.50003	36.29817	76.34001	119.04704
19.11	-148.00000	-122.10767	-95.28282	-67.44538	-38.50218	-8.34370	23.16040	56.16758	90.87152	127.51593	166.41584
19.10	-148.00000	-122.10767	-95.28282	-67.44538	-38.50218	-8.34370	23.16040	56.16758	90.87154	127.51629	166.42021
20.10	-70.00000	-46.10938	-21.39533	4.21248	30.79560	58.44940	87.28674	117.44291	149.08263	182.41038	217.68606
20.9	-70.00000	-46.10938	-21.39533	4.21248	30.79560	58.44940	87.28674	117.44291	149.08263	182.41041	217.68636
21.9	12.00000	33.79257	56.30720	79.60538	103.75817	128.84842	154.97379	182.25076	210.82027	240.85574	272.57504
21.8	12.00000	33.79257	56.30720	79.60538	103.75817	128.84842	154.97379	182.25076	210.82027	240.85574	272.57506
22.8	98.00000	117.59734	137.82138	158.72536	180.37069	202.82883	226.18378	250.53540	276.00403	302.73693	330.91781
22.7	98.00000	117.59734	137.82138	158.72536	180.37069	202.82883	226.18378	250.53540	276.00403	302.73693	330.91781
23.7	188.00000	205.30429	223.14454	241.56619	260.62161	280.37166	300.88773	322.25452	344.57382	367.96980	392.59662
23.6	188.00000	205.30429	223.14454	241.56619	260.62161	280.37166	300.88773	322.25452	344.57382	367.96980	392.59662
24.6	282.00000	296.91291	312.27454	328.12291	344.50178	361.46195	379.06295	397.37530	416.48363	436.49090	457.52464
24.5	282.00000	296.91291	312.27454	328.12291	344.50178	361.46195	379.06295	397.37530	416.48363	436.49090	457.52464
25.5	380.00000	392.42278	405.20970	418.39159	432.00391	446.08781	460.69147	475.87192	491.69750	508.25127	525.63590
25.4	380.00000	392.42278	405.20970	418.39159	432.00391	446.08781	460.69147	475.87192	491.69750	508.25127	525.63590
26.4	482.00000	491.83357	501.94863	512.36901	523.12210	534.23966	545.75888	557.72375	570.18693	583.21239	596.87912
26.3	482.00000	491.83357	501.94863	512.36901	523.12210	534.23966	545.75888	557.72375	570.18693	583.21239	596.87912
27.3	588.00000	595.14500	602.49020	610.05256	617.85153	625.90969	634.25347	642.91411	651.92899	661.34350	671.21367
27.2	588.00000	595.14500	602.49020	610.05256	617.85153	625.90969	634.25347	642.91411	651.92899	661.34350	671.21367
28.2	698.00000	702.35685	706.83350	711.44008	716.18825	721.09151	726.16566	731.42938	736.90502	742.61972	748.60692
28.1	698.00000	702.35685	706.83350	711.44008	716.18825	721.09151	726.16566	731.42938	736.90502	742.61972	748.60692
29.1	812.00000	813.46892	814.97773	816.52980	818.12900	819.77982	821.48754	823.25836	825.09972	827.02065	829.03228
29.0	812.00000	813.46892	814.97773	816.52980	818.12900	819.77982	821.48754	823.25836	825.09972	827.02065	829.03228

J = 30

0.30	-930.00000	-916.90397	-911.27162	-906.95806	-903.32626	-900.12992	-897.24282	-894.59002	-892.12268	-889.80691	-887.61802
1.30	-928.00000	-916.90368	-911.27162	-906.95806	-903.32626	-900.12992	-897.24282	-894.59002	-892.12268	-889.80691	-887.61802
1.29	-928.00000	-891.90020	-874.99148	-862.07811	-851.21940	-841.67133	-833.05338	-825.13983	-817.78375	-810.88313	-804.36375
2.29	-922.00000	-891.88698	-874.99140	-862.07811	-851.21940	-841.67133	-833.05338	-825.13983	-817.78375	-810.88313	-804.36375
2.28	-922.00000	-869.59492	-841.17163	-819.67590	-801.64472	-785.81470	-771.54399	-758.45297	-746.29495	-734.89887	-724.14035
3.28	-912.00000	-869.34736	-841.16891	-819.67585	-801.64472	-785.81470	-771.54399	-758.45297	-746.29495	-734.89887	-724.14035
3.27	-912.00000	-851.39781	-810.10334	-779.91871	-754.72029	-732.64848	-712.78280	-694.58264	-677.69801	-661.88678	-646.97317
4.27	-898.00000	-849.17905	-810.05228	-779.91739	-754.72025	-732.64848	-712.78280	-694.58264	-677.69801	-661.88678	-646.97317
4.26	-898.00000	-838.61427	-782.46996	-743.05702	-710.60581	-682.28774	-656.85639	-633.59537	-612.04451	-591.88688	-572.89311
5.26	-880.00000	-829.97413	-781.88175	-743.03398	-710.60488	-682.28770	-656.85639	-633.59537	-612.04451	-591.88688	-572.89311
5.25	-880.00000	-826.47548	-760.20561	-709.58205	-669.53474	-634.88788	-603.87770	-575.57607	-549.39927	-524.94895	-501.93827
6.25	-858.00000	-809.14921	-756.34577	-709.30752	-669.51939	-634.88705	-603.87766	-575.57606	-549.39927	-524.94895	-501.93827
6.24	-858.00000	-808.60499	-744.12335	-680.89471	-631.92988	-590.67405	-553.99948	-520.63588	-489.84545	-461.13585	-434.15626
7.24	-832.00000	-784.75157	-731.67213	-678.77178	-631.74941	-590.66125	-553.99865	-520.63583	-489.84544	-461.13585	-434.15626
7.23	-832.00000	-784.70337	-727.76023	-659.38091	-598.91732	-550.04037	-507.44286	-468.92575	-433.49242	-400.52868	-369.60761
8.23	-802.00000	-756.20990	-705.10229	-650.33470	-597.44983	-549.89408	-507.43065	-468.92484	-433.49236	-400.52867	-369.60761
8.22	-802.00000	-756.20699	-704.46009	-641.82552	-573.06691	-513.98758	-464.59051	-420.66501	-380.48955	-343.23474	-308.37084
9.22	-768.00000	-723.54737	-674.74972	-621.40048	-565.89047	-512.78759	-464.45515	-420.65234	-380.48852	-343.23467	-308.37083
9.21	-768.00000	-723.54724	-674.68374	-619.40650	-553.41394	-485.03353	-426.39196	-376.23377	-331.05534	-289.40153	-250.55056
10.21	-730.00000	-686.84447	-640.12910	-589.37643	-534.74585	-478.79124	-425.29157	-376.09768	-331.04158	-289.40036	-250.55048
10.20	-730.00000	-686.84447	-640.12420	-589.10485	-531.14648	-463.41240	-395.32009	-336.57069	-285.57216	-239.24675	-196.29163
11.20	-688.00000	-646.15212	-601.32123	-553.12502	-501.14072	-445.77020	-389.46920	-335.48274	-285.42849	-239.23152	-196.29031
11.19	-688.00000	-646.15212	-601.32095	-553.09901	-500.55627	-440.78066	-372.18418	-304.14441	-244.99088	-193.15697	-145.80999
12.19	-642.00000	-601.50022	-558.45201	-512.56733	-463.44991	-410.74235	-354.95243	-298.35853	-243.86462	-193.00182	-145.79323
12.18	-642.00000	-601.50022	-558.45200	-512.56543	-463.38492	-409.83835	-348.98453	-279.98564	-211.80756	-152.10717	-99.49576
13.18	-592.00000	-552.90714	-511.60535	-467.86775	-421.39043	-371.77693	-318.71280	-262.71797	-205.88697	-150.91411	-99.32811
13.17	-592.00000	-552.90714	-511.60535	-467.86764	-421.38493	-371.66677	-317.56062	-256.19807	-187.05994	-118.63520	-58.35009
14.17	-538.00000	-500.38472	-460.83412	-419.16172	-375.12916	-328.42315	-278.64825	-225.50468	-169.46936	-112.47020	-57.08161
14.16	-538.00000	-500.38472	-460.83412	-419.16171	-375.12879	-328.41298	-278.50002	-224.20599	-162.75418	-93.65054	-24.94909
15.16	-480.00000	-443.94090	-406.17289	-366.53809	-324.84159	-280.83632	-234.20115	-184.53700	-131.53453	-75.59764	-18.51113
15.15	-480.00000	-443.94090	-406.17289	-366.53809	-324.84157	-280.83558	-234.18673	-184.36506	-130.18486	-68.94427	0.00000
16.15	-418.00000	-383.58116	-347.64513	-310.05565	-270.64831	-229.22078	-185.51757	-139.20834	-89.88338	-37.20503	18.51113
16.14	-418.00000	-383.58116	-347.64513	-310.05565	-270.64831	-229.22073	-185.51646	-139.19113	-89.70276	-35.87527	24.94909
17.14	-352.00000	-319.30939	-285.26734	-249.75477	-212.63019	-173.72288	-132.82274	-89.66450	-43.90538	4.88303	57.08161
17.13	-352.00000	-319.30939	-285.26734	-249.75477	-212.63019	-173.72288	-132.82267	-89.66314	-43.88710	5.06064	58.35009
18.13	-282.00000	-251.12842	-219.05141	-185.66413	-150.84328	-114.44205	-76.28285	-36.14640	6.24485	51.25092	99.32811
18.12	-282.00000	-251.12842	-219.05141	-185.66413	-150.84328	-114.44205	-76.28284	-36.14632	6.24632	51.26887	99.49576
19.12	-208.00000	-179.04034	-149.00610	-117.80462	-85.32775	-51.44781	-16.01209	21.16514	60.31439	101.72920	145.79323
19.11	-208.00000	-179.04034	-149.00610	-117.80462	-85.32775	-51.44781	-16.01209	21.16515	60.31449	101.73064	145.80999
20.11	-130.00000	-103.04675	-75.13800	-46.19185	-16.11319	15.20958	47.90900	82.14464	118.11201	156.05575	196.29031
20.10	-130.00000	-103.04675	-75.13800	-46.19185	-16.11319	15.20958	47.90900	82.14464	118.11201	156.05585	196.29163
21.10	-48.00000	-23.14885	2.54789	29.16237	56.77813	85.49280	115.42170	146.70285	179.50390	214.03227	250.55048
21.9	-48.00000	-23.14885	2.54789	29.16237	56.77813	85.49280	115.42170	146.70285	179.50390	214.03227	250.55056
22.9	38.00000	60.65242	84.04765	108.24889	133.32915	159.37348	186.48205	214.77421	244.39417	275.51905	308.37083
22.8	38.00000	60.65242	84.04765	108.24889	133.32915	159.37348	186.48205	214.77421	244.39417	275.51905	308.37084
23.8	128.00000	148.35632	169.35824	191.06058	213.52656	236.82971	261.05639	286.30924	312.71182	340.41522	369.60761
23.7	128.00000	148.35632	169.35824	191.06058	213.52656	236.82971	261.05639	286.30924	312.71182	340.41522	369.60761
24.7	222.00000	239.96225	258.47722	277.59177	297.35989	317.84428	339.11850	361.26976	384.40286	408.64553	434.15626
24.6	222.00000	239.96225	258.47722	277.59177	297.35989	317.84428	339.11850	361.26976	384.40286	408.64553	434.15626
25.6	320.00000	335.46976	351.40266	367.83794	384.82076	402.40352	420.64764	439.62584	459.42538	480.15239	501.93827
25.5	320.00000	335.46976	351.40266	367.83794	384.82076	402.40352	420.64764	439.62584	459.42538	480.15239	501.93827
26.5	422.00000	434.87845	448.13298	461.79542	475.90241	490.49642	505.62720	521.35362	537.74623	554.89074	572.89311
26.4	422.00000	434.87845	448.13298	461.79542	475.90241	490.49642	505.62720	521.35362	537.74623	554.89074	572.89311
27.4	528.00000	538.18801	548.66688	559.46121	570.59932	582.11402	594.04372	606.43386	619.33885	632.82477	646.97317
27.3	528.00000	538.18801	548.66688	559.46121	570.59932	582.11402	594.04372	606.43386	619.33885	632.82477	646.97317
28.3	638.00000	645.39819	653.00331	660.83286	668.90697	677.24899	685.88620	694.85089	704.18171	713.92557	724.14035
28.2	638.00000	645.39819	653.00331	660.83286	668.90697	677.24899	685.88620	694.85089	704.18171	713.92557	724.14035
29.2	752.00000	756.50877	761.14138	765.90832	770.82162	775.89525	781.14554	786.59180	792.25714	798.16959	804.36375
29.1	752.00000	756.50877	761.14138	765.90832	770.82162	775.89525	781.14554	786.59180	792.25714	798.16959	804.36375
30.1	870.00000	871.51957	873.08036	874.68588	876.34013	878.04776	879.81420	881.64588	883.55050	885.53737	887.61802
30.0	870.00000	871.51957	873.08036	874.68588	876.34013	878.04776	879.81420	881.64588	883.55050	885.53737	887.61802

= 31

0.31	-992.00000	-978.45603	-972.63876	-968.18318	-964.43162	-961.12975	-958.14724	-955.40670	-952.85769	-950.46520	-948.20375
1.31	-990.00000	-978.45583	-972.63876	-968.18318	-964.43162	-961.12975	-958.14724	-955.40670	-952.85769	-950.46520	-948.20375
1.30	-990.00000	-952.54883	-935.08994	-921.75147	-910.53399	-900.66969	-891.76576	-883.58917	-875.98821	-868.85757	-862.12061
2.30	-984.00000	-952.53963	-935.08989	-921.75146	-910.53399	-900.66969	-891.76576	-883.58917	-875.98821	-868.85757	-862.12061
2.29	-984.00000	-929.28451	-899.98951	-877.79019	-859.16331	-842.80763	-828.06126	-814.53256	-801.96698	-790.18798	-779.06719
3.29	-974.00000	-929.10301	-899.98790	-877.79016	-859.16331	-842.80763	-828.06126	-814.53256	-801.96698	-790.18798	-779.06719
3.28	-974.00000	-909.90636	-867.60334	-836.45724	-810.43165	-787.62775	-767.09870	-748.28764	-730.83383	-714.48762	-699.06771
4.28	-960.00000	-908.13933	-867.57151	-836.45655	-810.43164	-787.62775	-767.09870	-748.28764	-730.83383	-714.48762	-699.06771
4.27	-960.00000	-896.11624	-838.49956	-797.98136	-764.48842	-735.23834	-708.95981	-684.91734	-662.63763	-641.79444	-622.15150
5.27	-942.00000	-888.45901	-838.10716	-797.96861	-764.48798	-735.23833	-708.95981	-684.91734	-662.63763	-641.79444	-622.15150
5.26	-942.00000	-884.03759	-814.34333	-762.77224	-721.54533	-685.78337	-653.74993	-624.50113	-597.43920	-572.15522	-548.35442
6.26	-920.00000	-867.53996	-811.46914	-762.60989	-721.53763	-685.78301	-653.74991	-624.50112	-597.43920	-572.15522	-548.35442
6.25	-920.00000	-866.76703	-796.82733	-731.92255	-681.95458	-639.46504	-601.60913	-567.14170	-535.31568	-505.62847	-477.72085
7.25	-894.00000	-843.22781	-786.20426	-730.53784	-681.85771	-639.45915	-601.60879	-567.14168	-535.31568	-505.62847	-477.72085
7.24	-894.00000	-843.15310	-780.81720	-707.91764	-646.55804	-596.60877	-552.73367	-512.97598	-476.36715	-442.28873	-410.30664
8.24	-864.00000	-814.78032	-759.59504	-701.02688	-645.69210	-596.53655	-552.72848	-512.97565	-476.36713	-442.28873	-410.30664
8.23	-864.00000	-814.77545	-758.57974	-689.66296	-617.62604	-557.94140	-507.43499	-462.19606	-420.72754	-382.23315	-346.18326
9.23	-830.00000	-782.18542	-729.44435	-671.76997	-612.67223	-557.28927	-507.37300	-462.19106	-420.72719	-382.23313	-346.18325
9.22	-830.00000	-782.18519	-729.32921	-668.60336	-596.25089	-525.49769	-466.38387	-415.10553	-368.58542	-325.59241	-285.44439
10.22	-792.00000	-745.53000	-695.02586	-639.95668	-580.91388	-521.52380	-465.82712	-415.04741	-368.58037	-325.59203	-285.44437
10.21	-792.00000	-745.52999	-695.01656	-639.46539	-575.18256	-501.33874	-431.46707	-372.34798	-320.23945	-272.55138	-228.21674
11.21	-750.00000	-704.87220	-656.36994	-604.03881	-547.46836	-487.64294	-427.98364	-371.83189	-320.18205	-272.54618	-228.21635
11.20	-750.00000	-704.87220	-656.36936	-603.98690	-546.37407	-479.62520	-405.16816	-335.73167	-276.32068	-223.40454	-174.68061
12.20	-704.00000	-660.24629	-613.61301	-563.75922	-510.21797	-452.71451	-392.41584	-332.47704	-275.81752	-223.34648	-174.67528
12.19	-704.00000	-660.24629	-613.61299	-563.75507	-510.08161	-450.95953	-382.65545	-307.92943	-238.60238	-178.77673	-125.12526
13.19	-654.00000	-611.67329	-566.85173	-519.27010	-468.56514	-414.27900	-356.23440	-295.64578	-235.43307	-178.27354	-125.06651
13.18	-654.00000	-611.67329	-566.85173	-519.26984	-468.55243	-414.03562	-353.90606	-284.70547	-209.84048	-140.42806	-80.16919
14.18	-600.00000	-559.16673	-516.14721	-470.72552	-422.62017	-371.45522	-316.78388	-258.47885	-197.72929	-137.27310	-79.66394
14.17	-600.00000	-559.16673	-516.14721	-470.72550	-422.61924	-371.43023	-316.43799	-255.73361	-186.08851	-111.14284	-41.56270
15.17	-542.00000	-502.73566	-461.53922	-418.22903	-372.57751	-324.29096	-272.98355	-218.21188	-159.85870	-99.05363	-38.40676
16.16	-542.00000	-502.73566	-461.53922	-418.22903	-372.57745	-324.28894	-272.94579	-217.78661	-156.85931	-87.07299	-12.08768
16.16	-480.00000	-442.38632	-403.05472	-361.84894	-318.57848	-273.00572	-224.82555	-173.64088	-119.00042	-60.76741	0.00000
16.15	-480.00000	-442.38632	-403.05472	-361.84894	-318.57848	-273.00558	-224.82231	-173.59270	-118.52512	-57.64648	12.08768
17.15	-414.00000	-378.12315	-340.71259	-301.63193	-260.71788	-217.77158	-172.54516	-124.72077	-73.88597	-19.56932	38.40676
17.14	-414.00000	-378.12315	-340.71259	-301.63193	-260.71788	-217.77158	-172.54493	-124.71648	-73.83104	-19.07007	41.56270
18.14	-344.00000	-309.94938	-274.52644	-237.61099	-199.06097	-158.70607	-116.33842	-71.69837	-24.45235	25.83357	79.66394
18.13	-344.00000	-309.94938	-274.52644	-237.61099	-199.06097	-158.70607	-116.33841	-71.69806	-24.44737	25.89170	80.16919
19.13	-270.00000	-237.86740	-204.50625	-169.81013	-133.65426	-95.89071	-56.34136	-14.78789	29.04374	75.50730	125.06651
19.12	-270.00000	-237.86740	-204.50625	-169.81013	-133.65426	-95.89071	-56.34136	-14.78787	29.04411	75.51260	125.12526
20.12	-192.00000	-161.87899	-130.65955	-98.24719	-64.53187	-29.38400	7.35095	45.86012	86.37449	129.18612	174.67528
20.11	-192.00000	-161.87899	-130.65955	-98.24719	-64.53187	-29.38400	7.35095	45.86012	86.37451	129.18651	174.68061
21.11	-110.00000	-81.98556	-52.99206	-22.93573	8.28061	40.77081	74.66979	110.13963	147.37822	186.63211	228.21635
21.10	-110.00000	-81.98556	-52.99206	-22.93573	8.28061	40.77081	74.66979	110.13963	147.37822	186.63214	228.21674
22.10.	-24.00000	1.81184	28.49179	56.11384	84.76361	114.54102	145.56402	177.97351	211.94034	247.67534	285.44437
22.9	-24.00000	1.81184	28.49179	56.11384	84.76361	114.54102	145.56402	177.97351	211.94034	247.67534	285.44439
23.9	66.00000	89.51237	113.78850	138.89335	164.90188	191.90141	219.99467	249.30397	279.97691	312.19453	346.18325
23.8	66.00000	89.51237	113.78850	138.89335	164.90188	191.90141	219.99467	249.30397	279.97691	312.19453	346.18326
24.8	160.00000	181.11535	202.89533	225.39635	248.68345	272.83224	297.93150	324.08668	351.42463	380.10034	410.30664
24.7	160.00000	181.11535	202.89533	225.39635	248.68345	272.83224	297.93150	324.08668	351.42463	380.10034	410.30664
25.7	258.00000	276.62024	295.81004	315.61767	336.09874	357.31783	379.35065	402.28699	426.23465	451.32498	477.72085
25.6	258.00000	276.62024	295.81004	315.61767	336.09874	357.31783	379.35065	402.28699	426.23465	451.32498	477.72085
26.6	360.00000	376.02662	392.53085	409.55313	427.14003	445.34557	464.23305	483.87741	504.36855	525.81579	548.35442
26.5	360.00000	376.02662	392.53085	409.55313	427.14003	445.34557	464.23305	483.87741	504.36855	525.81579	548.35442
27.5	466.00000	479.33412	493.05628	507.19933	521.80104	536.90525	552.56326	568.83581	585.79563	603.53112	622.15150
27.4	466.00000	479.33412	493.05628	507.19933	521.80104	536.90525	552.56326	568.83581	585.79563	603.53112	622.15150
28.4	576.00000	586.54246	597.38514	608.55345	620.07660	631.98847	644.32870	657.14418	670.49106	684.43751	699.06771
8.3	576.00000	586.54246	597.38514	608.55345	620.07660	631.98847	644.32870	657.14418	670.49106	684.43751	699.06771
9.3	690.00000	697.65138	705.51641	713.61317	721.96244	730.58831	739.51897	748.78775	758.43454	768.50777	779.06719
9.2	690.00000	697.65138	705.51641	713.61317	721.96244	730.58831	739.51897	748.78775	758.43454	768.50777	779.06719
0.2	808.00000	812.66069	817.44927	822.37656	827.45500	832.69900	838.12543	843.75424	849.60927	855.71949	862.12061
0.1	808.00000	812.66069	817.44927	822.37656	827.45500	832.69900	838.12543	843.75424	849.60927	855.71949	862.12061
1.1	930.00000	931.57021	933.18300	934.84197	936.55127	938.31570	940.14087	942.03341	944.00127	946.05409	948.20375
1.0	930.00000	931.57021	933.18300	934.84197	936.55127	938.31570	940.14087	942.03341	944.00127	946.05409	948.20375

J = 32

0.32	-1056.00000	-1042.00814	-1036.00593	-1031.40832	-1027.53699	-1024.12959	-1021.05168	-1018.22338	-1015.59270	-1013.12350	-1010.78949
1.32	-1054.00000	-1042.00802	-1036.00593	-1031.40832	-1027.53699	-1024.12959	-1021.05168	-1018.22338	-1015.59270	-1013.12350	-1010.78949
1.31	-1054.00000	-1015.19868	-997.18870	-983.42502	-971.84873	-961.66816	-952.47822	-944.03857	-936.19272	-928.83206	-921.87751
2.31	-1048.00000	-1015.19231	-997.18867	-983.42502	-971.84873	-961.66816	-952.47822	-944.03857	-936.19272	-928.83206	-921.87751
2.30	-1048.00000	-990.98648	-960.80917	-937.90544	-918.68258	-901.80107	-886.57892	-872.61246	-859.63925	-847.47728	-835.99417
3.30	-1038.00000	-990.85426	-960.80822	-937.90543	-918.68258	-901.80107	-886.57892	-872.61246	-859.63925	-847.47728	-835.99417
3.29	-1038.00000	-970.45620	-927.11280	-894.99917	-868.14521	-844.60861	-823.41580	-803.99357	-785.97038	-769.08902	-753.16269
4.29	-1024.00000	-969.06760	-927.09311	-894.99881	-868.14520	-844.60861	-823.41580	-803.99357	-785.97038	-769.08902	-753.16269
4.28	-1024.00000	-955.55086	-896.58045	-854.91732	-820.37697	-790.19303	-763.06625	-738.24160	-715.23251	-693.70336	-673.41094
5.28	-1006.00000	-948.85573	-896.32211	-854.91033	-820.37676	-790.19303	-763.06625	-738.24160	-715.23251	-693.70336	-673.41094
5.27	-1006.00000	-943.37733	-870.60085	-818.01111	-775.57208	-738.68840	-705.62890	-675.43120	-647.48292	-621.36440	-596.77278
6.27	-984.00000	-927.79891	-868.51445	-817.91650	-775.56826	-738.68825	-705.62890	-675.43120	-647.48292	-621.36440	-596.77278
6.26	-984.00000	-926.72134	-851.37750	-785.13201	-734.03332	-690.27863	-651.23303	-615.65772	-582.79349	-552.12681	-523.28976
7.26	-958.00000	-903.56452	-842.53613	-784.25289	-733.98220	-690.27596	-651.23289	-615.65772	-582.79349	-552.12681	-523.28976
7.25	-958.00000	-903.45072	-835.36718	-758.54666	-696.40303	-645.24211	-600.05536	-559.04633	-521.25631	-486.05946	-453.01363
8.25	-928.00000	-875.21834	-815.79797	-753.50800	-695.90660	-645.20716	-600.05320	-559.04621	-521.25630	-486.05946	-453.01363
8.24	-928.00000	-875.21035	-814.23844	-738.95271	-664.49702	-604.12579	-552.35978	-505.76812	-462.99259	-423.25092	-386.00985
9.24	-894.00000	-842.69810	-785.83389	-723.74104	-661.24829	-603.78258	-552.33206	-505.76619	-462.99247	-423.25092	-386.00985
9.23	-894.00000	-842.69769	-785.63792	-718.93705	-640.64192	-568.44025	-508.64310	-456.07704	-408.16834	-363.81823	-322.36302
10.23	-856.00000	-806.09516	-751.63360	-692.05358	-628.60432	-566.05557	-508.37147	-456.05290	-408.16653	-363.81811	-322.36301
10.22	-856.00000	-806.09515	-751.61637	-691.19490	-620.02369	-540.96612	-470.23588	-410.44054	-357.02962	-307.92212	-262.18534
11.22	-814.00000	-765.47549	-713.14631	-656.47660	-595.14291	-530.96851	-468.29645	-410.20571	-357.00741	-307.92039	-262.18522
11.21	-814.00000	-765.47549	-713.14516	-656.37620	-593.18721	-519.02147	-439.99711	-370.08216	-310.02437	-255.80122	-205.63157
12.21	-768.00000	-720.87823	-670.51411	-616.50419	-558.31082	-495.86928	-431.25926	-368.38567	-309.81008	-255.78027	-205.62993
12.20	-768.00000	-720.87823	-670.51405	-616.49543	-558.03565	-492.66192	-416.65360	-337.83823	-268.30879	-207.88778	-152.93201
13.20	-718.00000	-672.32736	-623.84740	-572.25009	-517.10211	-457.89958	-394.77237	-329.87070	-266.75334	-207.68691	-152.91228
13.19	-718.00000	-672.32736	-623.84740	-572.24949	-517.07381	-457.38658	-390.40673	-313.30703	-234.67592	-165.28948	-104.50308
14.19	-664.00000	-619.83826	-573.21660	-523.88528	-471.51161	-415.64697	-355.82542	-292.29513	-227.19032	-163.82721	-104.31412
14.18	-664.00000	-619.83826	-573.21660	-523.88525	-471.50933	-415.58822	-355.06215	-286.98938	-209.25653	-130.75192	-61.40781
15.18	-606.00000	-563.42122	-518.66738	-471.52989	-421.74306	-368.95735	-312.71395	-252.57280	-188.83911	-123.60167	-60.02585
15.17	-606.00000	-563.42122	-518.66738	-471.52989	-421.74291	-368.95211	-312.62003	-251.58739	-182.83497	-104.74340	-26.33576
16.17	-544.00000	-503.08333	-460.23059	-415.26312	-367.96048	-318.04350	-265.14738	-208.80197	-148.57528	-84.78757	-19.48391
16.16	-544.00000	-503.08333	-460.23059	-415.26312	-367.96047	-318.04312	-265.13841	-208.67481	-147.41494	-78.29037	0.00000
17.16	-478.00000	-438.82964	-397.92777	-355.13879	-310.27480	-263.10441	-213.33494	-160.58519	-104.36839	-44.24218	19.48391
17.15	-478.00000	-438.82964	-397.92777	-355.13879	-310.27480	-263.10439	-213.33425	-160.57250	-104.21405	-42.95399	26.33576
18.15	-408.00000	-370.66379	-331.77441	-291.19480	-248.76191	-204.27871	-157.50222	-108.12462	-55.74488	0.14848	60.02585
18.14	-408.00000	-370.66379	-331.77441	-291.19480	-248.76191	-204.27870	-157.50217	-108.12360	-55.72907	0.32294	61.40781
19.14	-334.00000	-298.58850	-261.78190	-223.45862	-183.47554	-141.66195	-97.81082	-51.66586	-2.90106	48.90996	104.31412
19.13	-334.00000	-298.58850	-261.78190	-223.45862	-183.47554	-141.66195	-97.81082	-51.66579	-2.89976	48.92808	104.50308
20.13	-256.00000	-222.60581	-187.95876	-151.95063	-114.45495	-75.32218	-34.37306	8.61098	53.90226	101.84950	152.91228
20.12	-256.00000	-222.60581	-187.95876	-151.95063	-114.45495	-75.32218	-34.37306	8.61099	53.90234	101.85101	152.93201
21.12	-174.00000	-142.71729	-110.31150	-76.68626	-41.72947	-5.30939	32.73065	72.57977	114.47037	158.69451	205.62993
21.11	-174.00000	-142.71729	-110.31150	-76.68626	-41.72947	-5.30939	32.73065	72.57977	114.47038	158.69462	205.63157
22.11	-88.00000	-58.92414	-28.84518	2.32260	34.67851	68.33877	103.44088	140.14970	178.66597	219.23886	262.18522
22.10	-88.00000	-58.92414	-28.84518	2.32260	34.67851	68.33877	103.44088	140.14970	178.66597	219.23887	262.18534
23.10	2.00000	28.77268	56.43627	85.06668	114.75160	145.59338	177.71263	211.25330	246.38967	283.33629	322.36301
23.9	2.00000	28.77268	56.43627	85.06668	114.75160	145.59338	177.71263	211.25330	246.38967	283.33630	322.36302
24.9	96.00000	120.37240	145.52971	171.53864	198.47654	226.43183	255.51106	285.83916	317.56724	350.88043	386.00985
24.8	96.00000	120.37240	145.52971	171.53864	198.47614	226.43183	255.51106	285.83916	317.56724	350.88043	386.00985
25.8	194.00000	215.87443	238.43262	261.73260	285.84123	310.83622	336.80880	363.86726	392.14180	421.79139	453.01363
25.7	194.00000	215.87443	238.43262	261.73260	285.84123	310.83622	336.80880	363.86726	392.14180	421.79139	453.01363
26.7	296.00000	315.27826	335.14297	355.64384	376.83809	398.79218	421.58402	445.30596	470.06884	496.00767	523.28976
26.6	296.00000	315.27826	335.14297	355.64384	376.83809	398.79218	421.58402	445.30596	470.06884	496.00767	523.28976
27.6	402.00000	418.58350	435.65909	453.26846	471.45957	490.28805	509.81910	530.12989	551.31296	573.48086	596.77278
27.5	402.00000	418.58350	435.65909	453.26846	471.45957	490.28805	509.81910	530.12989	551.31296	573.48086	596.77278
28.5	512.00000	525.78980	539.97962	554.60330	569.69980	585.31429	601.49964	618.31843	635.84562	654.17229	673.41094
28.4	512.00000	525.78980	539.97962	554.60330	569.69980	585.31429	601.49964	618.31843	635.84562	654.17229	673.41094
29.4	626.00000	636.89691	648.10341	659.64571	671.55394	683.86301	696.61381	709.85468	723.64352	738.05060	753.16269
29.3	626.00000	636.89691	648.10341	659.64571	671.55394	683.86301	696.61381	709.85468	723.64352	738.05060	753.16269
30.3	744.00000	751.90457	760.02952	768.39350	777.01793	785.92767	795.15178	804.72466	814.68744	825.09007	835.99417
30.2	744.00000	751.90457	760.02952	768.39350	777.01793	785.92767	795.15178	804.72466	814.68744	825.09007	835.99417
31.2	866.00000	870.81261	875.75715	880.84480	886.08838	891.50276	897.10533	902.91668	908.96142	915.26941	921.87751
31.1	866.00000	870.81261	875.75715	880.84480	886.08838	891.50276	897.10533	902.91668	908.96142	915.26941	921.87751
32.1	992.00000	993.62085	995.28563	996.99806	998.76241	1000.58364	1002.46753	1004.42094	1006.45205	1008.57081	1010.78949
32.0	992.00000	993.62085	995.28563	996.99806	998.76241	1000.58364	1002.46753	1004.42094	1006.45205	1008.57081	1010.78949

r = 33

0.33	-1122.00000	-1107.56032	-1101.37313	-1096.63348	-1092.64238	-1089.12945	-1085.95612	-1083.04007	-1080.32771	-1077.78180	-1075.37523
1.33	-1120.00000	-1107.56023	-1101.37313	-1096.63348	-1092.64238	-1089.12945	-1085.95612	-1083.04007	-1080.32771	-1077.78180	-1075.37523
1.32	-1120.00000	-1079.84951	-1061.28774	-1047.09875	-1035.16360	-1024.66674	-1015.19077	-1006.48803	-998.39729	-990.80658	-983.63444
2.32	-1114.00000	-1079.84510	-1061.28772	-1047.09875	-1035.16360	-1024.66674	-1015.19077	-1006.48803	-998.39729	-990.80658	-983.63444
2.31	-1114.00000	-1054.69833	-1023.63033	-1000.02156	-980.20246	-962.79497	-947.09693	-932.69263	-919.31173	-906.76675	-894.92129
3.31	-1104.00000	-1054.60257	-1023.62977	-1000.02155	-980.20246	-962.79497	-947.09693	-932.69263	-919.31173	-906.76675	-894.92129
3.30	-1104.00000	-1033.04782	-988.62961	-955.54401	-927.86070	-903.59089	-881.73398	-861.70033	-843.10757	-825.69093	-809.25806
4.30	-1090.00000	-1031.96972	-988.61751	-955.54383	-927.86070	-903.59089	-881.73398	-861.70033	-843.10757	-825.69093	-809.25806
4.29	-1090.00000	-1016.94369	-956.70052	-913.86259	-878.27066	-847.15131	-819.17536	-793.56789	-769.82896	-747.61349	-726.67131
5.29	-1072.00000	-1011.17221	-956.53239	-913.85879	-878.27057	-847.15131	-819.17536	-793.56789	-769.82896	-747.61349	-726.67131
5.28	-1072.00000	-1004.50887	-928.98196	-875.28566	-831.61200	-793.60165	-759.51377	-728.36567	-699.53000	-672.57614	-647.19310
6.28	-1050.00000	-989.92919	-927.50129	-875.23123	-831.61013	-793.60159	-759.51376	-728.36567	-699.53000	-672.57614	-647.19310
6.27	-1050.00000	-988.45461	-907.85873	-840.48606	-788.15156	-743.11072	-702.86913	-666.18259	-632.27792	-60).63014	-570.86246
7.27	-1024.00000	-965.76031	-900.69230	-839.94038	-788.12499	-743.10953	-702.86908	-666.18259	-632.27792	-600.63014	-570.86246
7.26	-1024.00000	-965.58982	-891.45599	-811.36819	-748.39572	-695.92303	-649.40228	-607.13377	-568.15790	-531.83942	-497.72753
8.26	-994.00000	-937.52186	-873.72115	-807.82619	-748.11786	-695.90641	-649.40139	-607.13372	-568.15790	-531.83942	-497.72753
8.25	-994.00000	-937.50896	-871.39622	-789.86262	-713.72412	-652.46895	-599.34328	-551.37344	-507.28046	-466.28523	-427.84863
9.25	-960.00000	-905.08369	-843.91489	-777.35274	-711.68705	-652.29305	-599.33113	-551.37271	-507.28042	-466.28523	-427.84863
9.24	-960.00000	-905.08298	-843.58943	-770.39993	-686.87953	-613.83698	-553.08061	-499.12092	-449.79375	-404.07327	-361.30265
10.24	-922.00000	-868.53862	-809.94619	-745.67360	-677.89375	-612.47401	-552.95200	-499.11112	-449.79311	-404.07323	-361.30265
10.23	-922.00000	-868.53859	-809.91505	-744.22530	-665.79430	-582.69918	-511.53938	-450.73788	-395.90792	-345.34551	-298.18997
11.23	-880.00000	-827.96092	-771.64517	-710.42838	-644.19392	-575.86535	-510.51560	-450.63461	-395.89956	-345.34494	-298.18993
11.22	-880.00000	-827.96092	-771.64293	-710.24012	-640.87090	-559.26763	-477.17523	-407.04993	-345.96487	-290.30329	-238.64579
12.22	-834.00000	-783.39517	-729.15119	-670.79119	-607.71879	-540.26926	-471.64710	-406.21629	-345.87709	-290.29597	-238.64530
12.21	-834.00000	-783.39517	-729.15106	-670.77325	-607.18471	-534.78753	-451.47015	-370.31593	-300.72278	-239.27144	-182.86204
13.21	-784.00000	-734.86860	-682.58901	-626.79875	-566.98339	-502.63445	-434.42933	-365.60817	-300.00806	-239.19472	-182.85565
13.20	-784.00000	-734.86860	-682.58900	-626.79741	-566.92267	-501.60428	-426.90094	-342.70329	-262.28201	-192.95576	-131.14651
14.20	-730.00000	-682.39868	-632.03951	-578.63381	-521.78803	-460.97336	-395.77831	-327.10140	-258.12524	-192.32627	-131.07929
14.19	-730.00000	-682.39868	-632.03951	-578.63373	-521.78268	-460.84118	-394.19070	-317.82578	-233.18531	-153.32329	-84.15721
15.19	-672.00000	-625.99703	-577.55500	-526.43470	-472.32587	-414.81220	-353.36030	-287.63762	-218.67356	-149.57900	-83.60010
15.18	-672.00000	-625.99703	-577.55500	-526.43470	-472.32549	-414.79918	-353.13829	-285.51002	-207.98429	-123.12829	-43.73506
16.18	-610.00000	-565.67171	-519.17068	-470.29311	-418.78416	-364.31540	-306.45082	-244.65244	-178.63985	-109.51801	-40.35002
16.17	-610.00000	-565.67171	-519.17068	-470.29311	-418.78414	-364.31437	-306.42723	-244.33599	-176.03261	-97.69924	-12.75499
17.17	-544.00000	-501.42841	-456.91107	-410.27099	-361.29239	-309.70610	-255.16597	-197.21556	-135.30587	-69.18265	0.00000
17.16	-544.00000	-501.42841	-456.91107	-410.27099	-361.29239	-309.70604	-255.16396	-197.18013	-134.90049	-66.16307	12.75499
18.16	-474.00000	-433.27125	-390.79374	-346.41174	-299.93871	-251.14715	-199.75303	-145.39073	-87.57963	-25.74929	40.35002
18.15	-474.00000	-433.27125	-390.79374	-346.41174	-299.93871	-251.14715	-199.75289	-145.38757	-87.53260	-25.26441	43.73506
19.15	-400.00000	-361.20331	-320.83160	-278.74672	-234.78514	-188.75053	-140.40275	-89.44092	-35.47644	22.00218	83.60010
19.14	-400.00000	-361.20331	-320.83160	-278.74672	-234.78514	-188.75053	-140.40274	-89.44069	-35.47210	22.05979	84.15721
20.14	-322.00000	-285.22689	-247.03433	-207.29913	-165.87672	-122.59535	-77.24785	-29.57955	20.73054	74.09946	131.07929
20.13	-322.00000	-285.22689	-247.03433	-207.29913	-165.87672	-122.59535	-77.24785	-29.57953	20.73086	74.10488	131.14651
21.13	-240.00000	-205.34375	-169.40929	-132.08652	-93.24701	-52.73931	-10.38248	34.04315	80.80964	130.26269	182.85565
21.12	-240.00000	-205.34375	-169.40929	-132.08652	-93.24701	-52.73931	-10.38248	34.04315	80.80966	130.26310	182.86204
22.12	-154.00000	-121.55527	-87.96219	-53.12237	-16.92154	20.77433	60.12437	101.32012	144.59608	190.24553	238.64530
22.11	-154.00000	-121.55527	-87.96219	-53.12237	-16.92154	20.77433	60.12437	101.32012	144.59608	190.24556	238.64579
23.11	-64.00000	-33.86252	-2.69749	29.58280	63.07990	97.91247	134.22075	172.17259	211.97192	253.87117	298.18993
23.10	-64.00000	-33.86252	-2.69749	29.58280	63.07990	97.91247	134.22075	172.17259	211.97192	253.87117	298.18997
24.10	30.00000	57.73363	86.38126	116.02069	146.74178	178.64930	211.86665	246.54094	282.85002	321.01251	361.30265
24.9	30.00000	57.73363	86.38126	116.02069	146.74178	178.64930	211.86665	246.54094	282.85002	321.01251	361.30265
25.9	128.00000	153.23251	179.27123	206.18464	234.05173	262.96441	293.03071	324.37905	357.16414	391.57531	427.84863
25.8	128.00000	153.23251	179.27123	206.18464	234.05173	262.96441	293.03071	324.37905	357.16414	391.57531	427.84863
26.8	230.00000	252.63355	275.97010	300.06928	324.99979	350.84148	377.68802	405.65058	434.86277	465.48760	497.72753
26.7	230.00000	252.63355	275.97010	300.06928	324.99979	350.84148	377.68802	405.65058	434.86277	465.48760	497.72753
27.7	336.00000	355.93630	376.47601	397.67024	419.57787	442.26725	465.81847	490.32646	515.90514	542.69320	570.86246
27.6	336.00000	355.93630	376.47601	397.67024	419.57787	442.26725	465.81847	490.32646	515.90514	542.69320	570.86246
28.6	446.00000	463.14039	480.78739	498.98392	517.77933	537.23090	557.40571	578.38317	600.25848	623.14741	647.19310
28.5	446.00000	463.14039	480.78739	498.98392	517.77933	537.23090	557.40571	578.38317	600.25848	623.14741	647.19310
29.5	560.00000	574.24549	588.90298	604.00734	619.59868	635.72350	652.43628	669.80144	687.89615	706.81417	726.67131
29.4	560.00000	574.24549	588.90298	604.00734	619.59868	635.72350	652.43628	669.80144	687.89615	706.81417	726.67131
30.4	678.00000	689.25136	700.82170	712.73800	725.03132	737.73763	750.89903	764.56535	778.79619	793.66398	809.25806
30.3	678.00000	689.25136	700.82170	712.73800	725.03132	737.73763	750.89903	764.56535	778.79619	793.66398	809.25806
31.3	800.00000	808.15776	816.54264	825.17383	834.07343	843.26705	852.78464	862.66163	872.94042	883.67248	894.92129
31.2	800.00000	808.15776	816.54264	825.17383	834.07343	843.26705	852.78464	862.66163	872.94042	883.67248	894.92129
32.2	926.00000	930.96453	936.06504	941.31304	946.72176	952.30652	958.08524	964.07914	970.31359	976.81936	983.63444
32.1	926.00000	930.96453	936.06504	941.31304	946.72176	952.30652	958.08524	964.07914	970.31359	976.81936	983.63444
33.1	1056.00000	1057.67149	1059.38826	1061.15415	1062.97355	1064.85158	1066.79420	1068.80847	1070.90283	1073.08754	1075.37523
33.0	1056.00000	1057.67149	1059.38826	1061.15415	1062.97355	1064.85158	1066.79420	1068.80847	1070.90283	1073.08754	1075.37523

J = 34

0.34	-1190.00000	-1175.11254	-1168.74035	-1163.85865	-1159.74777	-1156.12931	-1152.86056	-1149.85677	-1147.06274	-1144.44011	-1141.9609
1.34	-1188.00000	-1175.11248	-1168.74035	-1163.85865	-1159.74777	-1156.12931	-1152.86056	-1149.85677	-1147.06274	-1144.44011	-1141.9609
1.33	-1188.00000	-1146.50112	-1127.38702	-1112.77265	-1100.47859	-1089.66540	-1079.90338	-1070.93756	-1062.60190	-1054.78114	-1047.3913
2.33	-1182.00000	-1146.49809	-1127.38702	-1112.77265	-1100.47859	-1089.66540	-1079.90338	-1070.93756	-1062.60190	-1054.78114	-1047.3913
2.32	-1182.00000	-1120.41805	-1088.45279	-1064.13844	-1043.72289	-1025.78927	-1009.61526	-994.77304	-980.98441	-968.05637	-955.8485
3.32	-1172.00000	-1120.34905	-1088.45246	-1064.13844	-1043.72289	-1025.78927	-1009.61526	-994.77304	-980.98441	-968.05637	-955.8485
3.31	-1172.00000	-1097.67895	-1052.15227	-1018.09141	-989.57791	-964.57443	-942.05311	-921.40783	-902.24534	-884.29329	-867.3537
4.31	-1158.00000	-1096.85094	-1052.14488	-1018.09131	-989.57791	-964.57442	-942.05311	-921.40783	-902.24534	-884.29329	-867.3537
4.30	-1158.00000	-1080.31938	-1018.85032	-974.81553	-938.16885	-906.11277	-877.28683	-850.89599	-826.42681	-803.52470	-781.9325
5.30	-1140.00000	-1075.41660	-1018.74202	-974.81348	-938.16880	-906.11276	-877.28683	-850.89599	-826.42681	-803.52470	-781.9325
5.29	-1140.00000	-1067.45055	-989.47684	-934.58688	-889.66293	-850.52206	-815.40380	-783.30403	-753.58003	-725.79016	-699.6151
6.29	-1118.00000	-1053.93487	-988.44646	-934.55593	-889.66202	-850.52204	-815.40380	-783.30403	-753.58003	-725.79016	-699.6151
6.28	-1118.00000	-1051.95447	-966.34988	-897.94960	-844.29980	-797.95834	-756.51581	-718.71518	-683.76814	-651.13787	-620.4384
7.28	-1092.00000	-1029.81416	-960.69828	-897.61727	-844.28617	-797.95781	-756.51579	-718.71518	-683.76814	-651.13787	-620.4384
7.27	-1092.00000	-1029.56287	-949.14875	-866.42816	-802.49329	-748.64043	-700.77033	-657.23592	-617.07027	-579.62745	-544.4474
8.27	-1062.00000	-1001.68872	-933.37854	-864.02566	-802.34089	-748.63265	-700.76997	-657.23590	-617.07027	-579.62745	-544.4474
8.26	-1062.00000	-1001.66824	-930.01877	-842.57152	-765.27549	-702.92190	-648.37196	-599.00650	-553.58782	-511.33380	-471.6979
9.26	-1028.00000	-969.34040	-903.68574	-832.65073	-764.04492	-702.83377	-648.36673	-599.00622	-553.58781	-511.33380	-471.6979
9.25	-1028.00000	-969.33919	-903.15826	-823.04475	-735.22837	-661.59410	-599.63952	-544.22038	-493.45433	-446.35308	-402.2602
10.25	-990.00000	-932.85897	-869.95751	-800.83127	-728.86405	-660.84484	-599.58017	-544.21649	-493.45411	-446.35307	-402.2602
10.24	-990.00000	-932.85891	-869.90255	-798.47891	-712.71679	-626.83152	-555.23423	-493.17285	-436.85354	-384.81212	-336.2249
11.24	-948.00000	-892.32737	-831.86108	-765.88637	-694.66784	-622.45221	-554.71592	-493.12873	-436.85047	-384.81195	-336.2248
11.23	-948.00000	-892.32736	-831.85681	-765.54398	-689.32390	-600.77415	-516.98457	-446.44535	-384.06231	-326.88525	-273.7115
12.23	-902.00000	-847.79617	-789.51994	-726.60882	-658.44036	-586.00343	-513.73610	-446.05400	-384.02751	-326.88276	-273.7113
12.22	-902.00000	-847.79617	-789.51967	-726.57316	-657.44441	-577.28727	-487.71446	-405.61722	-335.60663	-272.83298	-214.8846
13.22	-852.00000	-799.29623	-743.07306	-682.90654	-618.19151	-548.49758	-475.34410	-403.05566	-335.29438	-272.80475	-214.8826
13.21	-852.00000	-799.29623	-743.07305	-682.90366	-618.06584	-546.53332	-463.45467	-373.71392	-292.88312	-223.13734	-159.9886
14.21	-798.00000	-746.84730	-692.61302	-634.96346	-573.43293	-507.41230	-436.67990	-363.09248	-290.77622	-222.88110	-159.9657
14.20	-798.00000	-746.84730	-692.61301	-634.96327	-573.42081	-507.12740	-433.58686	-348.46171	-258.91823	-179.04999	-109.4659
15.20	-740.00000	-690.46252	-638.19960	-582.93720	-524.31267	-461.83076	-394.89789	-323.47337	-249.61905	-177.27721	-109.2548
15.19	-740.00000	-690.46252	-638.19960	-582.93719	-524.31173	-461.79977	-394.39928	-319.23438	-232.70306	-143.50987	-64.4405
16.19	-678.00000	-630.15096	-579.87285	-526.93364	-471.03873	-411.80115	-348.70192	-281.16680	-209.29604	-135.28478	-62.9437
16.18	-678.00000	-630.15096	-579.87285	-526.93364	-471.03867	-411.79846	-348.64281	-280.42519	-203.97563	-116.46719	-27.7127
17.18	-612.00000	-565.91902	-517.66060	-467.02397	-413.76161	-357.56029	-298.00974	-234.56796	-166.67360	-94.53267	-20.44804
17.17	-612.00000	-565.91902	-517.66060	-467.02397	-413.76161	-357.56010	-298.00419	-234.47432	-165.68071	-88.21386	0.00000
18.17	-542.00000	-497.77138	-451.58275	-403.25782	-352.58364	-299.29780	-243.06792	-183.45887	-119.90193	-51.83687	20.44804
18.16	-542.00000	-497.77138	-451.58275	-403.25782	-352.58364	-299.29779	-243.06750	-183.44955	-119.77056	-50.59373	27.7127
19.16	-468.00000	-425.71148	-381.65389	-335.67089	-287.57631	-237.14486	-184.09820	-128.08264	-68.63415	-5.14982	62.9437
19.15	-468.00000	-425.71148	-381.65389	-335.67089	-287.57631	-237.14486	-184.09817	-128.08188	-68.62055	-4.97939	64.44055
20.15	-390.00000	-349.74191	-307.88494	-264.28955	-218.79122	-171.19341	-121.25732	-68.68643	-13.10191	45.99569	109.25484
20.14	-390.00000	-349.74191	-307.88494	-264.28955	-218.79122	-171.19341	-121.25732	-68.68638	-13.10076	46.01384	109.46593
21.14	-308.00000	-269.86466	-230.28423	-189.13367	-146.26669	-101.51004	-54.65562	-5.44904	46.42782	101.38461	159.96571
21.13	-308.00000	-269.86466	-230.28423	-189.13367	-146.26669	-101.51004	-54.65562	-5.44903	46.42790	101.38617	159.98864
22.13	-222.00000	-186.08129	-148.85816	-110.21850	-70.03177	-28.14435	15.62681	61.50313	109.75755	160.73458	214.88265
22.12	-222.00000	-186.08129	-148.85816	-110.21850	-70.03177	-28.14435	15.62681	61.50313	109.75755	160.73469	214.88465
23.12	-132.00000	-98.39299	-63.61179	-27.55594	9.89110	48.86583	89.53003	132.07799	176.74691	223.83222	273.71139
23.11	-132.00000	-98.39299	-63.61179	-27.55594	9.89110	48.86583	89.53003	132.07799	176.74691	223.83223	273.71153
24.11	-38.00000	-6.80074	25.45090	58.84464	93.48430	129.49111	167.00816	206.20645	247.29342	290.52517	336.22489
24.10	-38.00000	-6.80074	25.45090	58.84464	93.48430	129.49111	167.00816	206.20645	247.29342	290.52517	336.22490
25.10	60.00000	88.69470	118.32669	148.97572	180.73384	213.70831	248.02536	283.83537	321.31990	360.70186	402.26026
25.9	60.00000	88.69470	118.32669	148.97572	180.73384	213.70831	248.02536	283.83537	321.31990	360.70186	402.26026
26.9	162.00000	188.09269	215.01302	242.83126	271.62847	301.49888	332.55322	364.92305	398.76676	434.27800	471.69796
26.8	162.00000	188.09269	215.01302	242.83126	271.62847	301.49888	332.55322	364.92305	398.76676	434.27800	471.69796
27.8	268.00000	291.39272	315.50774	340.40633	366.15904	392.84784	420.56891	449.43631	479.58708	511.18834	544.44749
27.7	268.00000	291.39272	315.50774	340.40633	366.15904	392.84784	420.56891	449.43631	479.58708	511.18834	544.44749
28.7	378.00000	398.59437	419.80913	441.69686	464.31805	487.74294	512.05386	537.34831	563.74330	591.38125	620.43849
28.6	378.00000	398.59437	419.80913	441.69686	464.31805	487.74294	512.05386	537.34831	563.74330	591.38125	620.43849
29.6	492.00000	509.69729	527.91574	546.69949	566.09931	586.17409	606.99282	628.63716	651.20499	674.81529	699.61516
29.5	492.00000	509.69729	527.91574	546.69949	566.09931	586.17409	606.99282	628.63716	651.20499	674.81529	699.61516
30.5	610.00000	624.70119	639.82637	655.41143	671.49765	688.13289	705.37317	723.28479	741.94714	761.45668	781.93252
30.4	610.00000	624.70119	639.82637	655.41143	671.49765	688.13289	705.37317	723.28479	741.94714	761.45668	781.93252
31.4	732.00000	743.60581	755.53999	767.83032	780.50875	793.61231	807.18436	821.27616	835.94907	851.27762	867.35377
31.3	732.00000	743.60581	755.53999	767.83032	780.50875	793.61231	807.18436	821.27616	835.94907	851.27762	867.35377
32.3	858.00000	866.41096	875.05575	883.95417	893.12895	902.60645	912.41753	922.59865	933.19347	944.25497	955.84853
32.2	858.00000	866.41096	875.05575	883.95417	893.12895	902.60645	912.41753	922.59865	933.19347	944.25497	955.84853
33.2	988.00000	993.11645	998.37293	1003.78129	1009.35514	1015.11028	1021.06516	1027.24162	1033.66577	1040.36933	1047.39139
33.1	988.00000	993.11645	998.37293	1003.78129	1009.35514	1015.11028	1021.06516	1027.24162	1033.66577	1040.36933	1047.39139
34.1	1122.00000	1123.72213	1125.49089	1127.31024	1129.18469	1131.11952	1133.12087	1135.19600	1137.35361	1139.60427	1141.96098
34.0	1122.00000	1123.72213	1125.49089	1127.31024	1129.18469	1131.11952	1133.12087	1135.19600	1137.35361	1139.60427	1141.96098

J = 35

0.35	-1260.00000	-1244.66479	-1238.10758	-1233.08384	-1228.85318	-1225.12918	-1221.76502	-1218.67348	-1215.79776	-1213.09842	-1210.54672
1.35	-1258.00000	-1244.66476	-1238.10758	-1233.08384	-1228.85318	-1225.12918	-1221.76502	-1218.67348	-1215.79776	-1213.09842	-1210.54672
1.34	-1258.00000	-1215.15338	-1195.48653	-1180.44669	-1167.79369	-1156.66415	-1146.61606	-1137.38713	-1128.80655	-1120.75573	-1113.14837
2.34	-1252.00000	-1215.15129	-1195.48653	-1180.44669	-1167.79369	-1156.66415	-1146.61606	-1137.38713	-1128.80655	-1120.75573	-1113.14837
2.33	-1252.00000	-1188.14401	-1155.27638	-1130.25601	-1109.24380	-1090.78395	-1074.13388	-1058.85368	-1044.65727	-1031.34613	-1018.77587
3.33	-1242.00000	-1188.09453	-1155.27619	-1130.25601	-1109.24380	-1090.78395	-1074.13388	-1058.85368	-1044.65727	-1031.34613	-1018.77587
3.32	-1242.00000	-1164.34564	-1117.67967	-1082.64105	-1053.29666	-1027.55909	-1004.37311	-983.11600	-963.38363	-944.89605	-927.44980
4.32	-1228.00000	-1163.71583	-1117.67519	-1082.64100	-1053.29665	-1027.55909	-1004.37311	-983.11600	-963.38363	-944.89605	-927.44980
4.31	-1228.00000	-1145.70039	-1083.02277	-1037.77495	-1000.07099	-967.07704	-937.40041	-910.22570	-885.02591	-861.43688	-839.19446
5.31	-1210.00000	-1141.59708	-1082.95363	-1037.77385	-1000.07097	-967.07704	-937.40041	-910.22570	-885.02591	-861.43688	-839.19446
5.30	-1210.00000	-1132.22393	-1052.06884	-995.90862	-949.72326	-909.44875	-873.29842	-840.24583	-809.63270	-781.00620	-754.03877
6.30	-1188.00000	-1119.82095	-1051.36380	-995.89120	-949.72282	-909.44874	-873.29842	-840.24583	-809.63270	-781.00620	-754.03877
6.29	-1188.00000	-1117.21098	-1026.91603	-957.49417	-902.47163	-854.81924	-812.17171	-773.25456	-737.26347	-703.64949	-672.01748
7.29	-1162.00000	-1095.72556	-1022.57956	-957.29505	-902.46473	-854.81901	-812.17170	-773.25456	-737.26347	-703.64949	-672.01748
7.28	-1162.00000	-1095.36100	-1008.52407	-923.72480	-858.66589	-803.38697	-754.15639	-709.35083	-667.99207	-629.42254	-593.17277
8.28	-1132.00000	-1067.71679	-994.78791	-922.14473	-858.58377	-803.38338	-754.15625	-709.35082	-667.99207	-629.42254	-593.17277
8.27	-1132.00000	-1067.68480	-990.08603	-897.25339	-819.08606	-755.45299	-699.43690	-648.66308	-601.91199	-558.39474	-517.55650
9.27	-1098.00000	-1035.46637	-965.14734	-889.68478	-818.36498	-755.40970	-699.43469	-648.66298	-601.91199	-558.39474	-517.55650
9.26	-1098.00000	-1035.46435	-964.31317	-876.98233	-785.87863	-711.61176	-648.28487	-591.36449	-539.14460	-490.65412	-445.23337
10.26	-1060.00000	-999.05475	-931.66173	-857.55017	-781.59617	-711.21261	-648.25808	-591.36297	-539.14453	-490.65412	-445.23337
10.25	-1060.00000	-999.05465	-931.56689	-853.88073	-761.07444	-673.48699	-601.19643	-537.70572	-479.85305	-426.31498	-376.28561
11.25	-1018.00000	-958.57366	-893.78834	-822.84639	-746.62707	-670.83739	-600.94270	-537.68734	-479.85195	-426.31493	-376.28560
11.24	-1018.00000	-958.57366	-893.78038	-822.24248	-738.51989	-643.99364	-559.45462	-488.12210	-424.27103	-365.53083	-310.82014
12.24	-972.00000	-914.08027	-851.61579	-783.94565	-710.48574	-633.18164	-557.65971	-487.94500	-424.25763	-365.53000	-310.82010
12.23	-972.00000	-914.08027	-851.61526	-783.87680	-708.70540	-620.29482	-525.99391	-443.67667	-372.79067	-308.52001	-248.98055
13.23	-922.00000	-865.60945	-805.29588	-740.56333	-670.71099	-595.52693	-517.67890	-442.37141	-372.65965	-308.50995	-248.97994
13.22	-922.00000	-865.60945	-805.29586	-740.55730	-670.45992	-591.99021	-500.42042	-407.07996	-326.32257	-255.63987	-190.97010
14.22	-868.00000	-813.18346	-754.93409	-692.86613	-626.42889	-554.94862	-478.60707	-400.48706	-325.32716	-255.54012	-190.96255
14.21	-868.00000	-813.18346	-754.93409	-692.86570	-626.40239	-554.36077	-473.01297	-379.49796	-287.30992	-207.68498	-137.11490
15.21	-810.00000	-756.81710	-700.59859	-641.03078	-577.68921	-509.98746	-437.31731	-360.20486	-281.95515	-206.90676	-137.03877
15.20	-810.00000	-756.81710	-700.59859	-641.03075	-577.68699	-509.91662	-436.25548	-352.50429	-257.86293	-166.83724	-88.11207
16.20	-748.00000	-696.52055	-642.33488	-585.17914	-524.71270	-460.47881	-391.86679	-318.34683	-240.72475	-162.43329	-87.50192
16.19	-748.00000	-696.52055	-642.33488	-585.17914	-524.71255	-460.47208	-391.72550	-316.71692	-231.00724	-135.76071	-45.89141
17.19	-682.00000	-632.30103	-580.17443	-525.39296	-467.67293	-406.64935	-341.83528	-272.60042	-198.48975	-120.53772	-42.27651
17.18	-682.00000	-632.30103	-580.17443	-525.39296	-467.67292	-406.64883	-341.82061	-272.36614	-196.23722	-108.92694	-13.41811
18.18	-612.00000	-564.16377	-514.13973	-461.72889	-406.68858	-348.71614	-287.42195	-222.28777	-152.66213	-78.15424	0.00000
18.17	-612.00000	-564.16377	-514.13973	-461.72889	-406.68858	-348.71611	-287.42071	-222.26181	-152.31766	-75.24312	13.41811
19.17	-538.00000	-492.11263	-444.24723	-394.22749	-341.84199	-286.83266	-228.87677	-167.55762	-102.32158	-32.48928	42.27651
19.16	-538.00000	-492.11263	-444.24723	-394.22749	-341.84199	-286.83266	-228.87668	-167.55530	-102.28148	-32.02030	45.89141
20.16	-460.00000	-416.15054	-370.50922	-322.91863	-273.19225	-221.10577	-166.38432	-108.68253	-47.55216	17.60324	87.50192
20.15	-460.00000	-416.15054	-370.50922	-322.91863	-273.19225	-221.10577	-166.38431	-108.68236	-47.54840	17.65997	88.11207
21.15	-378.00000	-336.27973	-292.93506	-247.82478	-200.78299	-151.61226	-100.07399	-45.87406	11.35966	72.11378	137.03877
21.14	-378.00000	-336.27973	-292.93506	-247.82478	-200.78299	-151.61226	-100.07399	-45.87405	11.35995	72.11927	137.11490
22.14	-292.00000	-252.50192	-211.53197	-168.96315	-124.64720	-78.40898	-30.03887	20.71824	74.17933	130.74919	190.96255
22.13	-292.00000	-252.50192	-211.53197	-168.96315	-124.64720	-78.40898	-30.03887	20.71824	74.17935	130.74963	190.97010
23.13	-202.00000	-164.81848	-126.30560	-86.34714	-44.81030	-1.53908	43.65197	90.98659	140.73944	193.25536	248.97994
23.12	-202.00000	-164.81848	-126.30560	-86.34714	-44.81030	-1.53908	43.65197	90.98659	140.73944	193.25539	248.98055
24.12	-108.00000	-73.23049	-37.26045	0.01268	38.70781	78.96402	120.94592	164.85084	210.91912	259.44907	310.82010
24.11	-108.00000	-73.23049	-37.26045	0.01268	38.70781	78.96402	120.94592	164.85084	210.91912	259.44907	310.82014
25.11	-10.00000	22.26119	55.59989	90.10788	125.89132	163.07405	201.80209	242.24980	284.62831	329.19778	376.28560
25.10	-10.00000	22.26119	55.59989	90.10788	125.89132	163.07405	201.80209	242.24980	284.62831	329.19778	376.28561
26.10	92.00000	121.65585	152.27251	183.93165	216.72756	250.77002	286.18817	323.13570	361.79806	402.40257	445.23337
26.9	92.00000	121.65585	152.27251	183.93165	216.72756	250.77002	286.18817	323.13570	361.79806	402.40257	445.23337
27.9	198.00000	224.95292	252.75504	281.47845	311.20623	342.03501	374.07824	407.47066	442.37440	478.98753	517.55650
27.8	198.00000	224.95292	252.75504	281.47845	311.20623	342.03501	374.07824	407.47066	442.37440	478.98753	517.55650
28.8	308.00000	332.15192	357.04552	382.74371	409.31889	436.85520	465.45129	495.22416	526.31433	558.89306	593.17277
28.7	308.00000	332.15192	357.04552	382.74371	409.31889	436.85520	465.45129	495.22416	526.31433	558.89306	593.17277
29.7	422.00000	443.25246	465.14234	487.72366	511.05857	535.21920	560.29009	586.37136	613.58313	642.07153	672.01748
29.6	422.00000	443.25246	465.14234	487.72366	511.05857	535.21920	560.29009	586.37136	613.58313	642.07153	672.01748
30.6	540.00000	558.25420	577.04414	596.41516	616.41946	637.11757	658.58038	680.89179	704.15237	728.48433	754.03877
30.5	540.00000	558.25420	577.04414	596.41516	616.41946	637.11757	658.58038	680.89179	704.15237	728.48433	754.03877
31.5	662.00000	677.15689	692.74977	708.81557	725.39671	742.54241	760.31028	778.76845	797.99857	818.09976	839.19446
31.4	662.00000	677.15689	692.74977	708.81557	725.39671	742.54241	760.31028	778.76845	797.99857	818.09976	839.19446
32.4	788.00000	799.96027	812.25829	824.92265	837.98621	851.48706	865.46978	879.98710	895.10213	910.89151	927.44980
32.3	788.00000	799.96027	812.25829	824.92265	837.98621	851.48706	865.46978	879.98710	895.10213	910.89151	927.44980
33.3	918.00000	926.66415	935.56887	944.73452	954.18448	963.94588	974.05045	984.53572	995.44658	1006.83755	1018.77587
33.2	918.00000	926.66415	935.56887	944.73452	954.18448	963.94588	974.05045	984.53572	995.44658	1006.83755	1018.77587
34.2	1052.00000	1057.26837	1062.68082	1068.24954	1073.98853	1079.91405	1086.04509	1092.40410	1099.01797	1105.91931	1113.14837
34.1	1052.00000	1057.26837	1062.68082	1068.24954	1073.98853	1079.91405	1086.04509	1092.40410	1099.01797	1105.91931	1113.14837
35.1	1190.00000	1191.77277	1193.59352	1195.46632	1197.39583	1199.38746	1201.44753	1203.58353	1205.80440	1208.12100	1210.54672
35.0	1190.00000	1191.77277	1193.59352	1195.46632	1197.39583	1199.38746	1201.44753	1203.58353	1205.80440	1208.12100	1210.54672

J = 36

0.36	-1332.00000	-1316.21709	-1309.47484	-1304.30903	-1299.95860	-1296.12906	-1292.66948	-1289.49018	-1286.53279	-1283.75673	-1281.13247
1.36	-1330.00000	-1316.21706	-1309.47484	-1304.30903	-1299.95860	-1296.12906	-1292.66948	-1289.49018	-1286.53279	-1283.75673	-1281.13247
1.35	-1330.00000	-1285.80617	-1265.58624	-1250.12087	-1237.10889	-1225.66297	-1215.32880	-1205.83675	-1197.01123	-1188.73035	-1180.90537
2.35	-1324.00000	-1285.80474	-1265.58623	-1250.12087	-1237.10889	-1225.66297	-1215.32880	-1205.83675	-1197.01123	-1188.73035	-1180.90537
2.34	-1324.00000	-1257.87494	-1224.10096	-1198.37420	-1176.76516	-1157.77897	-1140.65275	-1124.93453	-1110.33028	-1096.63601	-1083.70331
3.34	-1314.00000	-1257.83962	-1224.10085	-1198.37420	-1176.76516	-1157.77897	-1140.65275	-1124.93453	-1110.33028	-1096.63601	-1083.70331
3.33	-1314.00000	-1233.04322	-1185.21103	-1149.19270	-1119.01677	-1092.54478	-1068.69389	-1046.82477	-1026.52239	-1007.49918	-989.54612
4.33	-1300.00000	-1232.56828	-1185.20832	-1149.19267	-1119.01677	-1092.54478	-1068.69389	-1046.82477	-1026.52239	-1007.49918	-989.54612
4.32	-1300.00000	-1213.10563	-1149.21261	-1102.73993	-1063.97663	-1030.04383	-999.51589	-971.55687	-945.62612	-921.34993	-898.45708
5.32	-1282.00000	-1209.72173	-1149.16885	-1102.73934	-1063.97662	-1030.04383	-999.51589	-971.55687	-945.62612	-921.34993	-898.45708
5.31	-1282.00000	-1198.85304	-1116.73975	-1059.24659	-1011.79174	-970.38099	-933.19709	-899.19070	-867.68772	-838.22404	-810.46377
6.31	-1260.00000	-1187.59332	-1116.26429	-1059.23688	-1011.79153	-970.38098	-933.19709	-899.19070	-867.68772	-838.22404	-810.46377
6.30	-1260.00000	-1184.21806	-1089.60184	-1019.09849	-962.66268	-913.69163	-869.83573	-829.79995	-792.76335	-758.16457	-725.59910
7.30	-1234.00000	-1163.49465	-1086.36060	-1018.98088	-962.65922	-913.69153	-869.83572	-829.79995	-792.76335	-758.16457	-725.59910
7.29	-1234.00000	-1162.97395	-1069.66906	-983.22735	-916.89351	-860.15753	-809.55802	-763.47690	-720.92215	-681.22387	-643.90275
8.29	-1204.00000	-1135.60399	-1057.97038	-982.21482	-916.84993	-860.15589	-809.55796	-763.47690	-720.92215	-681.22387	-643.90275
8.28	-1204.00000	-1135.55478	-1051.60007	-954.05669	-875.08956	-810.04206	-752.53191	-700.33990	-652.25079	-607.46651	-565.42310
9.28	-1170.00000	-1103.45965	-1028.30361	-948.50596	-874.67751	-810.02115	-752.53098	-700.33986	-652.25078	-607.46651	-565.42310
9.27	-1170.00000	-1103.45634	-1027.01721	-932.36648	-838.91593	-763.81127	-698.99504	-640.54557	-586.86026	-536.97350	-490.21996
10.27	-1132.00000	-1067.12445	-995.05355	-915.86299	-836.16441	-763.60410	-698.98319	-640.54499	-586.86024	-536.97350	-490.21996
10.26	-1132.00000	-1067.12428	-994.89348	-910.38013	-811.17483	-722.63960	-649.33970	-584.31247	-524.89708	-469.84865	-418.36846
11.26	-1090.00000	-1026.69860	-957.42100	-881.30952	-800.14673	-721.11095	-649.21893	-584.30499	-524.89669	-469.84863	-418.36846
11.25	-1090.00000	-1026.69860	-957.40652	-880.27728	-788.54617	-689.34768	-604.46391	-531.98627	-466.56395	-406.22873	-349.96496
12.25	-1044.00000	-982.24649	-915.43394	-842.79078	-763.88000	-681.92524	-603.51918	-531.90855	-466.55891	-406.22846	-349.96495
12.24	-1044.00000	-982.24649	-915.43292	-842.66147	-760.84051	-664.12417	-566.77561	-484.29805	-412.17247	-346.30196	-285.13640
13.24	-994.00000	-933.80743	-869.25364	-799.75840	-724.53056	-643.78780	-561.60144	-483.66573	-412.11933	-346.29848	-285.13622
13.23	-994.00000	-933.80743	-869.25358	-799.74613	-724.04653	-637.81502	-538.36032	-443.25013	-362.34200	-290.35197	-224.05680
14.23	-940.00000	-881.40644	-818.99956	-752.33319	-680.75797	-603.57951	-521.67733	-439.49888	-361.89619	-290.31456	-224.05439
14.22	-940.00000	-881.40644	-818.99955	-752.33226	-680.70191	-602.42046	-512.36191	-411.75186	-318.74538	-238.91499	-166.98444
15.22	-882.00000	-825.06014	-764.74931	-700.70844	-632.44021	-559.25738	-480.63453	-398.01091	-315.94178	-238.59230	-166.95809
15.21	-882.00000	-825.06014	-764.74931	-700.70838	-632.43509	-559.10181	-478.49991	-385.35310	-284.54030	-193.39832	-114.38746
16.21	-820.00000	-764.77995	-706.55445	-645.02381	-579.79379	-510.32479	-435.91531	-356.23742	-273.17400	-191.26977	-114.15347
16.20	-820.00000	-764.77995	-706.55445	-645.02380	-579.79341	-510.30859	-435.59322	-352.90440	-257.37240	-156.91425	-67.45007
17.20	-754.00000	-700.57394	-644.45055	-585.37302	-523.01625	-456.95423	-386.60937	-311.28029	-230.83800	-147.52271	-65.83707
17.19	-754.00000	-700.57394	-644.45055	-585.37302	-523.01622	-456.95286	-386.57237	-310.72598	-226.16156	-128.81965	-29.08062
18.19	-684.00000	-632.44798	-578.46292	-521.82065	-462.24499	-399.38668	-332.78790	-261.83370	-185.82844	-104.83429	-21.40412
18.18	-684.00000	-632.44798	-578.46292	-521.82065	-462.24498	-399.38659	-332.78447	-261.76500	-184.98234	-98.71176	0.00000
19.18	-610.00000	-560.40640	-508.61005	-454.41272	-397.57482	-337.80092	-274.71646	-207.82917	-136.48374	-59.98931	21.40412
19.17	-610.00000	-560.40640	-508.61005	-454.41272	-397.57482	-337.80091	-274.71620	-207.82236	-136.37230	-58.79373	29.08062
20.17	-532.00000	-484.45245	-434.90575	-383.18301	-329.07333	-272.32125	-212.61052	-149.53828	-82.57273	-11.01116	65.83707
20.16	-532.00000	-484.45245	-434.90575	-383.18301	-329.07333	-272.32124	-212.61050	-149.53772	-82.56108	-10.84544	67.45007
21.16	-450.00000	-404.58864	-357.36052	-308.15684	-256.79016	-203.03620	-146.62195	-87.20746	-24.35665	42.50982	114.15347
21.15	-450.00000	-404.58864	-357.36052	-308.15684	-256.79016	-203.03620	-146.62195	-87.20742	-24.35565	42.52787	114.38746
22.15	-364.00000	-320.81689	-275.98244	-229.35359	-180.76268	-130.01091	-76.85901	-21.01375	37.89286	100.33729	166.95809
22.14	-364.00000	-320.81689	-275.98244	-229.35359	-180.76268	-130.01091	-76.85901	-21.01375	37.89294	100.33889	166.98444
23.14	-274.00000	-233.13873	-190.77787	-146.78832	-101.01964	-53.29453	-3.40137	48.91644	103.97612	162.17980	224.05439
23.13	-274.00000	-233.13873	-190.77787	-146.78832	-101.01964	-53.29453	-3.40137	48.91644	103.97612	162.17992	224.05680
24.13	-180.00000	-141.55537	-101.75181	-60.47290	-17.58347	27.07502	· 73.69073	122.49007	173.75015	227.81729	285.13622
24.12	-180.00000	-141.55537	-101.75181	-60.47290	-17.58347	27.07502	73.69073	122.49007	173.75015	227.81730	285.13640
25.12	-82.00000	-46.06778	-8.90830	29.58320	69.52804	111.06800	154.37066	199.63661	247.10967	297.09166	349.96495
25.11	-82.00000	-46.06778	-8.90830	29.58320	69.52804	111.06800	154.37066	199.63661	247.10967	297.09166	349.96496
26.11	20.00000	53.32325	87.74941	123.37237	160.30063	198.66072	238.60172	280.30139	323.97480	369.88645	418.36846
26.10	20.00000	53.32325	87.74941	123.37237	160.30063	198.66072	238.60172	280.30139	323.97480	369.88645	418.36846
27.10	126.00000	156.61709	188.21866	220.88837	254.72274	289.83411	326.35457	364.44122	404.28346	446.11321	490.21996
27.9	126.00000	156.61709	188.21866	220.88837	254.72274	289.83411	326.35457	364.44122	404.28346	446.11321	490.21996
28.9	236.00000	263.81320	292.49727	322.12612	352.78489	384.57261	417.60547	452.02145	487.98645	525.70306	565.42310
28.8	236.00000	263.81320	292.49727	322.12612	352.78489	384.57261	417.60547	452.02145	487.98645	525.70306	565.42310
29.8	350.00000	374.91115	400.58343	427.08138	454.47928	482.86343	512.33499	543.01391	575.04420	608.60131	643.90275
29.7	350.00000	374.91115	400.58343	427.08138	454.47928	482.86343	512.33499	543.01391	575.04420	608.60131	643.90275
30.7	468.00000	489.91056	512.47562	535.75063	559.79939	584.69596	610.52707	637.39549	665.42443	694.76380	725.59910
30.6	468.00000	489.91056	512.47562	535.75063	559.79939	584.69596	610.52707	637.39549	665.42443	694.76380	725.59910
31.6	590.00000	608.81113	628.17257	648.13092	668.73979	690.06133	712.16834	735.14700	759.10054	784.15444	810.46377
31.5	590.00000	608.81113	628.17257	648.13092	668.73979	690.06133	712.16834	735.14700	759.10054	784.15444	810.46377
32.5	716.00000	731.61259	747.67320	764.21976	781.29586	798.95207	817.24758	836.25240	856.05038	876.74335	898.45708
32.4	716.00000	731.61259	747.67320	764.21976	781.29586	798.95207	817.24758	836.25240	856.05038	876.74335	898.45708
33.4	846.00000	858.31473	870.97660	884.01501	897.46371	911.36187	925.75529	940.69816	956.25535	972.50561	989.54612
33.3	846.00000	858.31473	870.97660	884.01501	897.46371	911.36187	925.75529	940.69816	956.25535	972.50561	989.54612
34.3	980.00000	988.91735	998.08200	1007.51487	1017.24002	1027.28532	1037.68340	1048.47282	1059.69975	1071.42021	1083.70331
34.2	980.00000	988.91735	998.08200	1007.51487	1017.24002	1027.28532	1037.68340	1048.47282	1059.69975	1071.42021	1083.70331
35.2	1118.00000	1123.42029	1128.98871	1134.71778	1140.62193	1146.71783	1153.02502	1159.56659	1166.37018	1173.46932	1180.90537
35.1	1118.00000	1123.42029	1128.98871	1134.71778	1140.62193	1146.71783	1153.02502	1159.56659	1166.37018	1173.46932	1180.90537
36.1	1260.00000	1261.82341	1263.69616	1265.62241	1267.60697	1269.65540	1271.77420	1273.97106	1276.25518	1278.63773	1281.13247
36.0	1260.00000	1261.82341	1263.69616	1265.62241	1267.60697	1269.65540	1271.77420	1273.97106	1276.25518	1278.63773	1281.13247

J = 37

0.37	-1406.00000	-1389.76941	-1382.84211	-1377.53424	-1373.06402	-1369.12895	-1365.57394	-1362.30690	-1359.26782	-1356.41504	-1353.71822
1.37	-1404.00000	-1389.76940	-1382.84211	-1377.53424	-1373.06402	-1369.12895	-1365.57394	-1362.30690	-1359.26782	-1356.41504	-1353.71822
1.36	-1404.00000	-1358.45941	-1337.68612	-1321.79517	-1308.42418	-1296.66187	-1286.04160	-1276.28641	-1267.21595	-1258.70500	-1250.66239
2.36	-1398.00000	-1358.45843	-1337.68612	-1321.79517	-1308.42418	-1296.66187	-1286.04160	-1276.28641	-1267.21595	-1258.70500	-1250.66239
2.35	-1398.00000	-1329.60986	-1294.92643	-1268.49295	-1246.28692	-1226.77428	-1209.17186	-1193.01556	-1178.00344	-1163.92601	-1150.63084
3.35	-1388.00000	-1329.58475	-1294.92636	-1268.49295	-1246.28692	-1226.77428	-1209.17186	-1193.01556	-1178.00344	-1163.92601	-1150.63084
3.34	-1388.00000	-1303.76697	-1254.74572	-1217.74613	-1186.73813	-1159.53138	-1135.01536	-1112.53409	-1091.66158	-1072.10265	-1053.64269
4.34	-1374.00000	-1303.41156	-1254.74409	-1217.74612	-1186.73813	-1159.53138	-1135.01536	-1112.53409	-1091.66158	-1072.10265	-1053.64269
4.33	-1374.00000	-1282.54943	-1217.41605	-1169.70974	-1129.88541	-1095.01288	-1063.63306	-1034.88934	-1008.22736	-983.26377	-959.72031
5.33	-1356.00000	-1279.79837	-1217.38855	-1169.70943	-1129.88541	-1095.01288	-1063.63306	-1034.88934	-1008.22736	-983.26377	-959.72031
5.32	-1356.00000	-1267.36373	-1183.47286	-1124.59778	-1075.86735	-1033.31815	-995.09940	-960.13833	-927.74486	-897.44352	-868.89002
6.32	-1334.00000	-1257.25867	-1183.15625	-1124.59241	-1075.86725	-1033.31815	-995.09940	-960.13833	-927.74486	-897.44352	-868.89002
6.31	-1334.00000	-1252.97483	-1154.42903	-1082.74739	-1024.86978	-974.57410	-929.50691	-888.35067	-850.26727	-814.68274	-781.18308
7.31	-1308.00000	-1233.12226	-1152.06402	-1082.67880	-1024.86807	-974.57406	-929.50691	-888.35067	-850.26727	-814.68274	-781.18308
7.30	-1308.00000	-1232.39003	-1132.67483	-1044.89506	-977.16281	-918.94839	-866.97321	-819.61281	-775.85956	-735.03073	-696.63692
8.30	-1278.00000	-1205.34832	-1122.94977	-1044.26019	-977.14001	-918.94765	-866.97319	-819.61281	-775.85956	-735.03073	-696.63692
8.29	-1278.00000	-1205.27374	-1114.58897	-1013.08318	-933.23243	-866.67602	-807.65237	-754.03430	-704.60238	-658.54779	-615.29684
9.29	-1244.00000	-1173.31825	-1093.16201	-1009.16376	-933.00189	-866.66608	-807.65199	-754.03428	-704.60238	-658.54779	-615.29684
9.28	-1244.00000	-1173.31289	-1091.22946	-989.37661	-894.33213	-818.13820	-751.75616	-691.75794	-636.59785	-585.30886	-537.21835
10.28	-1206.00000	-1137.06649	-1060.12868	-975.81066	-892.63210	-818.03304	-751.75101	-691.75772	-636.59784	-585.30886	-537.21835
10.27	-1206.00000	-1137.06620	-1059.86427	-967.97520	-863.31181	-774.18978	-699.60916	-632.97778	-571.97866	-515.40878	-462.47049
11.27	-1164.00000	-1096.70093	-1022.75298	-941.28370	-855.30956	-773.34081	-699.55305	-632.97479	-571.97852	-515.40877	-462.47049
11.26	-1164.00000	-1096.70092	-1022.72718	-939.57657	-839.60233	-737.14070	-651.86100	-577.98134	-510.92382	-448.97058	-391.14070
12.26	-1118.00000	-1052.29381	-980.96935	-903.13459	-818.66499	-732.35583	-651.38247	-577.94813	-510.92197	-448.97049	-391.14069
12.25	-1118.00000	-1052.29381	-980.96740	-902.89841	-813.73192	-709.20560	-610.27129	-527.29872	-453.69291	-386.15948	-323.34236
13.25	-1068.00000	-1003.88931	-934.94228	-860.48058	-779.64616	-693.37195	-607.26652	-527.00438	-453.67198	-386.15830	-323.34230
13.24	-1068.00000	-1003.88931	-934.94216	-860.45628	-778.74655	-683.96205	-577.92169	-482.29629	-400.73693	-327.21151	-259.22721
14.24	-1014.00000	-951.51552	-884.80611	-813.35549	-736.40242	-653.32089	-566.04151	-480.31096	-400.54552	-327.19793	-259.22646
14.23	-1014.00000	-951.51552	-884.80610	-813.35350	-736.28769	-651.14502	-551.79450	-446.05816	-353.14802	-272.51896	-199.00971
15.23	-956.00000	-895.19102	-830.64896	-761.96281	-688.54932	-609.61887	-524.90069	-437.10659	-351.78611	-272.39107	-199.00091
15.22	-956.00000	-895.19102	-830.64896	-761.96268	-688.53794	-609.29060	-520.87965	-418.23108	-313.70960	-222.97359	-143.03407
16.22	-894.00000	-834.92860	-772.52919	-706.46153	-636.26885	-561.31406	-480.82728	-394.93898	-306.92637	-222.02150	-142.94861
16.21	-894.00000	-834.92860	-772.52919	-706.46152	-636.26794	-561.27651	-480.12786	-388.67037	-283.85582	-180.97811	-92.03606
17.21	-828.00000	-770.73728	-710.48688	-646.95894	-579.78083	-508.45433	-432.29763	-350.59609	-263.88101	-175.84007	-91.37178
17.20	-828.00000	-770.73728	-710.48688	-646.95894	-579.78077	-508.45088	-432.20834	-349.35941	-255.13075	-149.04074	-48.03286
18.20	-758.00000	-702.62360	-644.55047	-583.52860	-519.24385	-451.29277	-379.13611	-302.05304	-219.40654	-132.11473	-44.18739
18.19	-758.00000	-702.62360	-644.55047	-583.52860	-519.24385	-451.29251	-379.12703	-301.88038	-217.47087	-120.75223	-14.07731
19.19	-684.00000	-630.59239	-574.74071	-516.22264	-454.76706	-390.03583	-321.59341	-248.85709	-171.06871	-87.68284	0.00000
19.18	-684.00000	-630.59239	-574.74071	-516.22264	-454.76706	-390.03581	-321.59265	-248.83813	-170.77694	-84.88526	14.07731
20.18	-606.00000	-554.64728	-501.07307	-445.07917	-386.42771	-324.82806	-259.91629	-191.22128	-118.11156	-39.78906	44.18739
20.17	-606.00000	-554.64728	-501.07307	-445.07917	-386.42771	-324.82806	-259.91624	-191.21958	-118.07749	-39.33711	48.03286
21.17	-524.00000	-474.79107	-423.55928	-370.12670	-314.28221	-255.77163	-194.28287	-129.42280	-60.67887	12.63785	91.37178
21.16	-524.00000	-474.79107	-423.55928	-370.12670	-314.28221	-255.77163	-194.28287	-129.42267	-60.67563	12.69342	92.03606
22.16	-438.00000	-391.02591	-342.20840	-291.38698	-238.37288	-182.94110	-124.81924	-63.67059	0.93241	69.55232	142.94861
22.15	-438.00000	-391.02591	-342.20840	-291.38698	-238.37288	-182.94110	-124.81924	-63.67059	0.93266	69.55784	143.03407
23.15	-348.00000	-303.35349	-257.02749	-208.87691	-158.73211	-106.39244	-51.61732	5.88673	66.48559	130.64871	199.00091
23.14	-348.00000	-303.35349	-257.02749	-208.87691	-158.73211	-106.39244	-51.61732	5.88673	66.48561	130.64916	199.00971
24.14	-254.00000	-211.77516	-168.02218	-122.60978	-75.38516	-26.16861	25.25386	79.14094	135.81110	195.66572	259.22646
24.13	-254.00000	-211.77516	-168.02218	-122.60978	-75.38516	-26.16861	25.25386	79.14094	135.81110	195.66575	259.22721
25.13	-156.00000	-116.29201	-75.19695	-32.59617	11.64799	57.69677	105.74120	156.01076	208.78549	264.41418	323.34230
25.12	-156.00000	-116.29201	-75.19695	-32.59617	11.64799	57.69677	105.74120	156.01076	208.78549	264.41418	323.34236
25.12	-54.00000	-16.90491	21.44456	61.15538	102.35135	145.17702	189.80308	236.43358	285.31609	336.75641	391.14069
26.11	-54.00000	-16.90491	21.44456	61.15538	102.35135	145.17702	189.80308	236.43358	285.31609	336.75641	391.14070
27.11	52.00000	86.38543	121.89940	158.63794	196.71195	236.25069	277.40633	320.36020	365.33142	412.58910	462.47049
27.10	52.00000	86.38543	121.89940	158.63794	196.71195	236.25069	277.40633	320.36020	365.33142	412.58910	462.47049
28.10	162.00000	193.57840	226.16511	259.84579	294.71920	330.90030	368.52415	407.75131	448.77523	491.83256	537.21835
28.9	162.00000	193.57840	226.16511	259.84579	294.71920	330.90030	368.52415	407.75131	448.77523	491.83256	537.21835
29.9	276.00000	304.67353	334.23970	364.77422	396.36436	429.11151	463.13467	498.57506	535.60242	574.42391	615.29684
29.8	276.00000	304.67353	334.23970	364.77422	396.36436	429.11151	463.13467	498.57506	535.60242	574.42391	615.29684
30.8	394.00000	419.67040	446.12145	473.41931	501.64016	530.87244	561.21987	592.80535	625.77640	660.31271	696.63692
30.7	394.00000	419.67040	446.12145	473.41931	501.64016	530.87244	561.21987	592.80535	625.77640	660.31271	696.63692
31.7	516.00000	538.56868	561.80897	585.77775	610.54050	636.17316	662.76473	690.42058	719.26706	749.45786	781.18308
31.6	516.00000	538.56868	561.80897	585.77775	610.54050	636.17316	662.76473	690.42058	719.26706	749.45786	781.18308
32.6	642.00000	661.36806	681.30104	701.84677	723.06026	745.00532	767.75667	791.40273	816.04942	841.82549	868.89002
32.5	642.00000	661.36806	681.30104	701.84677	723.06026	745.00532	767.75667	791.40273	816.04942	841.82549	868.89002
33.5	772.00000	788.06830	804.59665	821.62398	839.19508	857.36185	876.18506	895.73660	916.10253	937.38741	959.72031
33.4	772.00000	788.06830	804.59665	821.62398	839.19508	857.36185	876.18506	895.73660	916.10253	937.38741	959.72031
34.4	906.00000	918.66919	931.69492	945.10738	958.94124	973.23672	988.04087	1003.40933	1019.40873	1036.11991	1053.64269
34.3	906.00000	918.66919	931.69492	945.10738	958.94124	973.23672	988.04087	1003.40933	1019.40873	1036.11991	1053.64269
35.3	1044.00000	1053.17054	1062.59512	1072.29523	1082.29557	1092.62479	1103.31638	1114.40997	1125.95297	1138.00293	1150.63084
35.2	1044.00000	1053.17054	1062.59512	1072.29523	1082.29557	1092.62479	1103.31638	1114.40997	1125.95297	1138.00293	1150.63084
36.2	1186.00000	1191.57221	1197.29660	1203.18603	1209.25532	1215.52161	1222.00496	1228.72909	1235.72241	1243.01934	1250.66239
36.1	1186.00000	1191.57221	1197.29660	1203.18603	1209.25532	1215.52161	1222.00496	1228.72909	1235.72241	1243.01934	1250.66239
37.1	1332.00000	1333.87405	1335.79879	1337.77850	1339.81812	1341.92335	1344.10087	1346.35859	1348.70597	1351.15446	1353.71822
37.0	1332.00000	1333.87405	1335.79879	1337.77850	1339.81812	1341.92335	1344.10087	1346.35859	1348.70597	1351.15446	1353.71822

J = 38

0.38	-1482.00000	-1465.32177	-1458.20940	-1452.75946	-1448.16946	-1444.12884	-1440.47841	-1437.12361	-1434.00286	-1431.07336	-1428.30398
1.38	-1480.00000	-1465.32176	-1458.20940	-1452.75946	-1448.16946	-1444.12884	-1440.47841	-1437.12361	-1434.00286	-1431.07336	-1428.30398
1.37	-1480.00000	-1433.11304	-1411.78617	-1395.46958	-1381.73955	-1369.66082	-1358.75444	-1348.73611	-1339.42070	-1330.67967	-1322.41943
2.37	-1474.00000	-1433.11237	-1411.78617	-1395.46958	-1381.73955	-1369.66082	-1358.75444	-1348.73611	-1339.42070	-1330.67967	-1322.41943
2.36	-1474.00000	-1403.34799	-1367.75269	-1340.61220	-1317.80905	-1297.76988	-1279.69119	-1263.09675	-1247.67673	-1233.21611	-1219.55845
3.36	-1464.00000	-1403.33021	-1367.75265	-1340.61220	-1317.80905	-1297.76988	-1279.69119	-1263.09675	-1247.67673	-1233.21611	-1219.55845
3.35	-1464.00000	-1376.51246	-1326.28328	-1288.30116	-1256.46061	-1228.51882	-1203.33748	-1180.24390	-1158.80116	-1138.70642	-1119.73950
4.35	-1450.00000	-1376.24833	-1326.28231	-1288.30116	-1256.46061	-1228.51882	-1203.33748	-1180.24390	-1158.80116	-1138.70642	-1119.73950
4.34	-1450.00000	-1354.04097	-1287.63030	-1238.68382	-1197.79699	-1161.98397	-1129.75177	-1100.22299	-1072.82951	-1047.17831	-1022.98409
5.34	-1432.00000	-1351.83440	-1287.61315	-1238.68366	-1197.79699	-1161.98397	-1129.75177	-1100.22299	-1072.82951	-1047.17831	-1022.98409
5.33	-1432.00000	-1337.78315	-1252.25424	-1191.95999	-1141.94926	-1098.25972	-1059.00496	-1023.08842	-989.80390	-958.66446	-929.31740
6.33	-1410.00000	-1328.82437	-1252.04574	-1191.95705	-1141.94921	-1098.25972	-1059.00496	-1023.08842	-989.80390	-958.66446	-929.31740
6.32	-1410.00000	-1323.48625	-1221.39914	-1148.43022	-1089.09060	-1037.46544	-991.18448	-948.90616	-909.77481	-873.20368	-838.76918
7.32	-1384.00000	-1304.61006	-1219.70999	-1148.39066	-1089.08975	-1037.46542	-991.18448	-948.90616	-909.77481	-873.20368	-838.76918
7.31	-1384.00000	-1303.59634	-1197.63132	-1108.68923	-1039.46485	-979.75670	-926.40035	-877.75741	-832.80349	-790.84252	-751.37483
8.31	-1354.00000	-1276.94798	-1189.75169	-1108.29842	-1039.45307	-979.75638	-926.40034	-877.75741	-832.80349	-790.84252	-751.37483
8.30	-1354.00000	-1276.83657	-1179.10545	-1074.37771	-993.47585	-925.34611	-864.79474	-809.74407	-758.96526	-711.63750	-667.17690
9.30	-1320.00000	-1245.04010	-1159.73389	-1071.70357	-993.34920	-925.34146	-864.79459	-809.74407	-758.96526	-711.63750	-667.17690
9.29	-1320.00000	-1245.03155	-1156.90933	-1048.20206	-952.06639	-874.55694	-806.55887	-744.99718	-688.35455	-635.65823	-586.22712
10.29	-1282.00000	-1208.87922	-1126.88427	-1037.44057	-951.04926	-874.50458	-806.55666	-744.99710	-688.35454	-635.65823	-586.22712
10.28	-1282.00000	-1208.87873	-1126.45685	-1026.72714	-917.71878	-828.03045	-751.97073	-683.69119	-621.09237	-562.99185	-508.58924
11.28	-1240.00000	-1168.57935	-1089.77814	-1002.78518	-912.19973	-827.57326	-751.94521	-683.69002	-621.09233	-562.99185	-508.58924
11.27	-1240.00000	-1168.57933	-1089.73311	-1000.05989	-891.96682	-787.50074	-701.52400	-626.07329	-557.33884	-493.74987	-434.34309
12.27	-1194.00000	-1124.22115	-1048.21668	-964.96996	-874.89926	-784.58443	-701.28907	-626.05943	-557.33817	-493.74984	-434.34309
12.26	-1194.00000	-1124.22115	-1048.21304	-964.55033	-867.31964	-756.01176	-656.45111	-572.55182	-497.31710	-428.07920	-363.59078
13.26	-1144.00000	-1075.85420	-1002.35758	-922.71838	-836.06441	-744.39208	-654.80196	-572.41923	-497.30906	-428.07881	-363.59076
13.25	-1144.00000	-1075.85420	-1002.35734	-922.67150	-834.45580	-730.57458	-619.70769	-524.05360	-441.37838	-366.18231	-296.46638
14.25	-1090.00000	-1023.50993	-952.35029	-875.92325	-793.34602	-704.21304	-611.86901	-523.06037	-441.29903	-366.17752	-296.46616
14.24	-1090.00000	-1023.50993	-952.35028	-875.91913	-793.11876	-700.34583	-591.75533	-483.04937	-390.25381	-308.36724	-233.15327
15.24	-1032.00000	-967.20909	-898.29466	-824.78607	-745.99924	-661.05764	-570.20792	-477.71597	-389.62878	-308.31843	-233.15043
15.23	-1032.00000	-967.20909	-898.29466	-824.78578	-745.97471	-660.39268	-563.17599	-451.90485	-345.92636	-255.22946	-173.92341
16.23	-970.00000	-906.96595	-840.25662	-769.48587	-694.12378	-613.42092	-526.60130	-434.60991	-342.25666	-254.82768	-173.89345
16.22	-970.00000	-906.96595	-840.25662	-769.48585	-694.12164	-613.33713	-525.15869	-423.88360	-311.54097	-208.34361	-119.27043
17.22	-904.00000	-842.79055	-778.28126	-710.14531	-637.95528	-561.12737	-478.86689	-390.57268	-297.85468	-205.80991	-119.01285
17.21	-904.00000	-842.79055	-778.28126	-710.14531	-637.95512	-561.11897	-478.66075	-387.98736	-283.23506	-170.96937	-70.43795
18.21	-834.00000	-774.69018	-712.40049	-646.84807	-577.67567	-504.41649	-426.43447	-342.90797	-253.46248	-160.31823	-68.70760
18.20	-834.00000	-774.69018	-712.40049	-646.84807	-577.67566	-504.41579	-426.41143	-342.49605	-249.38183	-141.79799	-30.44009
19.20	-760.00000	-702.67021	-642.63754	-579.65316	-513.41061	-443.52265	-369.48164	-290.59821	-206.03875	-115.69354	-22.35272
19.19	-760.00000	-702.67021	-642.63754	-579.65316	-513.41060	-443.52260	-369.47953	-290.54798	-205.32034	-109.78130	0.00000
20.19	-682.00000	-626.73468	-569.00968	-508.60350	-445.24834	-378.61376	-308.28050	-233.69594	-154.11359	-68.69966	22.35272
20.18	-682.00000	-626.73468	-569.00968	-508.60350	-445.24834	-378.61376	-308.28034	-233.69097	-154.01934	-67.55333	30.44009
21.18	-600.00000	-546.88671	-491.53000	-433.73112	-373.25294	-309.80778	-243.03904	-172.49132	-97.56023	-17.43492	68.70760
21.17	-600.00000	-546.88671	-491.53000	-433.73112	-373.25294	-309.80778	-243.03903	-172.49091	-97.55030	-17.27445	70.43795
22.17	-514.00000	-463.12867	-410.20859	-355.06041	-297.47222	-237.19010	-173.90434	-107.22840	-36.66440	38.45383	119.01285
22.16	-514.00000	-463.12867	-410.20859	-355.06041	-297.47222	-237.19010	-173.90434	-107.22837	-36.66354	38.47165	119.27043
23.16	-424.00000	-375.46247	-325.05336	-272.61026	-217.94269	-160.82439	-100.98259	-38.08217	28.29890	98.70995	173.89345
23.15	-424.00000	-375.46247	-325.05336	-272.61026	-217.94269	-160.82439	-100.98259	-38.08217	28.29897	98.71158	173.92341
24.15	-330.00000	-283.88960	-236.07053	-186.39552	-134.69271	-80.75932	-24.35288	34.82122	97.12832	163.03356	233.15043
24.14	-330.00000	-283.88960	-236.07053	-186.39552	-134.69271	-80.75932	-24.35288	34.82122	97.12832	163.03369	233.15327
25.14	-232.00000	-188.41126	-143.26512	-96.42802	-47.74468	2.96721	55.92434	111.38796	169.67864	231.19841	296.46616
25.13	-232.00000	-188.41126	-143.26512	-96.42802	-47.74468	2.96721	55.92434	111.38796	169.67864	231.19842	296.46638
26.13	-130.00000	-89.02842	-46.64116	-2.71725	42.88351	90.32517	139.80187	191.54635	245.84210	303.04104	363.59076
26.12	-130.00000	-89.02842	-46.64116	-2.71725	42.88351	90.32517	139.80187	191.54635	245.84210	303.04104	363.59078
27.12	-24.00000	14.25812	53.79805	94.72902	137.17736	181.29047	227.24225	275.24034	325.53633	378.44038	434.34309
27.11	-24.00000	14.25812	53.79805	94.72902	137.17736	181.29047	227.24225	275.24034	325.53633	378.44038	434.34309
28.11	86.00000	121.44770	158.04980	195.90446	235.12504	275.84355	318.21533	362.42537	408.69695	457.30400	508.58924
28.10	86.00000	121.44770	158.04980	195.90446	235.12504	275.84355	318.21533	362.42537	408.69695	457.30400	508.58924
29.10	200.00000	232.53978	266.11183	300.80382	336.71681	373.96835	412.69653	453.06545	495.27263	539.55960	586.22712
29.9	200.00000	232.53978	266.11183	300.80382	336.71681	373.96835	412.69653	453.06545	495.27263	539.55960	586.22712
30.9	318.00000	347.53390	377.98229	409.42272	441.94454	475.65158	510.66563	547.13119	585.22188	625.14949	667.17690
30.8	318.00000	347.53390	377.98229	409.42272	441.94454	475.65158	510.66563	547.13119	585.22188	625.14949	667.17690
31.8	440.00000	466.42969	493.65958	521.75748	550.80147	580.88216	612.10581	644.59830	678.51069	714.02692	751.37483
31.7	440.00000	466.42969	493.65958	521.75748	550.80147	580.88216	612.10581	644.59830	678.51069	714.02692	751.37483
32.7	566.00000	589.22681	613.14237	637.80501	663.28186	689.65078	717.00300	745.44654	775.11089	806.15351	838.76918
32.6	566.00000	589.22681	613.14237	637.80501	663.28186	689.65078	717.00300	745.44654	775.11089	806.15351	838.76918
33.6	696.00000	715.92499	736.42953	757.56268	779.38087	801.94954	825.34533	849.65892	874.99894	901.49740	929.31740
33.5	696.00000	715.92499	736.42953	757.56268	779.38087	801.94954	825.34533	849.65892	874.99894	901.49740	929.31740
34.5	830.00000	846.52401	863.52011	881.02824	899.09436	917.77173	937.12270	957.22102	978.15500	1000.03189	1022.98409
34.4	830.00000	846.52401	863.52011	881.02824	899.09436	917.77173	937.12270	957.22102	978.15500	1000.03189	1022.98409
35.4	968.00000	981.02365	994.41325	1008.19977	1022.41880	1037.11162	1052.32652	1068.12060	1084.56223	1101.73440	1119.73950
35.3	968.00000	981.02365	994.41325	1008.19977	1022.41880	1037.11162	1052.32652	1068.12060	1084.56223	1101.73440	1119.73950
36.3	1110.00000	1119.42374	1129.10825	1139.07559	1149.35113	1159.96426	1170.94938	1182.34715	1194.20624	1206.58571	1219.55845
36.2	1110.00000	1119.42374	1129.10825	1139.07559	1149.35113	1159.96426	1170.94938	1182.34715	1194.20624	1206.58571	1219.55845
37.2	1256.00000	1261.72413	1267.60449	1273.65429	1279.88872	1286.32539	1292.98490	1299.89160	1307.07464	1314.56938	1322.41943
37.1	1256.00000	1261.72413	1267.60449	1273.65429	1279.88872	1286.32539	1292.98490	1299.89160	1307.07464	1314.56938	1322.41943
38.1	1406.00000	1407.92469	1409.90142	1411.93459	1414.02926	1416.19129	1418.42754	1420.74613	1423.15676	1425.67120	1428.30398
38.0	1406.00000	1407.92469	1409.90142	1411.93459	1414.02926	1416.19129	1418.42754	1420.74613	1423.15676	1425.67120	1428.30398

J = 39

0.39	-1560.00000	-1542.87415	-1535.57670	-1529.98469	-1525.27490	-1521.12874	-1517.38288	-1513.94033	-1510.73789	-1507.73168	-1504.88973
1.39	-1558.00000	-1542.87414	-1535.57670	-1529.98469	-1525.27490	-1521.12874	-1517.38288	-1513.94033	-1510.73789	-1507.73168	-1504.88973
1.38	-1558.00000	-1509.76699	-1487.88637	-1471.14409	-1457.05499	-1444.65983	-1433.46733	-1423.18585	-1413.62548	-1404.65436	-1396.17649
2.38	-1552.00000	-1509.76654	-1487.88637	-1471.14409	-1457.05499	-1444.65983	-1433.46733	-1423.18585	-1413.62548	-1404.65436	-1396.17649
2.37	-1552.00000	-1479.08875	-1442.57967	-1414.73192	-1391.33151	-1370.76572	-1352.21071	-1335.17811	-1319.35015	-1304.50631	-1290.48613
3.37	-1542.00000	-1479.07620	-1442.57965	-1414.73192	-1391.33151	-1370.76572	-1352.21071	-1335.17811	-1319.35015	-1304.50631	-1290.48613
3.36	-1542.00000	-1451.27581	-1399.82336	-1360.85765	-1328.18410	-1299.50701	-1273.66018	-1249.95417	-1227.94109	-1207.31046	-1187.83652
4.36	-1528.00000	-1451.08073	-1399.82278	-1360.85765	-1328.18410	-1299.50701	-1273.66018	-1249.95417	-1227.94109	-1207.31046	-1187.83652
4.35	-1528.00000	-1427.58457	-1359.85334	-1309.66168	-1267.71110	-1230.95689	-1197.87187	-1167.55771	-1139.43249	-1113.09351	-1088.24836
5.35	-1510.00000	-1425.83670	-1359.84270	-1309.66160	-1267.71110	-1230.95689	-1197.87187	-1167.55771	-1139.43249	-1113.09351	-1088.24836
5.34	-1510.00000	-1410.13912	-1323.07290	-1261.33160	-1210.03678	-1165.20523	-1124.91346	-1088.04075	-1053.86465	-1021.88674	-991.74580
6.34	-1488.00000	-1402.29834	-1322.93695	-1261.33000	-1210.03676	-1165.20523	-1124.91346	-1088.04075	-1053.86465	-1021.88674	-991.74580
6.33	-1488.00000	-1395.76313	-1290.49996	-1216.13953	-1155.32329	-1102.36467	-1054.86775	-1011.46592	-971.28561	-933.72711	-898.35718
7.33	-1462.00000	-1377.96060	-1289.31576	-1216.11695	-1155.32288	-1102.36466	-1054.86775	-1011.46592	-971.28561	-933.72711	-898.35718
7.32	-1462.00000	-1376.57918	-1264.62046	-1174.57824	-1103.79344	-1042.58022	-987.83803	-937.90975	-891.75322	-848.65872	-808.11609
8.32	-1432.00000	-1350.40141	-1258.40261	-1174.34146	-1103.78742	-1042.58008	-987.83803	-937.90975	-891.75322	-848.65872	-808.11609
8.31	-1432.00000	-1350.23726	-1245.22182	-1137.93587	-1055.79314	-986.04621	-923.95619	-867.46740	-815.33813	-766.73471	-721.06259
9.31	-1398.00000	-1318.62309	-1228.03452	-1136.16474	-1055.72467	-986.04406	-923.95613	-867.46740	-815.33813	-766.73471	-721.06259
9.30	-1398.00000	-1318.60967	-1224.02371	-1109.02486	-1012.04508	-933.04447	-863.39646	-800.25976	-742.12800	-688.01996	-637.24508
10.30	-1360.00000	-1282.56093	-1195.31950	-1100.80412	-1011.45239	-933.01885	-863.39553	-800.25973	-742.12800	-688.01996	-637.24508
10.29	-1360.00000	-1282.56010	-1194.64341	-1086.75689	-974.52776	-884.07625	-806.40294	-736.44513	-672.23389	-612.59494	-556.72266
11.29	-1318.00000	-1242.33250	-1158.49043	-1065.83936	-970.89678	-883.83627	-806.39154	-736.44468	-672.23387	-612.59494	-556.72266
11.28	-1318.00000	-1242.33246	-1158.41342	-1061.65160	-945.95714	-840.40025	-753.37011	-676.24083	-605.80031	-540.56141	-479.56864
12.28	-1272.00000	-1198.02742	-1117.17035	-1028.29363	-932.65563	-838.70352	-753.25778	-676.23516	-605.80008	-540.56140	-479.56864
12.27	-1272.00000	-1198.02742	-1117.16369	-1027.56874	-921.65008	-804.98077	-705.16757	-619.97852	-543.02318	-472.05123	-405.87561
13.27	-1222.00000	-1149.70119	-1071.49507	-986.46043	-893.80593	-796.97237	-704.30178	-619.92044	-543.02017	-472.05110	-405.87560
13.26	-1222.00000	-1149.70119	-1071.49461	-986.37229	-891.04789	-777.98190	-664.13228	-568.31151	-484.19075	-407.24159	-335.76323
14.26	-1168.00000	-1097.38890	-1021.62848	-940.02616	-851.57609	-756.32382	-659.32904	-567.83588	-484.15883	-407.23994	-335.76316
14.25	-1168.00000	-1097.38890	-1021.62846	-940.01782	-851.14045	-749.86107	-632.86864	-522.98109	-429.82357	-346.38690	-269.39130
15.25	-1110.00000	-1041.11368	-967.68338	-889.16930	-804.77203	-713.57228	-616.68799	-520.04359	-429.54917	-346.36887	-269.39041
15.24	-1110.00000	-1041.11368	-967.68338	-889.16929	-804.72076	-712.28192	-605.35353	-487.25674	-381.22779	-289.91716	-206.98526
16.24	-1048.00000	-980.89140	-909.73416	-834.09008	-753.34344	-666.62056	-573.26500	-475.45511	-379.39448	-289.75521	-206.97512
16.23	-1048.00000	-980.89140	-909.73416	-834.09004	-753.33861	-666.44035	-570.45433	-458.80690	-341.49911	-238.83038	-148.90721
17.23	-982.00000	-916.73323	-847.83146	-774.92645	-697.52740	-614.94943	-526.28973	-431.28611	-333.04081	-237.67635	-148.81201
17.22	-982.00000	-916.73323	-847.83146	-774.92645	-697.52702	-614.92971	-525.83464	-426.27293	-311.14165	-195.75340	-95.93121
18.22	-912.00000	-848.64725	-782.01104	-711.77420	-637.53041	-558.73848	-474.64898	-384.37636	-288.13974	-189.80295	-95.21178
18.21	-912.00000	-848.64725	-782.01104	-711.77420	-637.53039	-558.73671	-474.59290	-383.44555	-280.33076	-162.96832	-50.16038
19.21	-838.00000	-776.63944	-712.29880	-644.70002	-573.49693	-498.24564	-418.35280	-333.00812	-241.38832	-144.25073	-46.08372
19.20	-838.00000	-776.63944	-712.29880	-644.70002	-573.49693	-498.24551	-418.34720	-332.88137	-239.73273	-133.17117	-14.73284
20.20	-760.00000	-700.71428	-638.71401	-573.75224	-505.52783	-433.66515	-357.68029	-276.92321	-190.52513	-97.76900	0.00000
20.19	-760.00000	-700.71428	-638.71401	-573.75224	-505.52783	-433.66514	-357.67982	-276.90940	-190.27872	-95.08828	14.73284
21.19	-678.00000	-620.87521	-561.27126	-498.96676	-433.69584	-365.13332	-292.87153	-216.38157	-134.94934	-47.64841	46.08372
21.18	-678.00000	-620.87521	-561.27126	-498.96676	-433.69584	-365.13333	-292.87150	-216.38033	-134.92049	-47.21429	50.16038
22.18	-592.00000	-537.12489	-479.98177	-420.37086	-358.05495	-292.74799	-224.09820	-151.66138	-74.85600	7.10697	95.21178
22.17	-592.00000	-537.12489	-479.98177	-420.37086	-358.05495	-292.74799	-224.09820	-151.66129	-74.85322	7.16113	95.93121
23.17	-502.00000	-449.46541	-394.85430	-337.98564	-278.64622	-216.58161	-151.48316	-82.96846	-10.55013	66.41688	148.81201
23.16	-502.00000	-449.46541	-394.85430	-337.98564	-278.64622	-216.58161	-151.48316	-82.96845	-10.54991	66.42238	148.90721
24.16	-408.00000	-357.89843	-305.89581	-251.82765	-195.50145	-136.68922	-75.11710	-10.45024	57.73015	129.96394	206.97512
24.15	-408.00000	-357.89843	-305.89581	-251.82765	-195.50145	-136.68922	-75.11710	-10.45024	57.73016	129.96441	206.98526
25.15	-310.00000	-262.42529	-213.11183	-161.91006	-108.64570	-53.11357	4.93112	65.78480	129.81349	197.48022	269.39041
25.14	-310.00000	-262.42529	-213.11183	-161.91006	-108.64570	-53.11357	4.93112	65.78480	129.81349	197.48026	269.39130
26.14	-208.00000	-163.04706	-116.50686	-68.24345	-18.09898	34.11164	88.60805	145.65447	205.57418	268.77101	335.76316
26.13	-208.00000	-163.04706	-116.50686	-68.24345	-18.09898	34.11164	88.60805	145.65447	205.57418	268.77101	335.76323
27.13	-102.00000	-59.76463	-16.08455	29.16357	76.12257	124.95940	175.87143	229.09494	284.91717	343.69381	405.87560
27.12	-102.00000	-59.76463	-16.08455	29.16357	76.12257	124.95940	175.87143	229.09494	284.91717	343.69381	405.87561
28.12	8.00000	47.42128	88.15209	130.30395	174.00575	219.40782	266.68735	316.05571	367.76869	422.14114	479.56864
28.11	8.00000	47.42128	88.15209	130.30395	174.00575	219.40782	266.68735	316.05571	367.76869	422.14114	479.56864
29.11	122.00000	158.51006	196.20056	235.17185	275.53971	317.43898	361.02823	406.49615	454.07033	504.02969	556.72266
29.10	122.00000	158.51006	196.20056	235.17185	275.53971	317.43898	361.02823	406.49615	454.07033	504.02969	556.72266
30.10	240.00000	273.50121	308.05879	343.76241	380.71543	419.03806	458.87143	500.38319	543.77505	589.29346	637.24508
30.9	240.00000	273.50121	308.05879	343.76241	380.71543	419.03806	458.87143	500.38319	543.77505	589.29346	637.24508
31.9	362.00000	392.39431	423.72503	456.07156	489.52537	524.19269	560.19816	597.68957	636.84446	677.87928	721.06259
31.8	362.00000	392.39431	423.72503	456.07156	489.52537	524.19269	560.19816	597.68957	636.84446	677.87928	721.06259
32.8	488.00000	515.18899	543.19779	572.09586	601.96316	632.89252	664.99270	698.39261	733.24685	769.74365	808.11609
32.7	488.00000	515.18899	543.19779	572.09586	601.96316	632.89252	664.99270	698.39261	733.24685	769.74365	808.11609
33.7	618.00000	641.88495	666.47583	691.83239	718.02344	745.12876	773.24182	802.47329	832.95579	864.85061	898.35718
33.6	618.00000	641.88495	666.47583	691.83239	718.02344	745.12876	773.24182	802.47329	832.95579	864.85061	898.35718
34.6	752.00000	772.48194	793.55806	815.27866	837.70160	860.89395	884.93428	909.91553	935.94904	963.17009	991.74580
34.5	752.00000	772.48194	793.55806	815.27866	837.70160	860.89395	884.93428	909.91553	935.94904	963.17009	991.74580
35.5	890.00000	906.97972	924.44358	942.43254	960.99371	980.18172	1000.06050	1020.70566	1042.20776	1064.67675	1088.24836
35.4	890.00000	906.97972	924.44358	942.43254	960.99371	980.18172	1000.06050	1020.70566	1042.20776	1064.67675	1088.24836
36.4	1032.00000	1045.37811	1059.13158	1073.29217	1087.89639	1102.98657	1118.61224	1134.83196	1151.71586	1169.34904	1187.83652
36.3	1032.00000	1045.37811	1059.13158	1073.29217	1087.89639	1102.98657	1118.61224	1134.83196	1151.71586	1169.34904	1187.83652
37.3	1178.00000	1187.67694	1197.62138	1207.85597	1218.40671	1229.30376	1240.58240	1252.28436	1264.45954	1277.16855	1290.48613
37.2	1178.00000	1187.67694	1197.62138	1207.85597	1218.40671	1229.30376	1240.58240	1252.28436	1264.45954	1277.16855	1290.48613
38.2	1328.00000	1333.87605	1339.91238	1346.12254	1352.52212	1359.12918	1365.96485	1373.05412	1380.42688	1388.11942	1396.17649
38.1	1328.00000	1333.87605	1339.91238	1346.12254	1352.52212	1359.12918	1365.96485	1373.05412	1380.42688	1388.11942	1396.17649
39.1	1482.00000	1483.97533	1486.00405	1488.09068	1490.24040	1492.45923	1494.75421	1497.13366	1499.60754	1502.18793	1504.88973
39.0	1482.00000	1483.97533	1486.00405	1488.09068	1490.24040	1492.45923	1494.75421	1497.13366	1499.60754	1502.18793	1504.88973

J = 40

0.40	-1640.00000	-1622.42655	-1614.94401	-1609.20993	-1604.38035	-1600.12864	-1596.28736	-1592.75706	-1589.47293	-1586.39000	-1583.47549
1.40	-1638.00000	-1622.42654	-1614.94401	-1609.20993	-1604.38035	-1600.12864	-1596.28736	-1592.75706	-1589.47293	-1586.39000	-1583.47549
1.39	-1638.00000	-1588.42124	-1565.98670	-1548.81869	-1534.37051	-1521.65890	-1510.18025	-1499.63561	-1489.83028	-1480.62907	-1471.93356
2.39	-1632.00000	-1588.42093	-1565.98670	-1548.81869	-1534.37051	-1521.65890	-1510.18025	-1499.63561	-1489.83028	-1480.62907	-1471.93356
2.38	-1632.00000	-1556.83168	-1519.40730	-1490.85205	-1466.85428	-1445.76180	-1426.73041	-1409.25960	-1393.02367	-1377.79659	-1363.41389
3.38	-1622.00000	-1556.82284	-1519.40728	-1490.85205	-1466.85428	-1445.76180	-1426.73041	-1409.25960	-1393.02367	-1377.79659	-1363.41389
3.37	-1622.00000	-1528.05371	-1475.36565	-1435.41545	-1401.90852	-1372.49589	-1345.98340	-1321.66485	-1299.08135	-1277.91476	-1257.93374
4.37	-1608.00000	-1527.91045	-1475.36531	-1435.41545	-1401.90852	-1372.49589	-1345.98340	-1321.66485	-1299.08135	-1277.91476	-1257.93374
4.36	-1608.00000	-1503.18037	-1434.08366	-1382.64294	-1339.62749	-1301.93148	-1267.99323	-1236.89341	-1208.03623	-1181.00929	-1155.51310
5.36	-1590.00000	-1501.81155	-1434.07710	-1382.64289	-1339.62748	-1301.93148	-1267.99323	-1236.89341	-1208.03623	-1181.00929	-1155.51310
5.35	-1590.00000	-1484.45927	-1395.92048	-1332.71136	-1280.12932	-1234.15430	-1192.82460	-1154.99510	-1119.92697	-1087.11021	-1056.17512
6.35	-1568.00000	-1477.68887	-1395.83262	-1332.71049	-1280.12931	-1234.15430	-1192.82460	-1154.99510	-1119.92697	-1087.11021	-1056.17512
6.34	-1568.00000	-1469.82160	-1361.71256	-1285.87011	-1223.56642	-1169.27093	-1120.55614	-1076.02952	-1034.79934	-996.25280	-959.94691
7.34	-1542.00000	-1453.17738	-1360.89558	-1285.85734	-1223.56622	-1169.27092	-1120.55614	-1076.02952	-1034.79934	-996.25280	-959.94691
7.33	-1542.00000	-1451.32496	-1333.70828	-1242.53822	-1170.14418	-1107.41711	-1051.28511	-1000.06902	-952.70818	-908.47889	-866.86036
8.33	-1512.00000	-1425.70739	-1328.92881	-1242.39674	-1170.14114	-1107.41704	-1051.28510	-1000.06902	-952.70818	-908.47889	-866.86036
8.32	-1512.00000	-1425.46880	-1313.02387	-1203.72358	-1120.16631	-1048.77179	-985.13438	-927.20273	-873.71991	-823.83860	-776.95333
9.32	-1478.00000	-1394.06508	-1298.08294	-1202.57954	-1120.12982	-1048.77081	-985.13436	-927.20273	-873.71991	-823.83860	-776.95333
9.31	-1478.00000	-1394.04428	-1292.55582	-1171.99810	-1074.20372	-993.58574	-922.26391	-857.54276	-797.91625	-742.39264	-690.27122
10.31	-1440.00000	-1358.10983	-1265.43619	-1165.95382	-1073.86590	-993.57339	-922.26352	-857.54275	-797.91625	-742.39264	-690.27122
10.30	-1440.00000	-1358.10847	-1264.39024	-1148.22936	-1033.76373	-942.26743	-862.89168	-791.23378	-725.39964	-664.21564	-606.86905
11.30	-1398.00000	-1317.95898	-1228.88420	-1130.48075	-1031.47063	-942.14418	-862.88668	-791.23361	-725.39964	-664.21564	-606.86905
11.29	-1398.00000	-1317.95892	-1228.75507	-1124.30362	-1001.89128	-895.73416	-807.34668	-728.46981	-656.30155	-589.40096	-526.81446
12.29	-1352.00000	-1273.71149	-1187.82455	-1093.10789	-992.01666	-894.78318	-807.29418	-728.46753	-656.30147	-589.40096	-526.81446
12.28	-1352.00000	-1273.71149	-1187.81259	-1091.89167	-976.89150	-856.42766	-756.26570	-669.53116	-590.79640	-518.06792	-450.19192
13.28	-1302.00000	-1225.42934	-1142.35005	-1051.69599	-952.90779	-851.23716	-755.82698	-669.50633	-590.79529	-518.06788	-450.19192
13.27	-1302.00000	-1225.42934	-1142.34917	-1051.53443	-948.39386	-826.63345	-711.33295	-614.90560	-529.12954	-450.37370	-377.10914
14.27	-1248.00000	-1173.15162	-1092.63691	-1005.65332	-911.08625	-809.74769	-708.57301	-614.68574	-529.11703	-450.37315	-377.10912
14.26	-1248.00000	-1173.15162	-1092.63685	-1005.63686	-910.27885	-799.64835	-675.80748	-565.77299	-471.69781	-386.53549	-307.70723
15.26	-1190.00000	-1116.90410	-1038.81200	-955.10555	-864.84971	-767.18053	-664.50241	-564.25334	-471.58165	-386.52901	-307.70696
15.25	-1190.00000	-1116.90410	-1038.81200	-955.10420	-864.74581	-764.79096	-647.67389	-525.06945	-419.36601	-326.88717	-242.17875
16.25	-1128.00000	-1056.70435	-980.95915	-900.26700	-813.91164	-720.89175	-620.88396	-517.70228	-418.50399	-326.82433	-242.17542
16.24	-1128.00000	-1056.70435	-980.95915	-900.26691	-813.90107	-720.51843	-615.75385	-494.10218	-374.45525	-272.08898	-180.80916
17.24	-1062.00000	-992.56479	-919.13519	-841.29641	-758.48418	-669.89521	-574.55158	-472.87129	-369.72583	-271.59373	-180.77536
17.23	-1062.00000	-992.56479	-919.13519	-841.29640	-758.48329	-669.85058	-573.59275	-463.96291	-339.91242	-223.89454	-124.11733
18.23	-992.00000	-924.49435	-853.38007	-778.30189	-698.79741	-614.23780	-523.74514	-426.46572	-323.65949	-220.90212	-123.83545
18.22	-992.00000	-924.49435	-853.38007	-778.30189	-698.79735	-614.23346	-523.61422	-424.48308	-310.26034	-185.67873	-73.40557
19.22	-918.00000	-852.49967	-783.72269	-711.35882	-635.01707	-554.18789	-468.17610	-376.04550	-277.16671	-173.67367	-71.55677
19.21	-918.00000	-852.49967	-783.72269	-711.35882	-635.01707	-554.18754	-468.16181	-375.74091	-273.62890	-155.39903	-31.79165
20.21	-840.00000	-776.58569	-710.18448	-640.52150	-567.25848	-489.96820	-408.09083	-320.86072	-227.30349	-127.11137	-23.29431
20.20	-840.00000	-776.58569	-710.18448	-640.52150	-567.25848	-489.96818	-408.08953	-320.82411	-226.69547	-121.41992	0.00000
21.20	-758.00000	-696.75624	-632.78163	-565.83015	-495.60417	-421.73630	-343.76001	-261.05907	-172.79123	-77.96807	23.29431
21.19	-758.00000	-696.75624	-632.78163	-565.83015	-495.60417	-421.73630	-343.75991	-261.05545	-172.71174	-76.87197	31.79165
22.19	-672.00000	-613.01424	-551.52661	-487.31522	-420.11508	-349.60439	-275.38366	-196.94155	-113.59634	-24.42057	71.55677
22.18	-672.00000	-613.01424	-551.52661	-487.31522	-420.11508	-349.60439	-275.38365	-196.94125	-113.58790	-24.26575	73.40557
23.18	-582.00000	-525.36202	-466.42915	-405.00023	-340.83732	-273.65492	-203.10422	-128.74878	-50.02443	33.82897	123.83545
23.17	-582.00000	-525.36202	-466.42915	-405.00023	-340.83732	-273.65492	-203.10422	-128.74876	-50.02368	33.84648	124.11733
24.17	-488.00000	-433.80140	-377.49692	-318.90359	-257.80652	-193.95014	-127.02582	-56.65344	17.64725	96.50466	180.77536
24.16	-488.00000	-433.80140	-377.49692	-318.90359	-257.80652	-193.95014	-127.02582	-56.65344	17.64730	96.50630	180.80916
25.16	-390.00000	-338.33387	-284.73609	-229.03994	-171.05066	-110.53817	-47.22689	19.21875	89.21610	163.29880	242.17542
25.15	-390.00000	-338.33387	-284.73609	-229.03994	-171.05066	-110.53817	-47.22689	19.21875	89.21610	163.29893	242.17875
26.15	-288.00000	-238.96061	-188.15161	-135.42103	-80.59206	-23.45685	36.23204	98.77345	164.53500	233.97939	307.70696
26.14	-288.00000	-238.96061	-188.15161	-135.42103	-80.59206	-23.45685	36.23204	98.77345	164.53500	233.97940	307.70723
27.14	-182.00000	-135.68261	-87.74755	-38.05642	13.55130	67.26361	123.30331	181.93792	243.49397	308.37798	377.10912
27.13	-182.00000	-135.68261	-87.74755	-38.05642	13.55130	67.26361	123.30331	181.93792	243.49397	308.37798	377.10914
28.13	-72.00000	-28.50066	16.47278	63.04609	111.36476	161.59877	213.94883	268.65494	326.00839	386.36913	450.19192
28.12	-72.00000	-28.50066	16.47278	63.04609	111.36476	161.59877	213.94883	268.65494	326.00839	386.36913	450.19192
29.12	42.00000	82.58456	124.50662	167.88002	212.83626	259.52863	308.13770	358.87868	412.01173	467.85665	526.81446
29.11	42.00000	82.58456	124.50662	167.88002	212.83626	259.52863	308.13770	358.87868	412.01173	467.85665	526.81446
30.11	160.00000	197.57250	236.35166	276.43999	317.95578	361.03671	405.84460	452.57194	501.45069	552.76492	606.86905
30.10	160.00000	197.57250	236.35166	276.43999	317.95578	361.03671	405.84460	452.57194	501.45069	552.76492	606.86905
31.10	282.00000	316.46270	352.00596	388.72149	426.71497	466.10926	507.04857	549.70416	594.28194	641.03341	690.27122
31.9	282.00000	316.46270	352.00596	388.72149	426.71497	466.10926	507.04857	549.70416	594.28194	641.03341	690.27122
32.9	408.00000	439.25475	471.46790	504.72072	539.10677	574.73475	611.73210	650.24997	690.46983	732.61286	776.95333
32.8	408.00000	439.25475	471.46790	504.72072	539.10677	574.73475	611.73210	650.24997	690.46983	732.61286	776.95333
33.8	538.00000	565.94831	594.73608	624.43443	655.12522	686.90344	719.88046	754.18814	789.98470	827.46266	866.86036
33.7	538.00000	565.94831	594.73608	624.43443	655.12522	686.90344	719.88046	754.18814	789.98470	827.46266	866.86036
34.7	672.00000	696.54311	721.80933	747.85989	774.76522	802.60707	831.48114	861.50074	892.80167	925.54902	959.94691
34.6	672.00000	696.54311	721.80933	747.85989	774.76522	802.60707	831.48114	861.50074	892.80167	925.54902	959.94691
35.6	810.00000	831.03889	852.68662	874.99471	898.02244	921.83855	946.52351	972.17253	998.89967	1026.84349	1056.17512
35.5	810.00000	831.03889	852.68662	874.99471	898.02244	921.83855	946.52351	972.17253	998.89967	1026.84349	1056.17512
36.5	952.00000	969.43544	987.36707	1005.83686	1024.89311	1044.59179	1064.99842	1086.19049	1108.26078	1131.32195	1155.51310
36.4	952.00000	969.43544	987.36707	1005.83686	1024.89311	1044.59179	1064.99842	1086.19049	1108.26078	1131.32195	1155.51310
37.4	1098.00000	1111.73258	1125.84992	1140.38458	1155.37400	1170.86155	1186.89801	1203.54341	1220.86961	1238.96384	1257.93374
37.3	1098.00000	1111.73258	1125.84992	1140.38458	1155.37400	1170.86155	1186.89801	1203.54341	1220.86961	1238.96384	1257.93374
38.3	1248.00000	1257.93013	1268.13451	1278.63634	1289.46229	1300.64327	1312.21545	1324.22160	1336.71289	1349.75145	1363.41399
38.2	1248.00000	1257.93013	1268.13451	1278.63634	1289.46229	1300.64327	1312.21545	1324.22160	1336.71289	1349.75145	1363.41389
39.2	1402.00000	1408.02797	1414.22027	1420.59079	1427.15552	1433.93296	1440.94480	1448.21664	1455.77914	1463.66949	1471.93356
39.1	1402.00000	1408.02797	1414.22027	1420.59079	1427.15552	1433.93296	1440.94480	1448.21664	1455.77914	1463.66949	1471.93356
40.1	1560.00000	1562.02597	1564.10669	1566.24677	1568.45154	1570.72717	1573.08088	1575.52120	1578.05833	1580.70467	1583.47549
40.0	1560.00000	1562.02597	1564.10669	1566.24677	1568.45154	1570.72717	1573.08088	1575.52120	1578.05833	1580.70467	1583.47549

Appendix V

SUM RULES RELATING ENERGY LEVELS AND INERTIAL CONSTANTS OF A RIGID ASYMMETRIC ROTOR

$$\sum E_{ee} = \pi_J(A + B + C)$$
$$\sum E_{oo} = \sigma_J(A + C) + \pi_J B$$
$$\sum E_{oe} = \sigma_J(A + B) + \pi_J C$$
$$\sum E_{eo} = \sigma_J(B + C) + \pi_J A$$

J	π_J	σ_J
0	0	0
1	0	1
2	4	1
3	4	10
4	20	10
5	20	35
6	56	35
7	56	84
8	120	84
9	120	165
10	220	165
11	220	286
12	364	286

This form and extension of the sum rules are due to K. K. Innes [58]. See also R. Mecke [80].

SUM RULES INCLUDING CENTRIFUGAL DISTORTION CONSTANTS*

E^+ J even ee

$$(\tfrac{1}{6})J(J+1)(J+2)[A+B+C+(\tfrac{1}{5})D_K-2R_6]$$
$$-(\tfrac{1}{30})J^2(J+1)^2(J+2)(15D_J+3D_K+5D_{JK})$$
$$+J(J+1)[J(J+1)-2][R_6-(\tfrac{1}{10})D_K]$$

E^- J odd ee

$$(\tfrac{1}{6})J(J+1)(J-1)[A+B+C+(\tfrac{1}{5})D_K-2R_6]$$
$$-(\tfrac{1}{30})J^2(J+1)^2(J-1)(15D_J+3D_K+5D_{JK})$$
$$-J(J+1)[J(J+1)-2][R_6-(\tfrac{1}{10})D_K]$$

E^- J even eo

$$(\tfrac{1}{6})J(J+1)(J+2)[A+(\tfrac{1}{5})D_K-2R_6]+(\tfrac{1}{6})(J-1)J(J+1)(B+C)$$
$$-(\tfrac{1}{30})J^2(J+1)^2(J+2)(15D_J+3D_K+5D_{JK})$$
$$+J^2(J+1)^2[D_J-R_6-(\tfrac{1}{10})D_K]$$
$$+J(J+1)[6R_6+(\tfrac{1}{5})D_K]$$

E^+ J odd eo

$$(\tfrac{1}{6})\{J(J+1)(J-1)[A+(\tfrac{1}{5})D_K-2R_6]+J(J+1)(J+2)(B+C)\}$$
$$-(\tfrac{1}{30})J^2(J+1)^2(J-1)(15D_J+3D_K+5D_{JK})$$
$$-J^2(J+1)^2[D_J-R_6-(\tfrac{1}{10})D_K]$$
$$-J(J+1)[6R_6+(\tfrac{1}{5})D_K]$$

0^+ J even oe

$$(\tfrac{1}{6})\{J(J+1)(J+2)C+(J-1)J(J+1)[A+B+(\tfrac{1}{5})D_K-2R_6]\}$$
$$-(\tfrac{1}{30})(J-1)J^2(J+1)^2(15D_J+3D_K+5D_{JK})$$
$$-(\tfrac{1}{2})J^2(J+1)^2[D_J-2\delta_J-(\tfrac{1}{5})D_K]$$
$$+2J(J+1)[-R_5-R_6-\tfrac{1}{10})D_K]$$

0^- J odd oe

$$(\tfrac{1}{6})\{J(J+1)(J-1)C+(J+2)(J+1)J[A+B+(\tfrac{1}{5})D_K-2R_6]\}$$
$$-(\tfrac{1}{30})J^2(J+1)^2(J+2)(15D_J+3D_K+5D_{JK})$$
$$+(\tfrac{1}{2})J^2(J+1)^2[D_J-2\delta_J-(\tfrac{1}{5})D_K]$$
$$-2J(J+1)[-R_5-R_6-(\tfrac{1}{10})D_K]$$

0^- J even oo

$$(\tfrac{1}{6})\{J(J+1)(J+2)B+(J-1)(J)(J+1)[A+C+(\tfrac{1}{5})D_K-2R_6]\}$$
$$-(\tfrac{1}{30})(J-1)(J^2)(J+1)^2(15D_J+3D_K+5D_{JK})$$
$$-(\tfrac{1}{2})J^2(J+1)^2[D_J+2\delta_J-(\tfrac{1}{5})D_K]$$
$$-2J(J+1)[-R_5+R_6+(\tfrac{1}{10})D_K]$$

0^+ J odd oo

$$(\tfrac{1}{6})\{J(J+1)(J-1)B+J(J+1)(J+2)[A+C+(\tfrac{1}{5})D_K-2R_6]\}$$
$$-(\tfrac{1}{30})J^2(J+1)^2(J+2)(15D_K+3D_K+5D_{JK})$$
$$+(\tfrac{1}{2})J^2(J+1)^2[D_J+2\delta_J-(\tfrac{1}{5})D_K]$$
$$+2J(J+1)[-R_5+R_6+(\tfrac{1}{10})D_K].$$

* H. C. Allen, Jr., and W. B. Olsen, *J. Chem. Phys.*, **37,** 212 (1962).

Appendix VI

APPROXIMATE METHODS FOR CALCULATION OF RIGID ROTOR ENERGIES

VIa INTRODUCTION

In Chapter 2 the precise calculation of the rigid energy levels was discussed in detail. These procedures are the best when it is feasible to use them. Fortunately, the values of the energy levels are available at present up through $J = 40$ [8, 43, 51, 67] and are given in Appendix IV. These tabulated reduced energies are sufficient for most problems encountered in infrared bands. However, several approximate methods of calculating the energy levels have been developed and deserve mention here.

These methods can be divided into three general classes. In the first class the energies are calculated by perturbation techniques and are good only near the limits of the particular asymmetry parameter used to expand the energy. The second class is based on the solution of a simplified wave equation. By examining the wave equation it has been found that for certain assumptions it reduces to some well-known differential equation for which the mathematicians have already tabulated the solutions. Finally, it is sometimes possible to use Correspondence Principle arguments to calculate the energy levels. Several of these approximate methods are discussed in this Appendix.

VIb PERTURBATION ABOUT $b = 0$

Of the several asymmetry parameters which can be used as a basis for perturbation calculations near the symmetric rotor limits, the calculations using b are the most extensive [8]. For this reason this method is described first.

The Hamiltonian of a rigid asymmetric rotor 2j1 can be rewritten as

$$\mathbf{H} = \frac{B + C}{2}\mathbf{P}^2 + [A - \tfrac{1}{2}(B + C)][\mathbf{P}_a^{\ 2} + b(\mathbf{P}_c^{\ 2} - \mathbf{P}_b^{\ 2})], \tag{1}$$

in which $\mathbf{P}^2 = \mathbf{P}_a^{\ 2} + \mathbf{P}_b^{\ 2} + \mathbf{P}_c^{\ 2}$

and
$$b = \frac{C - B}{2A - B - C}.$$

$\mathbf{P}^2$ is a constant of the motion with the value $J(J + 1)\hbar^2$ and is not affected by the diagonalization of the second term. Thus $\mathbf{H}'$ the perturbation operator can be taken as

$$\mathbf{H}' = \mathbf{P}_a^{\ 2} + b(\mathbf{P}_c^{\ 2} - \mathbf{P}_b^{\ 2}) \tag{2}$$

and the energy may be expressed as

$$E = \frac{(B + C)}{2} J(J + 1) + [A - \tfrac{1}{2}(B + C)]E(b), \tag{3}$$

where E, $J(J + 1)$ and $E(b)$ are the eigenvalues of $\mathbf{H}$, $\mathbf{P}^2$, and $\mathbf{H}'$, respectively. In a I^r representation $\mathbf{H}'$ has the matrix elements

$$\begin{aligned} (J, K\,|\mathbf{H}'|\,J, K) &= K^2 \\ (J, K\,|\mathbf{H}'|\,J, K + 2) &= bf^{1/2}(J, K + 1) \end{aligned} \tag{4}$$

Subjecting $\mathbf{H}'$ to a Wang transformation results in four factors closely related to $E^{\pm}$ and $0^{\pm}$. Then with the help of conventional perturbation theory each of these matrices gives $E(b)$ as a power series in b with the nth order perturbation resulting in the coefficient of b^n.

Thus

$$E(b) = K_{-1}^{\ 2} + C_1 b + C_2 b^2 + C_3 b^3 + \ldots. \tag{5}$$

and general coefficients for the powers of b can be evaluated by standard perturbation techniques. These expressions are given explicitly in Table VI-1.

The energies for the nearly oblate rotor are readily obtained by expanding the energy $E(b^*)$ in a series such as (5). Using the definition of b^*, 2k5, and the coefficients of Table VI-1 and 2k6, we obtain the energies of the nearly oblate symmetric rotor.

VIc PERTURBATION ABOUT $\delta = 0$

A second method for obtaining the energies of a nearly symmetric rotor makes use of an asymmetry parameter closely related to κ:

$$\delta = \frac{\kappa + 1}{2} = \frac{(B - C)}{(A - C)}. \tag{1}$$

From the definition it is seen that $0 \leqslant \delta \leqslant 1$, $\delta = 0$ corresponding to the prolate symmetric rotor. The perturbation calculation is carried out about $\delta = 0$ and the energies of the near oblate rotor are then obtained as in 2j6*a*.

Table VI-1—General Values of the Coefficients C_i in the Expansion of $E(b)$

K	C_1	C_2	C_3	C_4	C_5
1	$\pm \frac{J(J+1)}{2}$ $0^{\pm}$		$\pm \frac{J(J+1)f(J,2)}{128}$ $0^{\mp}$	Same as value for all other K $-\frac{J^2(J+1)^2 f(J,2)}{2048}$	$\pm \frac{J(J+1)}{294{,}912}[108f(J,2) - 9J^2(J+1)^2 - 28f(J,4)]f(J,2)$ $0^{\pm}$
3			$\pm \frac{J(J+1)f(J,2)}{128}$ $0^{\pm}$	Same as value for all other K $+\frac{J^2(J+1)^2 f(J,2)}{2048}$	$\pm \frac{J(J+1)f(J,2)}{294{,}912}[108f(J,2) - 9J^2(J+1)^2 - 27f(J,4)]$ $0^{\mp}$
5					$\pm \frac{J(J+1)f(J,2)f(J,4)}{294{,}912}$ $0^{\pm}$
All other K	0	$\frac{f(J,K-1)}{4(K-1)} - \frac{f(J,K+1)}{4(K+1)}$	0	$\frac{f(J,K-1)}{128(K-1)^2}\left[\frac{2f(J,K+1)}{(K+1)} - \frac{2f(J,K-1)}{K-1} + \frac{f(J,K-3)}{K-2}\right] - \frac{f(J,K+1)}{128(K+1)^2}\left[\frac{2f(J,K-1)}{K-1} - \frac{2f(J,K+1)}{K+1} + \frac{f(J,K+3)}{K+2}\right]$	0

If no entry is found in a particular box the value of C is found in the last row.

$$C_0 = K_{-1}^2 \text{ for all } K.$$

The functions $f(J, n)$ appearing here are given by Eq. 2j12, except that here $f(J, 1) = \frac{1}{2}J(J+1)[J(J+1) - 2]$ or twice the value given by Eq. 2j12. The symbols $0\pm$ identify the submatrix to which the level belongs and are given here for a I^r representation (if $K_{-1} + K_{+1} = J$, the level is in an 0^- matrix; if $K_{-1} + K_{+1} = J + 1$, the level is in an 0^+ matrix). Since the row and column for $K = 0$ which appear in the E^+ matrix ($K_{-1} + K_{+1} = J$) is absent in the E^- matrix ($K_{-1} + K_{+1} = J + 1$), $f(J, 1)$ must be taken to be zero when computing C_2 or C_4 for the E^- levels. Furthermore, $f(J, 0)$ and $f(J, -1)$ should always be set equal to zero.

The Hamiltonian in the prolate symmetric limit is written as

$$\mathbf{H}^0 = [A\mathbf{P}_a^2 + C(\mathbf{P}_b^2 + \mathbf{P}_c^2)]. \tag{2}$$

The perturbation is now introduced as

$$\mathbf{H}' = \frac{(B - C)\mathbf{P}_b^2}{h^2} = (A - C)\,\delta\mathbf{P}_b^2, \tag{3}$$

so that the Hamiltonian

$$\mathbf{H} = \mathbf{H}^0 + \mathbf{H}' = [A\mathbf{P}_a^2 + B\mathbf{P}_b^2 + C\mathbf{P}_c^2], \tag{4}$$

which is the energy of any rigid asymmetric rotor and can be reduced to 2j9. Then, by standard perturbation techniques $E(\delta)$ can be expanded in the power series

$$E(\delta) = C_0 + C_1\delta + C_2\delta^2 + C_3\delta^3 + \dots. \tag{5}$$

General expressions for the C_1's can also be obtained for this case. They are given through C_3 in Table VI-2. The numerical values for these coefficients through $J = 12$ have been given by King, Hainer, and Cross [67].

Perturbations can also be carried out about $\kappa = 0$, the most asymmetric case. These calculations have been carried out to the second order by King, Hainer, and Cross, who have given numerical values of the coefficients in the expansion.

VId OTHER APPROXIMATE METHODS

The correspondence principle has been used to approximate the energies of a rigid asymmetric rotor [64]. In this method the integrals of the angular momentum components for each of the rotational coordinates (Eulerian angles) are integrated over a complete cycle of the coordinate and set equal to some integral multiple of n. The result is elliptical integrals which must then be solved for $E_{K_{-1}K_{+1}}$. This method does not remove the K degeneracy of the limiting symmetric rotor. For large values of J it yields very good energies in the regions A and A' of Fig. VI-1. The results get considerably poorer as one goes away from the line $E_\tau(\kappa)/J(J + 1) = \kappa$.

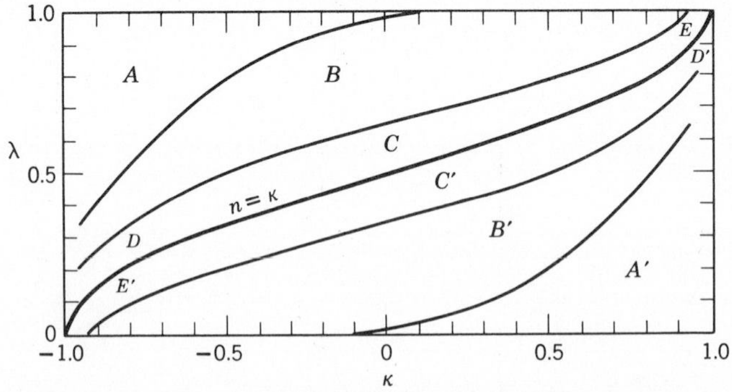

Fig. VI-1. Sketch of the regions of error expected in approximating the rigid rotor levels. The heavy line corresponds to $\dfrac{E_\tau(\kappa)}{J(J+1)} = \kappa$.

Table VI-2—Coefficients in the Expansion $E(\delta) = C_0 + C_1\delta + C_2\delta^2 + \dots$

K	C_0	C_1	C_2	C_3
0			$-\frac{1}{4}f(J, 1)$	$\frac{1}{4}C_2$
1		$\frac{3J(J+1)}{2} - 1 \quad 0^-$ $\frac{J(J+1)}{2} - 1 \quad 0^+$	$-\frac{1}{16}f(J, 2) \quad 0^{\pm}$	$-\frac{1}{512}f(J, 2)[16 \pm J(J+1)] \quad 0^{\pm}$
2			$\frac{1}{4}f(J, 1) - \frac{1}{24}f(J, 3) \quad E^+$ $- \frac{1}{24}f(J, 3) \quad E^-$	$\frac{1}{4}C_2$
3			$\frac{1}{16}f(J, 2) - \frac{1}{32}f(J, 4)$ $0^{\pm}$	$\frac{1}{512}f(J, 2)[16 \pm J(J+1)] - \frac{1}{64}f(J, 4)$ $0^{\mp}$
All other K	$-J(J+1) + 2K_{-1}^2$	$J(J+1) - K_{-1}^2$	$\frac{f(J, K-1)}{8(K-1)} - \frac{f(J, K+1)}{8(K+1)}$	$\frac{1}{4}C_2$

Another approximate method for obtaining the energy levels of the rigid asymmetric rotor is based on the similarity at high J between the energy matrix of the asymmetric rotor and that for the Mathieu equation [48]. The characteristic roots of the Mathieu equation are corrected by perturbation methods to give the approximate energies. The error is very small at low K values, that is, in the region D in Fig. VI-1. This is just the region where the correspondence principle approximation is the worst. Elsewhere in the figure the Mathieu approximation is not enough better to warrant the additional computation involved.

A complimentary technique [49] has been developed for the region of Fig. VI-1 where $|K|/[J(J+1)]^{1/2} \sim 1$. For this condition the energy matrix for high J approaches the matrix of an harmonic oscillator. This approximation is similar to the correspondence principle approximation in that it gives no splitting for the levels which are degenerate in the symmetric limit. The region of applicability is limited generally to the A and A' areas of Fig. VI-1. For more detail about these approximate methods the reader is referred to the original literature.

Appendix VII

THEORY OF CENTRIFUGAL DISTORTION CONSTANTS*

The centrifugal distortion constants are defined in 3e1 as

$$\tau_{gg'jj'} = \frac{\sum'[(V\,|\mu_{gg'}|\,V'')(V''\,|\mu_{jj'}|\,V)]}{h\nu_{VV''}}, \tag{1}$$

where the prime indicates that the state V is excluded from the sum. We now wish to reduce (1) to a more convenient form. The most convenient way to simplify the expression is to expand the μ_{gg}'s in terms of some suitable set of coordinates. We could use the set of normal coordinates [124] and neglect all the nonlinear terms.

This procedure is not convenient since the normal coordinates are seldom known. More often the force constants associated with some set of internal displacement coordinates are available; therefore such a set of coordinates is used here [70]. There are $(3N - 6)$ independent interatomic parameters where N is the number of atoms in the molecule. Let $\{\delta q_i\}$ be the corresponding set of internal displacement coordinates. These displacement coordinates might be changes in bond distances or valence angles. If $\mu^0_{gg'}$ is the equilibrium value of $\mu_{gg'}$ (all $\delta q_i = 0$), we can write

$$\mu_{gg'} = \mu^0_{gg'} + \sum_i \mu^i_{gg'}\,\delta q_i + 0(\delta q_i^2), \tag{2}$$

where

$$\mu^i_{gg'} = \left(\frac{\partial \mu_{gg'}}{\partial q_i}\right)_{\partial q_i = 0} \tag{3}$$

All terms in (2) quadratic or higher will be neglected. The basis functions in (1) are

*This Appendix is based entirely on the work of Kivelson and Wilson [70] and is reproduced here with their kind permission.

the harmonic-oscillator functions. $\mu^0_{gg'}$ and $\mu^i_{gg'}$ are both constants and since $V \neq V''$, the orthonormal property of the Hermite polynomials gives

$$(V\,|\mu_{gg'}|\,V'') = \sum_i \mu^i_{gg'}(V\,|\delta q_i|\,V''). \tag{4}$$

Substituting (4) into (1), we obtain

$$\tau_{gg'jj'} = \frac{\sum'_{V''} \sum_{ik} \mu^i_{gg'}\mu^k_{jj'}[(V\,|\delta q_i|\,V'')(V''\,|\delta q_k|\,V)]}{h\nu_{VV''}} \tag{5}$$

The internal displacement coordinates δq_i can be related to the normal coordinates Q_k by a linear transformation

$$\delta q_i = \sum_k b_{ik} Q_k \tag{6}$$

so that

$$(V\,|\delta q_i|\,V'') = (V\,|\sum_k b_{ik}Q_k|\,V'') = \sum_k b_{ik}(V\,|Q_k|\,V''). \tag{7}$$

For the harmonic-oscillator basis functions the quantity $(V\,|Q_k|\,V'')$ is nonvanishing only if all $v = v''$ except the single quantum number v_k associated with the normal coordinate Q_k. Here $v_k'' = v_k \pm 1$. By using this result, the τ's can be expressed as

$$\tau_{gg'jj'} = \sum_{ik} \mu^i_{gg'}\mu^k_{jj'} \sum_l b_{il}b_{kl}\{[(v_l\,|Q_l|\,v_l+1)$$
$$\times (v_l+1\,|Q_l|\,v_l)]/h\nu_{l,l+1} + [(v_l\,|Q_l|\,v_l-1)(v_l-1\,|Q_l|\,v_l)]/h\nu_{l,l-1}]\}. \tag{8}$$

The matrix elements are easily evaluated for the harmonic-oscillator to give

$$(v_l\,|Q_l|\,v_{l+1}) = \left[\frac{(v_l+1)h}{8\pi^2\nu_l^{\,0}}\right]^{1/2}$$
$$(v_l\,|Q_l|\,v_{l-1}) = \left(\frac{v_l h}{8\pi^2\nu_l^{\,0}}\right)^{1/2}. \tag{9}$$

The $\nu_l^{\,0}$ in (9) is given by $\nu_{l,l\pm1} = \mp\nu_l^{\,0}$. We must also make use of the elements of the inverse force constant matrix [9]. If f_{ik} is a potential constant in the expression for the potential energy when it is expressed in terms of the displacement coordinates δq_i, and $(f^{-1})_{ik}$ is an element of the matrix inverse to f, then [9]

$$(f^{-1})_{ik} = \frac{\sum_l b_{il}b_{kl}}{4\pi^2(\nu_l^{\,0})^2}. \tag{10}$$

Substituting (9) and (10) into (8) we arrive at a considerably simplified expression for the τ's,

$$\tau_{gg'jj'} = -\tfrac{1}{2}\sum_{ik} \mu^i_{gg'}\mu^k_{jj'}(f^{-1})_{ik}. \tag{11}$$

Wilson [123] has derived a similar expression (11) classically.

Additional simplification can be introduced by expressing the $\tau_{gg'jj'}$ in terms of the inertia tensor $\mathbf{I}$ rather than the inverse inertia tensor $\boldsymbol{\mu}$. We merely expand the inertia tensor in terms of the displacement coordinates δq_i; thus

$$\mathbf{I} = \mathbf{I}^0 + \sum_i [\mathbf{J}^i]_0\,\delta q_i + \ldots, \tag{12}$$

in which $\mathbf{I}^0$ is the equilibrium tensor and

$$\mathbf{J}^i = \left(\frac{\partial \mathrm{I}}{\partial q_i}\right), \tag{13}$$

and the zero subscript indicates the $\mathbf{J}^i$ is evaluated at $\delta\mathbf{q} = 0$. A useful relation is obtained by noting that $\boldsymbol{\mu} = \mathbf{I}^{-1}$, which leads to

$$\mathbf{I}\boldsymbol{\mu} = \mathbf{1} = \mathbf{I}^0\boldsymbol{\mu}^0 + I^0 \sum_i \boldsymbol{\mu}^i \, \delta q_i + \sum_i [\mathbf{J}^i]_0 \delta q_i \boldsymbol{\mu}^0 + \dots . \tag{14}$$

Since $\boldsymbol{\mu}$ is the inverse of $\mathbf{I}$ for all values of δq_i, $\mathbf{I}^0\boldsymbol{\mu}^0 = 1$. Since the δq_i are independent, (14) leads to the result

$$\boldsymbol{\mu}^i = -\boldsymbol{\mu}^0[\mathbf{J}^i]_0\boldsymbol{\mu}^0, \tag{15}$$

It is now convenient to use the principal axis system of $\mathbf{I}^0$ and $\boldsymbol{\mu}^0$ so that both of these tensors are diagonal. Then the diagonal elements of the former are the reciprocals of the diagonal elements of the latter. With this choice of axes (15) can be simplified to

$$\mu^i_{gg'} = -\frac{[J^i_{gg'}]_0}{I_{gg}{}^0 I^0_{g'g'}} . \tag{16}$$

As has been shown in Chapter 3, one cannot, in general, determine the equilibrium moments of inertia. As a result, the $I_{gg}{}^0$ and $I^0_{g'g'}$ must be replaced by the moments averaged over the ground vibrational state. Such a procedure is justified practically always.

If the averaged moments of inertia are known, we can then calculate the centrifugal distortion constants. For convenience a new quantity is defined which is made up of only those terms still to be evaluated. Thus

$$\begin{aligned} t_{gg'jj'} &= -2(I_{gg}{}^0 I^0_{g'g'} I_{jj}{}^0 I^0_{j'j'})\tau_{gg'jj'} \\ &= \sum_{ik} [J^i_{gg'}]_0 [J^k_{jj'}]_0 (f^{-1})_{ik}. \end{aligned} \tag{17}$$

Kivelson and Wilson [70] have given methods to evaluate **I** and **J** for various, often used internal coordinates. Their methods are presented here unchanged except for minor alterations of notation. However, their coordinate indices $\alpha\beta\gamma\delta$ are used in place of the $gg'jj'$ in preceding discussions.

Evaluation of *i* and *j*

Let $r = (\boldsymbol{\alpha}, \boldsymbol{\beta}, \boldsymbol{\gamma})$ be a $3N$-dimensional Cartesian position vector, where $\boldsymbol{\alpha}$, $\boldsymbol{\beta}$, $\boldsymbol{\gamma}$ are N-dimensional subvectors, each representing the positions of the N-atoms of the molecule along one of the three Cartesian axes. Furthermore, let $\boldsymbol{\alpha}$, $\boldsymbol{\beta}$, $\boldsymbol{\gamma}$ be taken in cyclic order. For example, if the ith element of $\boldsymbol{\alpha}$, α_i, stands for the x position of the ith atom, then β_i and γ_i stand for the y and z positions, respectively, of the same atom. The components of the inertia tensor can be written as

$$I_{\alpha\beta} = -\sum m_i \alpha_i \beta_i \tag{18}$$

$$I_{\alpha\alpha} = \sum m_i(\beta_i^2 + \gamma_i^2), \tag{19}$$

where m_i is the mass of the ith atom.

If the molecule is distorted from its equilibrium configuration, the coordinates are altered by increments $\delta\alpha_i$, $\delta\beta_i$, $\delta\gamma_i$. These increments cannot be made arbitrarily but must satisfy the conditions*

$$\sum m_i\,\delta\alpha_i = \sum m_i\,\delta\beta_i = \sum m_i\,\delta\gamma_i = 0 \tag{20}$$

$$\sum m_i(\alpha_i\,\delta\beta_i - \beta_i\,\delta\alpha_i) = 0, \text{ etc.} \tag{21}$$

However, it is possible to convert an unallowed set of increments $\delta\alpha_i'$, $\delta\beta_i'$, $\delta\gamma_i'$ into an allowed set by adding appropriate rigid translations and rigid rotations of the whole molecule. Thus

$$\delta\alpha_i = \delta\varepsilon_\alpha - \delta\eta_\gamma\cdot\beta_i + \delta\eta_\beta\cdot\gamma_i + \delta\alpha_i', \tag{22}$$

where $\delta\epsilon_\alpha$ is a translation along the α-axis, $\delta\eta_\gamma$ is a rotation about the γ-axis, etc. Insertion of (22) in conditions (20) and (21) shows that, if $\delta\alpha_i$ is to be allowable

$$\delta\varepsilon_\alpha = -\frac{\sum m_i\,\delta\alpha_i'}{M} \tag{23}$$

$$\delta\eta_\gamma = -\frac{\sum m_i(\alpha_i\,\delta\beta_i' - \beta_i\,\delta\alpha_i')}{I_{\gamma\gamma}} \tag{24}$$

since the variations are evaluated at equilibrium in the principal axis system and

$$\sum m_i\alpha_i = 0, \qquad \sum m_i\alpha_i\beta_i = 0. \tag{25}$$

M is the mass of the molecule. Consequently, we have the results

$$\begin{aligned}\delta I_{\alpha\beta} &= -\sum m_i(\alpha_i\,\delta\beta_i + \beta_i\,\delta\alpha_i)\\ &= -\sum m_i(\alpha_i\,\delta\beta_i' + \beta_i\,\delta\alpha_i' + \alpha_i^2\,\delta\eta_\gamma - \beta_i^2\,\delta\eta_\gamma\\ &= -2\,[\sum m_j\alpha_j^2 \sum m_i\beta_i\,\delta\alpha_i'\\ &\quad + \sum m_j\beta_j^2 \sum m_i\alpha_i\,\delta\beta_i']/I_{\gamma\gamma}\end{aligned} \tag{26}$$

$$\begin{aligned}\delta I_{\alpha\alpha} &= 2\sum m_i(\beta_i\,\delta\beta_i + \gamma_i\,\delta\gamma_i)\\ &= 2\sum m_i(\beta_i\,\delta\beta_i' + \gamma_i\,\delta\gamma_i').\end{aligned} \tag{27}$$

since all other terms vanish. Here the increments $\delta\alpha_i'$, etc., are entirely arbitrary. The coordinates α_i are equilibrium values measured from the center of mass.

To find a derivative $J_{\alpha\beta}^{(k)} = \partial I_{\alpha\beta}/\partial R_k$ it is only necessary to find any set of increments $\delta\alpha_i'$, etc., which produce an increment δR_k to the kth internal coordinate and leave the other internal coordinates with their equilibrium values. These increments do not need to satisfy the conditions (20) and (21). Without these results the required derivatives are often exceedingly difficult to obtain, as the allowed increments $\delta\alpha$ must be expressed in terms of the internal displacement coordinates, a process that involves the inversion of a $3N \times 3N$ matrix. The results just obtained have been used to derive rules for writing down the derivatives of **I** with respect to some commonly used internal displacement coordinates. These rules should greatly simplify the evaluation of the derivatives.

* These restrictions on the distortions may be considered as specifying the coordinate axes. These conditions are implied in the derivation of (3) to (17). See [125].

Bond stretching

The first application of the preceding results is to valence bond stretching, a commonly used internal displacement coordinate. The principal axis system of the molecule may be translated so that the origin passes through the atom A at one end of r_α, the valence bond to be stretched. (See Fig. VII-1.) All interatomic bonds are initially taken to be vectors directed in a path leading away from A. This path is unique, provided there are no rings, and only molecules without rings are considered here. Later, certain vectors are reversed as shown by dashed arrowheads in Fig. VII-1.

To evaluate one of the desired derivatives, we must consider the effect of a variation in r_α on $\Sigma\, m_i\alpha_i^2$, that is, $2\,\Sigma\, m_i\alpha_i\,\delta\alpha_i'$. By (27) $\delta\alpha_i'$ may be arbitrarily chosen. In this particular case we allow all atoms to the right of A to move a distance $\cos(r_\alpha \cdot \alpha)\,\delta r_\alpha$,

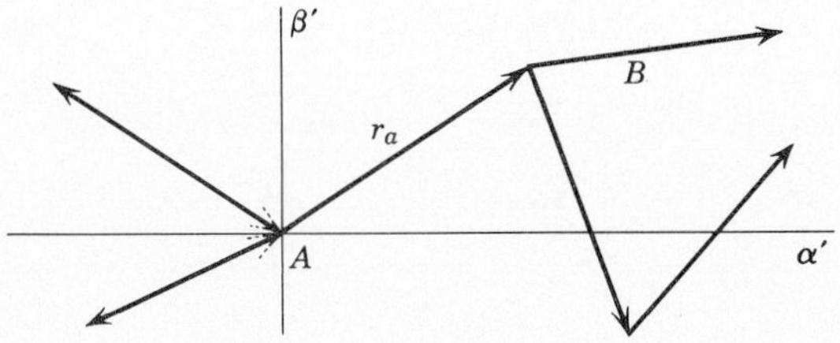

Fig. VII-1. Bond stretching.

whereas all atoms to the left of B are kept in place. All angles are taken between the positive direction of r_i and the positive α-axis.

If $\bar{\alpha}\bar{\beta}\bar{\gamma}$ is the position of the center of mass in the displaced coordinate system $\alpha'\beta'\gamma'$, then $\alpha_i = \alpha_i' - \bar{\alpha}$, where

$$M\bar{\alpha} = \sum^{A,B} m_i\alpha_i' = \sum^{A,B} M_j r_j \cos(r_j \cdot \alpha). \tag{28}$$

M is the total mass of the molecule, and M_j is the sum of the masses of all atoms "beyond" r_j;* $\sum^{A}$ indicates a sum over all atoms or bonds of the molecule on paths starting at A (including the atom at A) that do not include r_α, and $\sum^{B}$ is a sum over the rest of the molecule. With the aid of (28) we obtain

$$\begin{aligned} 2\sum m_i\alpha_i\,\delta\alpha_i' &= 2\sum^{B} m_i\alpha_i' \cos(r_\alpha \cdot \alpha)\,\delta r_\alpha - 2\sum^{B} m_i\bar{\alpha}\cos(r_\alpha \cdot \alpha)\,\delta r_\alpha \\ &= \frac{2}{M}\left[\sum_i^{A} m_i \sum_j^{B} M_j r_j \cos(r_j \cdot \alpha)\cos(r_\alpha \cdot \alpha)\right. \\ &\qquad \left. - \sum_i^{B} m_i \sum_j^{A} M_j r_j \cos(r_j \cdot \alpha)\cos(r_\alpha \cdot \alpha)\right]\delta r_\alpha. \end{aligned} \tag{29}$$

* "Beyond" r_j refers to all the atoms which are reached (from atom A) by paths that include r_j.

If the r_j's in $\sum^{A}$ are now redirected so that they point along paths leading towards A, then $\cos(r_j \cdot \alpha)$ goes to $-\cos(r_j \cdot \alpha)$ in the last term in (29). Thus

$$\frac{d(\sum m_i\alpha_i^2)}{dr_\alpha} = \frac{2}{M}[\sum_j (MM)_{\alpha j} r_j \cos(r_j \cdot \alpha)\cos(r_\alpha \cdot \alpha)], \tag{30}$$

where $(MM)_{\alpha j}$ is the product of the sum of all masses "beyond" r_j and "behind" r_α,* or vice versa, the choice made in such a way that $(MM)_{\alpha j}$ does not include the masses of any atoms on paths between m_j and m_α. β and γ may be substituted for α in this result.

Equation (26) may be used in a similar fashion with the consequence that

$$\frac{dI_{\alpha\beta}}{dr_\alpha} = -\frac{2}{MI_{\gamma\gamma}} \times [I_\alpha \sum_j (MM)_{\alpha j} r_j \cos(r_j \cdot \beta)\cos(r_\alpha \cdot \alpha) + I_\beta \sum_j (MM)_{\alpha j} r_j \cos(r_j \cdot \alpha)\cos(r_\alpha \cdot \beta)], \tag{31}$$

where the bonds are directed as above and

$$I_\alpha = \sum_i m_i\alpha_i^2 = \frac{1}{M}\sum_{j,k}(MM)_{jk} r_j r_k \cos(r_j \cdot \alpha)\cos(r_k \cdot \alpha). \tag{32}$$

I_β is defined similarly. The derivatives of $I_{\beta\gamma}$ and $I_{\gamma\alpha}$ are obtained by the usual permutations of indices.

Valence angle bending

The next application of these results is to valence angle bending where the angle lies in a principal plane, for example, the $\alpha\beta$ plane. The principal axis system may be translated so that the origin lies at the apex of the angle under consideration (see Fig. VII-2). All interatomic bonds are directed along paths leading away from the apex. Since the results above permit one to choose arbitrary displacements, it is convenient here to hold side B and the apex part C of the molecule rigid while part A of the molecule is moved out rigidly as θ changes. Part A of the molecule is distinguished from part B by the fact that an increase in θ_1 means a counterclockwise rotation about the positive γ axis, whereas an increase in θ_2 means a clockwise rotation. The required displacements are

$$\delta\alpha_i' = -\sum_j^{i} r_j \cos(r_j \cdot \beta)\,\delta\theta, \tag{33}$$

where the distances and angles refer to part A of the molecule and the sum is over the bonds forming a continuous path from the vertex to atom i. The bonds are ordered along such paths from the vertex to the given atom. No ordering exists for bonds on different branches. All other displacements are set equal to zero. Then

$$2\sum m_i\alpha_i\,\delta\alpha_i' = -2\sum_i^{A} m_i\alpha_i' \sum_j^{i} r_j\cos(r_j \cdot \beta)\,\delta\theta + 2\sum_i^{A} m_i\bar{\alpha}\sum_j^{i} r_j\cos(r_j \cdot \beta)\,\delta\theta, \tag{34}$$

* "Behind" r_α refers to all atoms which are not "beyond" r_α.

where $\sum^{A}$ is a sum over part A of the molecule. After suitable algebraic manipulations similar to those used in the previous case, we have

$$\frac{d(\sum m_i\alpha_i^2)}{d\theta} = -\frac{2}{M}\left\{\sum_{i\geqslant j}^{A'} M_iM^{(j)}r_ir_j\,[\cos(r_i\cdot\alpha)\cos(r_j\cdot\beta)\right.$$
$$+\cos(r_i\cdot\beta)\cos(r_j\cdot\alpha)]\frac{(1-\delta_{ij})}{2}$$
$$\left.-\sum_i^{A}\sum_j^{B,C,A''} M_iM_jr_ir_j\cos(r_i\cdot\beta)\cos(r_j\cdot\alpha)\right\}, \quad (35)$$

where $\delta_{ij} = 0$, if $i \neq j$, $\delta_{ij} = 1$ if $i = j$, and M_i is the sum of masses "beyond" r_i. In the first sum $\sum^{A'}$, $M^{(j)}$ is the sum of all masses "behind" r_j; r_i, and r_j are both in the A parts of the molecule, and r_j must be on the path from the vertex to r_i. In the second term $\sum_i^{A}\sum_j^{A''}$ are sums over the A part of the molecule in which r_i and r_j are not on the same path from the vertex.

An equivalent result may be obtained by keeping side A rigid and varying θ_2. The results have the same form as those stated except that the sums over A are substituted for sums over B, sums over B are substituted for sums over A, and the sign of the entire expression is changed.

$d(\sum m_i\beta_i^2)/d\theta$ is evaluated by the same formula except that α and β are interchanged, and the sign of the expression is changed. These results are valid for any of the principal planes, provided we remember that $\boldsymbol{\alpha\beta\gamma}$ are in cyclic order.

In a similar manner we may obtain

$$\frac{dI_{\alpha\beta}}{d\theta} = \left(\frac{2}{MI_{\gamma\gamma}}\right)\left\{2\sum_{i\geqslant j}^{A'} M_iM^{(j)}r_ir_j\,[\cos(r_i\cdot\alpha)\cos(r_j\cdot\alpha)I_\beta - \cos(r_j\cdot\beta)\right.$$
$$\times\cos(r_i\cdot\beta)I_\alpha]\left(1-\frac{\delta_{ij}}{2}\right) - \sum_i^{A}\sum_j^{B,C,A''} M_iM_jr_ir_j$$
$$\left.\times\,[\cos(r_i\cdot\alpha)\cos(r_j\cdot\alpha)I_\beta - \cos(r_i\cdot\beta)\cos(r_j\cdot\beta)I_\alpha]\right\}. \quad (36)$$

Where the valence angle does not lie in a principal plane, valence angle bending is more complicated. In this situation the formulas derived may be used as expressions for the projections of the molecular motions on the principal planes.

$$\frac{d(\sum m_i\alpha_i^2)}{d\theta} = \frac{d\theta'}{d\theta}\frac{dI_\alpha}{d\theta'} + \frac{d\theta''}{d\theta}\frac{dI_\alpha}{d\theta''}, \quad (37)$$

where θ' and θ'' are the projections of θ on the $\alpha\beta$ and $\gamma\alpha$ planes, respectively. The quantities just evaluated are really the derivatives with respect to θ'.

By direct but complicated trigonometric and vector manipulations, we obtain the result

$$\frac{d\theta'}{d\theta} = [\sin\theta\sin^2(r_a\cdot\gamma)]^{-1}\,[\sin^2(r_a\cdot\gamma)\sin^2\theta$$
$$-\{\cos(r_b\cdot\gamma)-\cos(r_a\cdot\gamma)\cos\theta\}^2]^{1/2}, \quad (38)$$

where θ is the angle between the bonds r_a and r_b; the other angles are taken with respect to the positive γ-axis, and r_a and r_b are defined in Fig. VII-2. Analogous expressions result for projections on the $\beta\gamma$ and $\gamma\alpha$ planes.

Plane-bond angle bending

Plane-bond bending is the change of angle between the normal (**N**) to a plane formed by two connecting bonds (r_1, r_2) and the bond r_a connected to the vertex of

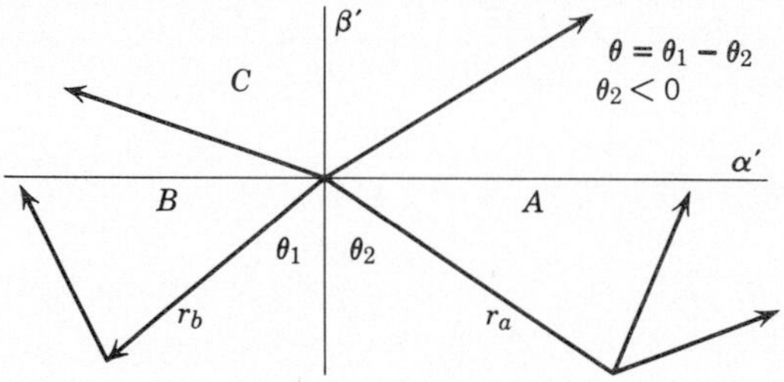

Fig. VII-2. Valence angle bending.

these two bonds where the change takes place in the (r_a, **N**) plane (see Fig. VII-3). If the angle lies in a principal plane ($\alpha\beta$) the coordinate system may be chosen as in Fig. VII-3, where the primed axes are parallel to the principal axes. The molecule is

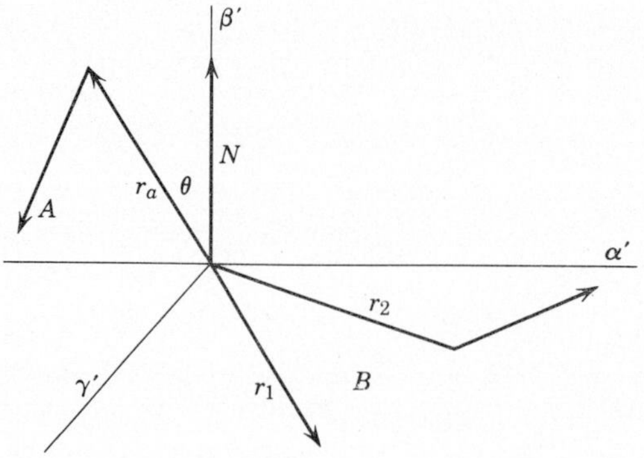

Fig. VII-3. Plane-bond angle bending.

divided into A and B parts as in valence angle bending. It is seen that the results are the same as in (35) and (36). For the special case of a planar molecule, the only non-vanishing derivatives are $dI_{\alpha\beta}/d\theta$ and $dI_{\gamma\beta}/d\theta$, where θ is the out-of-plane bending angle.

If the plane in question is not a principal plane, the same procedure must be carried out as in (37). Equation (38) holds here if $\mathbf{r}_b$ is the normal to the plane.

Torsion

Torsion is the change resulting from rotation about an interatomic bond (r_c) of the angle between the normal of a plane determined by r_e and a connecting bond r_a, and the normal of a plane determined by r_c and a connecting bond at the other end of r_c, that is, r_b (see Fig. VII-4). As for valence angle bending, we can consider the projections of the molecule on the various principal planes (see Fig. VII-5). The r_c bond

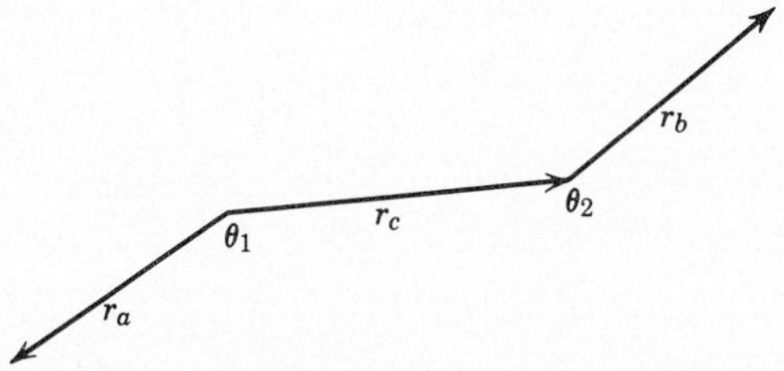

Fig. VII-4. Torsion.

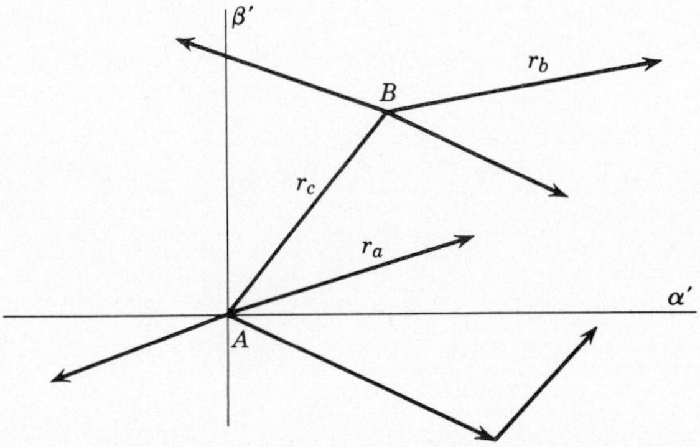

Fig. VII-5. Torsion. Projection on plane.

is the axis of rotation. The part of the molecule designated by B is kept fixed, whereas the part A rotates about r_c. In the projection in Fig. VII-5, the "center" of the A part is translated to the origin, and rotations of the A part about the γ' axis are considered. The bonds are directed away from the origin. With A and B defined in the manner described, the projections of the derivatives with respect to the torsion angle of the inertia tensor take on the same form as the corresponding projection for angle bending. However, here $d\theta'/d\theta$ is different. We must consider $\delta\theta$ as the change of the angle between normals—one where $\mathbf{r}_a$ changes but $\mathbf{r}_b$, $\mathbf{r}_c$, and θ_1 remain fixed (Fig. VII-4). The result is

$$\frac{d\theta'}{d\theta} = \frac{n_s'(n_t - n_s \cos\theta) + (1 - n_s^2)\sin\theta}{(1 - n_s^2)[(1 - n_s^2)\sin^2\theta - (n_t - n_s\cos\theta)^2]^{1/2}}, \tag{39}$$

where n_s and n_t are the γ-direction cosines of $\mathbf{r}_c \times \mathbf{r}_a$ and $\mathbf{r}_c \times \mathbf{r}_b$, respectively, θ is the angle between these two vectors, and $n_s' = dn_s/d\theta$. Furthermore,

$$n_s' = \frac{[\cos(\mathbf{r}_a \cdot \gamma) - \cos(\mathbf{r}_c \cdot \gamma)\cos\theta_1]\sin\theta_2 \sin\theta}{\mathbf{C}\cdot(\mathbf{A}\times\mathbf{B})} \tag{40}$$

where **C**, **A**, **B** are unit vectors in the direction of $\mathbf{r}_c$, $\mathbf{r}_a$, $\mathbf{r}_b$, respectively, and $|\mathbf{C}\times\mathbf{A}| = \sin\theta_1$ and $|\mathbf{C}\times\mathbf{B}| = \sin\theta_2$. These relations may be applied to projections on the $\beta\gamma$ and $\gamma\alpha$ planes by cyclic permutations.

In a symmetric rotor, the rules for torsion give $d\mathbf{I}/d\theta = 0$. In a molecule of the type CH_3AB_2, where CAB_2 lies in the $\gamma\alpha$ plane, the rule yields

$$\frac{dI_{\alpha\beta}}{d\theta} = (I_{\alpha\alpha} - I_{\beta\beta})\frac{I_{\gamma\gamma}^{(s)}}{I_{\gamma\gamma}} \tag{41}$$

as the only nonvanishing derivative with respect to the torsion angle where $I_{\gamma\gamma}^{(s)}$ is the $\gamma\gamma$ moment of the CH_3 part of the molecule.

Symmetry properties of $\mathbf{t}_{\alpha\beta\gamma\delta}$

If the molecule has any symmetry, it will usually be convenient to use internal symmetry coordinates [9]. Equation [17] holds if such coordinates are used throughout. The potential constant matrix **f** must be expressed in these internal symmetry coordinates as well. To obtain $J_{\alpha\beta}^{(k)}$, etc., one merely uses increments $\delta\alpha_i'$, etc., which produce an increment to a given internal symmetry coordinate.

The inertia tensor transforms as a tensor of the second rank. It has the same transformation properties as a linear combination of dyads formed from translation vectors. Thus if such a dyad is represented by **D**, which transforms as **rr**, the transformation properties of the inertia tensor may be related to it as follows:

$$\mathbf{I}_{\alpha\alpha} \rightarrow \mathbf{D}_{\beta\beta} + \mathbf{D}_{\gamma\gamma} \tag{42}$$

$$\mathbf{I}_{\alpha\beta} \rightarrow -\mathbf{D}_{\alpha\beta}. \tag{43}$$

The transformation properties of **D** under the symmetry operations of the point group to which the molecule belongs may be found in tables [4], since **D** transforms exactly as the polarizability tensor. The polarizability components, hence also those of **I**, do not always belong uniquely to a given irreducible representation (species), but linear combinations of them do.

If these linear combinations of the moments and products of inertia are expanded in powers of the normal coordinates or of any set of independent symmetry coordinates, each term must have the symmetry just outlined. The linear term therefore involves only coordinates of this symmetry species. If the problem is set up in terms of internal symmetry coordinates, the coefficients of the linear terms are the components of $\mathbf{J}^k$, and these vanish, except for symmetry coordinates of the appropriate species.

If normal coordinates or symmetry coordinates are used, the array of inverse potential constants $(f^{-1})_{ij}$ in (17) factors into submatrices corresponding to different species of the group. Thus there is no coupling of $[J_{\alpha\beta}^{(i)}]_0$ and $[J_{\gamma\delta}^{(j)}]_0$ if the ith and jth coordinates belong to different symmetry species. Such terms vanish and do not contribute to $t_{\alpha\beta\gamma\delta}$. The only nonvanishing combinations of t's are those made by products of linear combinations of components of $[\mathbf{J}^{(i)}]_0$ and $[\mathbf{J}^{(j)}]_0$, where the ith and jth coordinates belong to the same symmetry species. Such a product belongs to the completely symmetric species A_1. Since $(f^{-1})_{ij}$ is a constant and unaffected by the group operations on the internal displacement coordinates, it is seen that all nonvanishing linear combinations of t's belong to the A_1 species. This result is very useful in determining which t's vanish and in obtaining relations between the remaining ones. It is thus seen that the symmetry property of a linear combination of t's is determined by the symmetry properties of the product of two linear combinations of components of **I**, hence of **D**. Since $I_{\alpha\alpha}{}^0$ and $I_{\alpha\beta}{}^0$ are constants under the group operations, (17) can be used to relate the symmetry properties of the t's to those of the τ's.

The use of these results is illustrated in the following section for C_{2v} (asymmetric rotor) and C_{3v} (symmetric rotor) symmetry.

Molecules of C_{2v} symmetry

The species of the components of the polarizability may be obtained directly from tables. Noting (42) and (43), the relation between the transformation properties of **I** and **D** under the group operations, we obtain the results of Table VII-1 for a molecule

Table VII-1—Symmetry Species of the Components of the Moment of Inertia Tensor for Molecules of C_{2v} Symmetry

I_{xx}	A_1
I_{yy}	A_1
I_{zz}	A_1
I_{xy}	A_2
I_{xz}	B_1
I_{yz}	B_2

of C_{2v} symmetry where a right-handed coordinate system has been used and the z-axis has been chosen as the axis of twofold symmetry. It should be remembered that **I** is a symmetrical tensor. By the arguments of the preceding section it is seen that the only t's that do not vanish are those corresponding to a product of components of **I** which belongs to species A_1. Thus each component of the product must belong to the same species as the other. Once the symmetry properties of the t's have been determined it is easy to get those of the τ's. The result of this analysis is to give the following nonvanishing τ's: τ_{xxxx}, τ_{yyyy}, τ_{zzzz}, $\tau_{xxyy} = \tau_{yyxx}$, $\tau_{xxzz} = \tau_{zzxx}$, $\tau_{yyzz} = \tau_{zzyy}$, $\tau_{xyxy} = \tau_{xyyx} = \tau_{yxxy} = \tau_{yxyx}$, $\tau_{yzyz} = \tau_{zyyz} = \tau_{yzzy} = \tau_{zyzy}$, $\tau_{xzxz} = \tau_{xzzx} = \tau_{zxzx} = \tau_{zxxz}$.

Molecules of C_{3v} symmetry

As a second example of the symmetry arguments given consider a molecule of C_{3v} symmetry. Here not all the components of **D** belong uniquely to a species, and so linear combinations must be formed. The suitable linear combinations and their proper symmetry species are given in Table VII-2. It is necessary to take linear combinations of the products of the linear combinations given in Table VII-2 if the resulting sum of t's is to belong uniquely to a given species. These new combinations and their respective symmetry species are given in Table VII-3. Remembering to

Table VII-2—Symmetry Species of Linear Combinations of the Components of D for Molecules of C_{3v} Symmetry. The y-Axis Is Taken in a Plane of Symmetry

$(D_{xx} + D_{yy})/\sqrt{2} = D_a$	A_1
$D_{zz} = D_b$	A_1
$(D_{xy} + D_{yx})/\sqrt{2} = D_c$	$E(x)$
$(D_{xx} - D_{yy})/\sqrt{2} = D_d$	$E(y)$
$(D_{zx} + D_{xz})/\sqrt{2} = D_e$	$E(x)$
$(D_{zy} + D_{yz})/\sqrt{2} = D_f$	$E(y)$

Table VII-3—Symmetry Species of Products of Linear Combinations of the Components of D for Molecules of C_{3v} Symmetry. The Abbreviations of Table VII-2 Are Used. The y-Axis Lies in a Plane of Symmetry

A_1	A_2	$E(x)$	$E(y)$
D_a^2	$(D_dD_e - D_cD_f)/\sqrt{2}$	D_aD_c	D_aD_d
D_b^2		D_aD_e	D_aD_f
D_aD_b		D_bD_c	D_bD_d
$(D_c^2 + D_d^2)/\sqrt{2}$		D_bD_e	D_bD_f
$(D_e^2 + D_f^2)/\sqrt{2}$		$(D_cD_d + D_dD_c)/\sqrt{2}$	$(D_c^2 - D_d^2)/\sqrt{2}$
$(D_cD_e + D_dD_f)/\sqrt{2}$		$(D_eD_f + D_fD_e)/\sqrt{2}$	$(D_e^2 - D_f^2)/\sqrt{2}$
		$(D_cD_f + D_dD_e)/\sqrt{2}$	$(D_cD_e - D_dD_f)/\sqrt{2}$

identify the transformation properties of the components of **I** with those of **D** as shown in (42) and (43), we may now form the proper combination of t's. Since only combinations of t's belonging to species A_1 are permissible, the other products must vanish. If the y-axis lies in a symmetry plane, we get the following nonvanishing t's: $t_{xxxx} = t_{yyyy}$, $t_{xxyy} = t_{xxxx} - 2t_{xyxy}$, $t_{zzxx} = t_{zzyy}$, t_{zzzz}, $t_{zxzx} = t_{zyzy}$, $-t_{xyzx} = t_{yyyz} = -t_{xxyz}$.* In the molecules of C_{3v} symmetry, symmetric rotors $I^0_{xx} = I^0_{yy}$. Thus by the use of (17), it is seen that the same results hold for the τ's as for the t's.

Relations between τ's and force constants

The arguments given should simplify the calculation of the τ's. If the force constants and fairly good structural parameters are known for the molecule, the τ's may be

* It is easily seen that, in general, $t_{\mu\nu\mu\nu} = t_{\mu\nu\nu\mu} = t_{\nu\mu\mu\nu} = t_{\nu\mu\nu\mu}$.

calculated and used in the analysis of the rotational spectrum. However, because we usually do not have complete information concerning the force constants from vibrational analysis, the τ's cannot be calculated exactly. Then the distortion constants must be determined empirically from the rotational spectrum.* When this latter procedure can be followed, some knowledge of the force constants may be obtained from the empirically determined distortion constants. This knowledge may be used to supplement or check vibrational data.

In reference [69] it is shown that only nine independent τ's enter into a first-order treatment of the effect of centrifugal distortion on the rotational energy levels of polyatomic molecules. It is also shown that only six independent linear combinations of these enter into the final expression for the energy. For K doublets in an asymmetric molecule, only five linear combinations can be determined from the rotational spectrum [69], whereas for a symmetric rotor only three such linear combinations of τ's exist [113]. The force constant information that can be obtained from this analysis of centrifugal distortion is limited by the number of independent combinations of τ's.

A comparison of the distortion constants obtained empirically from rotational spectra with those calculated from the force constants obtained from vibrational spectra has been made for symmetric rotors [112]. The agreement was poor for some still undetermined reason. Such a comparison was also made for an asymmetric rotor (HDS) [114]. Only some of the constants could be determined unambiguously from the rotational spectrum, but they were in fairly good agreement with those calculated from vibrational data.

Example: nonlinear *ABA* molecule

The principles just discussed can be illustrated most simply for a nonlinear ABA molecule, Fig. VII-6. In this molecule $r_1 = r_2$ at equilibrium. The symmetry properties of such a molecule (C_{2v}) have already been discussed. Three symmetry coordinates that can be used are $S_1 = 2^{-1/2}(\delta r_1 + \delta r_2)$, $S_2 = \delta(2\theta)$, $S_3 = 2^{-1/2}(\delta r_1 - \delta r_2)$. They belong to the species A_1, A_1, and B_1, respectively. From Table VII-1 we find that δI_{xx}, δI_{yy}, δI_{zz} are functions of S_1 and S_2, whereas δI_{xz} is a function of S_3. Application of the rules given for bond stretching and valence angle bending yields

$$
\begin{aligned}
\frac{dI_{xx}}{dS_1} &= 2\sqrt{2}m_A r \sin^2\theta \\
\frac{dI_{zz}}{dS_1} &= 2\sqrt{2}m_A m_B r \frac{\cos^2\theta}{M} \\
\frac{dI_{xx}}{dS_2} &= 2m_A r^2 \sin\theta\cos\theta \\
\frac{dI_{zz}}{dS_2} &= -2m_A m_B r^2 \sin\theta\frac{\cos\theta}{M} \\
\frac{dI_{xz}}{dS_3} &= (2\sqrt{2}m_A r\cos\theta\sin\theta)(1 + 2m_A m_B^{-1}\sin^2\theta)^{-1}
\end{aligned}
\tag{44}
$$

* To evaluate these constants empirically different kinds of transitions are required. If the transitions are of the same type, the resulting equations in the constants may be nearly parallel.

It must be remembered that $I_{yy} = I_{xx} + I_{zz}$ in this molecule so that the derivatives of I_{yy} with respect to S_1 and S_2 exist and are easily obtainable. All other derivatives vanish as expected.

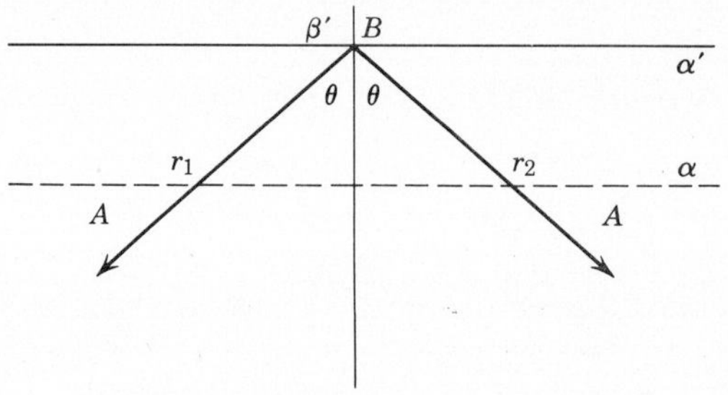

Fig. VII-6. Nonlinear *ABA* molecule.

The symmetry discussion given indicates which t's vanish. Furthermore, it is seen that in this case only four t's are linearly independent, the following relations holding among the seven nonvanishing ones:

$$\begin{aligned} t_{yyyy} &= t_{xxxx} + t_{zzzz} + 2t_{xxzz}, \\ t_{yyxx} &= t_{xxxx} + t_{zzxx}, \\ t_{yyzz} &= t_{zzzz} + t_{xxzz}. \end{aligned} \tag{45}$$

Appendix VIII

VIBRATIONAL-ROTATIONAL INTERACTION CONSTANTS

VIIIa INTRODUCTION

In Chapter 4, the vibrational-rotational interaction constants ζ, were discussed in considerable detail for symmetric rotor molecules. This interaction is particularly important for these molecules, for the correction to the energy appears in the first order for degenerate vibrational states. However, higher order interactions of this type can occur for nondegenerate states of symmetric rotors as well as for asymmetric rotors; hence the ζ constants are important for all molecules. Although the actual values of the ζ's depend upon the transformation coefficients between mass-weighted Cartesian coordinates and normal coordinates, hence on the force field, useful relations among the ζ's turn out to be independent of the force field. This problem has been treated by Meal and Polo [78, 79]. Therefore their work is summarized here and their notation related to that of previous workers and that used in this book.

From 4d9 and 1c13 we have the relationship that

$$(a) \qquad -\sum_{\alpha l} \frac{\partial(Q_k, Q_l)}{\partial(\zeta_\alpha, \eta_\alpha)} = -\sum_{\alpha l} (n_{\alpha k} m_{\alpha l} - m_{\alpha k} n_{\alpha l})$$

$$(b) \qquad -\sum_{\alpha l} \frac{\partial(Q_k, Q_l)}{\partial(\xi_\alpha, \zeta_\alpha)} = -\sum_{\alpha l} (l_{\alpha k} n_{\alpha l} - 1_{\alpha l} n_{\alpha k}) \qquad (1)$$

$$(c) \qquad -\sum_{\alpha l} \frac{\partial(Q_k, Q_l)}{\partial(\eta_\alpha, \xi_a)} = -\sum_{\alpha l} (m_{\alpha k} l_{\alpha l} - m_{\alpha l} l_{\alpha k}).$$

These quantities summed only over α for the two normal coordinates involved in the interaction have been defined as $\zeta_{lk}{}^x$, $\zeta_{lk}{}^y$, and $\zeta_{lk}{}^z$ by Meal and Polo.

The inverse of the transformation 1c10 may be written in matrix notation as

$$\mathbf{Q} = \mathbf{lq} \tag{2}$$

or in vector notation as

$$Q_i = \sum l_{i\alpha} q_\alpha.$$

From (2) we can see that the ζ^g, $g = x, y, z$ which have been defined by these authors are given by

$$\zeta_{ij}{}^g = \sum_\alpha (l_{i\alpha} x l_{y\alpha}) e_g. \tag{3}$$

If the matrices M^g are defined such that each consists of N identical (one for each atom) 3 × 3, blocks along the main diagonal, and each block having the form

$$M^x = \begin{bmatrix} 0 & 0 & 0 \\ 0 & 0 & 1 \\ 0 & -1 & 0 \end{bmatrix} \quad M^y = \begin{bmatrix} 0 & 0 & -1 \\ 0 & 0 & 0 \\ 1 & 0 & 0 \end{bmatrix} \quad M^z = \begin{bmatrix} 0 & 1 & 0 \\ -1 & 0 & 0 \\ 0 & 0 & 0 \end{bmatrix}, \tag{4}$$

the ζ_g can be defined in matrix notation as

$$\zeta^g = 1M^\alpha l^\dagger. \tag{5}$$

Thus the elements of ζ^g can be written down by inspection once the transformation matrix between the mass-weighted Cartesian coordinates and the normal coordinates have been evaluated for the molecule. There are, of course, $3N - 6$ vibrational coordinates ($3N - 5$ for linear molecules) plus three normal coordinates of translation and three normal coordinates of rotation (two for linear molecules). The latter six coordinates are defined by the Eckart conditions 1c5 and 1c7. These coordinates are given explicitly by the expressions

$$\begin{aligned} &(a) \qquad Q_i = R_g \qquad l_{ia} = \left(\frac{m_\alpha}{I_g}\right)^{1/2} (e_g \times r_\alpha) \\ &(b) \qquad Q_i = T_g \qquad l_{ia} = \left(\frac{m_\alpha}{M}\right)^{1/2} e_g. \end{aligned} \tag{6}$$

We readily verify that these definitions agree with those used in 4e8 and 4e10.

The M^g matrices have several very simple properties that can be useful.

$$\begin{aligned} &(a) \qquad M^{\dagger g} = -M^g \\ &(b) \qquad M^g M^{\dagger g} = M^{\dagger g} M^g \\ &(c) \qquad M^g M^{\dagger f} - M^f M^{g\dagger} = M^h \\ &(d) \qquad \sum_g M^g M^{\dagger g} = 2E. \end{aligned} \tag{7}$$

The diagonal blocks of the $M^g M^{\dagger g}$ matrices have the form

$$M^x M^{\dagger x} = \begin{bmatrix} 000 \\ 010 \\ 001 \end{bmatrix} \quad M^y M^{\dagger y} = \begin{bmatrix} 100 \\ 000 \\ 001 \end{bmatrix} \quad M^z M^{\dagger z} = \begin{bmatrix} 100 \\ 010 \\ 000 \end{bmatrix}. \tag{8}$$

The characteristic roots of these matrices are 1 ($2N$ times) and zero (N times).

The matrices $E - M^g M^{\dagger g}$ have an even simpler form and satisfy the same relations as a set of projection operators.

$$\begin{aligned} &(a) \qquad (E - M^g M^{\dagger g})^k = (E - M^g M^{\dagger g}) \\ &(b) \qquad (E - M^g M^{\dagger g})(E - M^f M^{\dagger f}) = 0 \\ &(c) \qquad \sum_g (E - M^g M^{\dagger g}) = E. \end{aligned} \tag{9}$$

The ζ^g matrices just defined then have the same properties as the M^g matrices since the two matrices are related through a similarity transformation (since l is real and orthogonal $l^\dagger = l^{-1}$), that is

$$\begin{aligned} &(a) \qquad \zeta^{\dagger g} = -\zeta^g \\ &(b \qquad \zeta^g \zeta^{\dagger f} - \zeta^f \zeta^{\dagger g} = \zeta^h \\ &(c) \qquad \sum_g \zeta^g \zeta^{\dagger g} = 2E. \end{aligned} \tag{10}$$

Each 3×3 block of the M^g matrices has the characteristic equation

$$\lambda(\lambda^2 + 1) = 0, \tag{11}$$

and since there are N of these blocks, the characteristic equation for M^g is

$$[\lambda(\lambda^2 + 1)]^N = 0. \tag{12}$$

The roots of this equation are 0, i, $-i$, each repeated N times. The characteristic equation of the ζ^g matrices is also given by (12) and has the same roots as the M^g matrices. Then it follows from the Cayley-Hamilton theorem that

$$\zeta^g(\zeta^g \zeta^g + E) = 0. \tag{13}$$

Thus the characteristic roots of the ζ^g are independent of the force field and depend only on the atomic masses and molecular geometry.

These ζ^g are of the greatest interest in the treatment of vibrational-rotational interaction. It is convenient to define the $3N \times 3N$ matrices

$$\boldsymbol{\zeta}^g = \mathbf{l}\mathbf{M}^g\mathbf{l}^\dagger = \mathbf{l}\mathbf{M}^g\mathbf{l}^{-1}, \tag{14}$$

which can be partitioned into the submatrices corresponding to the vibrational and rigid motions.

$$\zeta^g = \left[\begin{array}{c|c} \zeta_v^{\ g} & \zeta_{vr}^{\ g} \\ \hline -\zeta_{vr}^{\ g\dagger} & \zeta_r^{\ g} \end{array}\right] = \left[\begin{array}{c|c} \mathbf{l}_v M^g \mathbf{l}_v^{\ \dagger} & \mathbf{l}_v M^g \mathbf{l}_r^{\ \dagger} \\ \hline \mathbf{l}_r M^g \mathbf{l}_v^{\ \dagger} & \mathbf{l}_r M^g \mathbf{l}_r^{\ \dagger} \end{array}\right]. \tag{15}$$

The elements of ζ^g are readily given in terms of the $l_{i\alpha}$ vectors by (3). The $\zeta_r^{\ g}$ elements are determined for the rigid motions using (6*a*) and (6*b*). The only nonvanishing elements are found to be

$$\begin{aligned} &\zeta^g_{R_f R_h} = -\zeta^g_{R_h R_f} = \frac{J_g}{(I_f I_h)^{1/2}} \quad (a) \\ &\zeta^g_{T_f T_h} = -\zeta^g_{T_h T_f} = \quad 1, \quad (b) \end{aligned} \tag{16}$$

in which

$$J_x = \sum_\alpha m_\alpha x_\alpha^{\ 2}, \qquad J_y = \sum_\alpha M_\alpha y_\alpha^{\ 2}, \qquad J_z = \sum_\alpha m_\alpha z_\alpha^{\ 2}. \tag{17}$$

We can verify that the elements in 4e9 and 4e11 are given by these expressions.

The elements $\zeta_{ij}^{\ g} \neq 0$ only if the product $\Gamma(Q_i)\Gamma(Q_j)$ of the representations to which the coordinates Q_i and Q_j belong contains the representation $\Gamma(R_g)$. This is of course

Jahn's rule given earlier. This conclusion follows from the fact that $Q^{\dagger}\zeta^{g}\dot{Q}$ transforms as a rotation R_g under the operations of the symmetry group to which the molecule belongs. If R_g belongs to the totally symmetric representation, ζ^g breaks into blocks along the diagonal, one for each representation; otherwise there are only off-diagonal blocks connecting representations in pairs.

It should be noted that even if $\Gamma(Q_i)\Gamma(Q_j)$ contains R_g it still may be that $\zeta_{ij}{}^g$ is zero. This may happen because a cross-product of two parallel l-vectors will vanish.

VIIIb PROPERTIES OF THE $\zeta^g\zeta^{\dagger g}$ MATRICES

Relations among the ζ^g elements may be deduced from some of the properties of functions of these matrices such as $\zeta^g\zeta^{\dagger g}$, $E - \zeta^g\zeta^{\dagger g}$. These matrices have simpler properties and the advantage that their roots are real. The proofs of these relations are not given here, only the relations themselves. The proofs can be found in the original paper [78].

1. The matrices $\zeta_v{}^x\zeta_v{}^{\dagger x}$, $\zeta_v{}^y\zeta_v{}^{\dagger y}$, and $\zeta_v{}^z\zeta_v{}^{\dagger z}$ for molecules with no multiple axis of symmetry; $\zeta_v{}^x\zeta_v{}^{\dagger x} + \zeta_v{}^y\zeta_v{}^{\dagger y}$ and $\zeta_v{}^z\zeta_v{}^{\dagger z}$ for molecules with only one multiple axis of symmetry; and $\zeta_v{}^x\zeta_v{}^{\dagger x} + \zeta_v{}^y\zeta_v{}^{\dagger y} + \zeta_v{}^z\zeta_v{}^{\dagger z}$ for molecules with more than one multiple axis of symmetry can only have nonvanishing elements connecting normal coordinates belonging to the same representation.

2. The characteristic roots of the matrices $\zeta_v{}^g\zeta_v{}^{\dagger g}$ are $+1$ ($2N - 6$ times) J_fJ_h/I_fI_h (twice), and zero ($N - 2$ times). When the matrices are factored into diagonal blocks because of symmetry, the number of roots equal to $+1$ that a given submatrix has is $N_i^f + N_i^h$ (the number of degrees of freedom contributed by displacement along the f and h directions to the ith representation) decreased by a unit for every translation (different from T_g) or rotation belonging to the ith representation. If this is a totally symmetric representation, the number is further decreased by another unit. The number of roots equal to J_fJ_h/I_fI_h is given by the number of rotations (different from R_g) that belong to the ith representation. The number of roots equal to zero is N_i^g decreased by a unit for each of the rigid motions T_g, R_f, R_h, if these belong to the representation under consideration. If this is a totally symmetric representation, the number is increased by one.

A simplification is possible since for every pair of representations Γ_i and Γ_j whose product has the species R_g it is only necessary to determine the roots of $(\zeta_v{}^g\zeta_v{}^{\dagger g})_i$ with the smaller dimensions $n_i < n_j$. The roots of $(\zeta_v{}^g\zeta_v{}^{\dagger g})_j$ are the same plus $n_j - n_i$ zeros.

These results can now be applied to specific molecules. The H_2O molecule has the point group C_{2v}. The y-axis is chosen perpendicular to the plane of the molecule and the z-axis is taken as the twofold axis. The molecule has two vibrations of species A_1, ν_1, and ν_2 and a B_1 vibration, ν_3. Using Jahn's rule and the fact the R_y is of species B_1, we find the only nonvanishing $\zeta_v{}^g$ matrix is $\zeta_v{}^y$. This matrix has the form

$$\zeta_v{}^y = \begin{bmatrix} 0 & 0 & \zeta_{13}{}^y \\ 0 & 0 & \zeta_{23}{}^y \\ -\zeta_{13}{}^y & -\zeta_{23}{}^y & 0 \end{bmatrix}. \qquad (1)$$

From (1) the matrix $\zeta_{vv}{}^{y}\zeta_{v}{}^{\dagger y}$ is readily obtained. As would be expected from symmetry considerations it factors into two blocks along the main diagonal. These factors correspond to the A_1 representation

$$\begin{bmatrix} (\zeta_{13}{}^{y})^2 & \zeta_{13}{}^{y}\zeta_{23}{}^{y} \\ \zeta_{13}{}^{y}\zeta_{23}{}^{y} & (\zeta_{23}{}^{y})^2 \end{bmatrix}$$

and the B_1 representation

$$[(\zeta_{13}{}^{y})^2 + (\zeta_{23}{}^{y})^2].$$

For both of these representations $n_i^x + n_i^z = 3$. Considering the smaller factor B_1, the number of +1 roots is $3 - 1 - 1 = 1$ since T_x and R_y are of this species; thus

$$(\zeta_{13}{}^{y})^2 + (\zeta_{23}{}^{y})^2 = 1.$$

If we proceed to solve the other factor A_1, the number of roots +1 is $3 - 1 - 1 = 1$ since T_z belongs to this representation and it is totally symmetric. The other root is zero. From the properties of the trace of a matrix we find again

$$\zeta_{13}{}^{y^2} + \zeta_{23}{}^{y^2} = 1,$$

and no new information is obtained. This example clearly illustrates properties 1 and 2 of matrices of this type. The method applies in the same straightforward manner to all molecules although it becomes somewhat more tedious. The special case of symmetric rotor molecules was discussed in Chapter 4. If the $\zeta_v{}^{g}$'s are known with sufficient accuracy, they can furnish additional information concerning the vibrational potential function. It is convenient then to have the $\zeta_v{}^{g}$'s simply related to the parameters of the vibrational problem. This has been done [79] and the procedure is outlined below. It is not the intention to give a treatment of the vibrational problem here. Rather the reader is referred to Wilson, Decius, and Cross [9] for any further details.

VIIIc RELATION BETWEEN ζ^{g}'s AND POTENTIAL CONSTANTS

As is usual, it is convenient to set up the vibrational problem using a set of internal coordinates $R_1, R_2, \ldots, R_{n-6}$ plus the six coordinates of rigid motion. It is convenient to partition the **R** matrix and the transformation matrices **D** into submatrices for vibrational motion and rigid motion thus

$$\mathbf{R} = \begin{pmatrix} \mathbf{R}_v \\ \mathbf{R}_r \end{pmatrix} = \begin{pmatrix} \mathbf{L}_v & 0 \\ 0 & \mathbf{L}_R \end{pmatrix} \begin{pmatrix} \mathbf{Q}_v \\ \mathbf{Q}_r \end{pmatrix} = \begin{pmatrix} \mathbf{D}_v \\ \mathbf{D}_r \end{pmatrix} q. \tag{1}$$

The elements $\mathbf{R}_r$ can be defined (Eckart conditions) so that $\mathbf{R}_r = \mathbf{Q}_r = 0$ and therefore $\mathbf{L}_r = E_{6,6}$ and $\mathbf{D}_r = l_r$. Since the transformation matrix l is orthogonal,

$$DD^{\dagger} = LL^{\dagger} = G = \begin{pmatrix} G_v & 0 \\ 0 & G_r \end{pmatrix}, \tag{2}$$

where G_v is the familiar **G**-matrix [9] and G_r is a 6×6 identity matrix. Equation (1) may now be written in vector form

$$R_i = \sum_{\alpha} \mathbf{d}_{i\alpha} \cdot \mathbf{q}_{\alpha} = \sum_{\alpha} \mathbf{s}_{i\alpha} \cdot \boldsymbol{\rho}_{\alpha}, \tag{3}$$

where $\boldsymbol{\rho}_\alpha$ is the Cartesian displacement vector of atom α, and $s_{i\alpha}$ are the s-vectors. The components of the d-vectors are the elements of D and

$$d_{i\alpha} = m_\alpha^{1/2} s_{i\alpha}. \tag{4}$$

Meal and Polo [79] have then defined the C^g matrices. Starting from A.5 and making use of the definitions of $\mathbf{l}_v$ and $\mathbf{D}_v$, we can write

$$\zeta_v^g = L_v^{-1} C_v^g L_v^{\dagger -1} = L_v^\dagger \bar{C}_{vz}^g L_v, \tag{5}$$

in which C_v^g and C_v^{-g} are defined as

$$\begin{aligned} &(a) \qquad C_v^g = D_v M^g D_v^\dagger \\ &(b) \qquad C_v^{-g} = (D^{-1})_v^\dagger M^g (D^{-1})_v. \end{aligned} \tag{6}$$

Since D_v depends only on the masses and the geometry of the molecule, C_v^g and $\bar{C}_v^g$ also depend only on these parameters. Thus, if the transformation L_v has been determined from the normal coordinate analysis, the elements of ζ_v^g may be calculated.

The $\mathbf{D}$-matrix elements are quite readily obtained, but the elements of D^{-1} are generally quite complicated to obtain [95]. Thus in most of the subsequent discussion C_v^g is used. Thus using the definitions of C_v^g and M^g the elements of C_v^g may be written explicitly as

$$C_{ij}^g = \sum_\alpha (d_{i\alpha} \times d_{j\alpha}) \cdot e_g = \sum_\alpha m_\alpha^{-1} (s_{i\alpha} \times s_{j\alpha}) \cdot e_g \tag{7}$$

This expression is to be compared with that for the $\mathbf{G}$-matrix elements

$$G_{ij} = \sum_\alpha (d_{i\alpha} \cdot d_{j\alpha}) = \sum_\alpha m_\alpha^{-1} (s_{i\alpha} \cdot s_{j\alpha}), \tag{8}$$

thus the convenience of the C_v^g is apparent, since once the s- or d-vectors have been determined in the normal coordinate analysis, the elements of C_v^g are readily obtained.

Since $M^{\dagger g} = -M^g$, C_v^g is also skew-symmetric. By the use of symmetry coordinates C_v^g is considerably simplified since many elements vanish due to symmetry. The symmetry properties of C_v^g are the same as those of ζ_v^g. Equation (5) may be combined with the expressions

$$\begin{aligned} &(a) \qquad L_v^\dagger (G^{-1})_v L_v = E, \\ &\text{and} \\ &(b) \qquad L_v^\dagger F L_v = \Lambda \end{aligned} \tag{9}$$

to obtain the relations

$$\begin{aligned} &(a) \qquad L_v^\dagger (G^{-1})_v C_v^g (L_v^\dagger)^{-1} = \zeta_v^g \\ &\text{and} \\ &(b) \qquad L_v^\dagger F C_v^g (L_v^1)^{-1} = \Lambda \zeta_v^g. \end{aligned} \tag{10}$$

From (10a) we see that the matrices $(G^{-1})_v C_v^g$ and ζ_v^g are related by a similarity transformation and therefore have the same roots. Since $(G^{-1})_v C_v^g$ depends only on the masses of the molecule, it follows that the characteristic roots of ζ_v^g are independent of the force field as shown in A. Similarly from (10b) we see that FC_v^g and $\Lambda\zeta_v^g$ have the same roots.

By identification of the coefficients in the characteristic equations

$$\begin{aligned} &(a) \qquad |(G)_v^{-1} C_v^g - \lambda E| \equiv |\zeta_v^g - \lambda E| = 0 \\ &(b) \qquad |F C_v^g - \lambda' E| \equiv |\Lambda \zeta_v^g - \lambda' E| = 0. \end{aligned} \tag{11}$$

We can deduce useful relations among the ζ elements. Those obtained from (11a) are independent of the force field and merely give the sum rules again. However, (11b) relates the zetas with the force constants, and in some cases the need for the L_v^{-1} elements can be avoided.

Using these considerations Meal and Polo have deduced explicit formulas which relate the ζ's of a symmetric nonlinear triatomic molecule directly with the force constants,

$$\begin{aligned} &(a) \qquad (\zeta_{13}^{y})^2 = \frac{\lambda_1 - F_{11}/(G^{-1})_{11}}{\lambda_1 - \lambda_2} \\ &(b) \qquad (\zeta_{23}^{y})^2 = \frac{F_{11}/(G^{-1})_{11} - \lambda_2}{\lambda_1 - \lambda_2} \end{aligned} \tag{12}$$

In Eq. (12) $\lambda_1 = 4\pi^2\nu_1^2$ and $\lambda_2 = 4\pi^2\nu_2^2$, where ν_1 and ν_2 are the completely symmetric vibrational frequencies. These relations are based on the usual symmetry coordinates for this type of molecule, $R_1 = (\Delta r_1 + \Delta r_2)/\sqrt{2}$, $R_2 = \Delta\alpha$ and $R_3 = (\Delta r_1 - \Delta r_2)/\sqrt{2}$.

Application of these considerations to a symmetric top results in a useful isotope rule. We find

$$m_\alpha \sum_i \lambda_i(1 - \zeta_i) = k, \tag{13}$$

where k is a constant for isotopic molecules. The relation holds remarkably well for ζ's obtained from rather low resolution spectra. It must be remembered that these relations have all been deduced for an harmonic approximation. As a result it should not be expected that these relations hold exactly for constants derived from very high resolution work where the factors neglected in this treatment can be important. However, the considerations outlined in this appendix can be exceedingly useful in giving first approximations from which one can start in band analyses.

Appendix IX

TABLE OF LINE STRENGTHS

The line strengths listed here (see Tables IX-1 and IX-2) are the squares of the elements of the direction-cosine matrices in the representation which diagonalizes the energy matrix of the asymmetric rotor, summed over the Zeeman components and multiplied by three to account for the three equivalent space-fixed directions, that is, line strength for the component of the electric moment μ_g, parallel to the molecule-fixed axes $g = a, b, c$ is

$$\sum_{F=X,Y,Z} \sum_{M''} \sum_{M'} |(\Phi_{Fg}{}^A)_{J'',\tau'',M'',J',\tau',M'}|^2 = 3 \sum_{M''} \sum_{M'} \left| \int \psi^*_{J'',\tau'',M''} \Phi_{Fg}{}^A \psi_{J',\tau',M'} \, dv \right|^2$$

and thus is the corresponding prefactor on the right-hand side of 5c9 summed over X, Y, Z, M', and M''. Actual intensities can then be obtained provided the values of the integrals $\int \psi^*_{V'',e''} \mu_g \psi_{V',e'} \, dv$ are known. Relative intensities are obtained by putting such of these integrals that are nonvanishing equal to unity.

The entries have also been multiplied by 10^4 to eliminate decimal points. The parameter of asymmetry

$$\kappa = \frac{2b - a - c}{a - c}$$

where a, b, c equal $\hbar^2/2I_a, \hbar^2/2I_b, \hbar^2/2I_c$, respectively, and where the condition $I_a \leqslant I_b \leqslant I_c$ is applied in assigning the moments of inertia.

Transitions are classified by sub-branches which head the column in which the initial level can be found (identified by $JK_{-1},K_1;\tau$) and whose final level is in the adjacent column on the same row.

The lines in each sub-branch are listed in wings which can be identified by K_{-1} or K_1, whichever is held constant, as can be determined by the subscripts of the initial levels.

Sub-branches in adjacent columns have identical strengths. Those in columns 1 and 2 apply to the upper sign of κ; those in columns 8 and 9 apply to the lower sign.

The strengths found in columns 3 and 7 are those occurring in the prolate- and oblate-symmetric rotor. When no entry is given, the transition is forbidden. "High order forbidden" branches for which $|\Delta\tau| \geqslant 9$ have been omitted. When the entry is 0, the strength is less than 0.0001.

Table IX-1—Symmetric Rotor Sub-Branches

A. *a* AND *c* PROLATE-AND-OBLATE SUB-BRANCHES

Sub-branch ${}^{c,e}Q_{1,0}$	${}^{c,o}Q_{\bar{1},0}$	∓ 1	∓ 0.5	κ 0	± 0.5	± 1	Sub-branch ${}^{a,e}Q_{0,1}$	${}^{a,o}Q_{0,\bar{1}}$
$1_{0,1;-1}$	$1_{1,1;0}$	15000	15000	15000	15000	15000	$1_{1,0;1}$	$1_{1,1;0}$
$2_{0,2;-2}$	$2_{1,2;-1}$	25000	28223	31100	32845	33333	$2_{2,0;2}$	$2_{2,1;1}$
$3_{0,3;-3}$	$3_{1,3;-2}$	35000	45104	50431	52155	52500	$3_{3,0;3}$	$3_{3,1;2}$
$4_{0,4;-4}$	$4_{1,4;-3}$	45000	64494	70244	71708	72000	$4_{4,0;4}$	$4_{4,1;3}$
$5_{0,5;-5}$	$5_{1,5;-4}$	55000	84696	90073	91399	91667	$5_{5,0;5}$	$5_{5,1;4}$
$6_{0,6;-6}$	$6_{1,6;-5}$	65000	104928	109923	111174	111429	$6_{6,0;6}$	$6_{6,1;5}$
$7_{0,7;-7}$	$7_{1,7;-6}$	75000	125065	129799	131004	131250	$7_{7,0;7}$	$7_{7,1;6}$
$8_{0,8;-8}$	$8_{1,8;-7}$	85000	145135	149698	150871	151111	$8_{8,0;8}$	$8_{8,1;7}$
$9_{0,9;-9}$	$9_{1,9;-8}$	95000	165170	169614	170764	171000	$9_{9,0;9}$	$9_{9,1;8}$
$10_{0,10;-10}$	$10_{1,10;-9}$	105000	185187	189544	190677	190909	$10_{10,0;10}$	$10_{10,1;9}$
$11_{0,11;-11}$	$11_{1,11;-10}$	115000	205194	209484	210603	210834	$11_{11,0;11}$	$11_{11,1;10}$
$12_{0,12;-12}$	$12_{1,12;-11}$	125000	225195	229434	230542	230769	$12_{12,0;12}$	$12_{12,1;11}$
$2_{1,1;0}$	$2_{2,1;1}$	8333	8333	8333	8333	8333	$2_{1,1;0}$	$2_{1,2;-1}$
$3_{1,2;-1}$	$3_{2,2;0}$	14583	16278	18811	21875	23333	$3_{2,1;1}$	$3_{2,2;0}$
$4_{1,3;-2}$	$4_{2,3;-1}$	20250	26168	34242	39363	40500	$4_{3,1;2}$	$4_{3,2;1}$
$5_{1,4;-3}$	$5_{2,4;-2}$	25667	39338	52949	57742	58667	$5_{4,1;3}$	$5_{4,2;2}$
$6_{1,5;-4}$	$6_{2,5;-3}$	30952	56179	72319	76548	77381	$6_{5,1;4}$	$6_{5,2;3}$
$7_{1,6;-5}$	$7_{2,6;-4}$	36161	75597	91744	95646	96429	$7_{6,1;5}$	$7_{6,2;4}$
$8_{1,7;-6}$	$8_{2,7;-5}$	41319	95950	111231	114943	115694	$8_{7,1;6}$	$8_{7,2;5}$
$9_{1,8;-7}$	$9_{2,8;-6}$	46444	116333	130792	134381	135111	$9_{8,1;7}$	$9_{8,2;6}$
$10_{1,9;-8}$	$10_{2,9;-7}$	51545	136551	150418	153921	154636	$10_{9,1;8}$	$10_{9,2;7}$
$11_{1,10;-9}$	$11_{2,10;-8}$	56629	156642	170100	173540	174242	$11_{10,1;9}$	$11_{10,2;8}$
$12_{1,11;-10}$	$12_{2,11;-9}$	61699	176660	189825	193216	193910	$12_{11,1;10}$	$12_{11,2;9}$
$3_{2,1;1}$	$3_{3,1;2}$	8750	7403	6406	5944	5833	$3_{1,2;-1}$	$3_{1,3;-2}$
$4_{2,2;0}$	$4_{3,2;1}$	15750	13221	13196	15598	18000	$4_{2,2;0}$	$4_{2,3;-1}$
$5_{2,3;-1}$	$5_{3,3;0}$	22000	19105	23397	30662	33000	$5_{3,2;1}$	$5_{3,3;0}$
$6_{2,4;-2}$	$6_{3,4;-1}$	27857	26374	38620	47709	49524	$6_{4,2;2}$	$6_{4,3;1}$
$7_{2,5;-3}$	$7_{3,5;-2}$	33482	36237	57062	65399	66964	$7_{5,2;3}$	$7_{5,3;2}$
$8_{2,6;-4}$	$8_{3,6;-3}$	38958	49682	76155	83565	85000	$8_{6,2;4}$	$8_{6,3;3}$
$9_{2,7;-5}$	$9_{3,7;-4}$	44333	66864	95251	102089	103444	$9_{7,2;5}$	$9_{7,3;4}$
$10_{2,8;-6}$	$10_{3,8;-5}$	49636	86630	114393	120880	122183	$10_{8,2;6}$	$10_{8,3;5}$
$11_{2,9;-7}$	$11_{3,9;-6}$	54886	107332	133621	139873	141136	$11_{9,2;7}$	$11_{9,3;6}$
$12_{2,10;-8}$	$12_{3,10;-7}$	60096	128002	152940	159022	160256	$12_{10,2;8}$	$12_{10,3;7}$
$4_{3,1;2}$	$4_{4,1;3}$	9000	7587	6026	4847	4500	$4_{1,3;-2}$	$4_{1,4;-3}$
$5_{3,2;1}$	$5_{4,2;2}$	16500	13464	11058	11750	14667	$5_{2,3;-1}$	$5_{2,4;-2}$
$6_{3,3;0}$	$6_{4,3;1}$	23214	18339	17488	23981	27857	$6_{3,3;0}$	$6_{3,4;-1}$
$7_{3,4;-1}$	$7_{4,4;0}$	29464	22914	27745	39794	42857	$7_{4,3;1}$	$7_{4,4;0}$
$8_{3,5;-2}$	$8_{4,5;-1}$	35417	28185	43063	56506	59028	$8_{5,3;2}$	$8_{5,4;1}$
$9_{3,6;-3}$	$9_{4,6;-2}$	41167	35293	61523	73754	76000	$9_{6,3;3}$	$9_{6,4;2}$
$10_{3,7;-4}$	$10_{4,7;-3}$	46773	45350	80547	91464	93546	$10_{7,3;4}$	$10_{7,4;3}$
$11_{3,8;-5}$	$11_{4,8;-4}$	52273	59213	99473	109542	111515	$11_{8,3;5}$	$11_{8,4;4}$
$12_{3,9;-6}$	$12_{4,9;-5}$	57692	76888	118383	127913	129808	$12_{9,3;6}$	$12_{9,4;5}$
$5_{4,1;3}$	$5_{5,1;4}$	9167	7777	6127	4374	3667	$5_{1,4;-3}$	$5_{1,5;-4}$
$6_{4,2;2}$	$6_{5,2;3}$	17024	14084	10758	9464	12381	$6_{2,4;-2}$	$6_{2,5;-3}$
$7_{4,3;1}$	$7_{5,3;2}$	24107	19340	15156	18769	24107	$7_{3,4;-1}$	$7_{3,5;-2}$
$8_{4,4;0}$	$8_{5,4;1}$	30694	23768	21441	33034	37778	$8_{4,4;0}$	$8_{4,5;-1}$
$9_{4,5;-1}$	$9_{5,5;0}$	36944	27638	31860	49002	52778	$9_{5,4;1}$	$9_{5,5;0}$
$10_{4,6;-2}$	$10_{5,6;-1}$	42955	31542	47402	65474	68727	$10_{6,4;2}$	$10_{6,5;1}$
$11_{4,7;-3}$	$11_{5,7;-2}$	48788	36457	66028	82425	85379	$11_{7,4;3}$	$11_{7,5;2}$
$12_{4,8;-4}$	$12_{5,8;-3}$	54487	43527	85120	99805	102564	$12_{8,4;4}$	$12_{8,5;3}$
$6_{5,1;4}$	$6_{6,1;5}$	9286	7913	6271	4244	3095	$6_{1,5;-4}$	$6_{1,6;-5}$
$7_{5,2;3}$	$7_{6,2;4}$	17411	14552	11116	8220	10714	$7_{2,5;-3}$	$7_{2,6;-4}$
$8_{5,3;2}$	$8_{6,3;3}$	24792	20246	14956	14996	21250	$8_{3,5;-2}$	$8_{3,6;-3}$
$9_{5,4;1}$	$9_{6,4;2}$	31667	25157	18940	27035	33778	$9_{4,5;-1}$	$9_{4,6;-2}$
$10_{5,5;0}$	$10_{6,5;1}$	38182	29364	25162	42308	47727	$10_{5,5;0}$	$10_{5,6;-1}$
$11_{5,6;-1}$	$11_{6,6;0}$	44432	32945	35783	58220	62727	$11_{6,5;1}$	$11_{6,6;0}$
$12_{5,7;-2}$	$12_{6,7;-1}$	50481	36135	51610	74526	78526	$12_{7,5;2}$	$12_{7,6;1}$

Table IX-1 (Continued)

Sub-branch ${}^{c,e}Q_{1,0}$	${}^{c,o}Q_{\bar{1},0}$	κ ∓ 1	∓ 0.5	0	± 0.5	± 1	Sub-branch ${}^{a,e}Q_{0,1}$	${}^{a,o}Q_{0,\bar{1}}$
$7_{6,1;5}$	$7_{7,1;6}$	9375	8011	6383	4273	2679	$7_{1,6;-5}$	$7_{1,7;-6}$
$8_{6,2;4}$	$8_{7,2;5}$	17708	14899	11514	7682	9444	$8_{2,6;-4}$	$8_{2,7;-5}$
$9_{6,3;3}$	$9_{7,3;4}$	25333	20931	15562	12549	19000	$9_{3,6;-3}$	$9_{3,7;-4}$
$10_{6,4;2}$	$10_{7,4;3}$	32455	26254	18837	21925	30545	$10_{4,6;-2}$	$10_{4,7;-3}$
$11_{6,5;1}$	$11_{7,5;2}$	39205	30943	22510	36052	43561	$11_{5,6;-1}$	$11_{5,7;-2}$
$12_{6,6;0}$	$12_{7,6;1}$	45673	35027	28709	51607	57692	$12_{6,6;0}$	$12_{6,7;-1}$
$8_{7,1;6}$	$8_{8,1;7}$	9444	8087	6468	4346	2361	$8_{1,7;-6}$	$8_{1,8;-7}$
$9_{7,2;5}$	$9_{8,2;6}$	17944	15167	11832	7594	8444	$9_{2,7;-5}$	$9_{2,8;-6}$
$10_{7,3;4}$	$10_{8,3;5}$	25773	21462	16215	11172	17182	$10_{3,7;-4}$	$10_{3,8;-5}$
$11_{7,4;3}$	$11_{8,4;4}$	33106	27105	19675	18011	27879	$11_{4,7;-3}$	$11_{4,8;-4}$
$12_{7,5;2}$	$12_{8,5;3}$	40064	32174	22501	30163	40064	$12_{5,7;-2}$	$12_{5,8;-3}$
$9_{8,1;7}$	$9_{9,1;8}$	9500	8146	6535	4418	2111	$9_{1,8;-7}$	$9_{1,9;-8}$
$10_{8,2;6}$	$10_{9,2;7}$	18136	15380	12081	7731	7636	$10_{2,8;-6}$	$10_{2,9;-7}$
$11_{8,3;5}$	$11_{9,3;6}$	26136	21887	16748	10598	15682	$11_{3,8;-5}$	$11_{3,9;-6}$
$12_{8,4;4}$	$12_{9,4;5}$	33654	27788	20570	15383	25641	$12_{4,8;-4}$	$12_{4,9;-5}$
$10_{9,1;8}$	$10_{10,1;9}$	9545	8194	6588	4479	1909	$10_{1,9;-8}$	$10_{1,10;-9}$
$11_{9,2;7}$	$11_{10,2;8}$	18295	15554	12280	7933	6970	$11_{2,9;-7}$	$11_{2,10;-8}$
$12_{9,3;6}$	$12_{10,3;7}$	26442	22237	17176	10563	14423	$12_{3,9;-6}$	$12_{3,10;-7}$
$11_{10,1;9}$	$11_{11,1;10}$	9583	8234	6631	4530	1742	$11_{1,10;-9}$	$11_{1,11;-10}$
$12_{10,2;8}$	$12_{11,2;9}$	18429	15699	12443	8126	6410	$12_{2,10;-8}$	$12_{2,11;-9}$
$12_{11,1;10}$	$12_{12,1;11}$	9615	8268	6667	4571	1603	$12_{1,11;-10}$	$12_{1,12;-11}$
${}^{c,e}R_{1,0}$	${}^{c,e}P_{\bar{1},0}$	∓ 1	∓ 0.5	0	± 0.5	± 1	${}^{a,e}R_{0,1}$	${}^{a,e}P_{0,\bar{1}}$
$0_{0,0;0}$	$1_{1,0;1}$	10000	10000	10000	10000	10000	$0_{0,0;0}$	$1_{0,1;-1}$
$1_{1,0;1}$	$2_{2,0;2}$	15000	16934	18660	19707	20000	$1_{0,1;-1}$	$2_{0,2;-2}$
$2_{2,0;2}$	$3_{3,0;3}$	25000	25893	27201	29029	30000	$2_{0,2;-2}$	$3_{0,3;-3}$
$3_{3,0;3}$	$4_{4,0;4}$	35000	35773	36728	38312	40000	$3_{0,3;-3}$	$4_{0,4;-4}$
$4_{4,0;4}$	$5_{5,0;5}$	45000	45745	46619	47897	50000	$4_{0,4;-4}$	$5_{0,5;-5}$
$5_{5,0;5}$	$6_{6,0;6}$	55000	55730	56582	57727	60000	$5_{0,5;-5}$	$6_{0,6;-6}$
$6_{6,0;6}$	$7_{7,0;7}$	65000	65721	66562	67660	70000	$6_{0,6;-6}$	$7_{0,7;-7}$
$7_{7,0;7}$	$8_{8,0;8}$	75000	75714	76549	77628	80000	$7_{0,7;-7}$	$8_{0,8;-8}$
$8_{8,0;8}$	$9_{9,0;9}$	85000	85708	86539	87610	90000	$8_{0,8;-8}$	$9_{0,9;-9}$
$9_{9,0;9}$	$10_{10,0;10}$	95000	95704	96531	97597	100000	$9_{0,9;-9}$	$10_{0,10;-10}$
$10_{10,0;10}$	$11_{11,0;11}$	105000	105701	106525	107588	110000	$10_{0,10;-10}$	$11_{0,11;-11}$
$11_{11,0;11}$	$12_{12,0;12}$	115000	115698	116519	117580	120000	$11_{0,11;-11}$	$12_{0,12;-12}$
$1_{0,1;-1}$	$2_{1,1;0}$	15000	15000	15000	15000	15000	$1_{1,0;1}$	$2_{1,1;0}$
$2_{1,1;0}$	$3_{2,1;1}$	16667	22500	25581	26509	26667	$2_{1,1;0}$	$3_{1,2;-1}$
$3_{2,1;1}$	$4_{3,1;2}$	26250	29261	33801	36902	37500	$3_{1,2;-1}$	$4_{1,3;-2}$
$4_{3,1;2}$	$5_{4,1;3}$	36000	38400	41758	46530	48000	$4_{1,3;-2}$	$5_{1,4;-3}$
$5_{4,1;3}$	$6_{5,1;4}$	45833	48106	50867	55604	58333	$5_{1,4;-3}$	$6_{1,5;-4}$
$6_{5,1;4}$	$7_{6,1;5}$	55714	57930	60533	64605	68571	$6_{1,5;-4}$	$7_{1,6;-5}$
$7_{6,1;5}$	$8_{7,1;6}$	65625	67805	70356	73938	78750	$7_{1,6;-5}$	$8_{1,7;-6}$
$8_{7,1;6}$	$9_{8,1;7}$	75556	77710	80235	83593	88889	$8_{1,7;-6}$	$9_{1,8;-7}$
$9_{8,1;7}$	$10_{9,1;8}$	85500	87636	90142	93412	99000	$9_{1,8;-7}$	$10_{1,9;-8}$
$10_{9,1;8}$	$11_{10,1;9}$	95455	97576	100068	103301	109091	$10_{1,9;-8}$	$11_{1,10;-9}$
$11_{10,1;9}$	$12_{11,1;10}$	105416	107526	110008	113219	119166	$11_{1,10;-9}$	$12_{1,11;-10}$
$2_{0,2;-2}$	$3_{1,2;-1}$	20000	18636	17345	16724	16667	$2_{2,0;2}$	$3_{2,1;1}$
$3_{1,2;-1}$	$4_{2,2;0}$	18750	29055	30992	30230	30000	$3_{2,1;1}$	$4_{2,2;0}$
$4_{2,2;0}$	$5_{3,2;1}$	28000	34387	41441	42462	42000	$4_{2,2;0}$	$5_{2,3;-1}$
$5_{3,2;1}$	$6_{4,2;2}$	37500	41961	49227	53738	53333	$5_{2,3;-1}$	$6_{2,4;-2}$
$6_{4,2;2}$	$7_{5,2;3}$	47143	51182	56697	64087	64286	$6_{2,4;-2}$	$7_{2,5;-3}$
$7_{5,2;3}$	$8_{6,2;4}$	56875	60756	65450	73564	75000	$7_{2,5;-3}$	$8_{2,6;-4}$
$8_{6,2;4}$	$9_{7,2;5}$	66667	70451	74899	82413	85556	$8_{2,6;-4}$	$9_{2,7;-5}$
$9_{7,2;5}$	$10_{8,2;6}$	76500	80218	84567	91174	96000	$9_{2,7;-5}$	$10_{2,8;-6}$
$10_{8,2;6}$	$11_{9,2;7}$	86364	90031	94328	100297	106364	$10_{2,8;-6}$	$11_{2,9;-7}$
$11_{9,2;7}$	$12_{10,2;8}$	96250	99880	104134	109796	116666	$11_{2,9;-7}$	$12_{2,10;-8}$

Table IX-1 (Continued)

Sub-branch $^{c,e}R_{1,0}$	$^{c,e}P_{\bar{1},0}$	κ: ∓ 1	∓ 0.5	0	± 0.5	± 1	Sub-branch $^{a,e}R_{0,1}$	$^{a,e}P_{0,\bar{1}}$
$3_{0,3;-3}$	$4_{1,3;-2}$	25000	20331	18001	17567	17500	$3_{3,0;3}$	$4_{3,1;2}$
$4_{1,3;-2}$	$5_{2,3;-1}$	21000	34848	33475	32109	32000	$4_{3,1;2}$	$5_{3,2;1}$
$5_{2,3;-1}$	$6_{3,3;0}$	30000	41218	47032	45219	45000	$5_{3,2;1}$	$6_{3,3;0}$
$6_{3,3;0}$	$7_{4,3;1}$	39286	46575	57381	57683	57143	$6_{3,3;0}$	$7_{3,4;-1}$
$7_{4,3;1}$	$8_{5,3;2}$	48750	54876	64788	69691	68750	$7_{3,4;-1}$	$8_{3,5;-2}$
$8_{5,3;2}$	$9_{6,3;3}$	58333	64092	71834	80981	80000	$8_{3,5;-2}$	$9_{3,6;-3}$
$9_{6,3;3}$	$10_{7,3;4}$	68000	73557	80274	91312	91000	$9_{3,6;-3}$	$10_{3,7;-4}$
$10_{7,3;4}$	$11_{8,3;5}$	77727	83147	89526	100665	101818	$10_{3,7;-4}$	$11_{3,8;-5}$
$11_{8,3;5}$	$12_{9,3;6}$	87500	92819	90948	109320	112500	$11_{3,8;-5}$	$12_{3,9;-6}$
$4_{0,4;-4}$	$5_{1,4;-3}$	30000	20650	18478	18082	18000	$4_{4,0;4}$	$5_{4,1;3}$
$5_{1,4;-3}$	$6_{2,4;-2}$	23333	38686	34370	33475	33333	$5_{4,1;3}$	$6_{4,2;2}$
$6_{2,4;-2}$	$7_{3,4;-1}$	32143	48639	49439	47326	47143	$6_{4,2;2}$	$7_{4,3;1}$
$7_{3,4;-1}$	$8_{4,4;0}$	41250	52676	63082	60238	60000	$7_{4,3;1}$	$8_{4,4;0}$
$8_{4,4;0}$	$9_{5,4;1}$	50556	59229	73357	72645	72222	$8_{4,4;0}$	$9_{4,5;-1}$
$9_{5,4;1}$	$10_{6,4;2}$	60000	67888	80428	84877	84000	$9_{4,5;-1}$	$10_{4,6;-2}$
$10_{6,4;2}$	$11_{7,4;3}$	69546	77064	87093	96881	95455	$10_{4,6;-2}$	$11_{4,7;-3}$
$11_{7,4;3}$	$12_{8,4;4}$	79167	86444	95254	108226	106666	$11_{4,7;-3}$	$12_{4,8;-4}$
$5_{0,5;-5}$	$6_{1,5;-4}$	35000	20660	18847	18422	18333	$5_{5,0;5}$	$6_{5,1;4}$
$6_{1,5;-4}$	$7_{2,5;-3}$	25714	40254	35224	34447	34286	$6_{5,1;4}$	$7_{5,2;3}$
$7_{2,5;-3}$	$8_{3,5;-2}$	34375	54914	50352	48971	48750	$7_{5,2;3}$	$8_{5,3;2}$
$8_{3,5;-2}$	$9_{4,5;-1}$	43333	60334	65354	62490	62222	$8_{5,3;2}$	$9_{5,4;1}$
$9_{4,5;-1}$	$10_{5,5;0}$	52500	64543	79136	75309	75000	$9_{5,4;1}$	$10_{5,5;0}$
$10_{5,5;0}$	$11_{6,5;1}$	61818	72156	89354	87664	87273	$10_{5,5;0}$	$11_{5,6;-1}$
$11_{6,5;1}$	$12_{7,5;2}$	71250	80944	96120	99820	99167	$11_{5,6;-1}$	$12_{5,7;-2}$
$6_{0,6;-6}$	$7_{1,6;-5}$	40000	20793	19108	18664	18571	$6_{6,0;6}$	$7_{6,1;5}$
$7_{1,6;-5}$	$8_{2,6;-4}$	28125	40367	35988	35171	35000	$7_{6,1;5}$	$8_{6,2;4}$
$8_{2,6;-4}$	$9_{3,6;-3}$	36667	58807	51410	50241	50000	$8_{6,2;4}$	$9_{6,3;3}$
$9_{3,6;-3}$	$10_{4,6;-2}$	45500	68406	66193	64301	64000	$9_{6,3;3}$	$10_{6,4;2}$
$10_{4,6;-2}$	$11_{5,6;-1}$	54546	71334	81252	77627	77273	$10_{6,4;2}$	$11_{6,5;1}$
$11_{5,6;-1}$	$12_{6,6;0}$	63750	77023	95192	90399	90000	$11_{6,5;1}$	$12_{6,6;0}$
$7_{0,7;-7}$	$8_{1,7;-6}$	45000	20990	19300	18844	18750	$7_{7,0;7}$	$8_{7,1;6}$
$8_{1,7;-6}$	$9_{2,7;-5}$	30556	40255	36587	35733	35556	$8_{7,1;6}$	$9_{7,2;5}$
$9_{2,7;-5}$	$10_{3,7;-4}$	39000	60147	52457	51252	51000	$9_{7,2;5}$	$10_{7,3;4}$
$10_{3,7;-4}$	$11_{4,7;-3}$	47727	75043	67350	65775	65455	$10_{7,3;4}$	$11_{7,4;3}$
$11_{4,7;-3}$	$12_{5,7;-2}$	56667	79651	81971	79549	79167	$11_{7,4;3}$	$12_{7,5;2}$
$8_{0,8;-8}$	$9_{1,8;-7}$	50000	21170	19449	18985	18889	$8_{8,0;8}$	$9_{8,1;7}$
$9_{1,8;-7}$	$10_{2,8;-6}$	33000	40430	37059	36182	36000	$9_{8,1;7}$	$10_{8,2;6}$
$10_{2,8;-6}$	$11_{3,8;-5}$	41364	59963	53330	52078	51818	$10_{8,2;6}$	$11_{8,3;5}$
$11_{3,8;-5}$	$12_{4,8;-4}$	50000	78946	68596	66999	66667	$11_{8,3;5}$	$12_{8,4;4}$
$9_{0,9;-9}$	$10_{1,9;-8}$	55000	21315	19566	19097	19000	$9_{9,0;9}$	$10_{9,1;8}$
$10_{1,9;-8}$	$11_{2,9;-7}$	35454	40779	37443	36548	36364	$10_{9,1;8}$	$11_{9,2;7}$
$11_{2,9;-7}$	$12_{3,9;-6}$	43750	59640	54051	52766	52500	$11_{9,2;7}$	$12_{9,3;6}$
$10_{0,10;-10}$	$11_{1,10;-9}$	60000	21432	19662	19189	19091	$10_{10,0;10}$	$11_{10,1;9}$
$11_{1,10;-9}$	$12_{2,10;-8}$	37917	41138	37761	36854	36667	$11_{10,1;9}$	$12_{10,2;8}$
$11_{0,11;-11}$	$12_{1,11;-10}$	65000	21527	19742	19265	19167	$11_{11,0;11}$	$12_{11,1;10}$
$^{c,o}R_{1,0}$	$^{c,o}P_{\bar{1},0}$	∓ 1	∓ 0.5	0	± 0.5	± 1	$^{a,o}R_{0,1}$	$^{a,o}P_{0,\bar{1}}$
$1_{1,1;0}$	$2_{2,1;1}$	15000	15000	15000	15000	15000	$1_{1,1;0}$	$2_{1,2;-1}$
$2_{2,1;1}$	$3_{3,1;2}$	25000	25710	26243	26564	26667	$2_{1,2;-1}$	$3_{1,3;-2}$
$3_{3,1;2}$	$4_{4,1;3}$	35000	35758	36540	37210	37500	$3_{1,3;-2}$	$4_{1,4;-3}$
$4_{4,1;3}$	$5_{5,1;4}$	45000	45743	46583	47478	48000	$4_{1,4;-3}$	$5_{1,5;-4}$
$5_{5,1;4}$	$6_{6,1;5}$	55000	55730	56576	57578	58333	$5_{1,5;-4}$	$6_{1,6;-5}$
$6_{6,1;5}$	$7_{7,1;6}$	65000	65721	66561	67607	68571	$6_{1,6;-5}$	$7_{1,7;-6}$
$7_{7,1;6}$	$8_{8,1;7}$	75000	75714	76550	77609	78750	$7_{1,7;-6}$	$8_{1,8;-7}$
$8_{8,1;7}$	$9_{9,1;8}$	85000	85708	86539	87603	88889	$8_{1,8;-7}$	$9_{1,9;-8}$
$9_{9,1;8}$	$10_{10,1;9}$	95000	95704	96531	97595	99000	$9_{1,9;-8}$	$10_{1,10;-9}$
$10_{10,1;9}$	$11_{11,1;10}$	105000	105701	106525	107587	109091	$10_{1,10;-9}$	$11_{1,11;-10}$
$11_{11,1;10}$	$12_{12,1;11}$	115000	115698	116519	117580	119166	$11_{1,11;-10}$	$12_{1,12;-11}$

Table IX-1 (Continued)

Sub-branch $^{c,o}R_{1,0}$	Sub-branch $^{c,o}P_{\bar{1},0}$	κ: ∓1	κ: ∓0.5	κ: 0	κ: ±0.5	κ: ±1	Sub-branch $^{a,o}R_{0,1}$	Sub-branch $^{a,o}P_{0,\bar{1}}$
$2_{1,2;-1}$	$3_{2,2;0}$	16667	16667	16667	16667	16667	$2_{2,1;1}$	$3_{2,2;0}$
$3_{2,2;0}$	$4_{3,2;1}$	26250	28258	29391	29882	30000	$3_{2,2;0}$	$4_{2,3;-1}$
$4_{3,2;1}$	$5_{4,2;2}$	36000	38290	40354	41637	42000	$4_{2,3;-1}$	$5_{2,4;-2}$
$5_{4,2;2}$	$6_{5,2;3}$	45833	48094	50537	52600	53333	$5_{2,4;-2}$	$6_{2,5;-3}$
$6_{5,2;3}$	$7_{6,2;4}$	55714	57929	60461	63088	64286	$6_{2,5;-3}$	$7_{2,6;-4}$
$7_{6,2;4}$	$8_{7,2;5}$	65625	67805	70340	73291	75000	$7_{2,6;-4}$	$8_{2,7;-5}$
$8_{7,2;5}$	$9_{8,2;6}$	75556	77710	80231	83338	85556	$8_{2,7;-5}$	$9_{2,8;-6}$
$9_{8,2;6}$	$10_{9,2;7}$	85500	87636	90142	93314	96000	$9_{2,8;-6}$	$10_{2,9;-7}$
$10_{9,2;7}$	$11_{10,2;8}$	95455	97576	100068	103262	106364	$10_{2,9;-7}$	$11_{2,10;-8}$
$11_{10,2;8}$	$12_{11,2;9}$	105416	107526	110008	113205	116666	$11_{2,10;-8}$	$12_{2,11;-9}$
$3_{1,3;-2}$	$4_{2,3;-1}$	18750	18207	17796	17564	17500	$3_{3,1;2}$	$4_{3,2;1}$
$4_{2,3;-1}$	$5_{3,3;0}$	28000	31148	32063	32074	32000	$4_{3,2;1}$	$5_{3,3;0}$
$5_{3,3;0}$	$6_{4,3;1}$	37500	41486	44187	45001	45000	$5_{3,3;0}$	$6_{3,4;-1}$
$6_{4,3;1}$	$7_{5,3;2}$	47143	51127	54949	56948	57143	$6_{3,4;-1}$	$7_{3,5;-2}$
$7_{5,3;2}$	$8_{6,3;3}$	56875	60749	64999	68208	68750	$7_{3,5;-2}$	$8_{3,6;-3}$
$8_{6,3;3}$	$9_{7,3;4}$	66667	70450	74791	78959	80000	$8_{3,6;-3}$	$9_{3,7;-4}$
$9_{7,3;4}$	$10_{8,3;5}$	76500	80217	84543	89339	91000	$9_{3,7;-4}$	$10_{3,8;-5}$
$10_{8,3;5}$	$11_{9,3;6}$	86364	90031	94320	99469	101818	$10_{3,8;-5}$	$11_{3,9;-6}$
$11_{9,3;6}$	$12_{10,3;7}$	96250	99880	104133	109453	112500	$11_{3,9;-6}$	$12_{3,10;-7}$
$4_{1,4;-3}$	$5_{2,4;-2}$	21000	19363	18449	18082	18000	$4_{4,1;3}$	$5_{4,2;2}$
$5_{2,4;-2}$	$6_{3,4;-1}$	30000	33887	33934	33473	33333	$5_{4,2;2}$	$6_{4,3;1}$
$6_{3,4;-1}$	$7_{4,4;0}$	39286	45000	47370	47311	47143	$6_{4,3;1}$	$7_{4,4;0}$
$7_{4,4;0}$	$8_{5,4;1}$	48750	54655	59178	60145	60000	$7_{4,4;0}$	$8_{4,5;-1}$
$8_{5,4;1}$	$9_{6,4;2}$	58333	64063	69788	72255	72222	$8_{4,5;-1}$	$9_{4,6;-2}$
$9_{6,4;2}$	$10_{7,4;3}$	68000	73554	79716	83787	84000	$9_{4,6;-2}$	$10_{4,7;-3}$
$10_{7,4;3}$	$11_{8,4;4}$	77727	83147	89384	94830	95455	$10_{4,7;-3}$	$11_{4,8;-4}$
$11_{8,4;4}$	$12_{9,4;5}$	87500	92819	99012	105459	106666	$11_{4,8;-4}$	$12_{4,9;-5}$
$5_{1,5;-4}$	$6_{2,5;-3}$	23333	20137	18843	18422	18333	$5_{5,1;4}$	$6_{5,2;3}$
$6_{2,5;-3}$	$7_{3,5;-2}$	32143	36189	35151	34447	34286	$6_{5,2;3}$	$7_{5,3;2}$
$7_{3,5;-2}$	$8_{4,5;-1}$	41250	48511	49684	48970	48750	$7_{5,3;2}$	$8_{5,4;1}$
$8_{4,5;-1}$	$9_{5,5;0}$	50556	58512	62686	62483	62222	$8_{5,4;1}$	$9_{5,5;0}$
$9_{5,5;0}$	$10_{6,5;1}$	60000	67785	74286	75273	75000	$9_{5,5;0}$	$10_{5,6;-1}$
$10_{6,5;1}$	$11_{7,5;2}$	69546	77050	84774	87504	87273	$10_{5,6;-1}$	$11_{5,7;-2}$
$11_{7,5;2}$	$12_{8,5;3}$	79167	86442	94594	99259	99167	$11_{5,7;-2}$	$12_{5,8;-3}$
$6_{1,6;-5}$	$7_{2,6;-4}$	25714	20629	19107	18664	18571	$6_{6,1;5}$	$7_{6,2;4}$
$7_{2,6;-4}$	$8_{3,6;-3}$	34375	37948	35978	35171	35000	$7_{6,2;4}$	$8_{6,3;3}$
$8_{3,6;-3}$	$9_{4,6;-2}$	43333	51721	51284	50241	50000	$8_{6,3;3}$	$9_{6,4;2}$
$9_{4,6;-2}$	$10_{5,6;-1}$	52500	62496	65297	64301	64000	$9_{6,4;2}$	$10_{6,5;1}$
$10_{5,6;-1}$	$11_{6,6;0}$	61818	71831	78026	77624	77273	$10_{6,5;1}$	$11_{6,6;0}$
$11_{6,6;0}$	$12_{7,6;1}$	71250	80896	89476	90385	90000	$11_{6,6;0}$	$12_{6,7;-1}$
$7_{1,7;-6}$	$8_{2,7;-5}$	28125	20944	19300	18844	18750	$7_{7,1;6}$	$8_{7,2;5}$
$8_{2,7;-5}$	$9_{3,7;-4}$	36667	39200	36585	35733	35556	$8_{7,2;5}$	$9_{7,3;4}$
$9_{3,7;-4}$	$10_{4,7;-3}$	45500	54420	52436	51252	51000	$9_{7,3;4}$	$10_{7,4;3}$
$10_{4,7;-3}$	$11_{5,7;-2}$	54546	66365	67167	65775	65455	$10_{7,4;3}$	$11_{7,5;2}$
$11_{5,7;-2}$	$12_{6,7;-1}$	63750	76087	80851	79549	79167	$11_{7,5;2}$	$12_{7,6;1}$
$8_{1,8;-7}$	$9_{2,8;-6}$	30556	21157	19449	18985	18889	$8_{8,1;7}$	$9_{8,2;6}$
$9_{2,8;-6}$	$10_{3,8;-5}$	39000	40063	37061	36182	36000	$9_{8,2;6}$	$10_{8,3;5}$
$10_{3,8;-5}$	$11_{4,8;-4}$	47727	56523	53327	52078	51818	$10_{8,3;5}$	$11_{8,4;4}$
$11_{4,8;-4}$	$12_{5,8;-3}$	56667	69870	68563	66999	66667	$11_{8,4;4}$	$12_{8,5;3}$
$9_{1,9;-8}$	$10_{2,9;-7}$	33000	21311	19566	19097	19000	$9_{9,1;8}$	$10_{9,2;7}$
$10_{2,9;-7}$	$11_{3,9;-6}$	41364	40664	37443	36548	36364	$10_{9,2;7}$	$11_{9,3;6}$
$11_{3,9;-6}$	$12_{4,9;-5}$	50000	58070	54050	52766	52500	$11_{9,3;6}$	$12_{9,4;5}$
$10_{1,10;-9}$	$11_{2,10;-8}$	35454	21431	19662	19189	19091	$10_{10,1;9}$	$11_{10,2;8}$
$11_{2,10;-8}$	$12_{3,10;-7}$	43750	41102	37761	36854	36667	$11_{10,2;8}$	$12_{10,3;7}$
$11_{1,11;-10}$	$12_{2,11;-9}$	37917	21526	19742	19265	19167	$11_{11,1;10}$	$12_{11,2;9}$

Table IX-1 (Continued)

B. *a* AND *c* PROLATE-OR-OBLATE SUB-BRANCHES

Sub-branch $^{c,e}Q_{\bar{1},2}$	$^{c,o}Q_{1,\bar{2}}$	κ ∓ 1	∓ 0.5	0	± 0.5	± 1	Sub-branch $^{a,e}Q_{2,\bar{1}}$	$^{a,o}Q_{\bar{2},1}$
$2_{2,0;2}$	$2_{1,2;-1}$	8333	5110	2233	488		$2_{0,2;-2}$	$2_{2,1;1}$
$3_{2,1;1}$	$3_{1,3;-2}$	14583	5722	1328	165		$3_{1,2;-1}$	$3_{3,1;2}$
$4_{2,2;0}$	$4_{1,4;-3}$	20250	4363	650	78		$4_{2,2;0}$	$4_{4,1;3}$
$5_{2,3;-1}$	$5_{1,5;-4}$	25667	2859	374	54		$5_{3,2;1}$	$5_{5,1;4}$
$6_{2,4;-2}$	$6_{1,6;-5}$	30952	1843	266	43		$6_{4,2;2}$	$6_{6,1;5}$
$7_{2,5;-3}$	$7_{1,7;-6}$	36161	1262	218	35		$7_{5,2;3}$	$7_{7,1;6}$
$8_{2,6;-4}$	$8_{1,8;-7}$	41319	945	183	30		$8_{6,2;4}$	$8_{8,1;7}$
$9_{2,7;-5}$	$9_{1,9;-8}$	46444	770	160	26		$9_{7,2;5}$	$9_{9,1;8}$
$10_{2,8;-6}$	$10_{1,10;-9}$	51545	664	141	23		$10_{8,2;6}$	$10_{10,1;9}$
$11_{2,9;-7}$	$11_{1,11;-10}$	56629	590	125	21		$11_{9,2;7}$	$11_{11,1;10}$
$12_{2,10;-8}$	$12_{1,12;-11}$	61699	533	115	19		$12_{10,2;8}$	$12_{12,1;11}$
$3_{3,0;3}$	$3_{2,2;0}$	8750	7055	4522	1458		$3_{0,3;-3}$	$3_{2,2;0}$
$4_{3,1;2}$	$4_{2,3;-1}$	15750	11214	4568	638		$4_{1,3;-2}$	$4_{3,2;1}$
$5_{3,2;1}$	$5_{2,4;-2}$	22000	12576	2754	274		$5_{2,3;-1}$	$5_{4,2;2}$
$6_{3,3;0}$	$6_{2,5;-3}$	27857	11283	1492	171		$6_{3,3;0}$	$6_{5,2;3}$
$7_{3,4;-1}$	$7_{2,6;-4}$	33482	8559	925	132		$7_{4,3;1}$	$7_{6,2;4}$
$8_{3,5;-2}$	$8_{2,7;-5}$	38958	5932	685	108		$8_{5,3;2}$	$8_{7,2;5}$
$9_{3,6;-3}$	$9_{2,8;-6}$	44333	4077	567	92		$9_{6,3;3}$	$9_{8,2;6}$
$10_{3,7;-4}$	$10_{2,9;-7}$	49636	2945	490	80		$10_{7,3;4}$	$10_{9,2;7}$
$11_{3,8;-5}$	$11_{2,10;-8}$	54886	2294	433	71		$11_{8,3;5}$	$11_{10,2;8}$
$12_{3,9;-6}$	$12_{2,11;-9}$	60096	1917	387	64		$12_{9,3;6}$	$12_{11,2;9}$
$4_{4,0;4}$	$4_{3,2;1}$	9000	7558	5617	2547		$4_{0,4;-4}$	$4_{2,3;-1}$
$5_{4,1;3}$	$5_{3,3;0}$	16500	13242	7983	1599		$5_{1,4;-3}$	$5_{3,3;0}$
$6_{4,2;2}$	$6_{3,4;-1}$	23214	17320	6820	681		$6_{2,4;-2}$	$6_{4,3;1}$
$7_{4,3;1}$	$7_{3,5;-2}$	29464	19464	4223	374		$7_{3,4;-1}$	$7_{5,3;2}$
$8_{4,4;0}$	$8_{3,6;-3}$	35417	19178	2433	273		$8_{4,4;0}$	$8_{6,3;3}$
$9_{4,5;-1}$	$9_{3,7;-4}$	41167	16526	1579	222		$9_{5,4;1}$	$9_{7,3;4}$
$10_{4,6;-2}$	$10_{3,8;-5}$	46773	12665	1205	188		$10_{6,4;2}$	$10_{8,3;5}$
$11_{4,7;-3}$	$11_{3,9;-6}$	52273	9080	1014	163		$11_{7,4;3}$	$11_{9,3;6}$
$12_{4,8;-4}$	$12_{3,10;-7}$	57692	6485	888	144		$12_{8,4;4}$	$12_{10,3;7}$
$5_{5,0;5}$	$5_{4,2;2}$	9167	7775	6052	3368		$5_{0,5;-5}$	$5_{2,4;-2}$
$6_{5,1;4}$	$6_{4,3;1}$	17024	14062	9982	3054		$6_{1,5;-4}$	$6_{3,4;-1}$
$7_{5,2;3}$	$7_{4,4;0}$	24107	19225	11103	1459		$7_{2,5;-3}$	$7_{4,4;0}$
$8_{5,3;2}$	$8_{4,5;-1}$	30694	23287	9000	720		$8_{3,5;-2}$	$8_{5,4;1}$
$9_{5,4;1}$	$9_{4,6;-2}$	36944	26001	5708	481		$9_{4,5;-1}$	$9_{6,4;2}$
$10_{5,5;0}$	$10_{4,7;-3}$	42955	26852	3433	382		$10_{5,5;0}$	$10_{7,4;3}$
$11_{5,6;-1}$	$11_{4,8;-4}$	48788	25327	2306	321		$11_{6,5;1}$	$11_{8,4;4}$
$12_{5,7;-2}$	$12_{4,9;-5}$	54487	21546	1796	277		$12_{7,5;2}$	$12_{9,4;5}$
$6_{6,0;6}$	$6_{5,2;3}$	9286	7912	6257	3863		$6_{0,6;-6}$	$6_{2,5;-3}$
$7_{6,1;5}$	$7_{5,3;2}$	17411	14550	10952	4657		$7_{1,6;-5}$	$7_{3,5;-2}$
$8_{6,2;4}$	$8_{5,4;1}$	24792	20233	13841	2772		$8_{2,6;-4}$	$8_{4,5;-1}$
$9_{6,3;3}$	$9_{5,5;0}$	31667	25098	14023	1322		$9_{3,6;-3}$	$9_{5,5;0}$
$10_{6,4;2}$	$10_{5,6;-1}$	38182	29140	11121	785		$10_{4,6;-2}$	$10_{6,5;1}$
$11_{6,5;1}$	$11_{5,7;-2}$	44432	32202	7197	595		$11_{5,6;-1}$	$11_{7,5;2}$
$12_{6,6;0}$	$12_{5,8;-3}$	50481	33921	4472	494		$12_{6,6;0}$	$12_{8,5;3}$
$7_{7,0;7}$	$7_{6,2;4}$	9375	8011	6381	4141		$7_{0,7;-7}$	$7_{2,6;-4}$
$8_{7,1;6}$	$8_{6,3;3}$	17708	14899	11480	5982		$8_{1,7;-6}$	$8_{3,6;-3}$
$9_{7,2;5}$	$9_{6,4;2}$	25333	20930	15306	4603		$9_{2,7;-5}$	$9_{4,6;-2}$
$10_{7,3;4}$	$10_{6,5;1}$	32455	26247	17406	2348		$10_{3,7;-4}$	$10_{5,6;-1}$
$11_{7,4;3}$	$11_{6,6;0}$	39205	30913	16805	1258		$11_{4,7;-3}$	$11_{6,6;0}$
$12_{7,5;2}$	$12_{6,7;-1}$	45673	34922	13192	878		$12_{5,7;-2}$	$12_{7,6;1}$
$8_{8,0;8}$	$8_{7,2;5}$	9444	8087	6468	4302		$8_{0,8;-8}$	$8_{2,7;-5}$
$9_{8,1;7}$	$9_{7,3;4}$	17944	15167	11825	6888		$9_{1,8;-7}$	$9_{3,7;-4}$
$10_{8,2;6}$	$10_{7,4;3}$	25773	21462	16158	6602		$10_{2,8;-6}$	$10_{4,7;-3}$
$11_{8,3;5}$	$11_{7,5;2}$	33106	27105	19325	3955		$11_{3,8;-5}$	$11_{5,7;-2}$
$12_{8,4;4}$	$12_{7,6;1}$	40064	32170	20771	2027		$12_{4,8;-4}$	$12_{6,7;-1}$
$9_{9,0;9}$	$9_{8,2;6}$	9500	8146	6535	4403		$9_{0,9;-9}$	$9_{2,8;-6}$
$10_{9,1;8}$	$10_{8,3;5}$	18136	15380	12079	7464		$10_{1,9;-8}$	$10_{3,8;-5}$
$11_{9,2;7}$	$11_{8,4;4}$	26136	21887	16736	8319		$11_{2,9;-7}$	$11_{4,8;-4}$
$12_{9,3;6}$	$12_{8,5;3}$	33654	27788	20487	6108		$12_{3,9;-6}$	$12_{5,8;-3}$

Table IX-1 (Continued)

Sub-branch $^{c,e}Q\bar{1},2$	$^{c,o}Q1,\bar{2}$	∓1	∓0.5	κ 0	±0.5	±1	Sub-branch $^{a,e}Q2,\bar{1}$	$^{a,o}Q\bar{2},1$
$10_{10,0;10}$	$10_{9,2;7}$	9545	8194	6588	4474		$10_{0,10;-10}$	$10_{2,9;-7}$
$11_{10,1;9}$	$11_{9,3;6}$	18295	15554	12279	7835		$11_{1,10;-9}$	$11_{3,9;-6}$
$12_{10,2;8}$	$12_{9,4;5}$	26442	22237	17173	9567		$12_{2,10;-8}$	$12_{4,9;-5}$
$11_{11,0;11}$	$11_{10,2;8}$	9583	8234	6631	4528		$11_{0,11;-11}$	$11_{2,10;-8}$
$12_{11,1;10}$	$12_{10,3;7}$	18429	15699	12443	8090		$12_{1,11;-10}$	$12_{3,10;-7}$
$12_{12,0;12}$	$12_{11,2;9}$	9615	8268	6667	4571		$12_{0,12;-12}$	$12_{2,11;-9}$
$^{c,e}R\bar{1},2$	$^{c,e}P1,\bar{2}$	∓1	∓0.5	0	±0.5	±1	$^{a,e}R2,\bar{1}$	$^{a,e}P\bar{2},1$
$1_{1,0;1}$	$2_{0,2;-2}$	5000	3066	1340	293		$1_{0,1;-1}$	$2_{2,0;2}$
$2_{1,1;0}$	$3_{0,3;-3}$	10000	4167	1086	157		$2_{1,1;0}$	$3_{3,0;3}$
$3_{1,2;-1}$	$4_{0,4;-4}$	15000	3944	800	123		$3_{2,1;1}$	$4_{4,0;4}$
$4_{1,3;-2}$	$5_{0,5;-5}$	20000	3386	696	117		$4_{3,1;2}$	$5_{5,0;5}$
$5_{1,4;-3}$	$6_{0,6;-6}$	25000	2976	667	114		$5_{4,1;3}$	$6_{6,0;6}$
$6_{1,5;-4}$	$7_{0,7;-7}$	30000	2770	656	112		$6_{5,1;4}$	$7_{7,0;7}$
$7_{1,6;-5}$	$8_{0,8;-8}$	35000	2686	649	111		$7_{6,1;5}$	$8_{8,0;8}$
$8_{1,7;-6}$	$9_{0,9;-9}$	40000	2652	644	110		$8_{7,1;6}$	$9_{9,0;9}$
$9_{1,8;-7}$	$10_{0,10;-10}$	45000	2634	640	109		$9_{8,1;7}$	$10_{10,0;10}$
$10_{1,9;-8}$	$11_{0,11;-11}$	50000	2621	637	109		$10_{9,1;8}$	$11_{11,0;11}$
$11_{1,10;-9}$	$12_{0,12;-12}$	55000	2610	634	108		$11_{10,1;9}$	$12_{12,0;12}$
$2_{2,0;2}$	$3_{1,2;-1}$	1667	2062	1905	776		$2_{0,2;-2}$	$3_{2,1;1}$
$3_{2,1;1}$	$4_{1,3;-2}$	3750	5114	2884	480		$3_{1,2;-1}$	$4_{3,1;2}$
$4_{2,2;0}$	$5_{1,4;-3}$	6000	7788	2336	310		$4_{2,2;0}$	$5_{4,1;3}$
$5_{2,3;-1}$	$6_{1,5;-4}$	8333	8748	1768	268		$5_{3,2;1}$	$6_{5,1;4}$
$6_{2,4;-2}$	$7_{1,6;-5}$	10714	8172	1529	254		$6_{4,2;2}$	$7_{6,1;5}$
$7_{2,5;-3}$	$8_{1,7;-6}$	13125	7135	1445	246		$7_{5,2;3}$	$8_{7,1;6}$
$8_{2,6;-4}$	$9_{1,8;-7}$	15556	6332	1406	240		$8_{6,2;4}$	$9_{8,1;7}$
$9_{2,7;-5}$	$10_{1,9;-8}$	18000	5885	1380	235		$9_{7,2;5}$	$10_{9,1;8}$
$10_{2,8;-6}$	$11_{1,10;-9}$	20455	5673	1360	232		$10_{8,2;6}$	$11_{10,1;9}$
$11_{2,9;-7}$	$12_{1,11;-10}$	22917	5570	1344	229		$11_{9,2;7}$	$12_{11,1;10}$
$3_{3,0;3}$	$4_{2,2;0}$	1250	1176	1316	1061		$3_{0,3;-3}$	$4_{2,2;0}$
$4_{3,1;2}$	$5_{2,3;-1}$	3000	3166	3516	1032		$4_{1,3;-2}$	$5_{3,2;1}$
$5_{3,2;1}$	$6_{2,4;-2}$	5000	6089	4448	613		$5_{2,3;-1}$	$6_{4,2;2}$
$6_{3,3;0}$	$7_{2,5;-3}$	7143	9630	3653	469		$6_{3,3;0}$	$7_{5,2;3}$
$7_{3,4;-1}$	$8_{2,6;-4}$	9375	12493	2828	426		$7_{4,3;1}$	$8_{6,2;4}$
$8_{3,5;-2}$	$9_{2,7;-5}$	11667	13383	2447	404		$8_{5,3;2}$	$9_{7,2;5}$
$9_{3,6;-3}$	$10_{2,8;-6}$	14000	12500	2301	389		$9_{6,3;3}$	$10_{8,2;6}$
$10_{3,7;-4}$	$11_{2,9;-7}$	16364	11048	2226	378		$10_{7,3;4}$	$11_{9,2;7}$
$11_{3,8;-5}$	$12_{2,10;-8}$	18750	9884	2174	370		$11_{8,3;5}$	$12_{10,2;8}$
$4_{4,0;4}$	$5_{3,2;1}$	1000	882	849	963		$4_{0,4;-4}$	$5_{2,3;-1}$
$5_{4,1;3}$	$6_{3,3;0}$	2500	2272	2651	1677		$5_{1,4;-3}$	$6_{3,3;0}$
$6_{4,2;2}$	$7_{3,4;-1}$	4286	4145	5108	1110		$6_{2,4;-2}$	$7_{4,3;1}$
$7_{4,3;1}$	$8_{3,5;-2}$	6250	6720	6025	747		$7_{3,4;-1}$	$8_{5,3;2}$
$8_{4,4;0}$	$9_{3,6;-3}$	8333	10196	5011	636		$8_{4,4;0}$	$9_{6,3;3}$
$9_{4,5;-1}$	$10_{3,7;-4}$	10500	14160	3947	590		$9_{5,4;1}$	$10_{7,3;4}$
$10_{4,6;-2}$	$11_{3,8;-5}$	12727	17215	3427	561		$10_{6,4;2}$	$11_{8,3;5}$
$11_{4,7;-3}$	$12_{3,9;-6}$	15000	18047	3214	540		$11_{7,4;3}$	$12_{9,3;6}$
$5_{5,0;5}$	$6_{4,2;2}$	833	730	638	723		$5_{0,5;-5}$	$6_{2,4;-2}$
$6_{5,1;4}$	$7_{4,3;1}$	2143	1898	1863	2013		$6_{1,5;-4}$	$7_{3,4;-1}$
$7_{5,2;3}$	$8_{4,4;0}$	3750	3384	4016	1833		$7_{2,5;-3}$	$8_{4,4;0}$
$8_{5,3;2}$	$9_{4,5;-1}$	5556	5182	6701	1165		$8_{3,5;-2}$	$9_{5,4;1}$
$9_{5,4;1}$	$10_{4,6;-2}$	7500	7437	7610	901		$9_{4,5;-1}$	$10_{6,4;2}$
$10_{5,5;0}$	$11_{4,7;-3}$	9545	10443	6398	809		$10_{5,5;0}$	$11_{7,4;3}$
$11_{5,6;-1}$	$12_{4,8;-4}$	11667	14380	5109	758		$11_{6,5;1}$	$12_{8,4;4}$
$6_{6,0;6}$	$7_{5,2;3}$	714	625	533	530		$6_{0,6;-6}$	$7_{2,5;-3}$
$7_{6,1;5}$	$8_{5,3;2}$	1875	1652	1470	1860		$7_{1,6;-5}$	$8_{3,5;-2}$
$8_{6,2;4}$	$9_{5,4;1}$	3333	2968	2958	2593		$8_{2,6;-4}$	$9_{4,5;-1}$
$9_{6,3;3}$	$10_{5,5;0}$	5000	4516	5406	1801		$9_{3,6;-3}$	$10_{5,5;0}$
$10_{6,4;2}$	$11_{5,6;-1}$	6818	6288	8298	1256		$10_{4,6;-2}$	$11_{6,5;1}$
$11_{6,5;1}$	$12_{5,7;-2}$	8750	8348	9203	1069		$11_{5,6;-1}$	$12_{7,5;2}$

Table IX-1 (Continued)

Sub-branch ${}^{c,e}R\bar{1},2$	${}^{c,e}P1,\bar{2}$	∓ 1	∓ 0.5	κ 0	± 0.5	± 1	Sub-branch ${}^{a,e}R2,\bar{1}$	${}^{a,e}P\bar{2},1$
$7_{7,0;7}$	$8_{6,2;4}$	625	546	463	411		$7_{0,7;-7}$	$8_{2,6;-4}$
$8_{7,1;6}$	$9_{6,3;3}$	1667	1465	1267	1486		$8_{1,7;-6}$	$9_{3,6;-3}$
$9_{7,2;5}$	$10_{6,4;2}$	3000	2656	2402	2962		$9_{2,7;-5}$	$10_{4,6;-2}$
$10_{7,3;4}$	$11_{6,5;1}$	4545	4065	4106	2664		$10_{3,7;-4}$	$11_{5,6;-1}$
$11_{7,4;3}$	$12_{6,6;0}$	6250	5659	6816	1772		$11_{4,7;-3}$	$12_{6,6;0}$
$8_{8,0;8}$	$9_{7,2;5}$	556	485	411	341		$8_{0,8;-8}$	$9_{2,7;-5}$
$9_{8,1;7}$	$10_{7,3;4}$	1500	1316	1129	1154		$9_{1,8;-7}$	$10_{3,7;-4}$
$10_{8,2;6}$	$11_{7,4;3}$	2727	2407	2110	2759		$10_{2,8;-6}$	$11_{4,7;-3}$
$11_{8,3;5}$	$12_{7,5;2}$	4167	3705	3398	3519		$11_{3,8;-5}$	$12_{5,7;-2}$
$9_{9,0;9}$	$10_{8,2;6}$	500	436	369	298		$9_{0,9;-9}$	$10_{2,8;-6}$
$10_{9,1;8}$	$11_{8,3;5}$	1364	1195	1021	932		$10_{1,9;-8}$	$11_{3,8;-5}$
$11_{9,2;7}$	$12_{8,4;4}$	2500	2201	1910	2276		$11_{2,9;-7}$	$12_{4,8;-4}$
$10_{10,0;10}$	$11_{9,2;7}$	455	397	335	268		$10_{0,10;-10}$	$11_{2,9;-7}$
$11_{10,1;9}$	$12_{9,3;6}$	1250	1094	933	797		$11_{1,10;-9}$	$12_{3,9;-6}$
$11_{11,0;11}$	$12_{10,2;8}$	417	363	307	244		$11_{0,11;-11}$	$12_{2,10;-8}$
${}^{c,o}R\bar{1},2$	${}^{c,o}P1,\bar{2}$	∓ 1	∓ 0.5	0	± 0.5	± 1	${}^{a,o}R2,\bar{1}$	${}^{a,o}P\bar{2},1$
$2_{2,1;1}$	$3_{1,3;-2}$	1667	956	423	103		$2_{1,2;-1}$	$3_{3,1;2}$
$3_{2,2;0}$	$4_{1,4;-3}$	3750	1742	609	118		$3_{2,2;0}$	$4_{4,1;3}$
$4_{2,3;-1}$	$5_{1,5;-4}$	6000	2228	657	116		$4_{3,2;1}$	$5_{5,1;4}$
$5_{2,4;-2}$	$6_{1,6;-5}$	8333	2480	661	114		$5_{4,2;2}$	$6_{6,1;5}$
$6_{2,5;-3}$	$7_{1,7;-6}$	10714	2590	655	112		$6_{5,2;3}$	$7_{7,1;6}$
$7_{2,6;-4}$	$8_{1,8;-7}$	13125	2627	649	111		$7_{6,2;4}$	$8_{8,1;7}$
$8_{2,7;-5}$	$9_{1,9;-8}$	15556	2633	644	110		$8_{7,2;5}$	$9_{9,1;8}$
$9_{2,8;-6}$	$10_{1,10;-9}$	18000	2627	640	109		$9_{8,2;6}$	$10_{10,1;9}$
$10_{2,9;-7}$	$11_{1,11;-10}$	20455	2619	637	109		$10_{9,2;7}$	$11_{11,1;10}$
$11_{2,10;-8}$	$12_{1,12;-11}$	22917	2610	634	108		$11_{10,2;8}$	$12_{12,1;11}$
$3_{3,1;2}$	$4_{2,3;-1}$	1250	1025	643	213		$3_{1,3;-2}$	$4_{3,2;1}$
$4_{3,2;1}$	$5_{2,4;-2}$	3000	2317	1159	269		$4_{2,3;-1}$	$5_{4,2;2}$
$5_{3,3;0}$	$6_{2,5;-3}$	5000	3522	1389	265		$5_{3,3;0}$	$6_{5,2;3}$
$6_{3,4;-1}$	$7_{2,6;-4}$	7143	4450	1442	254		$6_{4,3;1}$	$7_{6,2;4}$
$7_{3,5;-2}$	$8_{2,7;-5}$	9375	5049	1429	246		$7_{5,3;2}$	$8_{7,2;5}$
$8_{3,6;-3}$	$9_{2,8;-6}$	11667	5372	1403	240		$8_{6,3;3}$	$9_{8,2;6}$
$9_{3,7;-4}$	$10_{2,9;-7}$	14000	5507	1379	235		$9_{7,3;4}$	$10_{9,2;7}$
$10_{3,8;-5}$	$11_{2,10;-8}$	16364	5539	1360	232		$10_{8,3;5}$	$11_{10,2;8}$
$11_{3,9;-6}$	$12_{2,11;-9}$	18750	5523	1344	229		$11_{9,3;6}$	$12_{11,2;9}$
$4_{4,1;3}$	$5_{3,3;0}$	1000	869	664	300		$4_{1,4;-3}$	$5_{3,3;0}$
$5_{4,2;2}$	$6_{3,4;-1}$	2500	2168	1455	440		$5_{2,4;-2}$	$6_{4,3;1}$
$6_{4,3;1}$	$7_{3,5;-2}$	4286	3662	2018	447		$6_{3,4;-1}$	$7_{5,3;2}$
$7_{4,4;0}$	$8_{3,6;-3}$	6250	5157	2266	424		$7_{4,4;0}$	$8_{6,3;3}$
$8_{4,5;-1}$	$9_{3,7;-4}$	8333	6471	2309	404		$8_{5,4;1}$	$9_{7,3;4}$
$9_{4,6;-2}$	$10_{3,8;-5}$	10500	7475	2272	389		$9_{6,4;2}$	$10_{8,3;5}$
$10_{4,7;-3}$	$11_{3,9;-6}$	12727	8130	2220	378		$10_{7,4;3}$	$11_{9,3;6}$
$11_{4,8;-4}$	$12_{3,10;-7}$	15000	8481	2173	370		$11_{8,4;4}$	$12_{10,3;7}$
$5_{5,1;4}$	$6_{4,3;1}$	833	729	601	346		$5_{1,5;-4}$	$6_{3,4;-1}$
$6_{5,2;3}$	$7_{4,4;0}$	2143	1889	1489	603		$6_{2,5;-3}$	$7_{4,4;0}$
$7_{5,3;2}$	$8_{4,5;-1}$	3750	3329	2360	656		$7_{3,5;-2}$	$8_{5,4;1}$
$8_{5,4;1}$	$9_{4,6;-2}$	5556	4947	2953	626		$8_{4,5;-1}$	$9_{6,4;2}$
$9_{5,5;0}$	$10_{4,7;-3}$	7500	6629	3209	589		$9_{5,5;0}$	$10_{7,4;3}$
$10_{5,6;-1}$	$11_{4,8;-4}$	9545	8239	3235	561		$10_{6,5;1}$	$11_{8,4;4}$
$11_{5,7;-2}$	$12_{4,9;-5}$	11667	9627	3171	540		$11_{7,5;2}$	$12_{9,4;5}$
$6_{6,1;5}$	$7_{5,3;2}$	714	625	527	357		$6_{1,6;-5}$	$7_{3,5;-2}$
$7_{6,2;4}$	$8_{5,4;1}$	1875	1651	1386	726		$7_{2,6;-4}$	$8_{4,5;-1}$
$8_{6,3;3}$	$9_{5,5;0}$	3333	2962	2400	873		$8_{3,6;-3}$	$9_{5,5;0}$
$9_{6,4;2}$	$10_{5,6;-1}$	5000	4489	3326	858		$9_{4,6;-2}$	$10_{6,5;1}$
$10_{6,5;1}$	$11_{5,7;-2}$	6818	6181	3940	804		$10_{5,6;-1}$	$11_{7,5;2}$
$11_{6,6;0}$	$12_{5,8;-3}$	8750	7975	4195	758		$11_{6,6;0}$	$12_{8,5;3}$

Table IX-I (Continued)

Sub-branch $^{c,o}R\bar{1},2$	$^{c,o}P1,\bar{2}$	κ: ∓1	∓0.5	0	±0.5	±1	Sub-branch $^{a,o}R2,\bar{1}$	$^{a,o}P\bar{2},1$
$7_{7,1;6}$	$8_{6,3;3}$	625	546	462	341		$7_{1,7;-6}$	$8_{3,6;-3}$
$8_{7,2;5}$	$9_{6,4;2}$	1667	1465	1251	793		$8_{2,7;-5}$	$9_{4,6;-2}$
$9_{7,3;4}$	$10_{6,5;1}$	3000	2656	2267	1069		$9_{3,7;-4}$	$10_{5,6;-1}$
$10_{7,4;3}$	$11_{6,6;0}$	4545	4062	3368	1110		$10_{4,7;-3}$	$11_{6,6;0}$
$11_{7,5;2}$	$12_{6,7;-1}$	6250	5646	4336	1050		$11_{5,7;-2}$	$12_{7,6;1}$
$8_{8,1;7}$	$9_{7,3;4}$	556	485	411	317		$8_{1,8;-7}$	$9_{3,7;-4}$
$9_{8,2;6}$	$10_{7,4;3}$	1500	1316	1126	808		$9_{2,8;-6}$	$10_{4,7;-3}$
$10_{8,3;5}$	$11_{7,5;2}$	2727	2407	2082	1216		$10_{3,8;-5}$	$11_{5,7;-2}$
$11_{8,4;4}$	$12_{7,6;1}$	4167	3704	3209	1361		$11_{4,8;-4}$	$12_{6,7;-1}$
$9_{9,1;8}$	$10_{8,3;5}$	500	436	369	290		$9_{1,9;-8}$	$10_{3,8;-5}$
$10_{9,2;7}$	$11_{8,4;4}$	1364	1195	1021	786		$10_{2,9;-7}$	$11_{4,8;-4}$
$11_{9,3;6}$	$12_{8,5;3}$	2500	2201	1904	1296		$11_{3,9;-6}$	$12_{5,8;-3}$
$10_{10,1;9}$	$11_{9,3;6}$	455	397	335	265		$10_{1,10;-9}$	$11_{3,9;-6}$
$11_{10,2;8}$	$12_{9,4;5}$	1250	1094	931	741		$11_{2,10;-8}$	$12_{4,9;-5}$
$11_{11,1;10}$	$12_{10,3;7}$	417	363	307	243		$11_{1,11;-10}$	$12_{3,10;-7}$

C. *b* prolate-and-oblate sub-branches

$^{b,e}Q\bar{1},1$	$^{b,e}Q1,\bar{1}$	∓1	∓0.5	0	±0.5	±1	$^{b,e}Q1,\bar{1}$	$^{b,e}Q\bar{1},1$
$1_{1,0;1}$	$1_{0,1;-1}$	15000	15000	15000	15000	15000	$1_{0,1;-1}$	$1_{1,0;1}$
$2_{1,1;0}$	$2_{0,2;-2}$	25000	21289	16667	12044	8333	$2_{1,1;0}$	$2_{2,0;2}$
$3_{1,2;-1}$	$3_{0,3;-3}$	35000	23196	14583	10583	8750	$3_{2,1;1}$	$3_{3,0;3}$
$4_{1,3;-2}$	$4_{0,4;-4}$	45000	22157	13527	10617	9000	$4_{3,1;2}$	$4_{4,0;4}$
$5_{1,4;-3}$	$5_{0,5;-5}$	55000	20634	13413	10753	9167	$5_{4,1;3}$	$5_{5,0;5}$
$6_{1,5;-4}$	$6_{0,6;-6}$	65000	19779	13484	10861	9286	$6_{5,1;4}$	$6_{6,0;6}$
$7_{1,6;-5}$	$7_{0,7;-7}$	75000	19511	13559	10943	9375	$7_{6,1;5}$	$7_{7,0;7}$
$8_{1,7;-6}$	$8_{0,8;-8}$	85000	19487	13620	11008	9444	$8_{7,1;6}$	$8_{8,0;8}$
$9_{1,8;-7}$	$9_{0,9;-9}$	95000	19524	13669	11060	9500	$9_{8,1;7}$	$9_{9,0;9}$
$10_{1,9;-8}$	$10_{0,10;-10}$	105000	19565	13710	11103	9545	$10_{9,1;8}$	$10_{10,0;10}$
$11_{1,10;-9}$	$11_{0,11;-11}$	115000	19604	13744	11139	9583	$11_{10,1;9}$	$11_{11,0;11}$
$12_{1,11;-10}$	$12_{0,12;-12}$	125000	19633	13774	11170	9615	$12_{11,1;10}$	$12_{12,0;12}$
$2_{2,0;2}$	$2_{1,1;0}$	8333	12044	16667	21289	25000	$2_{0,2;-2}$	$2_{1,1;0}$
$3_{2,1;1}$	$3_{1,2;-1}$	14583	24417	28872	24417	14583	$3_{1,2;-1}$	$3_{2,1;1}$
$4_{2,2;0}$	$4_{1,3;-2}$	20250	36119	31154	20622	15750	$4_{2,2;0}$	$4_{3,1;2}$
$5_{2,3;-1}$	$5_{1,4;-3}$	25667	43650	28164	20038	16500	$5_{3,2;1}$	$5_{4,1;3}$
$6_{2,4;-2}$	$6_{1,5;-4}$	30952	45529	26402	20356	17024	$6_{4,2;2}$	$6_{5,1;4}$
$7_{2,5;-3}$	$7_{1,6;-5}$	36161	43602	26163	20670	17411	$7_{5,2;3}$	$7_{6,1;5}$
$8_{2,6;-4}$	$8_{1,7;-6}$	41319	41002	26300	20926	17708	$8_{6,2;4}$	$8_{7,1;6}$
$9_{2,7;-5}$	$9_{1,8;-7}$	46444	39408	26465	21134	17944	$9_{7,2;5}$	$9_{8,1;7}$
$10_{2,8;-6}$	$10_{1,9;-8}$	51545	38815	26611	21307	18136	$10_{8,2;6}$	$10_{9,1;8}$
$11_{2,9;-7}$	$11_{1,10;-9}$	56629	38701	26737	21452	18295	$11_{9,2;7}$	$11_{10,1;9}$
$12_{2,10;-8}$	$12_{1,11;-10}$	61699	38736	26846	21576	18429	$12_{10,2;8}$	$12_{11,1;10}$
$3_{3,0;3}$	$3_{2,1;1}$	8750	10583	14583	23196	35000	$3_{0,3;-3}$	$3_{1,2;-1}$
$4_{3,1;2}$	$4_{2,2;0}$	15750	20622	31154	36119	20250	$4_{1,3;-2}$	$4_{2,2;0}$
$5_{3,2;1}$	$5_{2,3;-1}$	22000	32340	44017	32340	22000	$5_{2,3;-1}$	$5_{3,2;1}$
$6_{3,3;0}$	$6_{2,4;-2}$	27857	45986	45920	29422	23214	$6_{3,3;0}$	$6_{4,2;2}$
$7_{3,4;-1}$	$7_{2,5;-3}$	33482	58783	41862	29481	24107	$7_{4,3;1}$	$7_{5,2;3}$
$8_{3,5;-2}$	$8_{2,6;-4}$	38958	66715	39333	29932	24792	$8_{5,3;2}$	$8_{6,2;4}$
$9_{3,6;-3}$	$9_{2,7;-5}$	44333	68174	38859	30348	25333	$9_{6,3;3}$	$9_{7,2;5}$
$10_{3,7;-4}$	$10_{2,8;-6}$	49636	65282	38980	30705	25773	$10_{7,3;4}$	$10_{8,2;6}$
$11_{3,8;-5}$	$11_{2,9;-7}$	54886	61636	39182	31011	26136	$11_{8,3;5}$	$11_{9,2;7}$
$12_{3,9;-6}$	$12_{2,10;-8}$	60096	59285	39377	31275	26442	$12_{9,3;6}$	$12_{10,2;8}$
$4_{4,0;4}$	$4_{3,1;2}$	9000	10617	13527	22157	45000	$4_{0,4;-4}$	$4_{1,3;-2}$
$5_{4,1;3}$	$5_{3,2;1}$	16500	20038	28164	43650	25667	$5_{1,4;-3}$	$5_{2,3;-1}$
$6_{4,2;2}$	$6_{3,3;0}$	23214	29422	45920	45986	27857	$6_{2,4;-2}$	$6_{3,3;0}$
$7_{4,3;1}$	$7_{3,4;-1}$	29464	39987	59402	39987	29464	$7_{3,4;-1}$	$7_{4,3;1}$
$8_{4,4;0}$	$8_{3,5;-2}$	35417	52950	60829	38601	30694	$8_{4,4;0}$	$8_{5,3;2}$
$9_{4,5;-1}$	$9_{3,6;-3}$	41167	67954	55712	38960	31667	$9_{5,4;1}$	$9_{6,3;3}$
$10_{4,6;-2}$	$10_{3,7;-4}$	46773	81732	52398	39466	32455	$10_{6,4;2}$	$10_{7,3;4}$
$11_{4,7;-3}$	$11_{3,8;-5}$	52273	89952	51626	39938	33106	$11_{7,4;3}$	$11_{8,3;5}$
$12_{4,8;-4}$	$12_{3,9;-6}$	57692	90961	51673	40360	33654	$12_{8,4;4}$	$12_{9,3;6}$

Table IX-1 (Continued)

Sub-branch $^{b,e}Q_{\bar{1},1}$	$^{b,e}Q_{1,\bar{1}}$	κ ∓ 1	∓ 0.5	0	± 0.5	± 1	Sub-branch $^{b,e}Q_{1,\bar{1}}$	$^{b,e}Q_{\bar{1},1}$
$5_{5,0;5}$	$5_{4,1;3}$	9167	10753	13413	20634	55000	$5_{0,5;-5}$	$5_{1,4;-3}$
$6_{5,1;4}$	$6_{4,2;2}$	17024	20356	26402	45529	30952	$6_{1,5;-4}$	$6_{2,4;-2}$
$7_{5,2;3}$	$7_{4,3;1}$	24107	29481	41862	58783	33482	$7_{2,5;-3}$	$7_{3,4;-1}$
$8_{5,3;2}$	$8_{4,4;0}$	30694	38601	60829	52950	35417	$8_{3,5;-2}$	$8_{4,4;0}$
$9_{5,4;1}$	$9_{4,5;-1}$	36944	48332	74882	48332	36944	$9_{4,5;-1}$	$9_{5,4;1}$
$10_{5,5;0}$	$10_{4,6;-2}$	42955	59745	75829	47998	38182	$10_{5,5;0}$	$10_{6,4;2}$
$11_{5,6;-1}$	$11_{4,7;-3}$	48788	73909	69690	48463	39205	$11_{6,5;1}$	$11_{7,4;3}$
$12_{5,7;-2}$	$12_{4,8;-4}$	54487	90148	65598	48989	40064	$12_{7,5;2}$	$12_{8,4;4}$
$6_{6,0;6}$	$6_{5,1;4}$	9286	10861	13484	19779	65000	$6_{0,6;-6}$	$6_{1,5;-4}$
$7_{6,1;5}$	$7_{5,2;3}$	17411	20670	26163	43602	36161	$7_{1,6;-5}$	$7_{2,5;-3}$
$8_{6,2;4}$	$8_{5,3;2}$	24792	29932	39333	66715	38958	$8_{2,6;-4}$	$8_{3,5;-2}$
$9_{6,3;3}$	$9_{5,4;1}$	31667	38960	55712	67954	41167	$9_{3,6;-3}$	$9_{4,5;-1}$
$10_{6,4;2}$	$10_{5,5;0}$	38182	47998	75829	59745	42955	$10_{4,6;-2}$	$10_{5,5;0}$
$11_{6,5;1}$	$11_{5,6;-1}$	44432	57343	90410	57343	44432	$11_{5,6;-1}$	$11_{6,5;1}$
$12_{6,6;0}$	$12_{5,7;-2}$	50481	67590	90893	57486	45673	$12_{6,6;-0}$	$12_{7,5;2}$
$7_{7,0;7}$	$7_{6,1;5}$	9375	10943	13559	19511	75000	$7_{0,7;-7}$	$7_{1,6;-5}$
$8_{7,1;6}$	$8_{6,2;4}$	17708	20926	26300	41002	41319	$8_{1,7;-6}$	$8_{2,6;-4}$
$9_{7,2;5}$	$9_{6,3;3}$	25333	30348	38859	68174	44333	$9_{2,7;-5}$	$9_{3,6;-3}$
$10_{7,3;4}$	$10_{6,4;2}$	32455	39466	52398	81732	46773	$10_{3,7;-4}$	$10_{4,6;-2}$
$11_{7,4;3}$	$11_{6,5;1}$	39205	48463	69690	73909	48788	$11_{4,7;-3}$	$11_{5,6;-1}$
$12_{7,5;2}$	$12_{6,6;0}$	45673	57486	90893	67590	50481	$12_{5,7;-2}$	$12_{6,6;0}$
$8_{8,0;8}$	$8_{7,1;6}$	9444	11008	13620	19487	85000	$8_{0,8;-8}$	$8_{1,7;-6}$
$9_{8,1;7}$	$9_{7,2;5}$	17944	21134	26465	39408	46444	$9_{1,8;-7}$	$9_{2,7;-5}$
$10_{8,2;6}$	$10_{7,3;4}$	25773	30705	38980	65282	49636	$10_{2,8;-6}$	$10_{3,7;-4}$
$11_{8,3;5}$	$11_{7,4;3}$	33106	39938	51626	89952	52273	$11_{3,8;-5}$	$11_{4,7;-3}$
$12_{8,4;4}$	$12_{7,5;2}$	40064	48989	65598	90148	54487	$12_{4,8;-4}$	$12_{5,7;-2}$
$9_{9,0;9}$	$9_{8,1;7}$	9500	11060	13669	19524	95000	$9_{0,9;-9}$	$9_{1,8;-7}$
$10_{9,1;8}$	$10_{8,2;6}$	18136	21307	26611	38815	51545	$10_{1,9;-8}$	$10_{2,8;-6}$
$11_{9,2;7}$	$11_{8,3;5}$	26136	31011	39182	61636	54886	$11_{2,9;-7}$	$11_{3,8;-5}$
$12_{9,3;6}$	$12_{8,4;4}$	33654	40360	51673	90961	57692	$12_{3,9;-6}$	$12_{4,8;-4}$
$10_{10,0;10}$	$10_{9,1;8}$	9545	11103	13710	19565	105000	$10_{0,10;-10}$	$10_{1,9;-8}$
$11_{10,1;9}$	$11_{9,2;7}$	18295	21452	26737	38701	56629	$11_{1,10;-9}$	$11_{2,9;-7}$
$12_{10,2;8}$	$12_{9,3;6}$	26442	31275	39377	59285	60096	$12_{2,10;-8}$	$12_{3,9;-6}$
$11_{11,0;11}$	$11_{10,1;9}$	9583	11139	13744	19604	115000	$11_{0,11;-11}$	$11_{1,10;-9}$
$12_{11,1;10}$	$12_{10,2;8}$	18429	21576	26846	38736	61699	$12_{1,11;-10}$	$12_{2,10;-8}$
$12_{12,0;12}$	$12_{11,1;10}$	9615	11170	13774	19633	125000	$12_{0,12;-12}$	$12_{1,11;-10}$
$^{b,e}Q_{\bar{1},1}$	$^{b,e}Q_{1,\bar{1}}$	∓ 1	∓ 0.5	0	± 0.5	± 1	$^{b,e}Q_{1,\bar{1}}$	$^{b,e}Q_{\bar{1},1}$
$2_{2,1;1}$	$2_{1,2;-1}$	8333	8333	8333	8333	8333	$2_{1,2;-1}$	$2_{2,1;1}$
$3_{2,2;0}$	$3_{1,3;-2}$	14583	13160	11667	10173	8750	$3_{2,2;0}$	$3_{3,1;2}$
$4_{2,3;-1}$	$4_{1,4;-3}$	20250	16126	12886	10584	9000	$4_{3,2;1}$	$4_{4,1;3}$
$5_{2,4;-2}$	$5_{1,5;-4}$	25667	17823	13300	10751	9167	$5_{4,2;2}$	$5_{5,1;4}$
$6_{2,5;-3}$	$6_{1,6;-5}$	30952	18716	13464	10860	9286	$6_{5,2;3}$	$6_{6,1;5}$
$7_{2,6;-4}$	$7_{1,7;-6}$	36161	19158	13555	10943	9375	$7_{6,2;4}$	$7_{7,1;6}$
$8_{2,7;-5}$	$8_{1,8;-7}$	41319	19374	13619	11008	9444	$8_{7,2;5}$	$8_{8,1;7}$
$9_{2,8;-6}$	$9_{1,9;-8}$	46444	19487	13669	11060	9500	$9_{8,2;6}$	$9_{9,1;8}$
$10_{2,9;-7}$	$10_{1,10;-9}$	51545	19553	13710	11103	9545	$10_{9,2;7}$	$10_{10,1;9}$
$11_{2,10;-8}$	$11_{1,11;-10}$	56629	19598	13744	11139	9583	$11_{10,2;8}$	$11_{11,1;10}$
$12_{2,11;-9}$	$12_{1,12;-11}$	61699	19632	13774	11170	9615	$12_{11,2;9}$	$12_{12,1;11}$
$3_{3,1;2}$	$3_{2,2;0}$	8750	10173	11667	13160	14583	$3_{1,3;-2}$	$3_{2,2;0}$
$4_{3,2;1}$	$4_{2,3;-1}$	15750	18280	19208	18280	15750	$4_{2,3;-1}$	$4_{3,2;1}$
$5_{3,3;0}$	$5_{2,4;-2}$	22000	24936	23333	19781	16500	$5_{3,3;0}$	$5_{4,2;2}$
$6_{3,4;-1}$	$6_{2,5;-3}$	27857	30089	25173	20331	17024	$6_{4,3;1}$	$6_{5,2;3}$
$7_{3,5;-2}$	$7_{2,6;-4}$	33482	33722	25914	20668	17411	$7_{5,3;2}$	$7_{6,2;4}$
$8_{3,6;-3}$	$8_{2,7;-5}$	38958	36030	26251	20926	17708	$8_{6,3;3}$	$8_{7,2;5}$
$9_{3,7;-4}$	$9_{2,8;-6}$	44333	37360	26455	21134	17944	$9_{7,3;4}$	$9_{8,2;6}$
$10_{3,8;-5}$	$10_{2,9;-7}$	49636	38072	26609	21307	18136	$10_{8,3;5}$	$10_{9,2;7}$
$11_{3,9;-6}$	$11_{2,10;-8}$	54886	38443	26737	21452	18295	$11_{9,3;6}$	$11_{10,2;8}$
$12_{3,10;-7}$	$12_{2,11;-9}$	60096	38646	26846	21576	18429	$12_{10,3;7}$	$12_{11,2;9}$

Table IX-1 (Continued)

Sub-branch ${}^{b,o}Q_{\bar{1},1}$	${}^{b,o}Q_{1,\bar{1}}$	κ ∓ 1	∓ 0.5	0	± 0.5	± 1	Sub-branch ${}^{b,o}Q_{1,\bar{1}}$	${}^{b,o}Q_{\bar{1},1}$
$4_{4,1;3}$	$4_{3,2;1}$	9000	10584	12886	16126	20250	$4_{1,4;-3}$	$4_{2,3;-1}$
$5_{4,2;2}$	$5_{3,3;0}$	16500	19781	23333	24936	22000	$5_{2,4;-2}$	$5_{3,3;0}$
$6_{4,3;1}$	$6_{3,4;-1}$	23214	28237	30910	28237	23214	$6_{3,4;-1}$	$6_{4,3;1}$
$7_{4,4;0}$	$7_{3,5;-2}$	29464	35974	35396	29347	24107	$7_{4,4;0}$	$7_{5,3;2}$
$8_{4,5;-1}$	$8_{3,6;-3}$	35417	42717	37550	29917	24792	$8_{5,4;1}$	$8_{6,3;3}$
$9_{4,6;-2}$	$9_{3,7;-4}$	41167	48149	38467	30347	25333	$9_{6,4;2}$	$9_{7,3;4}$
$10_{4,7;-3}$	$10_{3,8;-5}$	46773	52121	38896	30705	25773	$10_{7,4;3}$	$10_{8,3;5}$
$11_{4,8;-4}$	$11_{3,9;-6}$	52273	54738	39163	31011	26136	$11_{8,4;4}$	$11_{9,3;6}$
$12_{4,9;-5}$	$12_{3,10;-7}$	57692	56299	39373	31275	26442	$12_{9,4;5}$	$12_{10,3;7}$
$5_{5,1;4}$	$5_{4,2;2}$	9167	10751	13300	17823	25667	$5_{1,5;-4}$	$5_{2,4;-2}$
$6_{5,2;3}$	$6_{4,3;1}$	17024	20331	25173	30089	27857	$6_{2,5;-3}$	$6_{3,4;-1}$
$7_{5,3;2}$	$7_{4,4;0}$	24107	29347	35396	35974	29464	$7_{3,5;-2}$	$7_{4,4;0}$
$8_{5,4;1}$	$8_{4,5;-1}$	30694	38050	43064	38050	30694	$8_{4,5;-1}$	$8_{5,4;1}$
$9_{5,5;0}$	$9_{4,6;-2}$	36944	46432	47757	38893	31667	$9_{5,5;0}$	$9_{6,4;2}$
$10_{5,6;-1}$	$10_{4,7;-3}$	42955	54266	50083	39458	32455	$10_{6,5;1}$	$10_{7,4;3}$
$11_{5,7;-2}$	$11_{4,8;-4}$	48788	61181	51087	39937	33106	$11_{7,5;2}$	$11_{8,4;4}$
$12_{5,8;-3}$	$12_{4,9;-5}$	54487	66823	51551	40360	33654	$12_{8,5;3}$	$12_{9,4;5}$
$6_{6,1;5}$	$6_{5,2;3}$	9286	10860	13464	18716	30952	$6_{1,6;-5}$	$6_{2,5;-3}$
$7_{6,2;4}$	$7_{5,3;2}$	17411	20668	25914	33722	33482	$7_{2,6;-4}$	$7_{3,5;-2}$
$8_{6,3;3}$	$8_{5,4;1}$	24792	29917	37550	42717	35417	$8_{3,6;-3}$	$8_{4,5;-1}$
$9_{6,4;2}$	$9_{5,5;0}$	31667	38893	47757	46432	36944	$9_{4,6;-2}$	$9_{5,5;0}$
$10_{6,5;1}$	$10_{5,6;-1}$	38182	47745	55515	47745	38182	$10_{5,6;-1}$	$10_{6,5;1}$
$11_{6,6;0}$	$11_{5,7;-2}$	44432	56495	60341	48430	39205	$11_{6,6;0}$	$11_{7,5;2}$
$12_{6,7;-1}$	$12_{5,8;-3}$	50481	65013	62764	48985	40064	$12_{7,6;1}$	$12_{8,5;3}$
$7_{7,1;6}$	$7_{6,2;4}$	9375	10943	13555	19158	36161	$7_{1,7;-6}$	$7_{2,6;-4}$
$8_{7,2;5}$	$8_{6,3;3}$	17708	20926	26251	36030	38958	$8_{2,7;-5}$	$8_{3,6;-3}$
$9_{7,3;4}$	$9_{6,4;2}$	25333	30347	38467	48149	41167	$9_{3,7;-4}$	$9_{4,6;-2}$
$10_{7,4;3}$	$10_{6,5;1}$	32455	39458	50083	54266	42955	$10_{4,7;-3}$	$10_{5,6;-1}$
$11_{7,5;2}$	$11_{6,6;0}$	39205	48430	60341	56495	44432	$11_{5,7;-2}$	$11_{6,6;0}$
$12_{7,6;1}$	$12_{6,7;-1}$	45673	57370	68182	57370	45673	$12_{6,7;-1}$	$12_{7,6;1}$
$8_{8,1;7}$	$8_{7,2;5}$	9444	11008	13619	19374	41319	$8_{1,8;-7}$	$8_{2,7;-5}$
$9_{8,2;6}$	$9_{7,3;4}$	17944	21134	26455	37360	44333	$9_{2,8;-6}$	$9_{3,7;-4}$
$10_{8,3;5}$	$10_{7,4;3}$	25773	30705	38896	52121	46773	$10_{3,8;-5}$	$10_{4,7;-3}$
$11_{8,4;4}$	$11_{7,5;2}$	33106	39937	51087	61181	48788	$11_{4,8;-4}$	$11_{5,7;-2}$
$12_{8,5;3}$	$12_{7,6;1}$	40064	48985	62764	65013	50481	$12_{5,8;-3}$	$12_{6,7;-1}$
$9_{9,1;8}$	$9_{8,2;6}$	9500	11060	13669	19487	46444	$9_{1,9;-8}$	$9_{2,8;-6}$
$10_{9,2;7}$	$10_{8,3;5}$	18136	21307	26609	38072	49636	$10_{2,9;-7}$	$10_{3,8;-5}$
$11_{9,3;6}$	$11_{8,4;4}$	26136	31011	39163	54738	52273	$11_{3,9;-6}$	$11_{4,8;-4}$
$12_{9,4;5}$	$12_{8,5;3}$	33654	40360	51551	66823	54487	$12_{4,9;-5}$	$12_{5,8;-3}$
$10_{10,1;9}$	$10_{9,2;7}$	9545	11103	13710	19553	51545	$10_{1,10;-9}$	$10_{2,9;-7}$
$11_{10,2;8}$	$11_{9,3;6}$	18295	21452	26737	38443	54886	$11_{2,10;-8}$	$11_{3,9;-6}$
$12_{10,3;7}$	$12_{9,4;5}$	26442	31275	39373	56299	57692	$12_{3,10;-7}$	$12_{4,9;-5}$
$11_{11,1;10}$	$11_{10,2;8}$	9583	11139	13744	19598	56629	$11_{1,11;-10}$	$11_{2,10;-8}$
$12_{11,2;9}$	$12_{10,3;7}$	18429	21576	26846	38646	60096	$12_{2,11;-9}$	$12_{3,10;-7}$
$12_{12,1;11}$	$12_{11,2;9}$	9615	11170	13774	19632	61699	$12_{1,12;-11}$	$12_{2,11;-9}$
${}^{b,e}R_{1,1}$	${}^{b,o}P_{\bar{1},\bar{1}}$	∓ 1	∓ 0.5	0	± 0.5	± 1	${}^{b,e}R_{1,1}$	${}^{b,o}P_{\bar{1},\bar{1}}$
$0_{0,0;0}$	$1_{1,1;0}$	10000	10000	10000	10000	10000	$0_{0,0;0}$	$1_{1,1;0}$
$1_{1,0;1}$	$2_{2,1;1}$	15000	15000	15000	15000	15000	$1_{0,1;-1}$	$2_{1,2;-1}$
$2_{2,0;2}$	$3_{3,1;2}$	25000	24086	22847	21383	20000	$2_{0,2;-2}$	$3_{1,3;-2}$
$3_{3,0;3}$	$4_{4,1;3}$	35000	34083	32533	29584	25000	$3_{0,3;-3}$	$4_{1,4;-3}$
$4_{4,0;4}$	$5_{5,1;4}$	45000	44117	42585	39100	30000	$4_{0,4;-4}$	$5_{1,5;-4}$
$5_{5,0;5}$	$6_{6,1;5}$	55000	54140	52653	49126	35000	$5_{0,5;-5}$	$6_{1,6;-5}$
$6_{6,0;6}$	$7_{7,1;6}$	65000	64155	62702	59250	40000	$6_{0,6;-6}$	$7_{1,7;-6}$
$7_{7,0;7}$	$8_{8,1;7}$	75000	74165	72737	69364	45000	$7_{0,7;-7}$	$8_{1,8;-7}$
$8_{8,0;8}$	$9_{9,1;8}$	85000	84173	82763	79453	50000	$8_{0,8;-8}$	$9_{1,9;-8}$
$9_{9,0;9}$	$10_{10,1;9}$	95000	94179	92782	89522	55000	$9_{0,9;-9}$	$10_{1,10;-9}$
$10_{10,0;10}$	$11_{11,1;10}$	105000	104184	102798	99576	60000	$10_{0,10;-10}$	$11_{1,11;-10}$
$11_{11,0;11}$	$12_{12,1;11}$	115000	114188	112810	109620	65000	$11_{0,11;-11}$	$12_{1,12;-11}$

Table IX-1 (Continued)

Sub-branch ${}^{b,e}R_{1,1}$	${}^{b,o}P_{\bar{1},\bar{1}}$	∓ 1	∓ 0.5	κ 0	± 0.5	± 1	Sub-branch ${}^{b,e}R_{1,1}$	${}^{b,o}P_{\bar{1},\bar{1}}$
$1_{0,1;-1}$	$2_{1,2;-1}$	15000	15000	15000	15000	15000	$1_{1,0;1}$	$2_{2,1;1}$
$2_{1,1;0}$	$3_{2,2;0}$	16667	16667	16667	16667	16667	$2_{1,1;0}$	$3_{2,2;0}$
$3_{2,1;1}$	$4_{3,2;1}$	26250	23549	21079	19563	18750	$3_{1,2;-1}$	$4_{2,3;-1}$
$4_{3,1;2}$	$5_{4,2;2}$	36000	33165	28748	23919	21000	$4_{1,3;-2}$	$5_{2,4;-2}$
$5_{4,1;3}$	$6_{5,2;3}$	45833	43122	38409	30161	23333	$5_{1,4;-3}$	$6_{2,5;-3}$
$6_{5,1;4}$	$7_{6,2;4}$	55714	53091	48508	38383	25714	$6_{1,5;-4}$	$7_{2,6;-4}$
$7_{6,1;5}$	$8_{7,2;5}$	65625	63059	58609	48001	28125	$7_{1,6;-5}$	$8_{2,7;-5}$
$8_{7,1;6}$	$9_{8,2;6}$	75556	73030	68678	58192	30556	$8_{1,7;-6}$	$9_{2,8;-6}$
$9_{8,1;7}$	$10_{9,2;7}$	85500	83002	78720	68479	33000	$9_{1,8;-7}$	$10_{2,9;-7}$
$10_{9,1;8}$	$11_{10,2;8}$	95455	92979	88749	78723	35454	$10_{1,9;-8}$	$11_{2,10;-8}$
$11_{10,1;9}$	$12_{11,2;9}$	105416	102958	98767	88911	37917	$11_{1,10;-9}$	$12_{2,11;-9}$
$2_{0,2;-2}$	$3_{1,3;-2}$	20000	21383	22847	24086	25000	$2_{2,0;2}$	$3_{3,1;2}$
$3_{1,2;-1}$	$4_{2,3;-1}$	18750	19563	21079	23549	26250	$3_{2,1;1}$	$4_{3,2;1}$
$4_{2,2;0}$	$5_{3,3;0}$	28000	23609	22028	23609	28000	$4_{2,2;0}$	$5_{3,3;0}$
$5_{3,2;1}$	$6_{4,3;1}$	37500	32338	26305	24633	30000	$5_{2,3;-1}$	$6_{3,4;-1}$
$6_{4,2;2}$	$7_{5,3;2}$	47143	42259	34093	27060	32143	$6_{2,4;-2}$	$7_{3,5;-2}$
$7_{5,2;3}$	$8_{6,3;3}$	56875	52226	43935	31293	34375	$7_{2,5;-3}$	$8_{3,6;-3}$
$8_{6,2;4}$	$9_{7,3;4}$	66667	62172	54199	37664	36667	$8_{2,6;-4}$	$9_{3,7;-4}$
$9_{7,2;5}$	$10_{8,3;5}$	76500	72110	64411	46127	39000	$9_{2,7;-5}$	$10_{3,8;-5}$
$10_{8,2;6}$	$11_{9,3;6}$	86364	82050	74550	56030	41364	$10_{2,8;-6}$	$11_{3,9;-6}$
$11_{9,2;7}$	$12_{10,3;7}$	96250	91993	84638	66512	43750	$11_{2,9;-7}$	$12_{3,10;-7}$
$3_{0,3;-3}$	$4_{1,4;-3}$	25000	29584	32533	34083	35000	$3_{3,0;3}$	$4_{4,1;3}$
$4_{1,3;-2}$	$5_{2,4;-2}$	21000	23919	28748	33165	36000	$4_{3,1;2}$	$5_{4,2;2}$
$5_{2,3;-1}$	$6_{3,4;-1}$	30000	24633	26305	32338	37500	$5_{3,2;1}$	$6_{4,3;1}$
$6_{3,3;0}$	$7_{4,4;0}$	39286	31500	26801	31500	39286	$6_{3,3;0}$	$7_{4,4;0}$
$7_{4,3;1}$	$8_{5,4;1}$	48750	41277	31054	31018	41250	$7_{3,4;-1}$	$8_{4,5;-1}$
$8_{5,3;2}$	$9_{6,4;2}$	58333	51336	39046	31553	43333	$8_{3,5;-2}$	$9_{4,6;-2}$
$9_{6,3;3}$	$10_{7,4;3}$	68000	61325	49147	33704	45500	$9_{3,6;-3}$	$10_{4,7;-3}$
$10_{7,3;4}$	$11_{8,4;4}$	77727	71271	59638	37919	47727	$10_{3,7;-4}$	$11_{4,8;-4}$
$11_{8,3;5}$	$12_{9,4;5}$	87500	81200	70013	44472	50000	$11_{3,8;-5}$	$12_{4,9;-5}$
$4_{0,4;-4}$	$5_{1,5;-4}$	30000	39100	42585	44117	45000	$4_{4,0;4}$	$5_{5,1;4}$
$5_{1,4;-3}$	$6_{2,5;-3}$	23333	30161	38409	43122	45833	$5_{4,1;3}$	$6_{5,2;3}$
$6_{2,4;-2}$	$7_{3,5;-2}$	32143	27060	34093	42259	47143	$6_{4,2;2}$	$7_{5,3;2}$
$7_{3,4;-1}$	$8_{4,5;-1}$	41250	31018	31054	41277	48750	$7_{4,3;1}$	$8_{5,4;1}$
$8_{4,4;0}$	$9_{5,5;0}$	50556	40057	31211	40057	50556	$8_{4,4;0}$	$9_{5,5;0}$
$9_{5,4;1}$	$10_{6,5;1}$	60000	50269	35484	38709	52500	$9_{4,5;-1}$	$10_{5,6;-1}$
$10_{6,4;2}$	$11_{7,5;2}$	69546	60393	43709	37729	54546	$10_{4,6;-2}$	$11_{5,7;-2}$
$11_{7,4;3}$	$12_{8,5;3}$	79167	70407	54102	37878	56667	$11_{4,7;-3}$	$12_{5,8;-3}$
$5_{0,5;-5}$	$6_{1,6;-5}$	35000	49126	52653	54140	55000	$5_{5,0;5}$	$6_{6,1;5}$
$6_{1,5;-4}$	$7_{2,6;-4}$	25714	38383	48508	53091	55714	$6_{5,1;4}$	$7_{6,2;4}$
$7_{2,5;-3}$	$8_{3,6;-3}$	34375	31293	43935	52226	56875	$7_{5,2;3}$	$8_{6,3;3}$
$8_{3,5;-2}$	$9_{4,6;-2}$	43333	31553	39046	51336	58333	$8_{5,3;2}$	$9_{6,4;2}$
$9_{4,5;-1}$	$10_{5,6;-1}$	52500	38709	35484	50269	60000	$9_{5,4;1}$	$10_{6,5;1}$
$10_{5,5;0}$	$11_{6,6;0}$	61818	48913	35365	48913	61818	$10_{5,5;0}$	$11_{6,6;0}$
$11_{6,5;1}$	$12_{7,6;1}$	71250	59273	39679	47228	63750	$11_{5,6;-1}$	$12_{6,7;-1}$
$6_{0,6;-6}$	$7_{1,7;-6}$	40000	59250	62702	64155	65000	$6_{6,0;6}$	$7_{7,1;6}$
$7_{1,6;-5}$	$8_{2,7;-5}$	28125	48001	58609	63059	65625	$7_{6,1;5}$	$8_{7,2;5}$
$8_{2,6;-4}$	$9_{3,7;-4}$	36667	37664	54199	62172	66667	$8_{6,2;4}$	$9_{7,3;4}$
$9_{3,6;-3}$	$10_{4,7;-3}$	45500	33704	49147	61325	68000	$9_{6,3;3}$	$10_{7,4;3}$
$10_{4,6;-2}$	$11_{5,7;-2}$	54546	37729	43709	60393	69546	$10_{6,4;2}$	$11_{7,5;2}$
$11_{5,6;-1}$	$12_{6,7;-1}$	63750	47228	39679	59273	71250	$11_{6,5;1}$	$12_{7,6;1}$
$7_{0,7;-7}$	$8_{1,8;-7}$	45000	69364	72737	74165	75000	$7_{7,0;7}$	$8_{8,1;7}$
$8_{1,7;-6}$	$9_{2,8;-6}$	30556	58192	68678	73030	75556	$8_{7,1;6}$	$9_{8,2;6}$
$9_{2,7;-5}$	$10_{3,8;-5}$	39000	46127	64411	72110	76500	$9_{7,2;5}$	$10_{8,3;5}$
$10_{3,7;-4}$	$11_{4,8;-4}$	47727	37919	59638	71271	77727	$10_{7,3;4}$	$11_{8,4;4}$
$11_{4,7;-3}$	$12_{5,8;-3}$	56667	37878	54102	70407	79167	$11_{7,4;3}$	$12_{8,5;3}$
$8_{0,8;-8}$	$9_{1,9;-8}$	50000	79453	82763	84173	85000	$8_{8,0;8}$	$9_{9,1;8}$
$9_{1,8;-7}$	$10_{2,9;-7}$	33000	68479	78720	83002	85500	$9_{8,1;7}$	$10_{9,2;7}$
$10_{2,8;-6}$	$11_{3,9;-6}$	41364	56030	74550	82050	86364	$10_{8,2;6}$	$11_{9,3;6}$
$11_{3,8;-5}$	$12_{4,9;-5}$	50000	44472	70013	81200	87500	$11_{8,3;5}$	$12_{9,4;5}$

Table IX-1 (Continued)

Sub-branch				κ			Sub-branch	
${}^{b,e}R_{1,1}$	${}^{b,o}P_{\bar{1},\bar{1}}$	∓1	∓0.5	0	±0.5	±1	${}^{b,e}R_{1,1}$	${}^{b,o}P_{\bar{1},\bar{1}}$
$9_{0,9;-9}$	$10_{1,10;-9}$	55000	89522	92782	94179	95000	$9_{9,0;9}$	$10_{10,1;9}$
$10_{1,9;-8}$	$11_{2,10;-8}$	35454	78723	88749	92979	95455	$10_{9,1;8}$	$11_{10,2;8}$
$11_{2,9;-7}$	$12_{3,10;-7}$	43750	66512	84638	91993	96250	$11_{9,2;7}$	$12_{10,3;7}$
$10_{0,10;-10}$	$11_{1,11;-10}$	60000	99576	102798	104184	105000	$10_{10,0;10}$	$11_{11,1;10}$
$11_{1,10;-9}$	$12_{2,11;-9}$	37917	88911	98767	102958	105416	$11_{10,1;9}$	$12_{11,2;9}$
$11_{0,11;-11}$	$12_{1,12;-11}$	65000	109620	112810	114188	115000	$11_{11,0;11}$	$12_{12,1;11}$
${}^{b,o}R_{\bar{1},1}$	${}^{b,e}P_{1,\bar{1}}$	∓1	∓0.5	0	±0.5	±1	${}^{b,o}R_{1,\bar{1}}$	${}^{b,e}P_{\bar{1},1}$
$1_{1,1;0}$	$2_{0,2;-2}$	5000	7226	10000	12774	15000	$1_{1,1;0}$	$2_{2,0;2}$
$2_{1,2;-1}$	$3_{0,3;-3}$	10000	16667	21498	23874	25000	$2_{2,1;1}$	$3_{3,0;3}$
$3_{1,3;-2}$	$4_{0,4;-4}$	15000	27406	32266	34065	35000	$3_{3,1;2}$	$4_{4,0;4}$
$4_{1,4;-3}$	$5_{0,5;-5}$	20000	38266	42535	44115	45000	$4_{4,1;3}$	$5_{5,0;5}$
$5_{1,5;-4}$	$6_{0,6;-6}$	25000	48829	52643	54140	55000	$5_{5,1;4}$	$6_{6,0;6}$
$6_{1,6;-5}$	$7_{0,7;-7}$	30000	59146	62700	64155	65000	$6_{6,1;5}$	$7_{7,0;7}$
$7_{1,7;-6}$	$8_{0,8;-8}$	35000	69327	72736	74165	75000	$7_{7,1;6}$	$8_{8,0;8}$
$8_{1,8;-7}$	$9_{0,9;-9}$	40000	79440	82762	84173	85000	$8_{8,1;7}$	$9_{9,0;9}$
$9_{1,9;-8}$	$10_{0,10;-10}$	45000	89517	92782	94179	95000	$9_{9,1;8}$	$10_{10,0;10}$
$10_{1,10;-9}$	$11_{0,11;-11}$	50000	99574	102798	104184	105000	$10_{10,1;9}$	$11_{11,0;11}$
$11_{1,11;-10}$	$12_{0,12;-12}$	55000	109619	112810	114188	115000	$11_{11,1;10}$	$12_{12,0;12}$
$2_{2,1;1}$	$3_{1,2;-1}$	1667	2792	5168	10000	16667	$2_{1,2;-1}$	$3_{2,1;1}$
$3_{2,2;0}$	$4_{1,3;-2}$	3750	7602	15000	22398	26250	$3_{2,2;0}$	$4_{3,1;2}$
$4_{2,3;-1}$	$5_{1,4;-3}$	6000	14796	26797	33039	36000	$4_{3,2;1}$	$5_{4,1;3}$
$5_{2,4;-2}$	$6_{1,5;-4}$	8333	24389	37946	43109	45833	$5_{4,2;2}$	$6_{5,1;4}$
$6_{2,5;-3}$	$7_{1,6;-5}$	10714	35443	48405	53090	55714	$6_{5,2;3}$	$7_{6,1;5}$
$7_{2,6;-4}$	$8_{1,7;-6}$	13125	46736	58587	63059	65625	$7_{6,2;4}$	$8_{7,1;6}$
$8_{2,7;-5}$	$9_{1,8;-7}$	15556	57689	68672	73029	75556	$8_{7,2;5}$	$9_{8,1;7}$
$9_{2,8;-6}$	$10_{1,9;-8}$	18000	68283	78719	83002	85500	$9_{8,2;6}$	$10_{9,1;8}$
$10_{2,9;-7}$	$11_{1,10;-9}$	20455	78648	88749	92979	95455	$10_{9,2;7}$	$11_{10,1;9}$
$11_{2,10;-8}$	$12_{1,11;-10}$	22917	88882	98767	102958	105416	$11_{10,2;8}$	$12_{11,1;10}$
$3_{3,1;2}$	$4_{2,2;0}$	1250	1537	2692	6941	18750	$3_{1,3;-2}$	$4_{2,2;0}$
$4_{3,2;1}$	$5_{2,3;-1}$	3000	4022	8877	19900	28000	$4_{2,3;-1}$	$5_{3,2;1}$
$5_{3,3;0}$	$6_{2,4;-2}$	5000	7698	19335	31792	37500	$5_{3,3;0}$	$6_{4,2;2}$
$6_{3,4;-1}$	$7_{2,5;-3}$	7143	13138	31685	42193	47143	$6_{4,3;1}$	$7_{5,2;3}$
$7_{3,5;-2}$	$8_{2,6;-4}$	9375	20912	43306	52219	56875	$7_{5,3;2}$	$8_{6,2;4}$
$8_{3,6;-3}$	$9_{2,7;-5}$	11667	31041	54046	62172	66667	$8_{6,3;3}$	$9_{7,2;5}$
$9_{3,7;-4}$	$10_{2,8;-6}$	14000	42620	64375	72111	76500	$9_{7,3;4}$	$10_{8,2;6}$
$10_{3,8;-5}$	$11_{2,9;-7}$	16364	54434	74542	82050	86364	$10_{8,3;5}$	$11_{9,2;7}$
$11_{3,9;-6}$	$12_{2,10;-8}$	18750	65840	84638	91993	96250	$11_{9,3;6}$	$12_{10,2;8}$
$4_{4,1;3}$	$5_{3,2;1}$	1000	1162	1666	4522	21000	$4_{1,4;-3}$	$5_{2,3;-1}$
$5_{4,2;2}$	$6_{3,3;0}$	2500	2920	5238	16127	30000	$5_{2,4;-2}$	$6_{3,3;0}$
$6_{4,3;1}$	$7_{3,4;-1}$	4286	5148	12183	29700	39286	$6_{3,4;-1}$	$7_{4,3;1}$
$7_{4,4;0}$	$8_{3,5;-2}$	6250	8062	23299	41022	48750	$7_{4,4;0}$	$8_{5,3;2}$
$8_{4,5;-1}$	$9_{3,6;-3}$	8333	12161	36249	51302	58333	$8_{5,4;1}$	$9_{6,3;3}$
$9_{4,6;-2}$	$10_{3,7;-4}$	10500	18094	48371	61321	68000	$9_{6,4;2}$	$10_{7,3;4}$
$10_{4,7;-3}$	$11_{3,8;-5}$	12727	26418	59438	71271	77727	$10_{7,4;3}$	$11_{8,3;5}$
$11_{4,8;-4}$	$12_{3,9;-6}$	15000	37104	69962	81200	87500	$11_{8,4;4}$	$12_{9,3;6}$
$5_{5,1;4}$	$6_{4,2;2}$	833	966	1253	2984	23333	$5_{1,5;-4}$	$6_{2,4;-2}$
$6_{5,2;3}$	$7_{4,3;1}$	2143	2475	3549	11918	32143	$6_{2,5;-3}$	$7_{3,4;-1}$
$7_{5,3;2}$	$8_{4,4;0}$	3750	4309	7685	26263	41250	$7_{3,5;-2}$	$8_{4,4;0}$
$8_{5,4;1}$	$9_{4,5;-1}$	5556	6389	15266	39231	50556	$8_{4,5;-1}$	$9_{5,4;1}$
$9_{5,5;0}$	$10_{4,6;-2}$	7500	8819	27015	50150	60000	$9_{5,5;0}$	$10_{6,4;2}$
$10_{5,6;-1}$	$11_{4,7;-3}$	9545	11954	40562	60377	69546	$10_{6,5;1}$	$11_{7,4;3}$
$11_{5,7;-2}$	$12_{4,8;-4}$	11667	16365	53190	70405	79167	$11_{7,5;2}$	$12_{8,4;4}$
$6_{6,1;5}$	$7_{5,2;3}$	714	829	1052	2102	25714	$6_{1,6;-5}$	$7_{2,5;-3}$
$7_{6,2;4}$	$8_{5,3;2}$	1875	2170	2825	8386	34375	$7_{2,6;-4}$	$8_{3,5;-2}$
$8_{6,3;3}$	$9_{5,4;1}$	3333	3839	5485	21522	43333	$8_{3,6;-3}$	$9_{4,5;-1}$
$9_{6,4;2}$	$10_{5,5;0}$	5000	5717	10057	36378	52500	$9_{4,6;-2}$	$10_{5,5;0}$
$10_{6,5;1}$	$11_{5,6;-1}$	6818	7736	18198	48537	61818	$10_{5,6;-1}$	$11_{6,5;1}$
$11_{6,6;0}$	$12_{5,7;-2}$	8750	9912	30549	59217	71250	$11_{6,6;0}$	$12_{7,5;2}$

Table IX-I (Continued)

Sub-branch $^{b,o}R\bar{1},1$	$^{b,e}P1,\bar{1}$	κ ∓ 1	∓ 0.5	0	± 0.5	± 1	Sub-branch $^{b,o}R1,\bar{1}$	$^{b,e}P\bar{1},1$
$7_{7,1;6}$	$8_{6,2;4}$	625	726	918	1615	28125	$7_{1,7;-6}$	$8_{2,6;-4}$
$8_{7,2;5}$	$9_{6,3;3}$	1667	1932	2461	5953	36667	$8_{2,7;-5}$	$9_{3,6;-3}$
$9_{7,3;4}$	$10_{6,4;2}$	3000	3467	4517	16387	45500	$9_{3,7;-4}$	$10_{4,6;-2}$
$10_{7,4;3}$	$11_{6,5;1}$	4545	5226	7432	32088	54546	$10_{4,7;-3}$	$11_{5,6;-1}$
$11_{7,5;2}$	$12_{6,6;0}$	6250	7134	12369	46152	63750	$11_{5,7;-2}$	$12_{6,6;0}$
$8_{8,1;7}$	$9_{7,2;5}$	556	645	816	1342	30556	$8_{1,8;-7}$	$9_{2,7;-5}$
$9_{8,2;6}$	$10_{7,3;4}$	1500	1741	2210	4454	39000	$9_{2,8;-6}$	$10_{3,7;-4}$
$10_{8,3;5}$	$11_{7,4;3}$	2727	3158	4030	12003	47727	$10_{3,8;-5}$	$11_{4,7;-3}$
$11_{8,4;4}$	$12_{7,5;2}$	4167	4808	6264	26503	56667	$11_{4,8;-4}$	$12_{5,7;-2}$
$9_{9,1;8}$	$10_{8,2;6}$	500	581	734	1176	33000	$9_{1,9;-8}$	$10_{2,8;-6}$
$10_{9,2;7}$	$11_{8,3;5}$	1364	1583	2009	3581	41364	$10_{2,9;-7}$	$11_{3,8;-5}$
$11_{9,3;6}$	$12_{8,4;4}$	2500	2898	3688	8874	50000	$11_{3,9;-6}$	$12_{4,8;-4}$
$10_{10,1;9}$	$11_{9,2;7}$	455	528	667	1059	35454	$10_{1,10;-9}$	$11_{2,9;-7}$
$11_{10,2;8}$	$12_{9,3;6}$	1250	1452	1841	3076	43750	$11_{2,10;-8}$	$12_{3,9;-6}$
$11_{11,1;10}$	$12_{10,2;8}$	417	484	611	967	37917	$11_{1,11;-10}$	$12_{2,10;-8}$

D. *b* prolate-or-oblate sub-branches.

$^{b,e}R\bar{1},3$	$^{b,o}P1,\bar{3}$	∓ 1	∓ 0.5	0	± 0.5	± 1	$^{b,e}R3,\bar{1}$	$^{b,o}P\bar{3},1$
$2_{2,0;2}$	$3_{1,3;-2}$	1667	1097	486	101		$2_{0,2;-2}$	$3_{3,1;2}$
$3_{2,1;1}$	$4_{1,4;-3}$	3750	1452	297	32		$3_{1,2;-1}$	$4_{4,1;3}$
$4_{2,2;0}$	$5_{1,5;-4}$	6000	1159	140	14		$4_{2,2;0}$	$5_{5,1;4}$
$5_{2,3;-1}$	$6_{1,6;-5}$	8333	758	77	9		$5_{3,2;1}$	$6_{6,1;5}$
$6_{2,4;-2}$	$7_{1,7;-6}$	10714	481	54	7		$6_{4,2;2}$	$7_{7,1;6}$
$7_{2,5;-3}$	$8_{1,8;-7}$	13125	323	42	6		$7_{5,2;3}$	$8_{8,1;7}$
$8_{2,6;-4}$	$9_{1,9;-8}$	15556	238	35	5		$8_{6,2;4}$	$9_{9,1;8}$
$9_{2,7;-5}$	$10_{1,10;-9}$	18000	191	30	4		$9_{7,2;5}$	$10_{10,1;9}$
$10_{2,8;-6}$	$11_{1,11;-10}$	20455	163	27	4		$10_{8,2;6}$	$11_{11,1;10}$
$11_{2,9;-7}$	$12_{1,12;-11}$	22917	144	24	3		$11_{9,2;7}$	$12_{12,1;11}$
$3_{3,0;3}$	$4_{2,3;-1}$	1250	1323	1091	416		$3_{0,3;-3}$	$4_{3,2;1}$
$4_{3,1;2}$	$5_{2,4;-2}$	3000	2753	1252	163		$4_{1,3;-2}$	$5_{4,2;2}$
$5_{3,2;1}$	$6_{2,5;-3}$	5000	3538	737	62		$5_{2,3;-1}$	$6_{5,2;3}$
$6_{3,3;0}$	$7_{2,6;-4}$	7143	3362	375	35		$6_{3,3;0}$	$7_{6,2;4}$
$7_{3,4;-1}$	$8_{2,7;-5}$	9375	2573	219	26		$7_{4,3;1}$	$8_{7,2;5}$
$8_{3,5;-2}$	$9_{2,8;-6}$	11667	1754	155	20		$8_{5,3;2}$	$9_{8,2;6}$
$9_{3,6;-3}$	$10_{2,9;-7}$	14000	1174	124	17		$9_{6,3;3}$	$10_{9,2;7}$
$10_{3,7;-4}$	$11_{2,10;-8}$	16364	826	104	14		$10_{7,3;4}$	$11_{10,2;8}$
$11_{3,8;-5}$	$12_{2,11;-9}$	18750	628	90	12		$11_{8,3;5}$	$12_{11,2;9}$
$4_{4,0;4}$	$5_{3,3;0}$	1000	1144	1259	855		$4_{0,4;-4}$	$5_{3,3;0}$
$5_{4,1;3}$	$6_{3,4;-1}$	2500	2771	2305	514		$5_{1,4;-3}$	$6_{4,3;1}$
$6_{4,2;2}$	$7_{3,5;-2}$	4286	4433	2107	189		$6_{2,4;-2}$	$7_{5,3;2}$
$7_{4,3;1}$	$8_{3,6;-3}$	6250	5663	1258	92		$7_{3,4;-1}$	$8_{6,3;3}$
$8_{4,4;0}$	$9_{3,7;-4}$	8333	6007	676	62		$8_{4,4;0}$	$9_{7,3;4}$
$9_{4,5;-1}$	$10_{3,8;-5}$	10500	5335	411	47		$9_{5,4;1}$	$10_{8,3;5}$
$10_{4,6;-2}$	$11_{3,9;-6}$	12727	4082	298	38		$10_{6,4;2}$	$11_{9,3;6}$
$11_{4,7;-3}$	$12_{3,10;-7}$	15000	2866	241	32		$11_{7,4;3}$	$12_{10,3;7}$
$5_{5,0;5}$	$6_{4,3;1}$	833	965	1174	1186		$5_{0,5;-5}$	$6_{3,4;-1}$
$6_{5,1;4}$	$7_{4,4;0}$	2143	2461	2707	1158		$6_{1,5;-4}$	$7_{4,4;0}$
$7_{5,2;3}$	$8_{4,5;-1}$	3750	4229	3560	488		$7_{2,5;-3}$	$8_{5,4;1}$
$8_{5,3;2}$	$9_{4,6;-2}$	5556	6040	3008	209		$8_{3,5;-2}$	$9_{6,4;2}$
$9_{5,4;1}$	$10_{4,7;-3}$	7500	7603	1831	125		$9_{4,5;-1}$	$10_{7,4;3}$
$10_{5,5;0}$	$11_{4,8;-4}$	9545	8531	1026	92		$10_{5,5;0}$	$11_{8,4;4}$
$11_{5,6;-1}$	$12_{4,9;-5}$	11667	8452	644	73		$11_{6,5;1}$	$12_{9,4;5}$
$6_{6,0;6}$	$7_{5,3;2}$	714	829	1039	1321		$6_{0,6;-6}$	$7_{3,5;-2}$
$7_{6,1;5}$	$8_{5,4;1}$	1875	2169	2640	1942		$7_{1,6;-5}$	$8_{4,5;-1}$
$8_{6,2;4}$	$9_{5,5;0}$	3333	3830	4213	1086		$8_{2,6;-4}$	$9_{5,5;0}$
$9_{6,3;3}$	$10_{5,6;-1}$	5000	5676	4833	447		$9_{3,6;-3}$	$10_{6,5;1}$
$10_{6,4;2}$	$11_{5,7;-2}$	6818	7575	3937	234		$10_{4,6;-2}$	$11_{7,5;2}$
$11_{6,5;1}$	$12_{5,8;-3}$	8750	9349	2441	161		$11_{5,6;-1}$	$12_{8,5;3}$

Table IX-1 (Continued)

Sub-branch $^{b,o}R\bar{1},3$	$^{b,o}P1,\bar{3}$	∓ 1	∓ 0.5	κ 0	± 0.5	± 1	Sub-branch $^{b,o}R3,\bar{1}$	$^{b,o}P\bar{3},1$
$7_{7,0;7}$	$8_{6,3;3}$	625	726	916	1313		$7_{0,7;-7}$	$8_{3,6;-3}$
$8_{7,1;6}$	$9_{6,4;2}$	1667	1932	2426	2561		$8_{1,7;-6}$	$9_{4,6;-2}$
$9_{7,2;5}$	$10_{6,5;1}$	3000	3466	4214	2032		$9_{2,7;-5}$	$10_{5,6;-1}$
$10_{7,3;4}$	$11_{6,6;0}$	4545	5221	5738	917		$10_{3,7;-4}$	$11_{6,6;0}$
$11_{7,4;3}$	$12_{6,7;-1}$	6250	7114	6113	427		$11_{4,7;-3}$	$12_{7,6;1}$
$8_{8,0;8}$	$9_{7,3;4}$	556	645	816	1238		$8_{0,8;-8}$	$9_{3,7;-4}$
$9_{8,1;7}$	$10_{7,4;3}$	1500	1741	2204	2870		$9_{1,8;-7}$	$10_{4,7;-3}$
$10_{8,2;6}$	$11_{7,5;2}$	2727	3158	3967	3126		$10_{2,8;-6}$	$11_{5,7;-2}$
$11_{8,3;5}$	$12_{7,6;1}$	4167	4807	5836	1752		$11_{3,8;-5}$	$12_{6,7;-1}$
$9_{9,0;9}$	$10_{8,3;5}$	500	581	734	1143		$9_{0,9;-9}$	$10_{3,8;-5}$
$10_{9,1;8}$	$11_{8,4;4}$	1364	1583	2008	2924		$10_{1,9;-8}$	$11_{4,8;-4}$
$11_{9,2;7}$	$12_{8,5;3}$	2500	2898	3676	4014		$11_{2,9;-7}$	$12_{5,8;-3}$
$10_{10,0;10}$	$11_{9,3;6}$	455	528	667	1049		$10_{0,10;-10}$	$11_{3,9;-6}$
$11_{10,1;9}$	$12_{9,4;5}$	1250	1452	1841	2831		$11_{1,10;-9}$	$12_{4,9;-5}$
$11_{11,0;11}$	$12_{10,3;7}$	417	484	611	964		$11_{0,11;-11}$	$12_{3,10;-7}$

Table IX-2—Forbidden Sub-Branches

A. *a* AND *c* SUB-BRANCHES

Sub-branch $^{c,o}Q\bar{3},2$	$^{c,e}Q3,\bar{2}$	∓ 1	∓ 0.5	κ 0	± 0.5	± 1	Sub-branch $^{a,o}Q2,\bar{3}$	$^{a,e}Q\bar{2},3$
$3_{3,1;2}$	$3_{0,3;-3}$		104	169	69		$3_{1,3;-2}$	$3_{3,0;3}$
$4_{3,2;1}$	$4_{0,4;-4}$		356	283	68		$4_{2,3;-1}$	$4_{4,0;4}$
$5_{3,3;0}$	$5_{0,5;-5}$		621	286	53		$5_{3,3;0}$	$5_{5,0;5}$
$6_{3,4;-1}$	$6_{0,6;-6}$		773	250	43		$6_{4,3;1}$	$6_{6,0;6}$
$7_{3,5;-2}$	$7_{0,7;-7}$		809	213	35		$7_{5,3;2}$	$7_{7,0;7}$
$8_{3,6;-3}$	$8_{0,8;-8}$		774	183	30		$8_{6,3;3}$	$8_{8,0;8}$
$9_{3,7;-4}$	$9_{0,9;-9}$		712	160	26		$9_{7,3;4}$	$9_{9,0;9}$
$10_{3,8;-5}$	$10_{0,10;-10}$		645	141	23		$10_{8,3;5}$	$10_{10,0;10}$
$11_{3,9;-6}$	$11_{0,11;-11}$		584	127	21		$11_{9,3;6}$	$11_{11,0;11}$
$12_{3,10;-7}$	$12_{0,12;-12}$		531	115	19		$12_{10,3;7}$	$12_{12,0;12}$
$4_{4,1;3}$	$4_{1,3;-2}$		31	164	153		$4_{1,4;-3}$	$4_{3,1;2}$
$5_{4,2;2}$	$5_{1,4;-3}$		174	504	193		$5_{2,4;-2}$	$5_{4,1;3}$
$6_{4,3;1}$	$6_{1,5;-4}$		527	708	163		$6_{3,4;-1}$	$6_{5,1;4}$
$7_{4,4;0}$	$7_{1,6;-5}$		1057	716	131		$7_{4,4;0}$	$7_{6,1;5}$
$8_{4,5;-1}$	$8_{1,7;-6}$		1557	642	108		$8_{5,4;1}$	$8_{7,1;6}$
$9_{4,6;-2}$	$9_{1,8;-7}$		1855	559	92		$9_{6,4;2}$	$9_{8,1;7}$
$10_{4,7;-3}$	$10_{1,9;-8}$		1940	489	80		$10_{7,4;3}$	$10_{9,1;8}$
$11_{4,8;-4}$	$11_{1,10;-9}$		1888	432	71		$11_{8,4;4}$	$11_{10,1;9}$
$12_{4,9;-5}$	$12_{1,11;-10}$		1770	387	64		$12_{9,4;5}$	$12_{11,1;10}$
$5_{5,1;4}$	$5_{2,3;-1}$		9	79	191		$5_{1,5;-4}$	$5_{3,2;1}$
$6_{5,2;3}$	$6_{2,4;-2}$		52	413	342		$6_{2,5;-3}$	$6_{4,2;2}$
$7_{5,3;2}$	$7_{2,5;-3}$		187	919	325		$7_{3,5;-2}$	$7_{5,2;3}$
$8_{5,4;1}$	$8_{2,6;-4}$		515	1216	268		$8_{4,5;-1}$	$8_{6,2;4}$
$9_{5,5;0}$	$9_{2,7;-5}$		1123	1236	222		$9_{5,5;0}$	$9_{7,2;5}$
$10_{5,6;-1}$	$10_{2,8;-6}$		1924	1128	188		$10_{6,5;1}$	$10_{8,2;6}$
$11_{5,7;-2}$	$11_{2,9;-7}$		2650	999	163		$11_{7,5;2}$	$11_{9,2;7}$
$12_{5,8,-3}$	$12_{2,10;-8}$		3098	885	144		$12_{8,5;3}$	$12_{10,2;8}$
$6_{6,1;5}$	$6_{3,3;0}$		4	33	168		$6_{1,6;-5}$	$6_{3,3;0}$
$7_{6,2;4}$	$7_{3,4;-1}$		22	212	457		$7_{2,6;-4}$	$7_{4,3;1}$
$8_{6,3;3}$	$8_{3,5;-2}$		68	714	524		$8_{3,6;-3}$	$8_{5,3;2}$
$9_{6,4;2}$	$9_{3,6;-3}$		184	1385	456		$9_{4,6;-2}$	$9_{6,3;3}$
$10_{6,5;1}$	$10_{3,7;-4}$		445	1777	379		$10_{5,6;-1}$	$10_{7,3;4}$
$11_{6,6;0}$	$11_{3,8;-5}$		971	1816	320		$11_{6,6;0}$	$11_{8,3;5}$
$12_{6,7;-1}$	$12_{3,9;-6}$		1829	1681	277		$12_{7,6;1}$	$12_{9,3;6}$

Table IX-2 (Continued)

Sub-branch				κ			Sub-branch	
$^{c,o}Q\bar{3},2$	$^{c,o}Q3,\bar{2}$	∓1	∓0.5	0	±0.5	±1	$^{a,o}Q2,\bar{3}$	$^{a,o}Q\bar{2},3$
$7_{7,1;6}$	$7_{4,3;1}$		3	17	114		$7_{1,7;-6}$	$7_{3,4;-1}$
$8_{7,2;5}$	$8_{4,4;0}$		13	98	480		$8_{2,7;-5}$	$8_{4,4;0}$
$9_{7,3;4}$	$9_{4,5;-1}$		36	385	714		$9_{3,7;-4}$	$9_{5,4;1}$
$10_{7,4;3}$	$10_{4,6;-2}$		84	1053	686		$10_{4,7;-3}$	$10_{6,4;2}$
$11_{7,5;2}$	$11_{4,7;-3}$		183	1885	584		$11_{5,7;-2}$	$11_{7,4;3}$
$12_{7,6;1}$	$12_{4,8;-4}$		384	2375	493		$12_{6,7;-1}$	$12_{8,4;4}$
$8_{8,1;7}$	$8_{5,3;2}$		2	10	69		$8_{1,8;-7}$	$8_{3,5;-2}$
$9_{8,2;6}$	$9_{5,4;1}$		8	52	403		$9_{2,8;-6}$	$9_{4,5;-1}$
$10_{8,3;5}$	$10_{5,5;0}$		23	190	830		$10_{3,8;-5}$	$10_{5,5;0}$
$11_{8,4;4}$	$11_{5,6;-1}$		52	588	931		$11_{4,8;-4}$	$11_{6,5;1}$
$12_{8,5;3}$	$12_{5,7;-2}$		102	1419	832		$12_{5,8;-3}$	$12_{7,5;2}$
$9_{9,1;8}$	$9_{6,3;3}$		1	7	41		$9_{1,9;-8}$	$9_{3,6;-3}$
$10_{9,2;7}$	$10_{6,4;2}$		6	34	282		$10_{2,9;-7}$	$10_{4,6;-2}$
$11_{9,3;6}$	$11_{6,5;1}$		17	106	817		$11_{3,9;-6}$	$11_{5,6;-1}$
$12_{9,4;5}$	$12_{6,6;0}$		36	304	1141		$12_{4,9;-5}$	$12_{6,6;0}$
$10_{10,1;9}$	$10_{7,3;4}$		1	5	25		$10_{1,10;-9}$	$10_{3,7;-4}$
$11_{10,2;8}$	$11_{7,4;3}$		5	24	180		$11_{2,10;-8}$	$11_{4,7;-3}$
$12_{10,3;7}$	$12_{7,5;2}$		12	70	677		$12_{3,10;-7}$	$12_{5,7;-2}$
$11_{11,1;10}$	$11_{8,3;5}$		1	4	17		$11_{1,11;-10}$	$11_{3,8;-5}$
$12_{11,2;9}$	$12_{8,4;4}$		4	19	112		$12_{2,11;-9}$	$12_{4,8;-4}$
$12_{12,1;11}$	$12_{9,3;6}$		0	3	13		$12_{1,12;-11}$	$12_{3,9;-6}$
$^{c,o}Q\bar{3},4$	$^{c,o}Q3,\bar{4}$	∓1	∓0.5	0	±0.5	±1	$^{a,o}Q4,\bar{3}$	$^{a,o}Q\bar{4},3$
$4_{4,0;4}$	$4_{1,4;-3}$		8	10	2		$4_{0,4;-4}$	$4_{4,1;3}$
$5_{4,1;3}$	$5_{1,5;-4}$		24	14	1		$5_{1,4;-3}$	$5_{5,1;4}$
$6_{4,2;2}$	$6_{1,6;-5}$		42	9	0		$6_{2,4;-2}$	$6_{6,1;5}$
$7_{4,3;1}$	$7_{1,7;-6}$		51	2	0		$7_{3,4;-1}$	$7_{7,1;6}$
$8_{4,4;0}$	$8_{1,8;-7}$		48	1	0		$8_{4,4;0}$	$8_{8,1;7}$
$9_{4,5;-1}$	$9_{1,9;-8}$		35	1	0		$9_{5,4;1}$	$9_{9,1;8}$
$10_{4,6;-2}$	$10_{1,10;-9}$		22	0	0		$10_{6,4;2}$	$10_{10,1;9}$
$11_{4,7;-3}$	$11_{1,11;-10}$		12	0	0		$11_{7,4;3}$	$11_{11,1;10}$
$12_{4,8;-4}$	$12_{1,12;-11}$		7	0	0		$12_{8,4;4}$	$12_{12,1;11}$
$5_{5,0;5}$	$5_{2,4;-2}$		7	17	7		$5_{0,5;-5}$	$5_{4,2;2}$
$6_{5,1;4}$	$6_{2,5;-3}$		27	39	4		$6_{1,5;-4}$	$6_{5,2;3}$
$7_{5,2;3}$	$7_{2,6;-4}$		62	44	1		$7_{2,5;-3}$	$7_{6,2;4}$
$8_{5,3;2}$	$8_{2,7;-5}$		106	28	0		$8_{3,5;-2}$	$8_{7,2;5}$
$9_{5,4;1}$	$9_{2,8;-6}$		142	13	0		$9_{4,5;-1}$	$9_{8,2;6}$
$10_{5,5;0}$	$10_{2,9;-7}$		154	5	0		$10_{5,5;0}$	$10_{9,2;7}$
$11_{5,6;-1}$	$11_{2,10;-8}$		137	3	0		$11_{6,5;1}$	$11_{10,2;8}$
$12_{5,7;-2}$	$12_{2,11;-9}$		102	1	0		$12_{7,5;2}$	$12_{11,2;9}$
$6_{6,0;6}$	$6_{3,4;-1}$		4	16	14		$6_{0;6;-6}$	$6_{4,3;1}$
$7_{6,1;5}$	$7_{3,5;-2}$		19	53	15		$7_{1,6;-5}$	$7_{5,3;2}$
$8_{6,2;4}$	$8_{3,6;-3}$		50	88	5		$8_{2,6;-4}$	$8_{6,3;3}$
$9_{6,3;3}$	$9_{3,7;-4}$		102	88	2		$9_{3,6;-3}$	$9_{7,3;4}$
$10_{6,4;2}$	$10_{3,8;-5}$		170	56	1		$10_{4,6;-2}$	$10_{8,3;5}$
$11_{6,5;1}$	$11_{3,9;-6}$		240	27	0		$11_{5,6;-1}$	$11_{9,3;6}$
$12_{6,6;0}$	$12_{3,10;-7}$		289	12	0		$12_{6,6;0}$	$12_{10,3;7}$
$7_{7,0;7}$	$7_{4,4;0}$		3	13	19		$7_{0,7;-7}$	$7_{4,4;0}$
$8_{7,1;6}$	$8_{4,5;-1}$		12	51	30		$8_{1,7;-6}$	$8_{5,4;1}$
$9_{7,2;5}$	$9_{4,6;-2}$		34	108	17		$9_{2,7;-5}$	$9_{6,4;2}$
$10_{7,3;4}$	$10_{4,7;-3}$		74	152	5		$10_{3,7;-4}$	$10_{7,4;3}$
$11_{7,4;3}$	$11_{4,8;-4}$		137	144	2		$11_{4,7;-3}$	$11_{8,4;4}$
$12_{7,5;2}$	$12_{4,9;-5}$		223	93	1		$12_{5,7;-2}$	$12_{9,4;5}$
$8_{8,0;8}$	$8_{5,4;1}$		2	10	20		$8_{0,8;-8}$	$8_{4,5;-1}$
$9_{8,1;7}$	$9_{5,5;0}$		8	41	46		$9_{1,8;-7}$	$9_{5,5;0}$
$10_{8,2;6}$	$10_{5,6;-1}$		23	102	39		$10_{2,8;-6}$	$10_{6,5;1}$
$11_{8,3;5}$	$11_{5,7;-2}$		51	179	15		$11_{3,8;-5}$	$11_{7,5;2}$
$12_{8,4;4}$	$12_{5,8;-3}$		97	229	5		$12_{4,8;-4}$	$12_{8,5;3}$

Table IX-2 (Continued)

Sub-branch ${}^{c,e}Q\bar{3},4$	${}^{c,e}Q3,\bar{4}$	∓ 1	∓ 0.5	κ 0	± 0.5	± 1	Sub-branch ${}^{a,e}Q4,\bar{3}$	${}^{a,e}Q\bar{4},3$
$9_{9,0;9}$	$9_{6,4;2}$		1	7	19		$9_{0,9;-9}$	$9_{4,6;-2}$
$10_{9,1;8}$	$10_{6,5;1}$		6	32	56		$10_{1,9;-8}$	$10_{5,6;-1}$
$11_{9,2;7}$	$11_{6,6;0}$		17	84	67		$11_{2,9;-7}$	$11_{6,6;0}$
$12_{9,3;6}$	$12_{6,7;-1}$		36	169	37		$12_{3,9;-6}$	$12_{7,6;1}$
$10_{10,0;10}$	$10_{7,4;3}$		1	5	17		$10_{0,10;-10}$	$10_{4,7;-3}$
$11_{10,1;9}$	$11_{7,5;2}$		5	24	59		$11_{1,10;-9}$	$11_{5,7;-2}$
$12_{10,2;8}$	$12_{7,6;1}$		12	66	93		$12_{2,10;-8}$	$12_{6,7;-1}$
$11_{11,0;11}$	$11_{8,4;4}$		1	4	14		$11_{0,11;-11}$	$11_{4,8;-4}$
$12_{11,1;10}$	$12_{8,5;3}$		4	19	57		$12_{1,11;-10}$	$12_{5,8;-3}$
$12_{12,0;12}$	$12_{9,4;5}$		0	3	12		$12_{0,12;-12}$	$12_{4,9;-5}$
${}^{c,e}R3,\bar{2}$	${}^{c,e}P\bar{3},2$	∓ 1	∓ 0.5	0	± 0.5	± 1	${}^{a,e}R\bar{2},3$	${}^{a,e}P2,\bar{3}$
$2_{0,2;-2}$	$3_{3,0;3}$		75	215	138		$2_{2,0;2}$	$3_{0,3,-3}$
$3_{0,3;-3}$	$4_{3,1;2}$		294	313	51		$3_{3,0;3}$	$4_{1,3;-2}$
$4_{0,4;-4}$	$5_{3,2;1}$		528	176	17		$4_{4,0;4}$	$5_{2,3;-1}$
$5_{0,5,-5}$	$6_{3,3;0}$		558	79	9		$5_{5,0;5}$	$6_{3,3;0}$
$6_{0,6;-6}$	$7_{3,4;-1}$		418	41	5		$6_{6,0;6}$	$7_{4,3;1}$
$7_{0,7;-7}$	$8_{3,5;-2}$		263	26	4		$7_{7,0;7}$	$8_{5,3;2}$
$8_{0,8;-8}$	$9_{3,6;-3}$		159	19	3		$8_{8,0;8}$	$9_{6,3;3}$
$9_{0,9;-9}$	$10_{3,7;-4}$		102	14	2		$9_{9,0;9}$	$10_{7,3;4}$
$10_{0,10;-10}$	$11_{3,8;-5}$		70	11	2		$10_{10,0;10}$	$11_{8,3;5}$
$11_{0,11;-11}$	$12_{3,9;-6}$		53	9	1		$11_{11,0;11}$	$12_{9,3;6}$
$3_{1,2;-1}$	$4_{4,0;4}$		24	146	272		$3_{2,1;1}$	$4_{0,4;-4}$
$4_{1,3;-2}$	$5_{4,1;3}$		122	538	212		$4_{3,1;2}$	$5_{1,4;-3}$
$5_{1,4;-3}$	$6_{4,2;2}$		377	665	76		$5_{4,1;3}$	$6_{2,4;-2}$
$6_{1,5;-4}$	$7_{4,3;1}$		803	409	34		$6_{5,1;4}$	$7_{3,4;-1}$
$7_{1,6;-5}$	$8_{4,4;0}$		1171	206	21		$7_{6,1;5}$	$8_{4,4;0}$
$8_{1,7;-6}$	$9_{4,5;-1}$		1200	115	15		$8_{7,1;6}$	$9_{5,4;1}$
$9_{1,8;-7}$	$10_{4,6;-2}$		938	76	11		$9_{8,1;7}$	$10_{6,4;2}$
$10_{1,9;-8}$	$11_{4,7;-3}$		631	57	9		$10_{9,1;8}$	$11_{7,4;3}$
$11_{1,10;-9}$	$12_{4,8;-4}$		409	45	7		$11_{10,1;9}$	$12_{8,4;4}$
$4_{2,2;0}$	$5_{5,0;5}$		11	70	262		$4_{2,2;0}$	$5_{0,5;-5}$
$5_{2,3;-1}$	$6_{5,1;4}$		46	359	464		$5_{3,2;1}$	$6_{1,5;-4}$
$6_{2,4;-2}$	$7_{5,2;3}$		142	894	220		$6_{4,2;2}$	$7_{2,5;-3}$
$7_{2,5;-3}$	$8_{5,3;2}$		364	1038	92		$7_{5,2;3}$	$8_{3,5;-2}$
$8_{2,6;-4}$	$9_{5,4;1}$		793	675	52		$8_{6,2;4}$	$9_{4,5;-1}$
$9_{2,7;-5}$	$10_{5,5;0}$		1395	365	36		$9_{7,2;5}$	$10_{5,5;0}$
$10_{2,8;-6}$	$11_{5,6;-1}$		1869	215	27		$10_{8,2;6}$	$11_{6,5;1}$
$11_{2,9;-7}$	$12_{5,7;-2}$		1887	148	21		$11_{9,2;7}$	$12_{7,5;2}$
$5_{3,2;1}$	$6_{6,0;6}$		7	39	184		$5_{2,3;-1}$	$6_{0,6;-6}$
$6_{3,3;0}$	$7_{6,1;5}$		27	187	624		$6_{3,3;0}$	$7_{1,6;-5}$
$7_{3,4;-1}$	$8_{6,2;4}$		69	610	487		$7_{4,3;1}$	$8_{2,6;-4}$
$8_{3,5;-2}$	$9_{6,3;3}$		157	1272	210		$8_{5,3;2}$	$9_{3,6;-3}$
$9_{3,6;-3}$	$10_{6,4;2}$		336	1426	108		$9_{6,3;3}$	$10_{4,6;-2}$
$10_{3,7;-4}$	$11_{6,5;1}$		684	963	71		$10_{7,3;4}$	$11_{5,6;-1}$
$11_{3,8;-5}$	$12_{6,6;0}$		1276	548	53		$11_{8,3;5}$	$12_{6,6;0}$
$6_{4,2;2}$	$7_{7,0;7}$		5	27	119		$6_{2,4;-2}$	$7_{0,7;-7}$
$7_{4,3;1}$	$8_{7,1;6}$		19	110	574		$7_{3,4;-1}$	$8_{1,7;-6}$
$8_{4,4;0}$	$9_{7,2;5}$		46	338	816		$8_{4,4;0}$	$9_{2,7;-5}$
$9_{4,5;-1}$	$10_{7,3;4}$		94	887	435		$9_{5,4;1}$	$10_{3,7;-4}$
$10_{4,6;-2}$	$11_{7,4;3}$		176	1662	205		$10_{6,4;2}$	$11_{4,7;-3}$
$11_{4,7;-3}$	$12_{7,5;2}$		321	1825	126		$11_{7,4;3}$	$12_{5,7;-2}$
$7_{5,2;3}$	$8_{8,0;8}$		4	21	81		$7_{2,5;-3}$	$8_{0,8;-8}$
$8_{5,3;2}$	$9_{8,1;7}$		15	78	424		$8_{3,5;-2}$	$9_{1,8;-7}$
$9_{5,4;1}$	$10_{8,2;6}$		36	206	995		$9_{4,5;-1}$	$10_{2,8;-6}$
$10_{5,5;0}$	$11_{8,3;5}$		69	514	790		$10_{5,5;0}$	$11_{3,8;-5}$
$11_{5,6;-1}$	$12_{8,4;4}$		121	1183	381		$11_{6,5;1}$	$12_{4,8;-4}$

Table IX-2 (Continued)

Sub-branch ${}^{c,e}R3,\bar{2}$	${}^{c,e}P\bar{3},2$	∓1	∓0.5	κ 0	±0.5	±1	Sub-branch ${}^{a,e}R\bar{2},3$	${}^{a,e}P2,\bar{3}$
$8_{6,2;4}$	$9_{9,0;9}$		4	18	60		$8_{2,6;-4}$	$9_{0,9;-9}$
$9_{6,3;3}$	$10_{9,1;8}$		13	62	292		$9_{3,6;-3}$	$10_{1,9;-8}$
$10_{6,4;2}$	$11_{9,2;7}$		29	149	910		$10_{4,6;-2}$	$11_{2,9;-7}$
$11_{6,5;1}$	$12_{9,3;6}$		55	324	1182		$11_{5,6;-1}$	$12_{3,9;-6}$
$9_{7,2;5}$	$10_{10,0;10}$		3	15	48		$9_{2,7;-5}$	$10_{0,10;-10}$
$10_{7,3;4}$	$11_{10,1;9}$		11	52	207		$10_{3,7;-4}$	$11_{1,10;-9}$
$11_{7,4;3}$	$12_{10,2;8}$		24	121	694		$11_{4,7;-3}$	$12_{2,10;-8}$
$10_{8,2;6}$	$11_{11,0;11}$		3	13	41		$10_{2,8;-6}$	$11_{0,11;-11}$
$11_{8,3;5}$	$12_{11,1;10}$		9	45	158		$11_{3,8;-5}$	$12_{1,11;-10}$
$11_{9,2;7}$	$12_{12,0;12}$		2	12	36		$11_{2,9;-7}$	$12_{0,12;-12}$
${}^{c,o}R3,\bar{2}$	${}^{c,o}P\bar{3},2$	∓1	∓0.5	0	±0.5	±1	${}^{a,o}R\bar{2},3$	${}^{a,o}P2,\bar{3}$
$3_{1,3;-2}$	$4_{4,1;3}$		10	21	13		$3_{3,1;2}$	$4_{1,4;-3}$
$4_{1,4;-3}$	$5_{4,2;2}$		31	38	12		$4_{4,1;3}$	$5_{2,4;-2}$
$5_{1,5;-4}$	$6_{4,3;1}$		56	39	8		$5_{5,1;4}$	$6_{3,4;-1}$
$6_{1,6;-5}$	$7_{4,4;0}$		73	32	5		$6_{6,1;5}$	$7_{4,4;0}$
$7_{1,7;-6}$	$8_{4,5;-1}$		80	24	4		$7_{7,1;6}$	$8_{5,4;1}$
$8_{1,8;-7}$	$9_{4,6;-2}$		77	18	3		$8_{8,1;7}$	$9_{6,4;2}$
$9_{1,9;-8}$	$10_{4,7;-3}$		68	14	2		$9_{9,1;8}$	$10_{7,4;3}$
$10_{1,10;-9}$	$11_{4,8;-4}$		58	11	2		$10_{10,1;9}$	$11_{8,4;4}$
$11_{1,11;-10}$	$12_{4,9;-5}$		49	9	1		$11_{11,1;10}$	$12_{9,4;5}$
$4_{2,3;-1}$	$5_{5,1;4}$		9	31	32		$4_{3,2;1}$	$5_{1,5;-4}$
$5_{2,4;-2}$	$6_{5,2;3}$		33	79	39		$5_{4,2;2}$	$6_{2,5;-3}$
$6_{2,5;-3}$	$7_{5,3;2}$		73	108	30		$6_{5,2;3}$	$7_{3,5;-2}$
$7_{2,6;-4}$	$8_{5,4;1}$		124	108	21		$7_{6,2;4}$	$8_{4,5;-1}$
$8_{2,7;-5}$	$9_{5,5;0}$		172	91	15		$8_{7,2;5}$	$9_{5,5;0}$
$9_{2,8;-6}$	$10_{5,6;-1}$		205	72	11		$9_{8,2;6}$	$10_{6,5;1}$
$10_{2,9;-7}$	$11_{5,7;-2}$		216	56	9		$10_{9,2;7}$	$11_{7,5;2}$
$11_{2,10;-8}$	$12_{5,8;-3}$		209	45	7		$11_{10,2;8}$	$12_{8,5;3}$
$5_{3,3;0}$	$6_{6,1;5}$		7	30	46		$5_{3,3;0}$	$6_{1,6;-5}$
$6_{3,4;-1}$	$7_{6,2;4}$		25	92	76		$6_{4,3;1}$	$7_{2,6;-4}$
$7_{3,5;-2}$	$8_{6,3;3}$		60	163	67		$7_{5,3;2}$	$8_{3,6;-3}$
$8_{3,6;-3}$	$9_{6,4;2}$		115	202	50		$8_{6,3;3}$	$9_{4,6;-2}$
$9_{3,7;-4}$	$10_{6,5;1}$		187	199	36		$9_{7,3;4}$	$10_{5,6;-1}$
$10_{3,8;-5}$	$11_{6,6;0}$		267	171	27		$10_{8,3;5}$	$11_{6,6;0}$
$11_{3,9;-6}$	$12_{6,7;-1}$		337	139	21		$11_{9,3;6}$	$12_{7,6;1}$
$6_{4,3;1}$	$7_{7,1;6}$		5	25	53		$6_{3,4;-1}$	$7_{1,7;-6}$
$7_{4,4;0}$	$8_{7,2;5}$		19	86	110		$7_{4,4;0}$	$8_{2,7;-5}$
$8_{4,5;-1}$	$9_{7,3;4}$		46	177	118		$8_{5,4;1}$	$9_{3,7;-4}$
$9_{4,6;-2}$	$10_{7,4;3}$		89	268	94		$9_{6,4;2}$	$10_{4,7;-3}$
$10_{4,7;-3}$	$11_{7,5;2}$		154	316	70		$10_{7,4;3}$	$11_{5,7;-2}$
$11_{4,8;-4}$	$12_{7,6;1}$		241	309	53		$11_{8,4;4}$	$12_{6,7;-1}$
$7_{5,3;2}$	$8_{8,1;7}$		4	21	53		$7_{3,5;-2}$	$8_{1,8;-7}$
$8_{5,4;1}$	$9_{8,2;6}$		15	73	134		$8_{4,5;-1}$	$9_{2,8;-6}$
$9_{5,5;0}$	$10_{8,3;5}$		35	163	173		$9_{5,5;0}$	$10_{3,8;-5}$
$10_{5,6;-1}$	$11_{8,4;4}$		69	281	155		$10_{6,5;1}$	$11_{4,8;-4}$
$11_{5,7;-2}$	$12_{8,5;3}$		119	390	120		$11_{7,5;2}$	$12_{5,8;-3}$
$8_{6,3;3}$	$9_{9,1;8}$		4	18	49		$8_{3,6;-3}$	$9_{1,9;-8}$
$9_{6,4;2}$	$10_{9,2;7}$		13	61	143		$9_{4,6;-2}$	$10_{2,9;-7}$
$10_{6,5;1}$	$11_{9,3;6}$		29	140	219		$10_{5,6;-1}$	$11_{3,9;-6}$
$11_{6,6;0}$	$12_{9,4;5}$		54	258	224		$11_{6,6;0}$	$12_{4,9;-5}$
$9_{7,3;4}$	$10_{10,1;9}$		3	15	44		$9_{3,7;-4}$	$10_{1,10;-9}$
$10_{7,4;3}$	$11_{10,2;8}$		11	52	140		$10_{4,7;-3}$	$11_{2,10;-8}$
$11_{7,5;2}$	$12_{10,3;7}$		24	119	249		$11_{5,7;-2}$	$12_{3,10;-7}$
$10_{8,3;5}$	$11_{11,1;10}$		3	13	40		$10_{3,8;-5}$	$11_{1,11;-10}$
$11_{8,4;4}$	$12_{11,2;9}$		9	45	131		$11_{4,8;-4}$	$12_{2,11;-9}$
$11_{9,3;6}$	$12_{12,1;11}$		2	12	36		$11_{3,9;-6}$	$12_{1,12;-11}$

Table IX-2 (Continued)

Sub-branch ${}^{c,e}R_{\bar{3},4}$	${}^{c,e}P_{3,\bar{4}}$	∓ 1	∓ 0.5	κ 0	± 0.5	± 1	Sub-branch ${}^{a,e}R_{4,\bar{3}}$	${}^{a,e}P_{\bar{4},3}$
$3_{3,0;3}$	$4_{0,4;-4}$		28	18	2		$3_{0,3;-3}$	$4_{4,0;4}$
$4_{3,1;2}$	$5_{0,5;-5}$		78	17	0		$4_{1,3;-2}$	$5_{5,0;5}$
$5_{3,2;1}$	$6_{0,6;-6}$		107	8	0		$5_{2,3;-1}$	$6_{6,0;6}$
$6_{3,3;0}$	$7_{0,7;-7}$		96	3	0		$6_{3,3;0}$	$7_{7,0;7}$
$7_{3,4;-1}$	$8_{0,8;-8}$		65	2·	0		$7_{4,3;1}$	$8_{8,0;8}$
$8_{3,5;-2}$	$9_{0,9;-9}$		39	1	0		$8_{5,3;2}$	$9_{9,0;9}$
$9_{3,6;-3}$	$10_{0,10;-10}$		23	1	0		$9_{6,3;3}$	$10_{10,0;10}$
$10_{3,7;-4}$	$11_{0,11,-11}$		14	1	0		$10_{7,3;4}$	$11_{11,0;11}$
$11_{3,8;-5}$	$12_{0,12,-12}$		9	0	0		$11_{8,3;5}$	$12_{12,0;12}$
$4_{4,0;4}$	$5_{1,4;-3}$		11	31	7		$4_{0,4;-4}$	$5_{4,1;3}$
$5_{4,1;3}$	$6_{1,5;-4}$		58	62	3		$5_{1,4;-3}$	$6_{5,1;4}$
$6_{4,2;2}$	$7_{1,6;-5}$		152	51	1		$6_{2,4;-2}$	$7_{6,1;5}$
$7_{4,3;1}$	$8_{1,7;-6}$		252	25	0		$7_{3,4;-1}$	$8_{7,1;6}$
$8_{4,4;0}$	$9_{1,8;-7}$		290	11	0		$8_{4,4;0}$	$9_{8,1;7}$
$9_{4,5;-1}$	$10_{1,9;-8}$		250	6	0		$9_{5,4;1}$	$10_{9,1;8}$
$10_{4,6;-2}$	$11_{1,10;-9}$		175	4	0		$10_{6,4;2}$	$11_{10,1;9}$
$11_{4,7;-3}$	$12_{1,11;-10}$		109	3	0		$11_{7,4;3}$	$12_{11,1;10}$
$5_{5,0;5}$	$6_{2,4;-2}$		3	21	15		$5_{0;5;-5}$	$6_{4,2;2}$
$6_{5,1;4}$	$7_{2,5;-3}$		17	83	12		$6_{1,5;-4}$	$7_{5,2;3}$
$7_{5,2;3}$	$8_{2,6;-4}$		65	124	4		$7_{2,5;-3}$	$8_{6,2;4}$
$8_{5,3;2}$	$9_{2,7;-5}$		173	96	1		$8_{3,5;-2}$	$9_{7,2;5}$
$9_{5,4;1}$	$10_{2,8;-6}$		335	50	1		$9_{4,5;-1}$	$10_{8,2;6}$
$10_{5,5;0}$	$11_{2,9;-7}$		477	23	0		$10_{5,5;0}$	$11_{9,2;7}$
$11_{5,6;-1}$	$12_{2,10;-8}$		517	13	0		$11_{6,5;1}$	$12_{10,2;8}$
$6_{6,0;6}$	$7_{3,4;-1}$		1	9	21		$6_{0,6;-6}$	$7_{4,3;1}$
$7_{6,1;5}$	$8_{3,5;-2}$		5	59	28		$7_{1,6;-5}$	$8_{5,3;2}$
$8_{6,2;4}$	$9_{3,6;-3}$		20	151	12		$8_{2,6;-4}$	$9_{6,3;3}$
$9_{6,3;3}$	$10_{3,7;-4}$		60	198	4		$9_{3,6;-3}$	$10_{7,3;4}$
$10_{6,4;2}$	$11_{3,8;-5}$		155	151	2		$10_{4,6;-2}$	$11_{8,3;5}$
$11_{6,5;1}$	$12_{3,9;-6}$		329	81	1		$11_{5,6;-1}$	$12_{9,3;6}$
$7_{7,0;7}$	$8_{4,4;0}$		0	4	20		$7_{0,7;-7}$	$8_{4,4;0}$
$8_{7,1;6}$	$9_{4,5;-1}$		2	26	46		$8_{1,7;-6}$	$9_{5,4;1}$
$9_{7,2;5}$	$10_{4,6;-2}$		8	107	31		$9_{2,7;-5}$	$10_{6,4;2}$
$10_{7,3;4}$	$11_{4,7;-3}$		22	230	11		$10_{3,7;-4}$	$11_{7,4;3}$
$11_{7,4;3}$	$12_{4,8;-4}$		55	282	4		$11_{4,7;-3}$	$12_{8,4;4}$
$8_{8,0;8}$	$9_{5,4;1}$		0	2	15		$8_{0,8;-8}$	$9_{4,5;-1}$
$9_{8,1;7}$	$10_{5,5;0}$		1	12	56		$9_{1,8;-7}$	$10_{5,5;0}$
$10_{8,2;6}$	$11_{5,6;-1}$		4	54	59		$10_{2,8;-6}$	$11_{6,5;1}$
$11_{8,3;5}$	$12_{5,7;-2}$		11	165	27		$11_{3,8;-5}$	$12_{7,5;2}$
$9_{9,0;9}$	$10_{6,4;2}$		0	1	10		$9_{0,9;-9}$	$10_{4,6;-2}$
$10_{9,1;8}$	$11_{6,5;1}$		1	6	53		$10_{1,9;-8}$	$11_{5,6;-1}$
$11_{9,2;7}$	$12_{6,6;0}$		3	25	87		$11_{2,9;-7}$	$12_{6,6;0}$
$10_{10,0;10}$	$11_{7,4;3}$		0	1	5		$10_{0,10;-10}$	$11_{4,7;-3}$
$11_{10,1;9}$	$12_{7,5;2}$		1	4	41		$11_{1,10;-9}$	$12_{5,7;-2}$
$11_{11,0;11}$	$12_{8,4;4}$		0	0	3		$11_{0,11;-11}$	$12_{4,8;-4}$
${}^{c,o}R_{\bar{3},4}$	${}^{c,o}P_{3,\bar{4}}$	∓ 1	∓ 0.5	0	± 0.5	± 1	${}^{a,o}R_{4,\bar{3}}$	${}^{a,o}P_{\bar{4},3}$
$4_{4,1;3}$	$5_{1,5;-4}$		2	1	0		$4_{1,4;-3}$	$5_{5,1;4}$
$5_{4,2;2}$	$6_{1,6;-5}$		5	2	0		$5_{2,4;-2}$	$6_{6,1;5}$
$6_{4,3;1}$	$7_{1,7;-6}$		8	2	0		$6_{3.4;-1}$	$7_{7,1;6}$
$7_{4,4;0}$	$8_{1,8;-7}$		11	1	0		$7_{4,4;0}$	$8_{8,1;7}$
$8_{4,5;-1}$	$9_{1,9;-8}$		12	1	0		$8_{5,4;1}$	$9_{9,1;8}$
$9_{4,6;-2}$	$10_{1,10;-9}$		11	1	0		$9_{6,4;2}$	$10_{10,1;9}$
$10_{4,7;-3}$	$11_{1,11;-10}$		9	1	0		$10_{7,4;3}$	$11_{11,1;10}$
$11_{4,8;-4}$	$12_{1,12;-11}$		8	0	0		$11_{8,4;4}$	$12_{12,1;11}$

Table IX-2 (Continued)

Sub-branch $^{c,o}R\bar{3},4$	$^{c,o}P3,\bar{4}$	∓1	∓0.5	κ 0	±0.5	±1	Sub-branch $^{a,o}R4,\bar{3}$	$^{a,o}P\bar{4},3$
$5_{5,1;4}$	$6_{2,5;-3}$		1	2	0		$5_{1,5;-4}$	$6_{5,2;3}$
$6_{5,2;3}$	$7_{2,6;-4}$		6	5	0		$6_{2,5;-3}$	$7_{6,2;4}$
$7_{5,3;2}$	$8_{2,7;-5}$		14	6	0		$7_{3,5;-2}$	$8_{7,2;5}$
$8_{5,4;1}$	$9_{2,8;-6}$		23	6	0		$8_{4,5;-1}$	$9_{8,2;6}$
$9_{5,5;0}$	$10_{2,9;-7}$		32	4	0		$9_{5,5;0}$	$10_{9,2;7}$
$10_{5,6;-1}$	$11_{2,10;-8}$		36	3	0		$10_{6.5;1}$	$11_{10,2;8}$
$11_{5,7;-2}$	$12_{2,11;-9}$		37	3	0		$11_{7,5;2}$	$12_{11,2;9}$
$6_{6,1;5}$	$7_{3,5;-2}$		1	2	1		$6_{1,6;-5}$	$7_{5,3;2}$
$7_{6,2;4}$	$8_{3,6;-3}$		4	8	1		$7_{2,6;-4}$	$8_{6,3;3}$
$8_{6,3;3}$	$9_{3,7;-4}$		11	13	1		$8_{3,6;-3}$	$9_{7,3;4}$
$9_{6,4;2}$	$10_{3,8,-5}$		23	14	1		$9_{4,6;-2}$	$10_{8,3;5}$
$10_{6,5;1}$	$11_{3,9;-6}$		39	13	0		$10_{5,6;-1}$	$11_{9,3;6}$
$11_{6,6;0}$	$12_{3,10;-7}$		56	10	0		$11_{6,6;0}$	$12_{10,3;7}$
$7_{7,1;6}$	$8_{4,5;-1}$		0	2	2		$7_{1,7;-6}$	$8_{5,4;1}$
$8_{7,2;5}$	$9_{4,6;-2}$		2	8	3		$8_{2,7;-5}$	$9_{6,4;2}$
$9_{7,3;4}$	$10_{4,7;-3}$		7	17	2		$9_{3,7;-4}$	$10_{7,4;3}$
$10_{7,4;3}$	$11_{4,8;-4}$		16	24	1		$10_{4,7;-3}$	$11_{8,4;4}$
$11_{7,5;2}$	$12_{4,9;-5}$		32	25	1		$11_{5,7;-2}$	$12_{9,4;5}$
$8_{8,1;7}$	$9_{5,5;0}$		0	1	2		$8_{1,8;-7}$	$9_{5,5;0}$
$9_{8,2;6}$	$10_{5,6;-1}$		1	6	4		$9_{2,8;-6}$	$10_{6,5;1}$
$10_{8,3;5}$	$11_{5,7;-2}$		4	17	4		$10_{3,8;-5}$	$11_{7,5;2}$
$11_{8,4;4}$	$12_{5,8;-3}$		10	29	3		$11_{4,8;-4}$	$12_{8,5;3}$
$9_{9,1;8}$	$10_{6,5;1}$		0	1	2		$9_{1,9;-8}$	$10_{5,6;-1}$
$10_{9,2;7}$	$11_{6,6;0}$		1	5	5		$10_{2,9;-7}$	$11_{6,6;0}$
$11_{9,3;6}$	$12_{6,7;-1}$		3	14	7		$11_{3,9;-6}$	$12_{7,6;1}$
$10_{10,1;9}$	$11_{7,5;2}$		0	1	2		$10_{1,10;-9}$	$11_{5,7;-2}$
$11_{10,2;8}$	$12_{7,6;1}$		1	3	6		$11_{2,10;-8}$	$12_{6,7;-1}$
$11_{11,1;10}$	$12_{8,5;3}$		0	0	2		$11_{1,11;-10}$	$12_{5,8;-3}$

B. *b* sub-branches.

$^{b,e}Q\bar{3},3$	$^{b,e}Q3,\bar{3}$	∓1	∓0.5	0	±0.5	±1	$^{b,e}Q3,\bar{3}$	$^{b,e}Q\bar{3},3$
$3_{3,0;3}$	$3_{0,3;-3}$		138	297	138		$3_{0,3;-3}$	$3_{3,0;3}$
$4_{3,1;2}$	$4_{0,4;-4}$		445	319	41		$4_{1,3;-2}$	$4_{4,0;4}$
$5_{3,2;1}$	$5_{0,5;-5}$		674	158	13		$5_{2,3;-1}$	$5_{5,0;5}$
$6_{3,3;0}$	$6_{0,6;-6}$		640	67	6		$6_{3,3;0}$	$6_{6,0;6}$
$7_{3,4;-1}$	$7_{0,7;-7}$		450	33	4		$7_{4,3;1}$	$7_{7,0;7}$
$8_{3,5;-2}$	$8_{0,8;-8}$		273	21	3		$8_{5,3;2}$	$8_{8,0;8}$
$9_{3,6;-3}$	$9_{0,9;-9}$		162	15	2		$9_{6,3;3}$	$9_{9,0;9}$
$10_{3,7;-4}$	$10_{0,10;-10}$		101	11	1		$10_{7,3;4}$	$10_{10,0;10}$
$11_{3,8;-5}$	$11_{0,11;-11}$		69	9	1		$11_{8,3;5}$	$11_{11,0;11}$
$12_{3,9;-6}$	$12_{0,12;-12}$		52	7	1		$12_{9,3;6}$	$12_{12,0;12}$
$4_{4,0;4}$	$4_{1,3;-2}$		41	319	445		$4_{0,4;-4}$	$4_{3,1;2}$
$5_{4,1;3}$	$5_{1,4;-3}$		230	846	230		$5_{1,4;-3}$	$5_{4,1;3}$
$6_{4,2;2}$	$6_{1,5;-4}$		684	793	70		$6_{2,4;-2}$	$6_{5,1;4}$
$7_{4,3;1}$	$7_{1,6;-5}$		1287	422	29		$7_{3,4;-1}$	$7_{6,1;5}$
$8_{4,4;0}$	$8_{1,7;-6}$		1640	197	17		$8_{4,4;0}$	$8_{7,1;6}$
$9_{4,5;-1}$	$9_{1,8;-7}$		1513	105	11		$9_{5,4;1}$	$9_{8,1;7}$
$10_{4,6;-2}$	$10_{1,9;-8}$		1105	67	8		$10_{6,4;2}$	$10_{9,1;8}$
$11_{4,7;-3}$	$11_{1,10;-9}$		713	49	6		$11_{7,4;3}$	$11_{10,1;9}$
$12_{4,8;-4}$	$12_{1,11;-10}$		449	38	5		$12_{8,4;4}$	$12_{11,1;10}$
$5_{5,0;5}$	$5_{2,3;-1}$		13	158	674		$5_{0,5;-5}$	$5_{3,2;1}$
$6_{5,1;4}$	$6_{2,4;-2}$		70	793	684		$6_{1,5;-4}$	$6_{4,2;2}$
$7_{5,2;3}$	$7_{2,5;-3}$		248	1513	248		$7_{2,5;-3}$	$7_{5,2;3}$
$8_{5,3;2}$	$8_{2,6;-4}$		681	1360	91		$8_{3,5;-2}$	$8_{6,2;4}$
$9_{5,4;1}$	$9_{2,7;-5}$		1444	763	48		$9_{4,5;-1}$	$9_{7,2;5}$
$10_{5,5;0}$	$10_{2,8;-6}$		2306	380	31		$10_{5,5;0}$	$10_{8,2;6}$
$11_{5,6;-1}$	$11_{2,9;-7}$		2749	212	22		$11_{6,5;1}$	$11_{9,2;7}$
$12_{5,7;-2}$	$12_{2,10;-8}$		2519	141	16		$12_{7,5;2}$	$12_{10,2;8}$

Table IX-2 (Continued)

Sub-branch ${}^{b,e}Q\bar{3},3$	${}^{b,e}Q3,\bar{3}$	∓ 1	∓ 0.5	κ 0	± 0.5	± 1	${}^{b,e}Q3,\bar{3}$	Sub-branch ${}^{b,e}Q\bar{3},3$
$6_{6,0;6}$	$6_{3,3;0}$		6	67	640		$6_{0,6;-6}$	$6_{3,3;0}$
$7_{6,1;5}$	$7_{3,4;-1}$		29	422	1287		$7_{1,6;-5}$	$7_{4,3;1}$
$8_{6,2;4}$	$8_{3,5;-2}$		91	1360	681		$8_{2,6;-4}$	$8_{5,3;2}$
$9_{6,3;3}$	$9_{3,6;-3}$		245	2252	245		$9_{3,6;-3}$	$9_{6,3;3}$
$10_{6,4;2}$	$10_{3,7;-4}$		592	1993	112		$10_{4,6;-2}$	$10_{7,3;4}$
$11_{6,5;1}$	$11_{3,8;-5}$		1278	1165	69		$11_{5,6;-1}$	$11_{8,3;5}$
$12_{6,6;0}$	$12_{3,9;-6}$		2337	609	48		$12_{6,6;0}$	$12_{9,3;6}$
$7_{7,0;7}$	$7_{4,3;1}$		4	33	450		$7_{0,7;-7}$	$7_{3,4;-1}$
$8_{7,1;6}$	$8_{4,4;0}$		17	197	1640		$8_{1,7;-6}$	$8_{4,4;0}$
$9_{7,2;5}$	$9_{4,5;-1}$		48	763	1444		$9_{2,7;-5}$	$9_{5,4;1}$
$10_{7,3;4}$	$10_{4,6;-2}$		112	1993	592		$10_{3,7;-4}$	$10_{6,4;2}$
$11_{7,4;3}$	$11_{4,7;-3}$		244	3040	244		$11_{4,7;-3}$	$11_{7,4;3}$
$12_{7,5;2}$	$12_{4,8;-4}$		512	2675	136		$12_{5,7;-2}$	$12_{8,4;4}$
$8_{8,0;8}$	$8_{5,3;2}$		3	21	273		$8_{0,8;-8}$	$8_{3,5;-2}$
$9_{8,1;7}$	$9_{5,4;1}$		11	105	1513		$9_{1,8;-7}$	$9_{4,5;-1}$
$10_{8,2;6}$	$10_{5,5;0}$		31	380	2306		$10_{2,8;-6}$	$10_{5,5;0}$
$11_{8,3;5}$	$11_{5,6;-1}$		69	1165	1278		$11_{3,8;-5}$	$11_{6,5;1}$
$12_{8,4;4}$	$12_{5,7;-2}$		136	2675	512		$12_{4,8;-4}$	$12_{7,5;2}$
$9_{9,0;9}$	$9_{6,3;3}$		2	15	162		$9_{0,9;-9}$	$9_{3,6;-3}$
$10_{9,1;8}$	$10_{6,4;2}$		8	67	1105		$10_{1,9;-8}$	$10_{4,6;-2}$
$11_{9,2;7}$	$11_{6,5;1}$		22	212	2749		$11_{2,9;-7}$	$11_{5,6;-1}$
$12_{9,3;6}$	$12_{6,6;0}$		48	609	2337		$12_{3,9;-6}$	$12_{6,6;-0}$
$10_{10,0;10}$	$10_{7,3;4}$		1	11	101		$10_{0,10;-10}$	$10_{3,7;-4}$
$11_{10,1;9}$	$11_{7,4;3}$		6	49	713		$11_{1,10;-9}$	$11_{4,7;-3}$
$12_{10,2;8}$	$12_{7,5;2}$		16	141	2519		$12_{2,10;-8}$	$12_{5,7;-2}$
$11_{11,0;11}$	$11_{8,3;5}$		1	9	69		$11_{0,11;-11}$	$11_{3,8;-5}$
$12_{11,1;10}$	$12_{8,4;4}$		5	38	449		$12_{1,11;-10}$	$12_{4,8;-4}$
$12_{12,0;12}$	$12_{9,3;6}$		1	7	52		$12_{0,12;-12}$	$12_{3,9;-6}$
${}^{b,o}Q\bar{3},3$	${}^{b,o}Q3,\bar{3}$	∓ 1	∓ 0.5	0	± 0.5	± 1	${}^{b,o}Q3,\bar{3}$	${}^{b,o}Q\bar{3},3$
$4_{4,1;3}$	$4_{1,4;-3}$		10	20	10		$4_{1,4;-3}$	$4_{4,1;3}$
$5_{4,2;2}$	$5_{1,5;-4}$		34	33	9		$5_{2,4;-2}$	$5_{5,1;4}$
$6_{4,3;1}$	$6_{1,6;-5}$		58	33	6		$6_{3,4;-1}$	$6_{6,1;5}$
$7_{4,4;0}$	$7_{1,7;-6}$		76	26	4		$7_{4,4;0}$	$7_{7,1;6}$
$8_{4,5;-1}$	$8_{1,8;-7}$		82	19	3		$8_{5,4;1}$	$8_{8,1;7}$
$9_{4,6;-2}$	$9_{1,9;-8}$		77	14	2		$9_{6,4;2}$	$9_{9,1;8}$
$10_{4,7;-3}$	$10_{1,10;-9}$		68	11	1		$10_{7,4;3}$	$10_{10,1;9}$
$11_{4,8;-4}$	$11_{1,11;-10}$		58	9	1		$11_{8,4;4}$	$11_{11,1;10}$
$12_{4,9;-5}$	$12_{1,12;-11}$		48	7	1		$12_{9,4;5}$	$12_{12,1;11}$
$5_{5,1;4}$	$5_{2,4;-2}$		9	33	34		$5_{1,5;-4}$	$5_{4,2;2}$
$6_{5,2;3}$	$6_{2,5;-3}$		36	83	36		$6_{2,5;-3}$	$6_{5,2;3}$
$7_{5,3;2}$	$7_{2,6;-4}$		84	108	25		$7_{3,5;-2}$	$7_{6,2;4}$
$8_{5,4;1}$	$8_{2,7;-5}$		144	102	16		$8_{4,5;-1}$	$8_{7,2;5}$
$9_{5,5;0}$	$9_{2,8;-6}$		199	83	11		$9_{5,5;0}$	$9_{8,2;6}$
$10_{5,6;-1}$	$10_{2,9;-7}$		233	63	8		$10_{6,5;1}$	$10_{9,2;7}$
$11_{5,7;-2}$	$11_{2,10;-8}$		241	48	6		$11_{7,5;2}$	$11_{10,2;8}$
$12_{5,8;-3}$	$12_{2,11;-9}$		228	38	5		$12_{8,5;3}$	$12_{11,2;9}$
$6_{6,1;5}$	$6_{3,4;-1}$		6	33	58		$6_{1,6;-5}$	$6_{4,3;1}$
$7_{6,2;4}$	$7_{3,5;-2}$		25	108	84		$7_{2,6;-4}$	$7_{5,3;2}$
$8_{6,3;3}$	$8_{3,6;-3}$		67	187	67		$8_{3,6;-3}$	$8_{6,3;3}$
$9_{6,4;2}$	$9_{3,7;-4}$		136	221	45		$9_{4,6;-2}$	$9_{7,3;4}$
$10_{6,5;1}$	$10_{3,8;-5}$		228	206	31		$10_{5,6;-1}$	$10_{8,3;5}$
$11_{6,6;0}$	$11_{3,9;-6}$		328	169	22		$11_{6,6;0}$	$11_{9,3;6}$
$12_{6,7;-1}$	$12_{3,10;-7}$		410	132	16		$12_{7,6;1}$	$12_{10,3;7}$

Table IX-2 (Continued)

Sub-branch $^{b,o}Q\bar{3},3$	$^{b,o}Q3,\bar{3}$	∓ 1	∓ 0.5	κ 0	± 0.5	± 1	Sub-branch $^{b,o}Q3,\bar{3}$	$^{b,o}Q\bar{3},3$
$7_{7,1;6}$	$7_{4,4;0}$		4	26	76		$7_{1,7;-6}$	$7_{4,4;0}$
$8_{7,2;5}$	$8_{4,5;-1}$		16	102	144		$8_{2,7;-5}$	$8_{5,4;1}$
$9_{7,3;4}$	$9_{4,6;-2}$		45	221	136		$9_{3,7;-4}$	$9_{6,4;2}$
$10_{7,4;3}$	$10_{4,7;-3}$		98	327	98		$10_{4,7;-3}$	$10_{7,4;3}$
$11_{7,5;2}$	$11_{4,8;-4}$		183	367	68		$11_{5,7;-2}$	$11_{8,4;4}$
$12_{7,6;1}$	$12_{4,9;-5}$		299	340	48		$12_{6,7;-1}$	$12_{9,4;5}$
$8_{8,1;7}$	$8_{5,4;1}$		3	19	82		$8_{1,8;-7}$	$8_{4,5;-1}$
$9_{8,2;6}$	$9_{5,5;0}$		11	83	199		$9_{2,8;-6}$	$9_{5,5;0}$
$10_{8,3;5}$	$10_{5,6;-1}$		31	206	228		$10_{3,8;-5}$	$10_{6,5;1}$
$11_{8,4;4}$	$11_{5,7;-2}$		68	367	183		$11_{4,8;-4}$	$11_{7,5;2}$
$12_{8,5;3}$	$12_{5,8;-3}$		129	498	129		$12_{5,8;-3}$	$12_{8,5;3}$
$9_{9,1;8}$	$9_{6,4;2}$		2	14	77		$9_{1,9;-8}$	$9_{4,6;-2}$
$10_{9,2;7}$	$10_{6,5;1}$		8	63	233		$10_{2,9;-7}$	$10_{5,6;-1}$
$11_{9,3;6}$	$11_{6,6;0}$		22	169	328		$11_{3,9;-6}$	$11_{6,6;0}$
$12_{9,4;5}$	$12_{6,7;-1}$		48	340	299		$12_{4,9;-5}$	$12_{7,6;1}$
$10_{10,1;9}$	$10_{7,4;3}$		1	11	68		$10_{1,10;-9}$	$10_{4,7;-3}$
$11_{10,2;8}$	$11_{7,5;2}$		6	48	241		$11_{2,10;-8}$	$11_{5,7;-2}$
$12_{10,3;7}$	$12_{7,6;1}$		16	132	410		$12_{3,10;-7}$	$12_{6,7;-1}$
$11_{11,1,10}$	$11_{8,4;4}$		1	9	58		$11_{1,11;-10}$	$11_{4,8;-4}$
$12_{11,2;9}$	$12_{8,5;3}$		5	38	228		$12_{2,11;-9}$	$12_{5,8;-3}$
$12_{12,1;11}$	$12_{9,4;5}$		1	7	48		$12_{1,12;-11}$	$12_{4,9;-5}$
$^{b,o}R\bar{3},3$	$^{b,e}P3,\bar{3}$	∓ 1	∓ 0.5	0	± 0.5	± 1	$^{b,o}R3,\bar{3}$	$^{b,e}P\bar{3},3$
$3_{3,1;2}$	$4_{0,4;-4}$		38	41	13		$3_{1,3;-2}$	$4_{4,0;4}$
$4_{3,2;1}$	$5_{0,5;-5}$		111	62	12		$4_{2,3;-1}$	$5_{5,0;5}$
$5_{3,3;0}$	$6_{0,6;-6}$		175	59	9		$5_{3,3;0}$	$6_{6,0;6}$
$6_{3,4;-1}$	$7_{0,7;-7}$		206	50	7		$6_{4,3;1}$	$7_{7,0;7}$
$7_{3,5;-2}$	$8_{0,8;-8}$		209	42	6		$7_{5,3;2}$	$8_{8,0;8}$
$8_{3,6;-3}$	$9_{0,9;-9}$		195	35	5		$8_{6,3;3}$	$9_{9,0;9}$
$9_{3,7;-4}$	$10_{0,10;-10}$		177	30	4		$9_{7,3;4}$	$10_{10,0;10}$
$10_{3,8;-5}$	$11_{0,11;-11}$		158	27	4		$10_{8,3;5}$	$11_{11,0;11}$
$11_{3,9;-6}$	$12_{0,12;-12}$		142	24	3		$11_{9,3;6}$	$12_{12,0;12}$
$4_{4,1;3}$	$5_{1,4;-3}$		14	63	41		$4_{1,4;-3}$	$5_{4,1;3}$
$5_{4,2;2}$	$6_{1,5;-4}$		77	149	44		$5_{2,4;-2}$	$6_{5,1;4}$
$6_{4,3;1}$	$7_{1,6;-5}$		209	182	34		$6_{3,4;-1}$	$7_{6,1;5}$
$7_{4,4;0}$	$8_{1,7;-6}$		370	170	26		$7_{4,4;0}$	$8_{7,1;6}$
$8_{4,5;-1}$	$9_{1,8;-7}$		493	146	20		$8_{5,4;1}$	$9_{8,1;7}$
$9_{4,6;-2}$	$10_{1,9;-8}$		549	122	17		$9_{6,4;2}$	$10_{9,1;8}$
$10_{4,7;-3}$	$11_{1,10;-9}$		550	104	14		$10_{7,4;3}$	$11_{10,1;9}$
$11_{4,8;-4}$	$12_{1,11;-10}$		519	90	12		$11_{8,4;4}$	$12_{11,1;10}$
$5_{5,1;4}$	$6_{2,4;-2}$		3	43	71		$5_{1,5;-4}$	$6_{4,2;2}$
$6_{5,2;3}$	$7_{2,5;-3}$		23	175	97		$6_{2,5;-3}$	$7_{5,2;3}$
$7_{5,3;2}$	$8_{2,6;-4}$		87	305	80		$7_{3,5;-2}$	$8_{6,2;4}$
$8_{5,4;1}$	$9_{2,7;-5}$		234	347	61		$8_{4,5;-1}$	$9_{7,2;5}$
$9_{5,5;0}$	$10_{2,8;-6}$		466	324	47		$9_{5,5;0}$	$10_{8,2;6}$
$10_{5,6;-1}$	$11_{2,9;-7}$		715	279	38		$10_{6,5;1}$	$11_{9,2;7}$
$11_{5,7;-2}$	$12_{2,10;-8}$		896	237	32		$11_{7,5;2}$	$12_{10,2;8}$
$6_{6,1;5}$	$7_{3,4;-1}$		1	19	87		$6_{1,6;-5}$	$7_{4,3;1}$
$7_{6,2;4}$	$8_{3,5;-2}$		7	118	162		$7_{2,6;-4}$	$8_{5,3;2}$
$8_{6,3;3}$	$9_{3,6;-3}$		26	321	153		$8_{3,6;-3}$	$9_{6,3;3}$
$9_{6,4;2}$	$10_{3,7;-4}$		80	495	119		$9_{4,6;-2}$	$10_{7,3;4}$
$10_{6,5;1}$	$11_{3,8;-5}$		208	547	91		$10_{5,6;-1}$	$11_{8,3;5}$
$11_{6,6;0}$	$12_{3,9;-6}$		446	511	73		$11_{6,6;0}$	$12_{9,3;6}$
$7_{7,1;6}$	$8_{4,4;0}$		1	7	82		$7_{1,7;-6}$	$8_{4,4;0}$
$8_{7,2;5}$	$9_{4,5;-1}$		3	55	218		$8_{2,7;-5}$	$9_{5,4;1}$
$9_{7,3;4}$	$10_{4,6;-2}$		11	218	247		$9_{3,7;-4}$	$10_{6,4;2}$
$10_{7,4;3}$	$11_{4,7;-3}$		29	492	205		$10_{4,7;-3}$	$11_{7,4;3}$
$11_{7,5;2}$	$12_{4,8;-4}$		73	711	158		$11_{5,7;-2}$	$12_{8,4;4}$

Table IX-2 (Continued)

Sub-branch $^{b,o}R_{\bar{3},3}$	$^{b,e}P_{3,\bar{3}}$	∓1	∓0.5	κ 0	±0.5	±1	Sub-branch $^{b,o}R_{3,\bar{3}}$	$^{b,e}P_{\bar{3},3}$
$8_{8,1;7}$	$9_{5,4;1}$		0	4	62		$8_{1,8;-7}$	$9_{4,5;-1}$
$9_{8,2;6}$	$10_{5,5;0}$		2	24	241		$9_{2,8;-6}$	$10_{5,5;0}$
$10_{8,3;5}$	$11_{5,6;-1}$		6	108	346		$10_{3,8;-5}$	$11_{6,5;1}$
$11_{8,4;4}$	$12_{5,7;-2}$		14	336	319		$11_{4,8;-4}$	$12_{7,5;2}$
$9_{9,1;8}$	$10_{6,4;2}$		0	2	38		$9_{1,9;-8}$	$10_{4,6;-2}$
$10_{9,2;7}$	$11_{6,5;1}$		1	12	219		$10_{2,9;-7}$	$11_{5,6;-1}$
$11_{9,3;6}$	$12_{6,6;0}$		4	50	420		$11_{3,9;-6}$	$12_{6,6;0}$
$10_{10,1;9}$	$11_{7,4;3}$		0	1	22		$10_{1,10;-9}$	$11_{4,7;-3}$
$11_{10,2;8}$	$12_{7,5;2}$		1	7	165		$11_{2,10;-8}$	$12_{5,7;-2}$
$11_{11,1;10}$	$12_{8,4;4}$		0	1	12		$11_{1,11;-10}$	$12_{4,8;-4}$
$^{b,e}R_{\bar{3},5}$	$^{b,o}P_{3,\bar{5}}$	∓1	∓0.5	0	±0.5	±1	$^{b,e}R_{5,\bar{3}}$	$^{b,o}P_{\bar{5},3}$
$4_{4,0;4}$	$5_{1,5;-4}$		2	2	0		$4_{0,4;-4}$	$5_{5,1;4}$
$5_{4,1;3}$	$6_{1,6;-5}$		6	3	0		$5_{1,4;-3}$	$6_{6,1;5}$
$6_{4,2;2}$	$7_{1,7;-6}$		11	2	0		$6_{2,4;-2}$	$7_{7,1;6}$
$7_{4,3;1}$	$8_{1,8;-7}$		13	1	0		$7_{3,4;-1}$	$8_{8,1;7}$
$8_{4,4;0}$	$9_{1,9;-8}$		12	0	0		$8_{4,4;0}$	$9_{9,1;8}$
$9_{4,5;-1}$	$10_{1,10;-9}$		9	0	0		$9_{5,4;1}$	$10_{10,1;9}$
$10_{4,6;-2}$	$11_{1,11;-10}$		5	0	0		$10_{6,4;2}$	$11_{11,1;10}$
$11_{4,7;-3}$	$12_{1,12;-11}$		3	0	0		$11_{7,4;3}$	$12_{12,1;11}$
$5_{5,0;5}$	$6_{2,5;-3}$		2	4	2		$5_{0,5;-5}$	$6_{5,2;3}$
$6_{5,1;4}$	$7_{2,6;-4}$		8	9	1		$6_{1,5;-4}$	$7_{6,2;4}$
$7_{5,2;3}$	$8_{2,7;-5}$		18	10	0		$7_{2,5;-3}$	$8_{7,2;5}$
$8_{5,3;2}$	$9_{2,8;-6}$		30	6	0		$8_{3,5;-2}$	$9_{8,2;6}$
$9_{5,4;1}$	$10_{2,9;-7}$		40	3	0		$9_{4,5;-1}$	$10_{9,2;7}$
$10_{5,5;0}$	$11_{2,10;-8}$		43	1	0		$10_{5,5;0}$	$11_{10,2;8}$
$11_{5,6;-1}$	$12_{2,11;-9}$		37	1	0		$11_{6,5;1}$	$12_{11,2;9}$
$6_{6,0;6}$	$7_{3,5;-2}$		1	5	4		$6_{0,6;-6}$	$7_{5,3;2}$
$7_{6,1;5}$	$8_{3,6;-3}$		5	15	4		$7_{1,6;-5}$	$8_{6,3;3}$
$8_{6,2;4}$	$9_{3,7;-4}$		15	24	1		$8_{2,6;-4}$	$9_{7,3;4}$
$9_{6,3;3}$	$10_{3,8;-5}$		31	23	0		$9_{3,6;-3}$	$10_{8,3;5}$
$10_{6,4;2}$	$11_{3,9;-6}$		52	14	0		$10_{4,6;-2}$	$11_{9,3;6}$
$11_{6,5;1}$	$12_{3,10;-7}$		73	6	0		$11_{5,6;-1}$	$12_{10,3;7}$
$7_{7,0;7}$	$8_{4,5;-1}$		1	4	6		$7_{0,7;-7}$	$8_{5,4;1}$
$8_{7,1;6}$	$9_{4,6;-2}$		3	16	9		$8_{1,7;-6}$	$9_{6,4;2}$
$9_{7,2;5}$	$10_{4,7;-3}$		9	33	4		$9_{2,7;-5}$	$10_{7,4;3}$
$10_{7,3;4}$	$11_{4,8;-4}$		22	44	1		$10_{3,7;-4}$	$11_{8,4;4}$
$11_{7,4;3}$	$12_{4,9;-5}$		42	39	0		$11_{4,7;-3}$	$12_{9,4;5}$
$8_{8,0;8}$	$9_{5,5;0}$		0	3	7		$8_{0,8;-8}$	$9_{5,5;0}$
$9_{8,1;7}$	$10_{5,6;-1}$		2	13	15		$9_{1,8;-7}$	$10_{6,5;1}$
$10_{8,2;6}$	$11_{5,7;-2}$		6	33	11		$10_{2,8;-6}$	$11_{7,5;2}$
$11_{8,3;5}$	$12_{5,8;-3}$		14	57	4		$11_{3,8;-5}$	$12_{8,5;3}$
$9_{9,0;9}$	$10_{6,5;1}$		0	2	8		$9_{0,9;-9}$	$10_{5,6;-1}$
$10_{9,1;8}$	$11_{6,6;0}$		1	9	21		$10_{1,9;-8}$	$11_{6,6;0}$
$11_{9,2;7}$	$12_{6,7;-1}$		4	28	22		$11_{2,9;-7}$	$12_{7,6;1}$
$10_{10,0;10}$	$11_{7,5;2}$		0	1	7		$10_{0,10;-10}$	$11_{5,7;-2}$
$11_{10,1;9}$	$12_{7,6;1}$		1	7	24		$11_{1,10;-9}$	$12_{6,7;-1}$
$11_{11,0;11}$	$12_{8,5;3}$		0	1	6		$11_{0,11;-11}$	$12_{5,8;-3}$

Bibliography

1. E. U. Condon and G. H. Shortley, *Theory of Atomic Spectra* (Cambridge University Press, Cambridge, England, 1935).
2. H. Eyring, J. Walter, and G. E. Kimball, *Quantum Chemistry* (John Wiley and Sons, New York, 1944).
3. W. Gordy, W. V. Smith, and R. F. Trambarulo, *Microwave Spectroscopy* (John Wiley and Sons, New York, 1953).
4. G. Herzberg, *Infrared and Raman Spectra* (D. Van Nostrand Co., New York, 1945).
5. E. C. Kemble, *The Fundamental Principles of Quantum Mechanics* (McGraw-Hill Book Co., New York, 1937).
6. H. Margenau and G. M. Murphy, *The Mathematics of Physics and Chemistry* (D. Van Nostrand Co., New York, 1943).
7. L. Pauling and E. B. Wilson, Jr., *Introduction to Quantum Mechanics* (McGraw-Hill Book Co., New York, 1935).
8. C. H. Townes and A. L. Schawlow, *Microwave Spectroscopy* (McGraw-Hill Book Co., New York, 1955).
9. E. B. Wilson, Jr., J. C. Decius, and P. C. Cross, *Molecular Vibrations* (McGraw-Hill Book Co., New York, 1955).
10. H. C. Allen, Jr., *J. Chem. Phys.*, **22,** 83 (1954).
11. ———, *Phil. Trans. Roy. Soc., London*, **253,** 335 (1961).
12. ———, and E. K. Plyler, *J. Am. Chem. Soc.*, **80,** 2673 (1958).
13. ———, *J. Research Nat. Bur. Standards*, **52,** 205 (1954).
14. ———, *J. Chem. Phys.*, **25,** 1132 (1956).
15. ———, *J. Chem. Phys.*, **31,** 1062 (1959).
16. ———, *J. Research Nat. Bur. Standards*, **63A,** 145 (1959).
17. H. C. Allen, Jr., L. R. Blaine, and E. K. Plyler, *J. Research Nat. Bur. Standards*, **56,** 279 (1956).
18. H. C. Allen, Jr., E. D. Tidwell, and E. K. Plyler, *J. Chem. Phys.*, **25,** 302 (1956).
19. ———, *J. Am. Chem. Soc.*, **78,** 3034 (1956),
20. ———, *J. Research. Nat. Bur. Standards*, **57,** 213 (1956).
21. W. S. Benedict, E. K. Plyler, and E. D. Tidwell, *J. Research Nat. Bur. Standards*, **61,** 123 (1958); *J. Chem. Phys.*, **29,** 829 (1958). W. S. Benedict and E. K. Plyler, *Can. J. Phys.*, **35,** 1235 (1957).
22. W. S. Benedict, N. Gailar, and E. K. Plyler, *J. Chem. Phys.*, **24,** 1139 (1956).
23. L. F. H. Bovey, *J. Chem. Phys.*, **21,** 830 (1953).
24. D. R. J. Boyd and H. C. Longuet-Higgins, *Proc. Roy. Soc., London*, **A213,** 55 (1952).
25. D. R. J. Boyd and H. W. Thompson, *Trans. Faraday Soc.*, **48,** 493 (1952).
26. ———, *Proc. Roy. Soc., London*, **A216,** 143 (1953).
27. D. G. Burkhard, *J. Chem. Phys.*, **21,** 1541 (1953).
28. ———, and D. M. Dennison, *Phys. Rev.*, **84,** 408 (1951).
29. D. G. Burkhard and J. C. Irvin, *J. Chem. Phys.*, **23,** 1405, 2469 (1955).
30. C. C. Costain and G. B. B. M. Sutherland, *J. Phys, Chem.*, **56,** 321 (1952).
31. C. C. Costain, *J. Chem. Phys.*, **29,** 864 (1958).
32. C. P. Courtoy, *Ann. Soc. Sci.*, Brussels, Ser. I **73,** 5 (1959).
33. B. L. Crawford, Jr., *J. Chem. Phys.*, **8,** 273 (1940).
34. ———, and P. C. Cross, *J. Chem. Phys.*, **5,** 621 (1937).

35. P. C. Cross, *Phys. Rev.*, **47,** 7 (1935).
36. ———, R. M. Hainer, and G. W. King, *J. Chem. Phys.*, **12,** 210 (1944).
37. B. T. Darling and D. M. Dennison, *Phys. Rev.*, **57,** 128 (1940).
38. D. M. Dennison, *Revs. Mod. Phys.*, **3,** 280 (1931).
39. ———, and M. Johnston, *Phys. Rev.*, **47,** 93 (1935); M. Johnston and D. M. Dennison, *Phys. Rev.*, **48,** 868 (1935).
40. F. P. Dickey and H. H. Nielsen, *Phys. Rev.*, **70,** 109 (1946).
41. C. Eckart, *Phys. Rev.*, **47,** 552 (1935).
42. A. M. Emerson, Thesis (University of Washington, 1958).
43. G. Erlandsson, *Arkiv Fysik*, **10,** 65 (1956).
44. E. Fermi, *Z. Physik*, **71,** 250 (1931).
45. G. W. Funke, *Z. Physik*, **99,** 341 (1936).
46. ———, *Z. Physik*, **104,** 169 (1937).
47. ———, and E. Lindholm, *Z. Physik*, **106,** 518 (1937).
48. S. Golden, *J. Chem. Phys.*, **16,** 78 (1948).
49. ———, and J. K. Bragg, *J. Chem. Phys.*, **17,** 439 (1949).
50. A. H. Guenther, T. A. Wiggins, and D. H. Rank, *J. Chem. Phys.*, **28,** 682 (1958).
51. R. M. Hainer, P. C. Cross, and G. W. King, *J. Chem. Phys.*, **17,** 826 (1949).
52. R. M. Hainer and G. W. King, *J. Chem. Phys.*, **15,** 89 (1947).
53. K. T. Hecht, *J. Mole. Spect.*, **5,** 355, 390 (1960).
54. R. C. Herman and W. H. Shaffer, *J. Chem. Phys.*, **16,** 453 (1948); **18,** 1207 (1950).
55. R. W. Kilb, Tables of Degenerate Mathieu Functions (Harvard University Press, Cambridge, 1956).
56. D. R. Herschbach, *J. Chem. Phys.*, **31,** 91 (1959).
57. J. B. Howard, *J. Chem. Phys.*, **5,** 451 (1937).
58. K. K. Innes, P. C. Cross, and P. A. Giguère, *J. Chem. Phys.*, **19,** 1086 (1951).
59. K. K. Innes, Thesis (University of Washington, 1951).
60. E. V. Ivash and D. M. Dennison, *J. Chem. Phys.*, **21,** 1804 (1954).
61. H. A. Jahn, *Proc. Roy. Soc. London*, **A168,** 469, 495 (1938); **A171,** 450 (1939).
62. ———, *Phys. Rev.*, **56,** 680 (1939).
63. O. M. Jordahl, *Phys. Rev.*, **45,** 87 (1934).
64. G. W. King, *J. Chem. Phys.*, **15,** 820 (1947).
65. ———, *J. Chem. Phys.*, **15,** 85 (1947).
66. ———, P. C. Cross, and G. B. Thomas, *J. Chem. Phys.*, **14,** 35 (1946).
67. G. W. King, R. M. Hainer, and P. C. Cross, *J. Chem. Phys.*, **11,** 27 (1943).
68. D. Kivelson, *J. Chem. Phys.*, **22,** 1733 (1954); **23,** 2230, 2236 (1955).
69. ———, and E. B. Wilson, Jr., *J. Chem. Phys.*, **20,** 1575 (1952).
70. ———, *J. Chem. Phys.*, **21,** 1229 (1953).
71. J. S. Koehler and D. M. Dennison, *Phys. Rev.*, **57,** 1006 (1940).
72. V. W. Laurie, *J. Chem. Phys.*, **28,** 704 (1958).
73. R. B. Lawrance and M. W. P. Strandberg, *Phys. Rev.*, **83,** 363 (1951).
74. D. R. Lide, Jr., *J. Chem. Phys.*, **20,** 1761 (1952).
75. ———, and D. Kivelson, *J. Chem. Phys.*, **23,** 2191 (1955).
76. D. R. Lide, Jr., and D. E. Mann, *J. Chem. Phys.* **28,** 572 (1958).
77. R. C. Lord and R. E. Merrifield, *J. Chem. Phys.*, **20,** 1348 (1952).
78. J. H. Meal and S. R. Polo, *J. Chem. Phys.*, **24,** 1119 (1956).
79. ———, *J. Chem. Phys.*, **24,** 1126 (1956).
80. R. Mecke, *Z. Physik*, **81,** 313 (1933).
81. ———, and R. Ziegler, *Z. Physik*, **101,** 405 (1936).
82. I. M. Mills and H. W. Thompson, *Proc. Roy. Soc. London*, **A226,** 306 (1954).
83. H. M. Mould, W. C. Price, and G. R. Wilkinson, *Spectrochim. Acta*, **15,** 313 (1959).
84. R. S. Mulliken, *Phys. Rev.*, **59,** 873 (1941).
85. H. H. Nielsen, *J. Chem. Phys.*, **5,** 818 (1937).
86. ———, *Phys. Rev.*, **60,** 794 (1941).
87. ———, *Phys. Rev.*, **66,** 282 (1944).
88. ———, *Phys. Rev.*, **77,** 130 (1950).

89. ———, *Rev. Mod. Phys.*, **23,** 90 (1951).
90. ———, and D. M. Dennison, *Phys. Rev.*, **72,** 1101 (1947).
91. G. Placzek and E. Teller, *Z. Physik*, **81,** 209 (1933).
92. E. K. Plyler and E. F. Barker, *Phys. Rev.*, **38,** 1827 (1931).
93. J. Pickworth and H. W. Thompson, *Proc. Roy. Soc., London*, **A222,** 443 (1954).
94. K. S. Pitzer and W. D. Gwinn, *J. Chem. Phys.*, **10,** 428 (1942).
95. S. R. Polo, *J. Chem. Phys.*, **24,** 1133 (1956).
96. D. A. Ramsay, Private Communication.
97. D. H. Rank, A. H. Guenther, and J. N. Shearer, *J. Opt. Soc. Am.*, **46,** 953 (1956).
98. D. H. Rank, J. N. Shearer, A. H. Guenther, and T. A. Wiggins, *J. Chem. Phys.*, **27,** 532 (1957).
99. B. S. Ray, *Z. Physik*, **78,** 74 (1932).
100. D. G. Rea and H. W. Thompson, *Trans. Faraday Soc.*, **52,** 1304 (1950).
101. A. Sayvetz, *J. Chem. Phys.*, **7,** 383 (1939).
102. W. H. Shaffer, *J. Chem. Phys.*, **9,** 607 (1941).
103. ———, *J. Chem. Phys.*, **10,** 1 (1942).
104. ———, and R. C. Herman, *J. Chem. Phys.*, **13,** 83 (1945).
105. W. H. Shaffer and A. H. Nielsen, *J. Chem. Phys.*, **9,** 847 (1941).
106. W. H. Shaffer and R. P. Schuman, *J. Chem. Phys.*, **12,** 504 (1944).
107. J. N. Shearer, T. A. Wiggins, A. H. Guenther, and D. H. Rank, *J. Chem. Phys.*, **25,** 724 (1956).
108. H. Y. Sheng, E. F. Barker, and D. M. Dennison, *Phys. Rev.*, **60,** 786 (1941).
109. S. Silver, *J. Chem. Phys.*, **10,** 565 (1942).
110. ———, and E. Ebers, *J. Chem. Phys.*, **10,** 559 (1942).
111. S. Silver and W. H. Shaffer, *J. Chem. Phys.*, **9,** 599 (1941).
112. J. W. Simmons and W. E. Anderson, *Phys. Rev.*, **80,** 338 (1950).
113. Z. I. Slawsky and D. M. Dennison, *J. Chem. Phys.*, **7,** 509 (1939).
114. R. E. Hillger and M. W. P. Strandberg, *Phys. Rev.*, **82,** 327 (1951).
115. E. Teller, *Hand- und Jahrbuch d Chem Phys.*, **IX** 2, 125 (1934).
116. J. H. Van Vleck, *Phys. Rev.*, **33,** 467 (1929).
117. ———, *Rev. Mod. Phys.*, **23,** 213 (1951).
118. S. C. Wang, *Phys. Rev.*, **34,** 243 (1929).
119. T. A. Wiggins, J. N. Shearer, E. R. Shull, and D. H. Rank, *J. Chem. Phys.*, **22,** 547 (1954).
120. T. A. Wiggins, E. R. Shull, J. M. Bennett, and D. H. Rank, *J. Chem. Phys.*, **21,** 1940 (1953).
121. E. B. Wilson, Jr., *J. Chem. Phys.*, **3,** 818 (1935).
122. ———, *J. Chem. Phys.*, **4,** 313 (1936).
123. ———, *J. Chem. Phys.*, **4,** 526 (1936).
124. ———, *J. Chem. Phys.*, **5,** 617 (1937).
125. ———, and J. B. Howard, *J. Chem. Phys.*, **4,** 260 (1936).
126. E. B. Wilson, Jr., C. C. Lin, and D. R. Lide, *J. Chem. Phys.*, **23,** 136 (1955).

AUTHOR INDEX

SUBJECT INDEX